NONLINEAR DYNAMICS AND CHAOS

NONLINEAR DYNAMICS AND CHAOS

With Applications to Physics, Biology, Chemistry, and Engineering

STEVEN H. STROGATZ

Westview Press
A Member of the Perseus Books Group

Printed in the United States of America.

Cataloging-in-Publication Data is available from the Library of Congress

ISBN-13 978-0-7382-0453-6
ISBN-10 0-7382-0453-6

Westview Press is a member of the Perseus Books Group.

Find us on the World Wide Web at http://www.westviewpress.com.

Westview Press books are available at special discounts for bulk purchases in the U.S. by corporations, institutions, and other organizations. For more information, please contact the Special Markets Department at The Perseus Books Group, 11 Cambridge Center, Cambridge, MA 02142, or call 1-612-252-5298, or e-mail special.sales@perseus-books.com.

Text design by Joyce C. Weston
Set in 10-point Times by Compset, Inc.

Cover art is a computer-generated picture of a twisted scroll ring, from Winfree and Strogatz (1984) with permission. Scroll rings are self-sustaining sources of waves in diverse excitable media, including heart muscle, neural tissue, and excitable chemical reactions (Winfree and Strogatz 1984, Winfree 1987b).

First paperback printing, December 2000

EBC 30 29 28 27

CONTENTS

PREFACE

This textbook is aimed at newcomers to nonlinear dynamics and chaos, especially students taking a first course in the subject. It is based on a one-semester course I've taught for the past several years at MIT and Cornell. My goal is to explain the mathematics as clearly as possible, and to show how it can be used to understand some of the wonders of the nonlinear world.

The mathematical treatment is friendly and informal, but still careful. Analytical methods, concrete examples, and geometric intuition are stressed. The theory is developed systematically, starting with first-order differential equations and their bifurcations, followed by phase plane analysis, limit cycles and their bifurcations, and culminating with the Lorenz equations, chaos, iterated maps, period doubling, renormalization, fractals, and strange attractors.

A unique feature of the book is its emphasis on applications. These include mechanical vibrations, lasers, biological rhythms, superconducting circuits, insect outbreaks, chemical oscillators, genetic control systems, chaotic waterwheels, and even a technique for using chaos to send secret messages. In each case, the scientific background is explained at an elementary level and closely integrated with the mathematical theory.

Prerequisites

The essential prerequisite is single-variable calculus, including curve-sketching, Taylor series, and separable differential equations. In a few places, multivariable calculus (partial derivatives, Jacobian matrix, divergence theorem) and linear algebra (eigenvalues and eigenvectors) are used. Fourier analysis is not assumed, and is developed where needed. Introductory physics is used throughout. Other scientific prerequisites would depend on the applications considered, but in all cases, a first course should be adequate preparation.

Possible Courses

The book could be used for several types of courses:

- A broad introduction to nonlinear dynamics, for students with no prior exposure to the subject. (This is the kind of course I have taught.) Here one goes straight through the whole book, covering the core material at the beginning of each chapter, selecting a few applications to discuss in depth and giving light treatment to the more advanced theoretical topics or skipping them altogether. A reasonable schedule is seven weeks on Chapters 1–8, and five or six weeks on Chapters 9–12. Make sure there's enough time left in the semester to get to chaos, maps, and fractals.
- A traditional course on nonlinear ordinary differential equations, but with more emphasis on applications and less on perturbation theory than usual. Such a course would focus on Chapters 1–8.
- A modern course on bifurcations, chaos, fractals, and their applications, for students who have already been exposed to phase plane analysis. Topics would be selected mainly from Chapters 3, 4, and 8–12.

For any of these courses, the students should be assigned homework from the exercises at the end of each chapter. They could also do computer projects; build chaotic circuits and mechanical systems; or look up some of the references to get a taste of current research. This can be an exciting course to teach, as well as to take. I hope you enjoy it.

Conventions

Equations are numbered consecutively within each section. For instance, when we're working in Section 5.4, the third equation is called (3) or Equation (3), but elsewhere it is called (5.4.3) or Equation (5.4.3). Figures, examples, and exercises are always called by their full names, e.g., Exercise 1.2.3. Examples and proofs end with a loud thump, denoted by the symbol ■.

Acknowledgments

Thanks to the National Science Foundation for financial support. For help with the book, thanks to Diana Dabby, Partha Saha, and Shinya Watanabe (students); Jihad Touma and Rodney Worthing (teaching assistants); Andy Christian, Jim Crutchfield, Kevin Cuomo, Frank DeSimone, Roger Eckhardt, Dana Hobson, and Thanos Siapas (for providing figures); Bob Devaney, Irv Epstein, Danny Kaplan, Willem Malkus, Charlie Marcus, Paul Matthews, Arthur Mattuck, Rennie Mirollo, Peter Renz, Dan Rockmore, Gil Strang, Howard Stone, John Tyson, Kurt Wiesenfeld, Art Winfree, and Mary Lou Zeeman (friends and colleagues who gave advice); and to my editor Jack Repcheck, Lynne Reed, Production Supervisor, and all the other helpful people at Addison-Wesley. Finally, thanks to my family and Elisabeth for their love and encouragement.

Steven H. Strogatz
Cambridge, Massachusetts

1

OVERVIEW

1.0 Chaos, Fractals, and Dynamics

There is a tremendous fascination today with chaos and fractals. James Gleick's book *Chaos* (Gleick 1987) was a bestseller for months—an amazing accomplishment for a book about mathematics and science. Picture books like *The Beauty of Fractals* by Peitgen and Richter (1986) can be found on coffee tables in living rooms everywhere. It seems that even nonmathematical people are captivated by the infinite patterns found in fractals (Figure 1.0.1). Perhaps most important of all, chaos and fractals represent hands-on mathematics that is alive and changing. You can turn on a home computer and create stunning mathematical images that no one has ever seen before.

Figure 1.0.1

The aesthetic appeal of chaos and fractals may explain why so many people have become intrigued by these ideas. But maybe you feel the urge to go deeper—to learn the mathematics behind the pictures, and to see how the ideas can be applied to problems in science and engineering. If so, this is a textbook for you.

The style of the book is informal (as you can see), with an emphasis on concrete examples and geometric thinking, rather than proofs and abstract arguments. It is also an extremely "applied" book—virtually every idea is illustrated by some application to science or engineering. In many cases, the applications are drawn from the recent research literature. Of course, one problem with such an applied approach is that not everyone is an expert in physics *and* biology *and* fluid mechanics . . . so the science as well as the mathematics will need to be explained from scratch. But that should be fun, and it can be instructive to see the connections among different fields.

Before we start, we should agree about something: chaos and fractals are part of an even grander subject known as ***dynamics***. This is the subject that deals with change, with systems that evolve in time. Whether the system in question settles down to equilibrium, keeps repeating in cycles, or does something more complicated, it is dynamics that we use to analyze the behavior. You have probably been exposed to dynamical ideas in various places—in courses in differential equations, classical mechanics, chemical kinetics, population biology, and so on. Viewed from the perspective of dynamics, all of these subjects can be placed in a common framework, as we discuss at the end of this chapter.

Our study of dynamics begins in earnest in Chapter 2. But before digging in, we present two overviews of the subject, one historical and one logical. Our treatment is intuitive; careful definitions will come later. This chapter concludes with a "dynamical view of the world," a framework that will guide our studies for the rest of the book.

1.1 Capsule History of Dynamics

Although dynamics is an interdisciplinary subject today, it was originally a branch of physics. The subject began in the mid-1600s, when Newton invented differential equations, discovered his laws of motion and universal gravitation, and combined them to explain Kepler's laws of planetary motion. Specifically, Newton solved the two-body problem—the problem of calculating the motion of the earth around the sun, given the inverse-square law of gravitational attraction between them. Subsequent generations of mathematicians and physicists tried to extend Newton's analytical methods to the three-body problem (e.g., sun, earth, and moon) but curiously this problem turned out to be much more difficult to solve. After decades of effort, it was eventually realized that the three-body problem was essentially *impossible* to solve, in the sense of obtaining explicit formulas for the motions of the three bodies. At this point the situation seemed hopeless.

The breakthrough came with the work of Poincaré in the late 1800s. He introduced a new point of view that emphasized qualitative rather than quantitative questions. For example, instead of asking for the exact positions of the planets at all times, he asked "Is the solar system stable forever, or will some planets eventually fly off to infinity?" Poincaré developed a powerful *geometric* approach to analyzing such questions. That approach has flowered into the modern subject of dynamics, with applications reaching far beyond celestial mechanics. Poincaré was also the first person to glimpse the possibility of ***chaos***, in which a deterministic system exhibits aperiodic behavior that depends sensitively on the initial conditions, thereby rendering long-term prediction impossible.

But chaos remained in the background in the first half of this century; instead dynamics was largely concerned with nonlinear oscillators and their applications in physics and engineering. Nonlinear oscillators played a vital role in the development of such technologies as radio, radar, phase-locked loops, and lasers. On the theoretical side, nonlinear oscillators also stimulated the invention of new mathematical techniques—pioneers in this area include van der Pol, Andronov, Littlewood, Cartwright, Levinson, and Smale. Meanwhile, in a separate development, Poincaré's geometric methods were being extended to yield a much deeper understanding of classical mechanics, thanks to the work of Birkhoff and later Kolmogorov, Arnol'd, and Moser.

The invention of the high-speed computer in the 1950s was a watershed in the history of dynamics. The computer allowed one to experiment with equations in a way that was impossible before, and thereby to develop some intuition about nonlinear systems. Such experiments led to Lorenz's discovery in 1963 of chaotic motion on a strange attractor. He studied a simplified model of convection rolls in the atmosphere to gain insight into the notorious unpredictability of the weather. Lorenz found that the solutions to his equations never settled down to equilibrium or to a periodic state—instead they continued to oscillate in an irregular, aperiodic fashion. Moreover, if he started his simulations from two slightly different initial conditions, the resulting behaviors would soon become totally different. The implication was that the system was *inherently* unpredictable—tiny errors in measuring the current state of the atmosphere (or any other chaotic system) would be amplified rapidly, eventually leading to embarrassing forecasts. But Lorenz also showed that there was structure in the chaos—when plotted in three dimensions, the solutions to his equations fell onto a butterfly-shaped set of points (Figure 1.1.1). He argued that this set had to be "an infinite complex of surfaces"—today we would regard it as an example of a fractal.

Lorenz's work had little impact until the 1970s, the boom years for chaos. Here are some of the main developments of that glorious decade. In 1971 Ruelle and Takens proposed a new theory for the onset of turbulence in fluids, based on abstract considerations about strange attractors. A few years later, May found examples of chaos in iterated mappings arising in population biology, and wrote an influential review article that stressed the pedagogical importance of studying simple nonlinear systems, to counterbalance the often misleading linear intuition fostered by traditional education. Next came the most surprising discovery of all, due to the physicist Feigenbaum. He discovered that there are certain universal laws governing the transition from regular to chaotic behavior; roughly speaking, completely different systems can go chaotic in the same way. His work established a link between chaos and

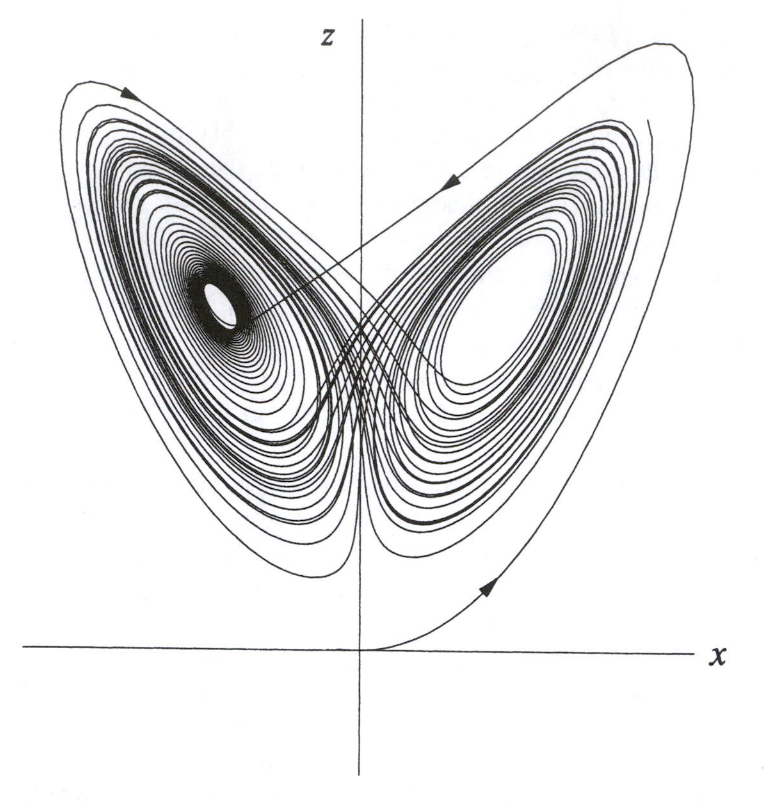

Figure 1.1.1

phase transitions, and enticed a generation of physicists to the study of dynamics. Finally, experimentalists such as Gollub, Libchaber, Swinney, Linsay, Moon, and Westervelt tested the new ideas about chaos in experiments on fluids, chemical reactions, electronic circuits, mechanical oscillators, and semiconductors.

Although chaos stole the spotlight, there were two other major developments in dynamics in the 1970s. Mandelbrot codified and popularized fractals, produced magnificent computer graphics of them, and showed how they could be applied in a variety of subjects. And in the emerging area of mathematical biology, Winfree applied the geometric methods of dynamics to biological oscillations, especially circadian (roughly 24-hour) rhythms and heart rhythms.

By the 1980s many people were working on dynamics, with contributions too numerous to list. Table 1.1.1 summarizes this history.

1.2 The Importance of Being Nonlinear

Now we turn from history to the logical structure of dynamics. First we need to introduce some terminology and make some distinctions.

Dynamics - A Capsule History

1666	Newton	Invention of calculus, explanation of planetary motion
1700s		Flowering of calculus and classical mechanics
1800s		Analytical studies of planetary motion
1890s	Poincaré	Geometric approach, nightmares of chaos
1920–1950		Nonlinear oscillators in physics and engineering, invention of radio, radar, laser
1920–1960	Birkhoff Kolmogorov Arnol'd Moser	Complex behavior in Hamiltonian mechanics
1963	Lorenz	Strange attractor in simple model of convection
1970s	Ruelle & Takens	Turbulence and chaos
	May	Chaos in logistic map
	Feigenbaum	Universality and renormalization, connection between chaos and phase transitions
		Experimental studies of chaos
	Winfree	Nonlinear oscillators in biology
	Mandelbrot	Fractals
1980s		Widespread interest in chaos, fractals, oscillators, and their applications

Table 1.1.1

There are two main types of dynamical systems: ***differential equations*** and ***iterated maps*** (also known as difference equations). Differential equations describe the evolution of systems in continuous time, whereas iterated maps arise in problems where time is discrete. Differential equations are used much more widely in science and engineering, and we shall therefore concentrate on them. Later in the book we will see that iterated maps can also be very useful, both for providing simple examples of chaos, and also as tools for analyzing periodic or chaotic solutions of differential equations.

Now confining our attention to differential equations, the main distinction is between ordinary and partial differential equations. For instance, the equation for a damped harmonic oscillator

$$m\frac{d^2x}{dt^2}+b\frac{dx}{dt}+kx=0 \qquad (1)$$

is an ordinary differential equation, because it involves only ordinary derivatives dx/dt and d^2x/dt^2. That is, there is only one independent variable, the time t. In contrast, the heat equation

$$\frac{\partial u}{\partial t} = \frac{\partial^2 u}{\partial x^2}$$

is a partial differential equation—it has both time t and space x as independent variables. Our concern in this book is with purely temporal behavior, and so we deal with ordinary differential equations almost exclusively.

A very general framework for ordinary differential equations is provided by the system

$$\begin{aligned} \dot{x}_1 &= f_1(x_1, \ldots, x_n) \\ &\vdots \\ \dot{x}_n &= f_n(x_1, \ldots, x_n). \end{aligned} \tag{2}$$

Here the overdots denote differentiation with respect to t. Thus $\dot{x}_i \equiv dx_i/dt$. The variables $x_1, \ldots, x_n$ might represent concentrations of chemicals in a reactor, populations of different species in an ecosystem, or the positions and velocities of the planets in the solar system. The functions $f_1, \ldots, f_n$ are determined by the problem at hand.

For example, the damped oscillator (1) can be rewritten in the form of (2), thanks to the following trick: we introduce new variables $x_1 = x$ and $x_2 = \dot{x}$. Then $\dot{x}_1 = x_2$, from the definitions, and

$$\begin{aligned} \dot{x}_2 = \ddot{x} &= -\tfrac{b}{m}\dot{x} - \tfrac{k}{m}x \\ &= -\tfrac{b}{m}x_2 - \tfrac{k}{m}x_1 \end{aligned}$$

from the definitions and the governing equation (1). Hence the equivalent system (2) is

$$\begin{aligned} \dot{x}_1 &= x_2 \\ \dot{x}_2 &= -\tfrac{b}{m}x_2 - \tfrac{k}{m}x_1. \end{aligned}$$

This system is said to be ***linear***, because all the x_i on the right-hand side appear to the first power only. Otherwise the system would be ***nonlinear***. Typical nonlinear terms are products, powers, and functions of the x_i, such as x_1x_2, $(x_1)^3$, or $\cos x_2$.

For example, the swinging of a pendulum is governed by the equation

$$\ddot{x} + \tfrac{g}{L}\sin x = 0,$$

where x is the angle of the pendulum from vertical, g is the acceleration due to gravity, and L is the length of the pendulum. The equivalent system is nonlinear:

$$\dot{x}_1 = x_2$$
$$\dot{x}_2 = -\tfrac{g}{L}\sin x_1 .$$

Nonlinearity makes the pendulum equation very difficult to solve analytically. The usual way around this is to fudge, by invoking the small angle approximation $\sin x \approx x$ for $x << 1$. This converts the problem to a linear one, which can then be solved easily. But by restricting to small x, we're throwing out some of the physics, like motions where the pendulum whirls over the top. Is it really necessary to make such drastic approximations?

It turns out that the pendulum equation *can* be solved analytically, in terms of elliptic functions. But there ought to be an easier way. After all, the motion of the pendulum is simple: at low energy, it swings back and forth, and at high energy it whirls over the top. There should be some way of extracting this information from the system directly. This is the sort of problem we'll learn how to solve, using geometric methods.

Here's the rough idea. Suppose we happen to know a solution to the pendulum system, for a particular initial condition. This solution would be a pair of functions $x_1(t)$ and $x_2(t)$, representing the position and velocity of the pendulum. If we construct an abstract space with coordinates (x_1, x_2), then the solution $(x_1(t), x_2(t))$ corresponds to a point moving along a curve in this space (Figure 1.2.1).

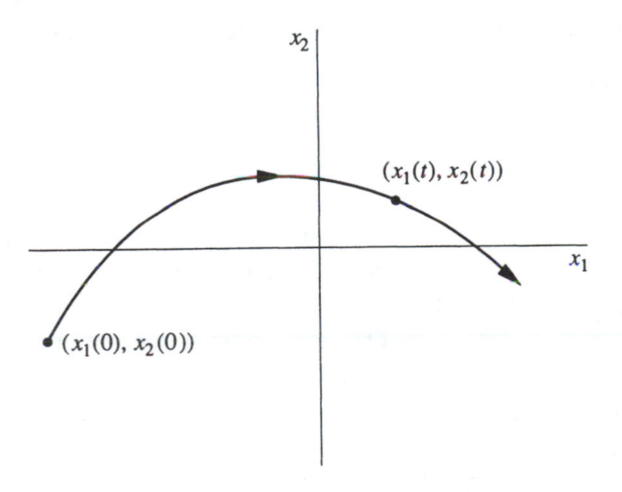

Figure 1.2.1

This curve is called a ***trajectory***, and the space is called the ***phase space*** for the system. The phase space is completely filled with trajectories, since each point can serve as an initial condition.

Our goal is to run this construction *in reverse:* given the system, we want to draw the trajectories, and thereby extract information about the solutions. In many cases, geometric reasoning will allow us to draw the trajectories *without actually solving the system*!

Some terminology: the phase space for the general system (2) is the space with coordinates $x_1, \ldots, x_n$. Because this space is n-dimensional, we will refer to (2) as an ***n*-dimensional system** or an ***n*th-order** system. Thus n represents the dimension of the phase space.

Nonautonomous Systems

You might worry that (2) is not general enough because it doesn't include any explicit *time dependence*. How do we deal with time-dependent or ***nonautonomous*** equations like the forced harmonic oscillator $m\ddot{x} + b\dot{x} + kx = F\cos t$? In this case too there's an easy trick that allows us to rewrite the system in the form (2). We let $x_1 = x$ and $x_2 = \dot{x}$ as before but now we introduce $x_3 = t$. Then $\dot{x}_3 = 1$ and so the equivalent system is

$$\begin{aligned}
\dot{x}_1 &= x_2 \\
\dot{x}_2 &= \tfrac{1}{m}\left(-kx_1 - bx_2 + F\cos x_3\right) \qquad (3) \\
\dot{x}_3 &= 1
\end{aligned}$$

which is an example of a *three*-dimensional system. Similarly, an nth-order time-dependent equation is a special case of an $(n+1)$-dimensional system. By this trick, we can always remove any time dependence by adding an extra dimension to the system.

The virtue of this change of variables is that it allows us to visualize a phase space with trajectories *frozen* in it. Otherwise, if we allowed explicit time dependence, the vectors and the trajectories would always be wiggling—this would ruin the geometric picture we're trying to build. A more physical motivation is that the ***state*** of the forced harmonic oscillator is truly three-dimensional: we need to know three numbers, x, $\dot{x}$, and t, to predict the future, given the present. So a three-dimensional phase space is natural.

The cost, however, is that some of our terminology is nontraditional. For example, the forced harmonic oscillator would traditionally be regarded as a second-order linear equation, whereas we will regard it as a third-order nonlinear system, since (3) is nonlinear, thanks to the cosine term. As we'll see later in the book, forced oscillators have many of the properties associated with nonlinear systems, and so there are genuine conceptual advantages to our choice of language.

Why Are Nonlinear Problems So Hard?

As we've mentioned earlier, most nonlinear systems are impossible to solve analytically. Why are nonlinear systems so much harder to analyze than linear ones? The essential difference is that *linear systems can be broken down into parts*. Then each part can be solved separately and finally recombined to get the answer. This idea allows a fantastic simplification of complex problems, and underlies such methods as normal modes, Laplace transforms, superposition arguments, and Fourier analysis. In this sense, a linear system is precisely equal to the sum of its parts.

But many things in nature don't act this way. Whenever parts of a system interfere, or cooperate, or compete, there are nonlinear interactions going on. Most of everyday life is nonlinear, and the principle of superposition fails spectacularly. If you listen to your two favorite songs at the same time, you won't get double the pleasure! Within the realm of physics, nonlinearity is vital to the operation of a laser, the formation of turbulence in a fluid, and the superconductivity of Josephson junctions.

1.3 A Dynamical View of the World

Now that we have established the ideas of nonlinearity and phase space, we can present a framework for dynamics and its applications. Our goal is to show the logical structure of the entire subject. The framework presented in Figure 1.3.1 will guide our studies thoughout this book.

The framework has two axes. One axis tells us the number of variables needed to characterize the state of the system. Equivalently, this number is the *dimension of the phase space*. The other axis tells us whether the system is linear or *nonlinear*.

For example, consider the exponential growth of a population of organisms. This system is described by the first-order differential equation

$$\dot{x} = rx$$

where x is the population at time t and r is the growth rate. We place this system in the column labeled "$n = 1$" because *one* piece of information—the current value of the population x—is sufficient to predict the population at any later time. The system is also classified as linear because the differential equation $\dot{x} = rx$ is linear in x.

As a second example, consider the swinging of a pendulum, governed by

$$\ddot{x} + \frac{g}{L}\sin x = 0.$$

In contrast to the previous example, the state of this system is given by *two* variables: its current angle x and angular velocity $\dot{x}$. (Think of it this way: we need the initial values of both x and $\dot{x}$ to determine the solution uniquely. For example, if we knew only x, we wouldn't know which way the pendulum was swinging.) Because two variables are needed to specify the state, the pendulum belongs in the $n = 2$ column of Figure 1.3.1. Moreover, the system is nonlinear, as discussed in the previous section. Hence the pendulum is in the lower, nonlinear half of the $n = 2$ column.

Figure 1.3.1

Number of variables →

Nonlinearity ↓	$n = 1$	$n = 2$	$n \geq 3$	$n >> 1$	Continuum
Linear	*Growth, decay, or equilibrium* Exponential growth RC circuit Radioactive decay	*Oscillations* Linear oscillator Mass and spring RLC circuit 2-body problem (Kepler, Newton)	Civil engineering, structures Electrical engineering	*Collective phenomena* Coupled harmonic oscillators Solid-state physics Molecular dynamics Equilibrium statistical mechanics	*Waves and patterns* Elasticity Wave equations Electromagnetism (Maxwell) Quantum mechanics (Schrödinger, Heisenberg, Dirac) Heat and diffusion Acoustics Viscous fluids
Nonlinear	Fixed points Bifurcations Overdamped systems, relaxational dynamics Logistic equation for single species	Pendulum Anharmonic oscillators Limit cycles Biological oscillators (neurons, heart cells) Predator-prey cycles Nonlinear electronics (van der Pol, Josephson)	***The frontier*** *Chaos* Strange attractors (Lorenz) 3-body problem (Poincaré) Chemical kinetics Iterated maps (Feigenbaum) Fractals (Mandelbrot) Forced nonlinear oscillators (Levinson, Smale) Practical uses of chaos Quantum chaos ?	Coupled nonlinear oscillators Lasers, nonlinear optics Nonequilibrium statistical mechanics Nonlinear solid-state physics (semiconductors) Josephson arrays Heart cell synchronization Neural networks Immune system Ecosystems Economics	*Spatio-temporal complexity* Nonlinear waves (shocks, solitons) Plasmas Earthquakes General relativity (Einstein) Quantum field theory Reaction-diffusion, biological and chemical waves Fibrillation Epilepsy Turbulent fluids (Navier-Stokes) Life

One can continue to classify systems in this way, and the result will be something like the framework shown here. Admittedly, some aspects of the picture are debatable. You might think that some topics should be added, or placed differently, or even that more axes are needed—the point is to think about classifying systems on the basis of their dynamics.

There are some striking patterns in Figure 1.3.1. All the simplest systems occur in the upper left-hand corner. These are the small linear systems that we learn about in the first few years of college. Roughly speaking, these linear systems exhibit growth, decay, or equilibrium when $n = 1$, or oscillations when $n = 2$. The italicized phrases in Figure 1.3.1 indicate that these broad classes of phenomena first arise in this part of the diagram. For example, an RC circuit has $n = 1$ and cannot oscillate, whereas an RLC circuit has $n = 2$ and can oscillate.

The next most familiar part of the picture is the upper right-hand corner. This is the domain of classical applied mathematics and mathematical physics where the linear partial differential equations live. Here we find Maxwell's equations of electricity and magnetism, the heat equation, Schrödinger's wave equation in quantum mechanics, and so on. These partial differential equations involve an infinite "continuum" of variables because each point in space contributes additional degrees of freedom. Even though these systems are large, they are tractable, thanks to such linear techniques as Fourier analysis and transform methods.

In contrast, the lower half of Figure 1.3.1—the nonlinear half—is often ignored or deferred to later courses. But no more! In this book we start in the lower left corner and systematically head to the right. As we increase the phase space dimension from $n = 1$ to $n = 3$, we encounter new phenomena at every step, from fixed points and bifurcations when $n = 1$, to nonlinear oscillations when $n = 2$, and finally chaos and fractals when $n = 3$. In all cases, a geometric approach proves to be very powerful, and gives us most of the information we want, even though we usually can't solve the equations in the traditional sense of finding a formula for the answer. Our journey will also take us to some of the most exciting parts of modern science, such as mathematical biology and condensed-matter physics.

You'll notice that the framework also contains a region forbiddingly marked "The frontier." It's like in those old maps of the world, where the mapmakers wrote, "Here be dragons" on the unexplored parts of the globe. These topics are not completely unexplored, of course, but it is fair to say that they lie at the limits of current understanding. The problems are very hard, because they are both large and nonlinear. The resulting behavior is typically complicated in *both space and time,* as in the motion of a turbulent fluid or the patterns of electrical activity in a fibrillating heart. Toward the end of the book we will touch on some of these problems—they will certainly pose challenges for years to come.

PART 1

ONE-DIMENSIONAL FLOWS

2

FLOWS ON THE LINE

2.0 Introduction

In Chapter 1, we introduced the general system

$$\dot{x}_1 = f_1(x_1, \ldots, x_n)$$
$$\vdots$$
$$\dot{x}_n = f_n(x_1, \ldots, x_n)$$

and mentioned that its solutions could be visualized as trajectories flowing through an n-dimensional phase space with coordinates $(x_1, \ldots, x_n)$. At the moment, this idea probably strikes you as a mind-bending abstraction. So let's start slowly, beginning here on earth with the simple case $n = 1$. Then we get a single equation of the form

$$\dot{x} = f(x).$$

Here $x(t)$ is a real-valued function of time t, and $f(x)$ is a smooth real-valued function of x. We'll call such equations ***one-dimensional*** or ***first-order systems***.

Before there's any chance of confusion, let's dispense with two fussy points of terminology:

1. The word *system* is being used here in the sense of a dynamical system, not in the classical sense of a collection of two or more equations. Thus a single equation can be a "system."
2. We do not allow f to depend explicitly on time. Time-dependent or "nonautonomous" equations of the form $\dot{x} = f(x,t)$ are more complicated, because one needs *two* pieces of information, x and t, to predict the future state of the system. Thus $\dot{x} = f(x,t)$ should really be regarded as a *two-dimensional* or *second-order* system, and will therefore be discussed later in the book.

2.1 A Geometric Way of Thinking

Pictures are often more helpful than formulas for analyzing nonlinear systems. Here we illustrate this point by a simple example. Along the way we will introduce one of the most basic techniques of dynamics: *interpreting a differential equation as a vector field.*

Consider the following nonlinear differential equation:

$$\dot{x} = \sin x. \tag{1}$$

To emphasize our point about formulas versus pictures, we have chosen one of the few nonlinear equations that can be solved in closed form. We separate the variables and then integrate:

$$dt = \frac{dx}{\sin x},$$

which implies

$$t = \int \csc x \, dx$$

$$= -\ln|\csc x + \cot x| + C.$$

To evaluate the constant C, suppose that $x = x_0$ at $t = 0$. Then $C = \ln|\csc x_0 + \cot x_0|$. Hence the solution is

$$t = \ln\left|\frac{\csc x_0 + \cot x_0}{\csc x + \cot x}\right|. \tag{2}$$

This result is exact, but a headache to interpret. For example, can you answer the following questions?

1. Suppose $x_0 = \pi/4$; describe the qualitative features of the solution $x(t)$ for all $t > 0$. In particular, what happens as $t \to \infty$?
2. For an *arbitrary* initial condition x_0, what is the behavior of $x(t)$ as $t \to \infty$?

Think about these questions for a while, to see that formula (2) is not transparent.

In contrast, a graphical analysis of (1) is clear and simple, as shown in Figure 2.1.1. We think of t as time, x as the position of an imaginary particle moving along the real line, and $\dot{x}$ as the velocity of that particle. Then the differential equation $\dot{x} = \sin x$ represents a ***vector field*** on the line: it dictates the velocity vector $\dot{x}$ at each x. To sketch the vector field, it is convenient to plot $\dot{x}$ versus x, and then draw arrows on the x-axis to indicate the corresponding velocity vector at each x. The arrows point to the right when $\dot{x} > 0$ and to the left when $\dot{x} < 0$.

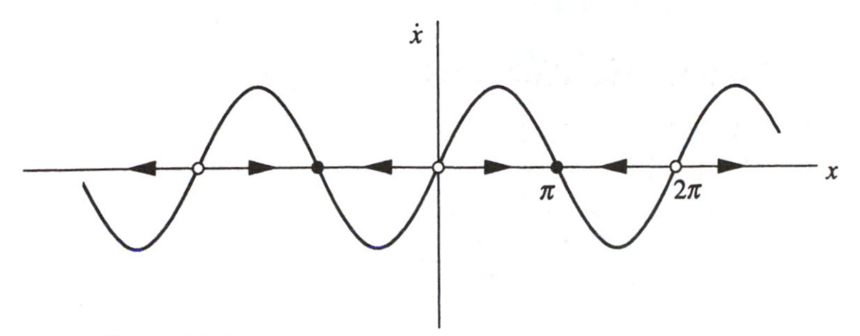

Figure 2.1.1

Here's a more physical way to think about the vector field: imagine that fluid is flowing steadily along the x-axis with a velocity that varies from place to place, according to the rule $\dot{x} = \sin x$. As shown in Figure 2.1.1, the ***flow*** is to the right when $\dot{x} > 0$ and to the left when $\dot{x} < 0$. At points where $\dot{x} = 0$, there is no flow; such points are therefore called ***fixed points***. You can see that there are two kinds of fixed points in Figure 2.1.1: solid black dots represent ***stable*** fixed points (often called *attractors* or *sinks,* because the flow is toward them) and open circles represent ***unstable*** fixed points (also known as *repellers* or *sources*).

Armed with this picture, we can now easily understand the solutions to the differential equation $\dot{x} = \sin x$. We just start our imaginary particle at x_0 and watch how it is carried along by the flow.

This approach allows us to answer the questions above as follows:

1. Figure 2.1.1 shows that a particle starting at $x_0 = \pi/4$ moves to the right faster and faster until it crosses $x = \pi/2$ (where $\sin x$ reaches its maximum). Then the particle starts slowing down and eventually approaches the stable fixed point $x = \pi$ from the left. Thus, the qualitative form of the solution is as shown in Figure 2.1.2.

   Note that the curve is concave up at first, and then concave down; this corresponds to the initial acceleration for $x < \pi/2$, followed by the deceleration toward $x = \pi$.

2. The same reasoning applies to any initial condition x_0. Figure 2.1.1 shows that if $\dot{x} > 0$ initially, the particle heads to the right and asymptotically approaches the nearest stable fixed point. Similarly, if $\dot{x} < 0$ initially, the particle approaches the nearest stable fixed point to its left. If $\dot{x} = 0$, then x remains constant. The qualitative form of the solution for any initial condition is sketched in Figure 2.1.3.

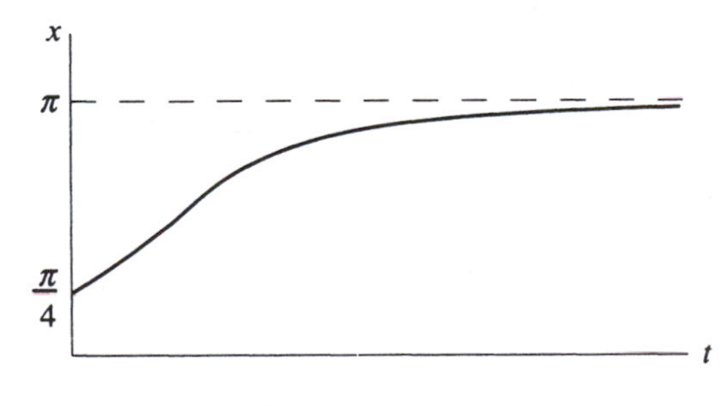

Figure 2.1.2

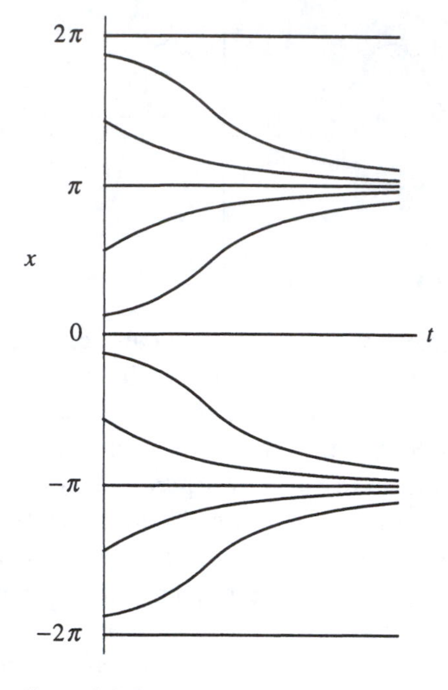

Figure 2.1.3

In all honesty, we should admit that a picture can't tell us certain *quantitative* things: for instance, we don't know the time at which the speed $|\dot{x}|$ is greatest. But in many cases *qualitative* information is what we care about, and then pictures are fine.

2.2 Fixed Points and Stability

The ideas developed in the last section can be extended to any one-dimensional system $\dot{x} = f(x)$. We just need to draw the graph of $f(x)$ and then use it to sketch the vector field on the real line (the x-axis in Figure 2.2.1).

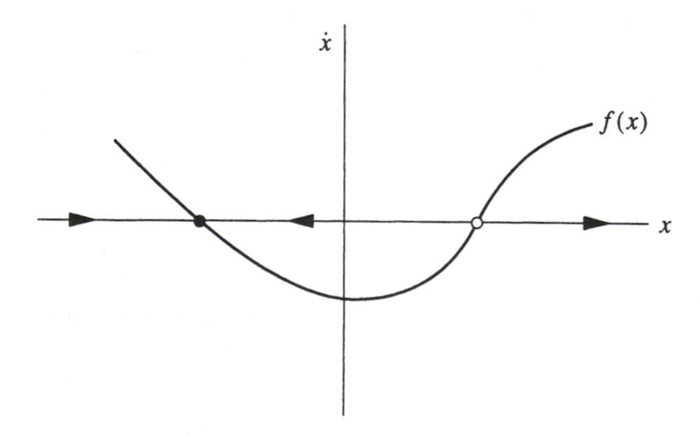

Figure 2.2.1

As before, we imagine that a fluid is flowing along the real line with a local velocity $f(x)$. This imaginary fluid is called the phase fluid, and the real line is the phase space. The flow is to the right where $f(x) > 0$ and to the left where $f(x) < 0$. To find the solution to $\dot{x} = f(x)$ starting from an arbitrary initial condition x_0, we place an imaginary particle (known as a ***phase point***) at x_0 and watch how it is carried along by the flow. As time goes on, the phase point moves along the x-axis according to some function $x(t)$. This function is called the ***trajectory*** based at x_0, and it represents the solution of the differential equation starting from the initial condition x_0. A picture like Figure 2.2.1, which shows all the qualitatively different trajectories of the system, is called a ***phase portrait.***

The appearance of the phase portrait is controlled by the fixed points x^*, defined by $f(x^*) = 0$; they correspond to stagnation points of the flow. In Figure 2.2.1, the solid black dot is a stable fixed point (the local flow is toward it) and the open dot is an unstable fixed point (the flow is away from it).

In terms of the original differential equation, fixed points represent ***equilibrium*** solutions (sometimes called steady, constant, or rest solutions, since if $x = x^*$ initially, then $x(t) = x^*$ for all time). An equilibrium is defined to be stable if all sufficiently small disturbances away from it damp out in time. Thus stable equilibria are represented geometrically by stable fixed points. Conversely, unstable equilibria, in which disturbances grow in time, are represented by unstable fixed points.

EXAMPLE 2.2.1:

Find all fixed points for $\dot{x} = x^2 - 1$, and classify their stability.

Solution: Here $f(x) = x^2 - 1$. To find the fixed points, we set $f(x^*) = 0$ and solve for x^*. Thus $x^* = \pm 1$. To determine stability, we plot $x^2 - 1$ and then sketch the vector field (Figure 2.2.2). The flow is to the right where $x^2 - 1 > 0$ and to the left where $x^2 - 1 < 0$. Thus $x^* = -1$ is stable, and $x^* = 1$ is unstable. ■

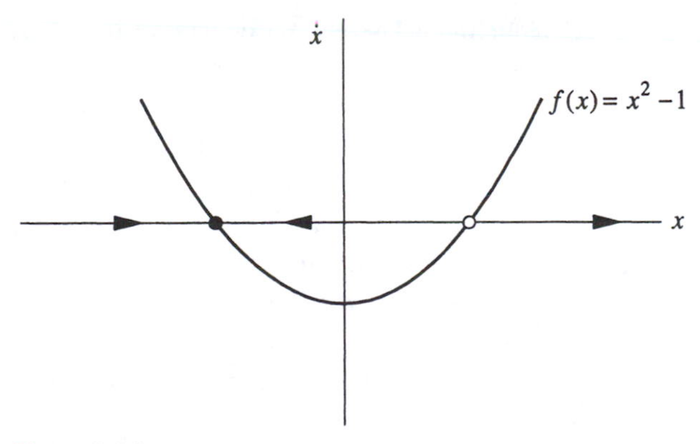

Figure 2.2.2

Note that the definition of stable equilibrium is based on *small* disturbances; certain large disturbances may fail to decay. In Example 2.2.1, all small disturbances to $x^* = -1$ will decay, but a large disturbance that sends x to the right of $x = 1$ will *not* decay—in fact, the phase point will be repelled out to $+\infty$. To emphasize this aspect of stability, we sometimes say that $x^* = -1$ is ***locally stable***, but not globally stable.

EXAMPLE 2.2.2:

Consider the electrical circuit shown in Figure 2.2.3. A resistor R and a capacitor C are in series with a battery of constant dc voltage V_0. Suppose that the switch is closed at $t = 0$, and that there is no charge on the capacitor initially. Let $Q(t)$ denote the charge on the capacitor at time $t \geq 0$. Sketch the graph of $Q(t)$.

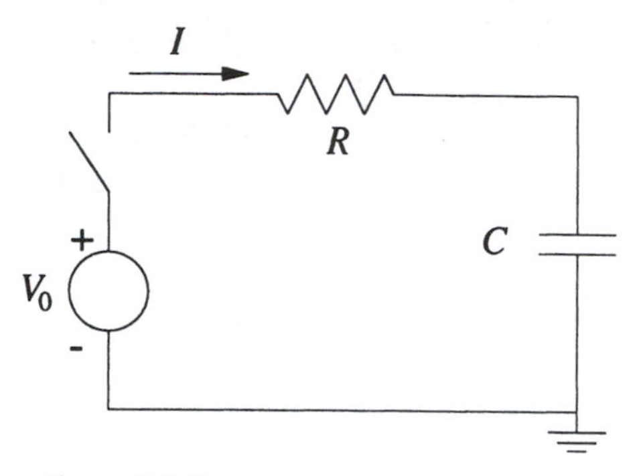

Figure 2.2.3

Solution: This type of circuit problem is probably familiar to you. It is governed by linear equations and can be solved analytically, but we prefer to illustrate the geometric approach.

First we write the circuit equations. As we go around the circuit, the total voltage drop must equal zero; hence $-V_0 + RI + Q/C = 0$, where I is the current flowing through the resistor. This current causes charge to accumulate on the capacitor at a rate $\dot{Q} = I$. Hence

$$-V_0 + R\dot{Q} + Q/C = 0 \quad \text{or}$$

$$\dot{Q} = f(Q) = \frac{V_0}{R} - \frac{Q}{RC}.$$

The graph of $f(Q)$ is a straight line with a negative slope (Figure 2.2.4). The corresponding vector field has a fixed point where $f(Q) = 0$, which occurs at $Q^* = CV_0$. The flow is to the right where $f(Q) > 0$ and to the left where $f(Q) < 0$. Thus the flow is always toward Q^*—it is a *stable* fixed point. In fact, it is ***globally stable***, in the sense that it is approached from *all* initial conditions.

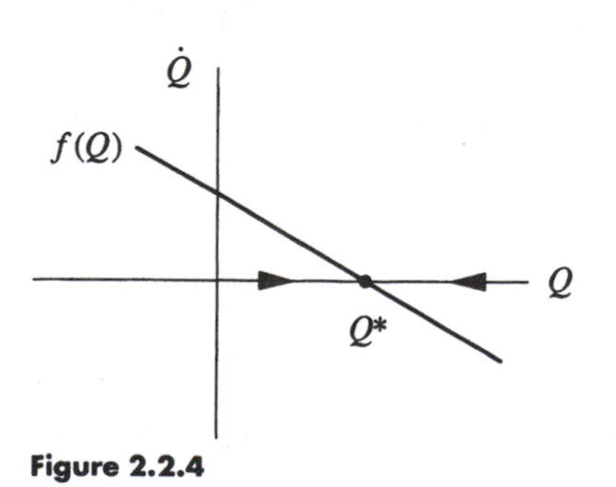

Figure 2.2.4

To sketch $Q(t)$, we start a phase point at the origin of Figure 2.2.4 and imagine how it would move. The flow carries the phase point monotonically toward Q^*. Its speed

$\dot{Q}$ decreases linearly as it approaches the fixed point; therefore $Q(t)$ is increasing and concave down, as shown in Figure 2.2.5. ■

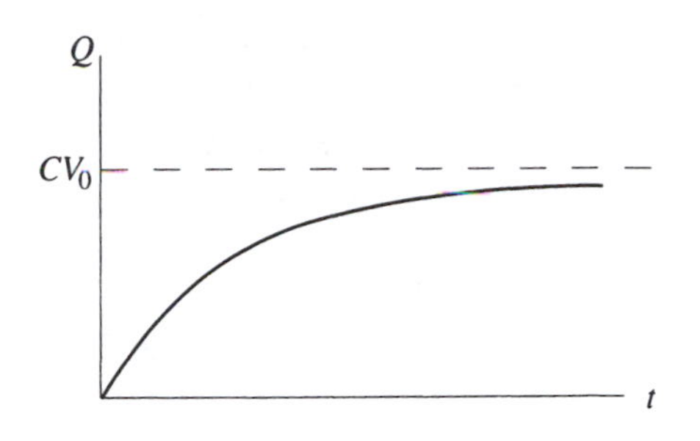

Figure 2.2.5

EXAMPLE 2.2.3:

Sketch the phase portrait corresponding to $\dot{x} = x - \cos x$, and determine the stability of all the fixed points.

Solution: One approach would be to plot the function $f(x) = x - \cos x$ and then sketch the associated vector field. This method is valid, but it requires you to figure out what the graph of $x - \cos x$ looks like.

There's an easier solution, which exploits the fact that we know how to graph $y = x$ and $y = \cos x$ *separately*. We plot both graphs on the same axes and then observe that they intersect in exactly one point (Figure 2.2.6).

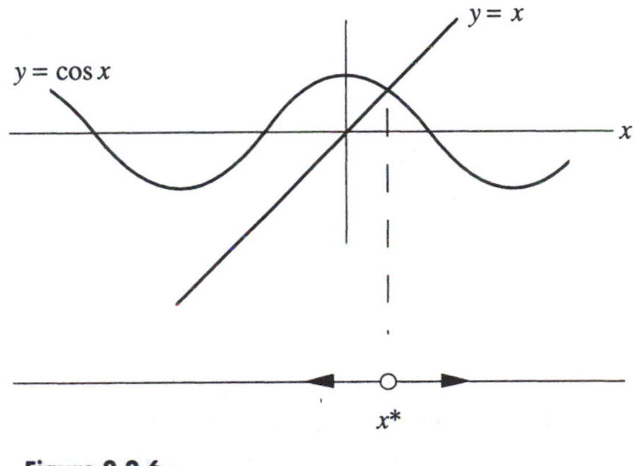

Figure 2.2.6

This intersection corresponds to a fixed point, since $x^* = \cos x^*$ and therefore $f(x^*) = 0$. Moreover, when the line lies above the cosine curve, we have $x > \cos x$ and so $\dot{x} > 0$: the flow is to the right. Similarly, the flow is to the left where the line is below the cosine curve. Hence x^* is the only fixed point, and it is unstable. Note that we can classify the stability of x^*, even though we don't have a formula for x^* itself! ■

2.3 Population Growth

The simplest model for the growth of a population of organisms is $\dot{N} = rN$, where $N(t)$ is the population at time t, and $r > 0$ is the growth rate. This model

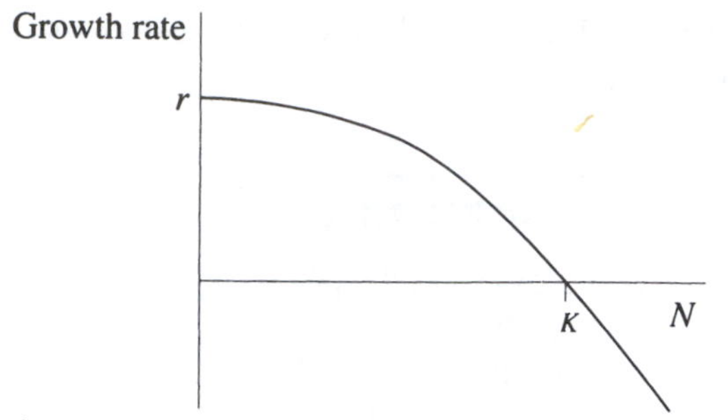

Figure 2.3.1

predicts exponential growth: $N(t) = N_0 e^{rt}$, where N_0 is the population at $t = 0$.

Of course such exponential growth cannot go on forever. To model the effects of overcrowding and limited resources, population biologists and demographers often assume that the per capita growth rate $\dot{N}/N$ decreases when N becomes sufficiently large, as shown in Figure 2.3.1. For small N, the growth rate equals r, just as before. However, for populations larger than a certain ***carrying capacity*** K, the growth rate actually becomes negative; the death rate is higher than the birth rate.

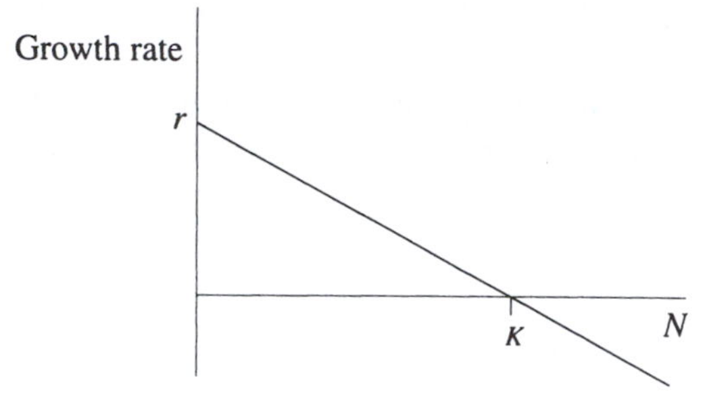

Figure 2.3.2

A mathematically convenient way to incorporate these ideas is to assume that the per capita growth rate $\dot{N}/N$ decreases *linearly* with N (Figure 2.3.2).

This leads to the ***logistic equation***

$$\dot{N} = rN\left(1 - \frac{N}{K}\right)$$

first suggested to describe the growth of human populations by Verhulst in 1838. This equation can be solved analytically (Exercise 2.3.1) but once again we prefer a graphical approach. We plot $\dot{N}$ versus N to see what the vector field looks like. Note that we plot only $N \geq 0$, since it makes no sense to think about a negative population (Figure 2.3.3). Fixed points occur at $N^* = 0$ and $N^* = K$, as found by setting $\dot{N} = 0$ and solving for N. By looking at the flow in Figure 2.3.3, we see that $N^* = 0$ is an unstable fixed point and $N^* = K$ is a stable fixed point. In biological terms, $N = 0$ is an unstable equilibrium: a small population will grow exponentially fast and run away from $N = 0$. On the other hand, if N is disturbed slightly from K, the disturbance will decay monotonically and $N(t) \to K$ as $t \to \infty$.

In fact, Figure 2.3.3 shows that if we start a phase point at *any* $N_0 > 0$, it will always flow toward $N = K$. Hence *the population always approaches the carrying capacity.*

The only exception is if $N_0 = 0$; then there's nobody around to start reproducing, and so $N = 0$ for all time. (The model does not allow for spontaneous generation!)

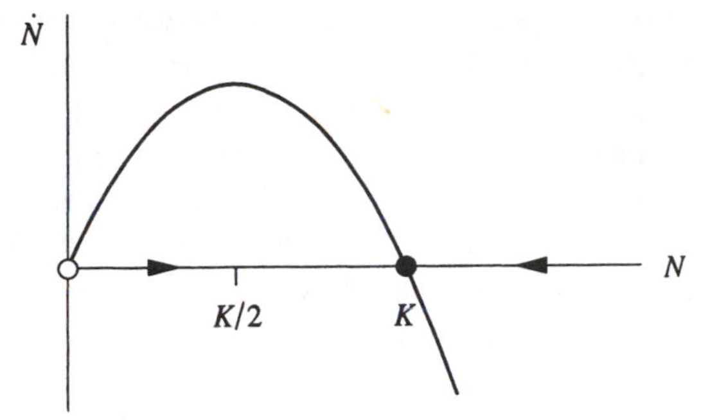

Figure 2.3.3

Figure 2.3.3 also allows us to deduce the qualitative shape of the solutions. For example, if $N_0 < K/2$, the phase point moves faster and faster until it crosses $N = K/2$, where the parabola in Figure 2.3.3 reaches its maximum. Then the phase point slows down and eventually creeps toward $N = K$. In biological terms, this means that the population initially grows in an accelerating fashion, and the graph of $N(t)$ is concave up. But after $N = K/2$, the derivative $\dot{N}$ begins to decrease, and so $N(t)$ is concave down as it asymptotes to the horizontal line $N = K$ (Figure 2.3.4). Thus the graph of $N(t)$ is S-shaped or *sigmoid* for $N_0 < K/2$.

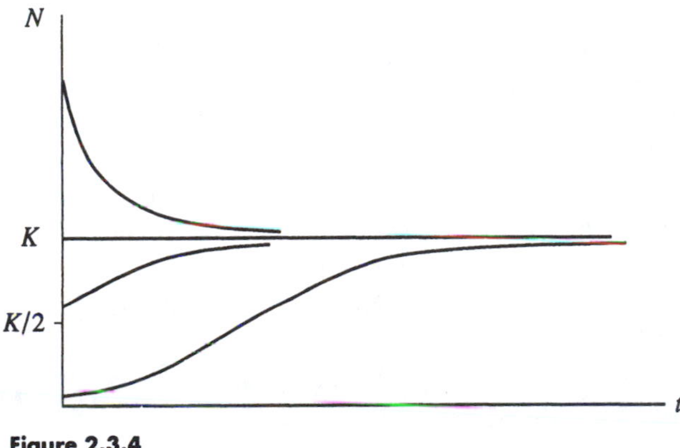

Figure 2.3.4

Something qualitatively different occurs if the initial condition N_0 lies between $K/2$ and K; now the solutions are decelerating from the start. Hence these solutions are concave down for all t. If the population initially exceeds the carrying capacity ($N_0 > K$), then $N(t)$ decreases toward $N = K$ and is concave up. Finally, if $N_0 = 0$ or $N_0 = K$, then the population stays constant.

Critique of the Logistic Model

Before leaving this example, we should make a few comments about the biological validity of the logistic equation. The algebraic form of the model is not to be taken literally. The model should really be regarded as a metaphor for populations that have a tendency to grow from zero population up to some carrying capacity K.

Originally a much stricter interpretation was proposed, and the model was argued to be a universal law of growth (Pearl 1927). The logistic equation was tested in laboratory experiments in which colonies of bacteria, yeast, or other simple organisms were grown in conditions of constant climate, food supply, and absence of predators. For a good review of this literature, see Krebs (1972, pp. 190–200). These experiments often yielded sigmoid growth curves, in some cases with an impressive match to the logistic predictions.

On the other hand, the agreement was much worse for fruit flies, flour beetles, and other organisms that have complex life cycles, involving eggs, larvae, pupae, and adults. In these organisms, the predicted asymptotic approach to a steady carrying capacity was never observed—instead the populations exhibited large, persistent fluctuations after an initial period of logistic growth. See Krebs (1972) for a discussion of the possible causes of these fluctuations, including age structure and time-delayed effects of overcrowding in the population.

For further reading on population biology, see Pielou (1969) or May (1981). Edelstein–Keshet (1988) and Murray (1989) are excellent textbooks on mathematical biology in general.

2.4 Linear Stability Analysis

So far we have relied on graphical methods to determine the stability of fixed points. Frequently one would like to have a more quantitative measure of stability, such as the rate of decay to a stable fixed point. This sort of information may be obtained by *linearizing* about a fixed point, as we now explain.

Let x^* be a fixed point, and let $\eta(t) = x(t) - x^*$ be a small perturbation away from x^*. To see whether the perturbation grows or decays, we derive a differential equation for η. Differentiation yields

$$\dot{\eta} = \tfrac{d}{dt}(x - x^*) = \dot{x},$$

since x^* is constant. Thus $\dot{\eta} = \dot{x} = f(x) = f(x^* + \eta)$. Now using Taylor's expansion we obtain

$$f(x^* + \eta) = f(x^*) + \eta f'(x^*) + O(\eta^2),$$

where $O(\eta^2)$ denotes quadratically small terms in η. Finally, note that $f(x^*) = 0$ since x^* is a fixed point. Hence

$$\dot{\eta} = \eta f'(x^*) + O(\eta^2).$$

Now if $f'(x^*) \neq 0$, the $O(\eta^2)$ terms are negligible and we may write the approximation

$$\dot{\eta} \approx \eta f'(x^*).$$

This is a linear equation in η, and is called the ***linearization about*** x^*. It shows that *the perturbation* $\eta(t)$ *grows exponentially if* $f'(x^*) > 0$ *and decays if* $f'(x^*) < 0$. If $f'(x^*) = 0$, the $O(\eta^2)$ terms are not negligible and a nonlinear analysis is needed to determine stability, as discussed in Example 2.4.3 below.

The upshot is that the slope $f'(x^*)$ at the fixed point determines its stability. If you look back at the earlier examples, you'll see that the slope was always negative at a stable fixed point. The importance of the *sign* of $f'(x^*)$ was clear from our graphical approach; the new feature is that now we have a measure of *how* stable a fixed point is—that's determined by the *magnitude* of $f'(x^*)$. This magnitude plays the role of an exponential growth or decay rate. Its reciprocal $1/|f'(x^*)|$ is a ***characteristic time scale***; it determines the time required for $x(t)$ to vary significantly in the neighborhood of x^*.

EXAMPLE 2.4.1:

Using linear stability analysis, determine the stability of the fixed points for $\dot{x} = \sin x$.

Solution: The fixed points occur where $f(x) = \sin x = 0$. Thus $x^* = k\pi$, where k is an integer. Then

$$f'(x^*) = \cos k\pi = \begin{cases} 1, & k \text{ even} \\ -1, & k \text{ odd}. \end{cases}$$

Hence x^* is unstable if k is even and stable if k is odd. This agrees with the results shown in Figure 2.1.1. ■

EXAMPLE 2.4.2:

Classify the fixed points of the logistic equation, using linear stability analysis, and find the characteristic time scale in each case.

Solution: Here $f(N) = rN\left(1 - \frac{N}{K}\right)$, with fixed points $N^* = 0$ and $N^* = K$. Then $f'(N) = r - \frac{2rN}{K}$ and so $f'(0) = r$ and $f'(K) = -r$. Hence $N^* = 0$ is unstable and $N^* = K$ is stable, as found earlier by graphical arguments. In either case, the characteristic time scale is $1/|f'(N^*)| = 1/r$. ■

EXAMPLE 2.4.3:

What can be said about the stability of a fixed point when $f'(x^*) = 0$?

Solution: Nothing can be said in general. The stability is best determined on a case-by-case basis, using graphical methods. Consider the following examples:

(a) $\dot{x} = -x^3$ (b) $\dot{x} = x^3$ (c) $\dot{x} = x^2$ (d) $\dot{x} = 0$

Each of these systems has a fixed point $x^* = 0$ with $f'(x^*) = 0$. However the stability is different in each case. Figure 2.4.1 shows that (a) is stable and (b) is unstable. Case (c) is a hybrid case we'll call ***half-stable***, since the fixed point is attracting from the left and repelling from the right. We therefore indicate this type of fixed point by a half-filled circle. Case (d) is a whole line of fixed points; perturbations neither grow nor decay.

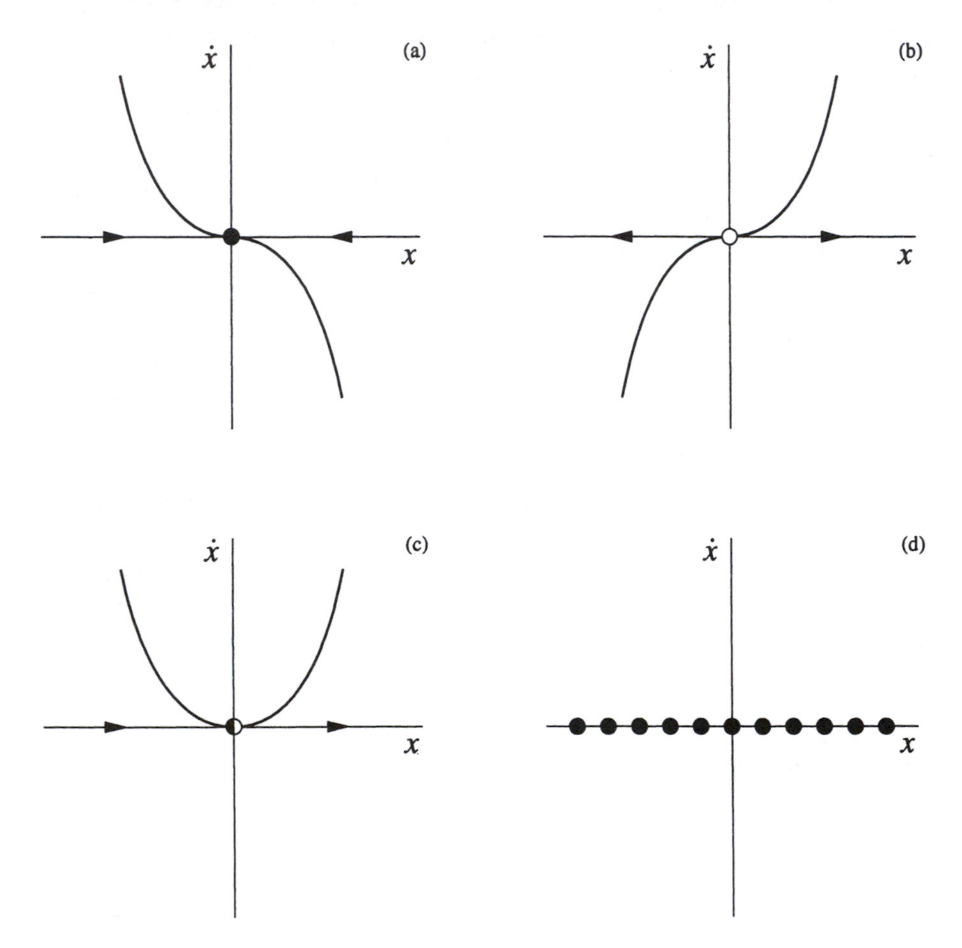

Figure 2.4.1

These examples may seem artificial, but we will see that they arise naturally in the context of *bifurcations*—more about that later. ∎

2.5 Existence and Uniqueness

Our treatment of vector fields has been very informal. In particular, we have taken a cavalier attitude toward questions of existence and uniqueness of solutions to the system $\dot{x} = f(x)$. That's in keeping with the "applied" spirit of this book. Nevertheless, we should be aware of what can go wrong in pathological cases.

EXAMPLE 2.5.1:

Show that the solution to $\dot{x} = x^{1/3}$ starting from $x_0 = 0$ is *not* unique.

Solution: The point $x = 0$ is a fixed point, so one obvious solution is $x(t) = 0$ for all t. The surprising fact is that there is *another* solution. To find it we separate variables and integrate:

$$\int x^{-1/3} dx = \int dt$$

so $\frac{3}{2}x^{2/3} = t + C$. Imposing the initial condition $x(0) = 0$ yields $C = 0$. Hence $x(t) = \left(\frac{2}{3}t\right)^{3/2}$ is also a solution! ■

When uniqueness fails, our geometric approach collapses because the phase point doesn't know how to move; if a phase point were started at the origin, would it stay there or would it move according to $x(t) = \left(\frac{2}{3}t\right)^{3/2}$? (Or as my friends in elementary school used to say when discussing the problem of the irresistible force and the immovable object, perhaps the phase point would explode!)

Actually, the situation in Example 2.5.1 is even worse than we've let on—there are *infinitely* many solutions starting from the same initial condition (Exercise 2.5.4).

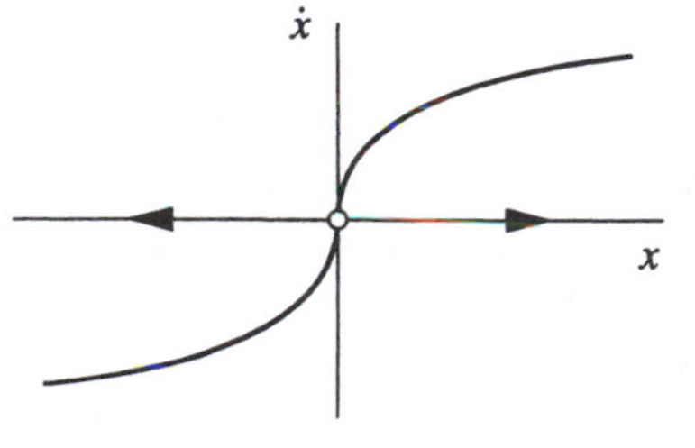

Figure 2.5.1

What's the source of the non-uniqueness? A hint comes from looking at the vector field (Figure 2.5.1). We see that the fixed point $x^* = 0$ is *very* unstable—the slope $f'(0)$ is infinite.

Chastened by this example, we state a theorem that provides sufficient conditions for existence and uniqueness of solutions to $\dot{x} = f(x)$.

Existence and Uniqueness Theorem: Consider the initial value problem

$$\dot{x} = f(x), \qquad x(0) = x_0.$$

Suppose that $f(x)$ and $f'(x)$ are continuous on an open interval R of the x-axis, and suppose that x_0 is a point in R. Then the initial value problem has a solution $x(t)$ on some time interval $(-\tau, \tau)$ about $t = 0$, and the solution is unique.

For proofs of the existence and uniqueness theorem, see Borrelli and Coleman (1987), Lin and Segel (1988), or virtually any text on ordinary differential equations.

This theorem says that *if $f(x)$ is smooth enough,* then solutions exist and are unique. Even so, there's no guarantee that solutions exist forever, as shown by the next example.

EXAMPLE 2.5.2:

Discuss the existence and uniqueness of solutions to the initial value problem $\dot{x} = 1 + x^2$, $x(0) = x_0$. Do solutions exist for all time?

Solution: Here $f(x) = 1 + x^2$. This function is continuous and has a continuous derivative for all x. Hence the theorem tells us that solutions exist and are unique for any initial condition x_0. But *the theorem does not say that the solutions exist for all time;* they are only guaranteed to exist in a (possibly very short) time interval around $t = 0$.

For example, consider the case where $x(0) = 0$. Then the problem can be solved analytically by separation of variables:

$$\int \frac{dx}{1 + x^2} = \int dt,$$

which yields

$$\tan^{-1} x = t + C$$

The initial condition $x(0) = 0$ implies $C = 0$. Hence $x(t) = \tan t$ is the solution. But notice that this solution exists only for $-\pi/2 < t < \pi/2$, because $x(t) \to \pm\infty$ as $t \to \pm\pi/2$. Outside of that time interval, there is no solution to the initial value problem for $x_0 = 0$. ■

The amazing thing about Example 2.5.2 is that the system has solutions that reach infinity *in finite time*. This phenomenon is called ***blow-up***. As the name suggests, it is of physical relevance in models of combustion and other runaway processes.

There are various ways to extend the existence and uniqueness theorem. One can allow f to depend on time t, or on several variables $x_1, \ldots, x_n$. One of the most useful generalizations will be discussed later in Section 6.2.

From now on, we will not worry about issues of existence and uniqueness—our vector fields will typically be smooth enough to avoid trouble. If we happen to come across a more dangerous example, we'll deal with it then.

2.6 Impossibility of Oscillations

Fixed points dominate the dynamics of first-order systems. In all our examples so far, all trajectories either approached a fixed point, or diverged to $\pm\infty$. In fact, those are the *only* things that can happen for a vector field on the real line. The reason is that trajectories are forced to increase or decrease monotonically, or remain constant (Figure 2.6.1). To put it more geometrically, the phase point never reverses direction.

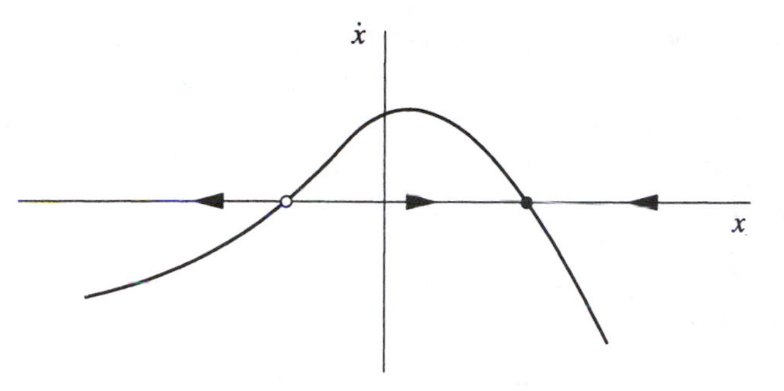

Figure 2.6.1

Thus, if a fixed point is regarded as an equilibrium solution, the approach to equilibrium is always *monotonic*—overshoot and damped oscillations can never occur in a first-order system. For the same reason, undamped oscillations are impossible. Hence *there are no periodic solutions to* $\dot{x} = f(x)$.

These general results are fundamentally topological in origin. They reflect the fact that $\dot{x} = f(x)$ corresponds to flow on a *line*. If you flow monotonically on a line, you'll never come back to your starting place—that's why periodic solutions are impossible. (Of course, if we were dealing with a *circle* rather than a line, we *could* eventually return to our starting place. Thus vector fields on the circle can exhibit periodic solutions, as we discuss in Chapter 4.)

Mechanical Analog: Overdamped Systems

It may seem surprising that solutions to $\dot{x} = f(x)$ can't oscillate. But this result becomes obvious if we think in terms of a mechanical analog. We regard $\dot{x} = f(x)$ as a limiting case of Newton's law, in the limit where the "inertia term" $m\ddot{x}$ is negligible.

For example, suppose a mass m is attached to a nonlinear spring whose restoring force is $F(x)$, where x is the displacement from the origin. Furthermore, suppose that the mass is immersed in a vat of very viscous fluid, like honey or motor oil (Figure 2.6.2), so that it is subject to a damping force $b\dot{x}$. Then Newton's law is

$$m\ddot{x} + b\dot{x} = F(x).$$

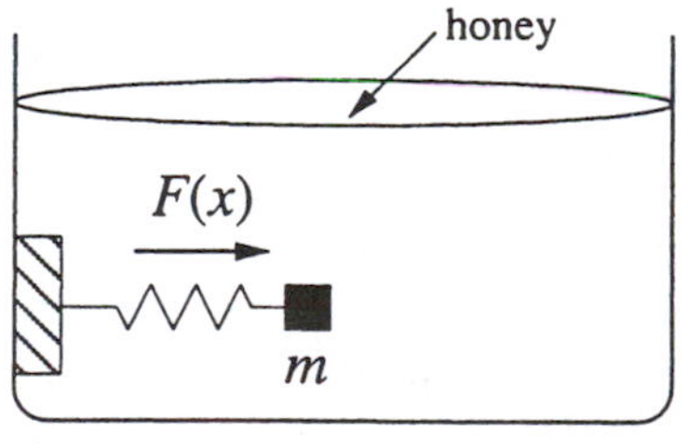

Figure 2.6.2

If the viscous damping is strong compared to the inertia term ($b\dot{x} >> m\ddot{x}$), the system should behave like $b\dot{x} = F(x)$, or equivalently $\dot{x} = f(x)$, where $f(x) = b^{-1}F(x)$. In this ***overdamped*** limit, the behavior of the mechanical system is clear. The mass prefers to sit at a stable equilibrium, where $f(x) = 0$ and $f'(x) < 0$. If displaced a bit, the mass is slowly dragged back to equilibrium by the restoring force. No overshoot can occur, because the damping is enormous. And undamped oscillations are out of the question! These conclusions agree with those obtained earlier by geometric reasoning.

Actually, we should confess that this argument contains a slight swindle. The neglect of the inertia term $m\ddot{x}$ is valid, but only after a rapid initial transient during which the inertia and damping terms are of comparable size. An honest discussion of this point requires more machinery than we have available. We'll return to this matter in Section 3.5.

2.7 Potentials

There's another way to visualize the dynamics of the first-order system $\dot{x} = f(x)$, based on the physical idea of potential energy. We picture a particle sliding down the walls of a potential well, where the ***potential*** $V(x)$ is defined by

$$f(x) = -\frac{dV}{dx}.$$

As before, you should imagine that the particle is heavily damped—its inertia is completely negligible compared to the damping force and the force due to the potential. For example, suppose that the particle has to slog through a thick layer of goo that covers the walls of the potential (Figure 2.7.1).

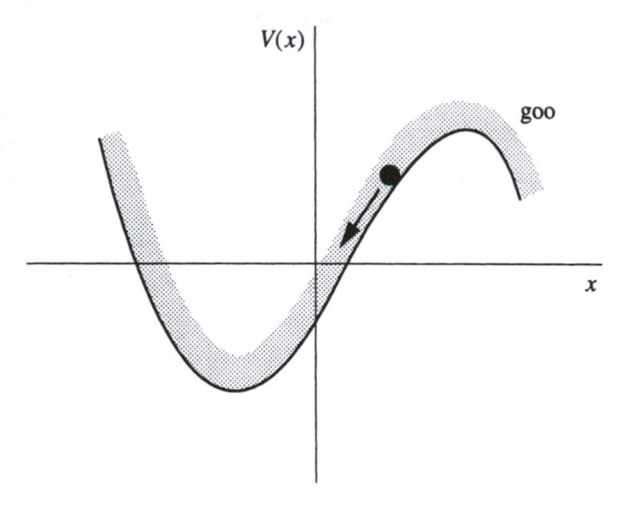

Figure 2.7.1

The negative sign in the definition of V follows the standard convention in physics; it implies that the particle always moves "downhill" as the motion proceeds. To see this, we think of x as a function of t, and then calculate the time-derivative of $V(x(t))$. Using the chain rule, we obtain

$$\frac{dV}{dt} = \frac{dV}{dx}\frac{dx}{dt}.$$

Now for a first-order system,

$$\frac{dx}{dt} = -\frac{dV}{dx},$$

since $\dot{x} = f(x) = -dV/dx$, by the definition of the potential. Hence,

$$\frac{dV}{dt} = -\left(\frac{dV}{dx}\right)^2 \leq 0.$$

Thus $V(t)$ *decreases along trajectories,* and so the particle always moves toward lower potential. Of course, if the particle happens to be at an ***equilibrium*** point where $dV/dx = 0$, then V remains constant. This is to be expected, since $dV/dx = 0$ implies $\dot{x} = 0$; equilibria occur at the fixed points of the vector field. Note that local minima of $V(x)$ correspond to *stable* fixed points, as we'd expect intuitively, and local maxima correspond to unstable fixed points.

EXAMPLE 2.7.1:

Graph the potential for the system $\dot{x} = -x$, and identify all the equilibrium points.

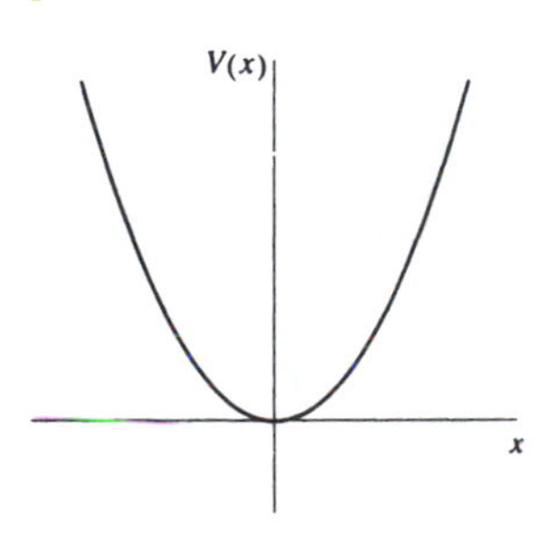

Figure 2.7.2

Solution: We need to find $V(x)$ such that $-dV/dx = -x$. The general solution is $V(x) = \frac{1}{2}x^2 + C$, where C is an arbitrary constant. (It always happens that the potential is only defined up to an additive constant. For convenience, we usually choose $C = 0$.) The graph of $V(x)$ is shown in Figure 2.7.2. The only equilibrium point occurs at $x = 0$, and it's stable. ■

EXAMPLE 2.7.2:

Graph the potential for the system $\dot{x} = x - x^3$, and identify all equilibrium points.

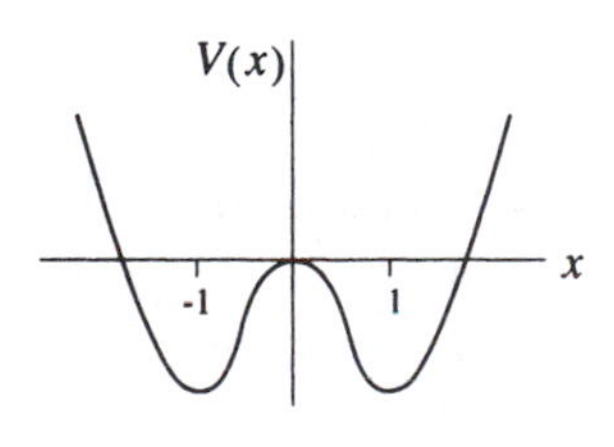

Figure 2.7.3

Solution: Solving $-dV/dx = x - x^3$ yields $V = -\frac{1}{2}x^2 + \frac{1}{4}x^4 + C$. Once again we set $C = 0$. Figure 2.7.3 shows the graph of V. The local minima at $x = \pm 1$ correspond to stable equilibria, and the local maximum at $x = 0$ corresponds to an unstable equilibrium. The potential shown in Figure 2.7.3 is often called a ***double-well potential***, and the system is said to be ***bistable***, since it has two stable equilibria. ■

2.8 Solving Equations on the Computer

Throughout this chapter we have used graphical and analytical methods to analyze first-order systems. Every budding dynamicist should master a third tool: numerical methods. In the old days, numerical methods were impractical because they required enormous amounts of tedious hand-calculation. But all that has changed, thanks to the computer. Computers enable us to approximate the solutions to analytically intractable problems, and also to visualize those solutions. In this section we take our first look at dynamics on the computer, in the context of ***numerical integration*** of $\dot{x} = f(x)$.

Numerical integration is a vast subject. We will barely scratch the surface. See Chapter 15 of Press et al. (1986) for an excellent treatment.

Euler's Method

The problem can be posed this way: given the differential equation $\dot{x} = f(x)$, subject to the condition $x = x_0$ at $t = t_0$, find a systematic way to approximate the solution $x(t)$.

Suppose we use the vector field interpretation of $\dot{x} = f(x)$. That is, we think of a fluid flowing steadily on the x-axis, with velocity $f(x)$ at the location x. Imagine we're riding along with a phase point being carried downstream by the fluid. Initially we're at x_0, and the local velocity is $f(x_0)$. If we flow for a short time Δt, we'll have moved a distance $f(x_0)\Delta t$, because distance = rate $\times$ time. Of course, that's not quite right, because our velocity was changing a little bit throughout the step. But over a sufficiently *small* step, the velocity will be nearly constant and our approximation should be reasonably good. Hence our new position $x(t_0 + \Delta t)$ is approximately $x_0 + f(x_0)\Delta t$. Let's call this approximation x_1. Thus

$$x(t_0 + \Delta t) \approx x_1 = x_0 + f(x_0)\Delta t.$$

Now we iterate. Our approximation has taken us to a new location x_1; our new velocity is $f(x_1)$; we step forward to $x_2 = x_1 + f(x_1)\Delta t$; and so on. In general, the update rule is

$$x_{n+1} = x_n + f(x_n)\Delta t.$$

This is the simplest possible numerical integration scheme. It is known as ***Euler's method***.

Euler's method can be visualized by plotting x versus t (Figure 2.8.1). The curve shows the exact solution $x(t)$, and the open dots show its values $x(t_n)$ at the discrete times $t_n = t_0 + n\Delta t$. The black dots show the approximate values given by the Euler method. As you can see, the approximation gets bad in a hurry unless Δt is extremely small. Hence Euler's method is not recommended in practice, but it contains the conceptual essence of the more accurate methods to be discussed next.

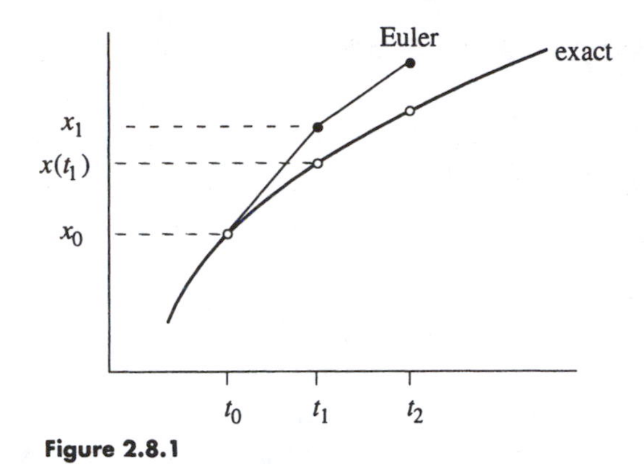

Figure 2.8.1

Refinements

One problem with the Euler method is that it estimates the derivative only at the left end of the time interval between t_n and t_{n+1}. A more sensible approach would be to use the *average* derivative across this interval. This is the idea behind the ***improved Euler method***. We first take a trial step across the interval, using the Euler method. This produces a trial value $\tilde{x}_{n+1} = x_n + f(x_n)\Delta t$; the tilde above the x indicates that this is a tentative step, used only as a probe. Now that we've estimated the derivative on both ends of the interval, we average $f(x_n)$ and $f(\tilde{x}_{n+1})$, and use that to take the *real* step across the interval. Thus the improved Euler method is

$$\tilde{x}_{n+1} = x_n + f(x_n)\Delta t \qquad \text{(the trial step)}$$

$$x_{n+1} = x_n + \tfrac{1}{2}\left[f(x_n) + f(\tilde{x}_{n+1})\right]\Delta t. \qquad \text{(the real step)}$$

This method is more accurate than the Euler method, in the sense that it tends to make a smaller ***error*** $E = |x(t_n) - x_n|$ for a given ***stepsize*** Δt. In both cases, the error $E \to 0$ as $\Delta t \to 0$, but the error decreases *faster* for the improved Euler method. One can show that $E \propto \Delta t$ for the Euler method, but $E \propto (\Delta t)^2$ for the improved Euler method (Exercises 2.8.7 and 2.8.8). In the jargon of numerical analysis, the Euler method is first order, whereas the improved Euler method is second order.

Methods of third, fourth, and even higher orders have been concocted, but you should realize that higher order methods are not necessarily superior. Higher order methods require more calculations and function evaluations, so there's a computational cost associated with them. In practice, a good balance is achieved by the ***fourth-order Runge–Kutta method***. To find x_{n+1} in terms of x_n, this method first requires us to calculate the following four numbers (cunningly chosen, as you'll see in Exercise 2.8.9):

$$
\begin{aligned}
k_1 &= f(x_n)\Delta t \\
k_2 &= f(x_n + \tfrac{1}{2}k_1)\Delta t \\
k_3 &= f(x_n + \tfrac{1}{2}k_2)\Delta t \\
k_4 &= f(x_n + k_3)\Delta t .
\end{aligned}
$$

Then x_{n+1} is given by

$$x_{n+1} = x_n + \tfrac{1}{6}(k_1 + 2k_2 + 2k_3 + k_4).$$

This method generally gives accurate results without requiring an excessively small stepsize Δt. Of course, some problems are nastier, and may require small steps in certain time intervals, while permitting very large steps elsewhere. In such cases, you may want to use a Runge–Kutta routine with an automatic stepsize control; see Press et al. (1986) for details.

Now that computers are so fast, you may wonder why we don't just pick a tiny Δt once and for all. The trouble is that excessively many computations will occur, and each one carries a penalty in the form of ***round-off error***. Computers don't have infinite accuracy—they don't distinguish between numbers that differ by some small amount δ. For numbers of order 1, typically $\delta \approx 10^{-7}$ for single precision and $\delta \approx 10^{-16}$ for double precision. Round-off error occurs during every calculation, and will begin to accumulate in a serious way if Δt is too small. See Hubbard and West (1991) for a good discussion.

Practical Matters

You have several options if you want to solve differential equations on the computer. If you like to do things yourself, you can write your own numerical integration routines, and plot the results using whatever graphics facilities are available. The information given above should be enough to get you started. For further guidance, consult Press et al. (1986); they provide sample routines written in Fortran, C, and Pascal.

A second option is to use existing packages for numerical methods. The software libraries by IMSL and NAG have a wide variety of state-of-the-art numerical integrators. These libraries are well documented, reliable, and flexible, and can be found at most university computing centers or networks. The packages *Matlab*, *Mathematica*, and *Maple* are more interactive and also have programs for solving ordinary differential equations.

The final option is for people who want to explore dynamics, not computing. Dynamical systems software has recently become available for personal computers. All you have to do is type in the equations and the parameters; the program solves the equations numerically and plots the results. Some recommended programs are *Phaser* (Kocak 1989) for the IBM PC or *MacMath* (Hubbard and West

1992) for the Macintosh. *MacMath* was used to generate many of the plots in this book.

These programs are easy to use, and they will help you build intuition about dynamical systems.

EXAMPLE 2.8.1:

Use *MacMath* to solve the system $\dot{x} = x(1-x)$ numerically.

Solution: This is a logistic equation (Section 2.3) with parameters $r = 1$, $K = 1$. Previously we gave a rough sketch of the solutions, based on geometric arguments; now we can draw a more quantitative picture.

As a first step, we plot the ***slope field*** for the system in the (t, x) plane (Figure 2.8.2). Here the equation $\dot{x} = x(1-x)$ is being interpreted in a new way: for each point (t, x), the equation gives the slope dx/dt of the solution passing through that point. The slopes are indicated by little line segments in Figure 2.8.2.

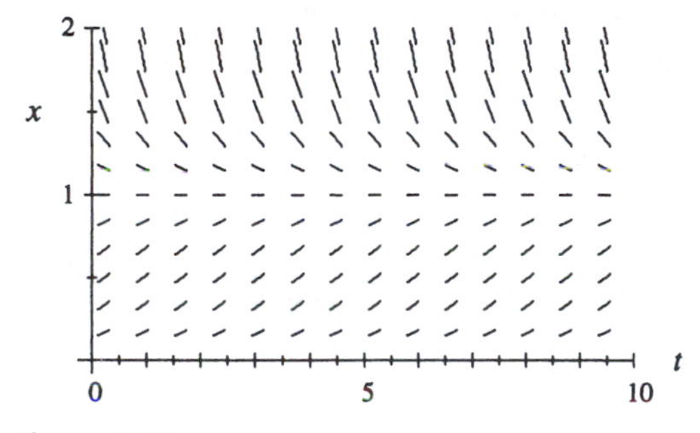

Figure 2.8.2

Finding a solution now becomes a problem of drawing a curve that is always tangent to the local slope. Figure 2.8.3 shows four solutions starting from various points in the (t, x) plane.

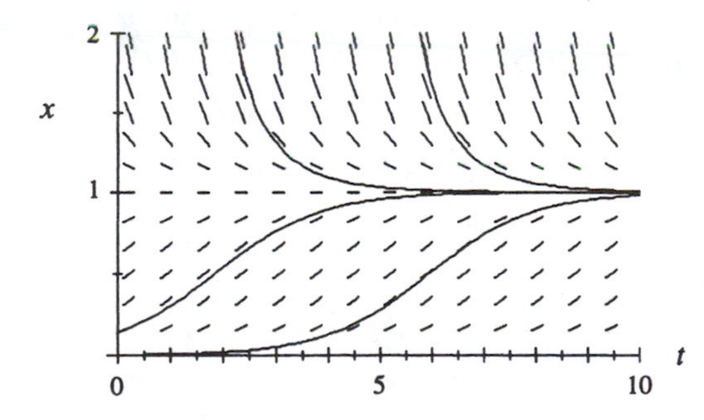

Figure 2.8.3

These numerical solutions were computed using the Runge–Kutta method with a stepsize $\Delta t = 0.1$. The solutions have the shape expected from Section 2.3. ■

Computers are indispensable for studying dynamical systems. We will use them liberally throughout this book, and you should do likewise.

EXERCISES FOR CHAPTER 2

2.1 A Geometric Way of Thinking

In the next three exercises, interpret $\dot{x} = \sin x$ as a flow on the line.

2.1.1 Find all the fixed points of the flow.

2.1.2 At which points x does the flow have greatest velocity to the right?

2.1.3

a) Find the flow's acceleration $\ddot{x}$ as a function of x.

b) Find the points where the flow has maximum positive acceleration.

2.1.4 (Exact solution of $\dot{x} = \sin x$) As shown in the text, $\dot{x} = \sin x$ has the solution $t = \ln|(\csc x_0 + \cot x_0)/(\csc x + \cot x)|$, where $x_0 = x(0)$ is the initial value of x.

a) Given the specific initial condition $x_0 = \pi/4$, show that the solution above can be inverted to obtain

$$x(t) = 2\tan^{-1}\left(\frac{e^t}{1+\sqrt{2}}\right).$$

Conclude that $x(t) \to \pi$ as $t \to \infty$, as claimed in Section 2.1. (You need to be good with trigonometric identities to solve this problem.)

b) Try to find the analytical solution for $x(t)$, given an *arbitrary* initial condition x_0.

2.1.5 (A mechanical analog)

a) Find a mechanical system that is approximately governed by $\dot{x} = \sin x$.

b) Using your physical intuition, explain why it now becomes obvious that $x^* = 0$ is an unstable fixed point and $x^* = \pi$ is stable.

2.2 Fixed Points and Stability

Analyze the following equations graphically. In each case, sketch the vector field on the real line, find all the fixed points, classify their stability, and sketch the graph of $x(t)$ for different initial conditions. Then try for a few minutes to obtain the analytical solution for $x(t)$; if you get stuck, don't try for too long since in several cases it's impossible to solve the equation in closed form!

2.2.1 $\dot{x} = 4x^2 - 16$

2.2.2 $\dot{x} = 1 - x^{14}$

2.2.3 $\dot{x} = x - x^3$

2.2.4 $\dot{x} = e^{-x} \sin x$

2.2.5 $\dot{x} = 1 + \frac{1}{2} \cos x$

2.2.6 $\dot{x} = 1 - 2\cos x$

2.2.7 $\dot{x} = e^x - \cos x$ (Hint: Sketch the graphs of e^x and $\cos x$ on the same axes, and look for intersections. You won't be able to find the fixed points explicitly, but you can still find the qualitative behavior.)

2.2.8 (Working backwards, from flows to equations) Given an equation $\dot{x} = f(x)$, we know how to sketch the corresponding flow on the real line. Here you are asked to solve the opposite problem: For the phase portrait shown in Figure 1, find an equation that is consistent with it. (There are an infinite number of correct answers—and wrong ones too.)

Figure 1

2.2.9 (Backwards again, now from solutions to equations) Find an equation $\dot{x} = f(x)$ whose solutions $x(t)$ are consistent with those shown in Figure 2.

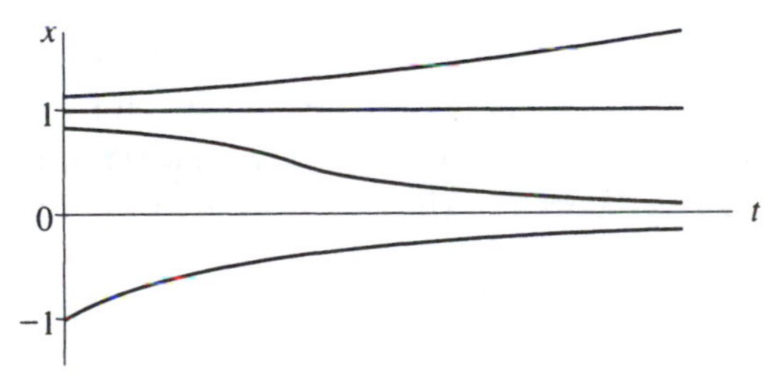

Figure 2

2.2.10 (Fixed points) For each of (a)–(e), find an equation $\dot{x} = f(x)$ with the stated properties, or if there are no examples, explain why not. (In all cases, assume that $f(x)$ is a smooth function.)

a) Every real number is a fixed point.

b) Every integer is a fixed point, and there are no others.

c) There are precisely three fixed points, and all of them are stable.

d) There are no fixed points.

e) There are precisely 100 fixed points.

2.2.11 (Analytical solution for charging capacitor) Obtain the analytical solution of the initial value problem $\dot{Q} = \frac{V_0}{R} - \frac{Q}{RC}$, with $Q(0) = 0$, which arose in Example 2.2.2.

2.2.12 (A nonlinear resistor) Suppose the resistor in Example 2.2.2 is replaced by a nonlinear resistor. In other words, this resistor does not have a linear

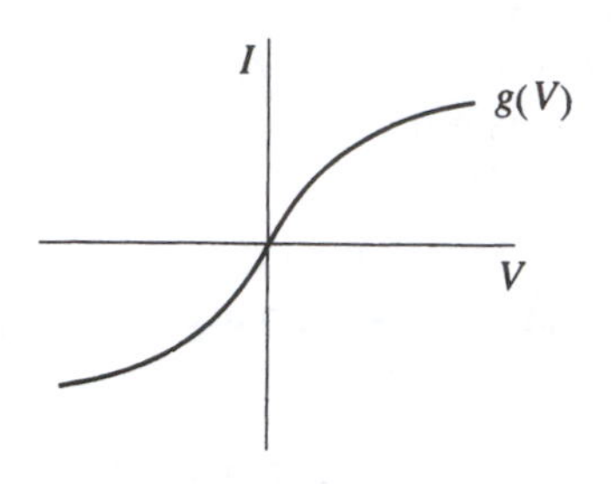

Figure 3

relation between voltage and current. Such nonlinearity arises in certain solid-state devices. Instead of $I_R = V/R$, suppose we have $I_R = g(V)$, where $g(V)$ has the shape shown in Figure 3.

Redo Example 2.2.2 in this case. Derive the circuit equations, find all the fixed points, and analyze their stability. What qualitative effects does the nonlinearity introduce (if any)?

2.2.13 (Terminal velocity) The velocity $v(t)$ of a skydiver falling to the ground is governed by $m\dot{v} = mg - kv^2$, where m is the mass of the skydiver, g is the acceleration due to gravity, and $k > 0$ is a constant related to the amount of air resistance.

a) Obtain the analytical solution for $v(t)$, assuming that $v(0) = 0$.

b) Find the limit of $v(t)$ as $t \to \infty$. This limiting velocity is called the *terminal velocity*. (Beware of bad jokes about the word *terminal* and parachutes that fail to open.)

c) Give a graphical analysis of this problem, and thereby re-derive a formula for the terminal velocity.

d) An experimental study (Carlson et al. 1942) confirmed that the equation $m\dot{v} = mg - kv^2$ gives a good quantitative fit to data on human skydivers. Six men were dropped from altitudes varying from 10,600 feet to 31,400 feet to a terminal altitude of 2,100 feet, at which they opened their parachutes. The long free fall from 31,400 to 2,100 feet took 116 seconds. The average weight of the men and their equipment was 261.2 pounds. In these units, $g = 32.2$ ft/sec^2. Compute the average velocity V_{avg}.

e) Using the data given here, estimate the terminal velocity, and the value of the drag constant k. (Hints: First you need to find an exact formula for $s(t)$, the distance fallen, where $s(0) = 0$, $\dot{s} = v$, and $v(t)$ is known from part (a). You should get $s(t) = \frac{V^2}{g} \ln\left(\cosh \frac{gt}{V}\right)$, where V is the terminal velocity. Then solve for V graphically or numerically, using $s = 29{,}300$, $t = 116$, and $g = 32.2$.)

A slicker way to estimate V is to suppose $V \approx V_{\text{avg}}$, as a rough first approximation. Then show that $gt/V \approx 15$. Since $gt/V >> 1$, we may use the approximation $\ln(\cosh x) \approx x - \ln 2$ for $x >> 1$. Derive this approximation and then use it to obtain an analytical estimate of V. Then k follows from part (b). This analysis is from Davis (1962).

2.3 Population Growth

2.3.1 (Exact solution of logistic equation) There are two ways to solve the logistic equation $\dot{N} = rN(1 - N/K)$ analytically for an arbitrary initial condition N_0.

a) Separate variables and integrate, using partial fractions.

b) Make the change of variables $x = 1/N$. Then derive and solve the resulting differential equation for x.

2.3.2 (Autocatalysis) Consider the model chemical reaction

$$A + X \underset{k_{-1}}{\overset{k_1}{\rightleftharpoons}} 2X$$

in which one molecule of X combines with one molecule of A to form two molecules of X. This means that the chemical X stimulates its own production, a process called *autocatalysis.* This positive feedback process leads to a chain reaction, which eventually is limited by a "back reaction" in which $2X$ returns to $A + X$.

According to the *law of mass action* of chemical kinetics, the rate of an elementary reaction is proportional to the product of the concentrations of the reactants. We denote the concentrations by lowercase letters $x = [X]$ and $a = [A]$. Assume that there's an enormous surplus of chemical A, so that its concentration a can be regarded as constant. Then the equation for the kinetics of x is

$$\dot{x} = k_1 a x - k_{-1} x^2$$

where k_1 and k_{-1} are positive parameters called *rate constants.*

a) Find all the fixed points of this equation and classify their stability.

b) Sketch the graph of $x(t)$ for various initial values x_0.

2.3.3 (Tumor growth) The growth of cancerous tumors can be modeled by the Gompertz law $\dot{N} = -aN\ln(bN)$, where $N(t)$ is proportional to the number of cells in the tumor, and $a, b > 0$ are parameters.

a) Interpret a and b biologically.

b) Sketch the vector field and then graph $N(t)$ for various initial values.

The predictions of this simple model agree surprisingly well with data on tumor growth, as long as N is not too small; see Aroesty et al. (1973) and Newton (1980) for examples.

2.3.4 (The Allee effect) For certain species of organisms, the effective growth rate $\dot{N}/N$ is highest at intermediate N. This is the called the Allee effect (Edelstein–Keshet 1988). For example, imagine that it is too hard to find mates when N is very small, and there is too much competition for food and other resources when N is large.

a) Show that $\dot{N}/N = r - a(N-b)^2$ provides an example of Allee effect, if r, a, and b satisfy certain constraints, to be determined.

b) Find all the fixed points of the system and classify their stability.

c) Sketch the solutions $N(t)$ for different initial conditions.

d) Compare the solutions $N(t)$ to those found for the logistic equation. What are the qualitative differences, if any?

2.4 Linear Stability Analysis

Use linear stability analysis to classify the fixed points of the following systems. If linear stability analysis fails because $f'(x^*) = 0$, use a graphical argument to decide the stability.

2.4.1 $\dot{x} = x(1-x)$

2.4.2 $\dot{x} = x(1-x)(2-x)$

2.4.3 $\dot{x} = \tan x$

2.4.4 $\dot{x} = x^2(6-x)$

2.4.5 $\dot{x} = 1 - e^{-x^2}$

2.4.6 $\dot{x} = \ln x$

2.4.7 $\dot{x} = ax - x^3$, where a can be positive, negative, or zero. Discuss all three cases.

2.4.8 Using linear stability analysis, classify the fixed points of the Gompertz model of tumor growth $\dot{N} = -aN\ln(bN)$. (As in Exercise 2.3.3, $N(t)$ is proportional to the number of cells in the tumor and $a, b > 0$ are parameters.)

2.4.9 (Critical slowing down) In statistical mechanics, the phenomenon of "critical slowing down" is a signature of a second-order phase transition. At the transition, the system relaxes to equilibrium much more slowly than usual. Here's a mathematical version of the effect:

a) Obtain the analytical solution to $\dot{x} = -x^3$ for an arbitrary initial condition. Show that $x(t) \to 0$ as $t \to \infty$, but that the decay is not exponential. (You should find that the decay is a much slower algebraic function of t.)

b) To get some intuition about the slowness of the decay, make a numerically accurate plot of the solution for the initial condition $x_0 = 10$, for $0 \le t \le 10$. Then, on the same graph, plot the solution to $\dot{x} = -x$ for the same initial condition.

2.5 Existence and Uniqueness

2.5.1 (Reaching a fixed point in a finite time) A particle travels on the half-line $x \ge 0$ with a velocity given by $\dot{x} = -x^c$, where c is real and constant.

a) Find all values of c such that the origin $x = 0$ is a stable fixed point.

b) Now assume that c is chosen such that $x = 0$ is stable. Can the particle ever reach the origin in a *finite* time? Specifically, how long does it take for the particle to travel from $x = 1$ to $x = 0$, as a function of c?

2.5.2 ("Blow-up": Reaching infinity in a finite time) Show that the solution to $\dot{x} = 1 + x^{10}$ escapes to $+\infty$ in a finite time, starting from any initial condition. (Hint: Don't try to find an exact solution; instead, compare the solutions to those of $\dot{x} = 1 + x^2$.)

2.5.3 Consider the equation $\dot{x} = rx + x^3$, where $r > 0$ is fixed. Show that $x(t) \to \pm\infty$ in finite time, starting from any initial condition $x_0 \neq 0$.

2.5.4 (Infinitely many solutions with the same initial condition) Show that the initial value problem $\dot{x} = x^{1/3}$, $x(0) = 0$, has an infinite number of solutions. (Hint:

Construct a solution that stays at $x = 0$ until some arbitrary time t_0, after which it takes off.)

2.5.5 (A general example of non-uniqueness) Consider the initial value problem $\dot{x} = |x|^{p/q}$, $x(0) = 0$, where p and q are positive integers with no common factors.

a) Show that there are an infinite number of solutions if $p < q$.

b) Show that there is a unique solution if $p > q$.

2.5.6 (The leaky bucket) The following example (Hubbard and West 1991, p. 159) shows that in some physical situations, non-uniqueness is natural and obvious, not pathological.

Consider a water bucket with a hole in the bottom. If you see an empty bucket with a puddle beneath it, can you figure out when the bucket was full? No, of course not! It could have finished emptying a minute ago, ten minutes ago, or whatever. The solution to the corresponding differential equation must be non-unique when integrated backwards in time.

Here's a crude model of the situation. Let $h(t)$ = height of the water remaining in the bucket at time t; a = area of the hole; A = cross-sectional area of the bucket (assumed constant); $v(t)$ = velocity of the water passing through the hole.

a) Show that $av(t) = A\dot{h}(t)$. What physical law are you invoking?

b) To derive an additional equation, use conservation of energy. First, find the change in potential energy in the system, assuming that the height of the water in the bucket decreases by an amount Δh and that the water has density ρ. Then find the kinetic energy transported out of the bucket by the escaping water. Finally, assuming all the potential energy is converted into kinetic energy, derive the equation $v^2 = 2gh$.

c) Combining (b) and (c), show $\dot{h} = -C\sqrt{h}$, where $C = \sqrt{2g}\left(\frac{a}{A}\right)$.

d) Given $h(0) = 0$ (bucket empty at $t = 0$), show that the solution for $h(t)$ is non-unique *in backwards time*, i.e., for $t < 0$.

2.6 Impossibility of Oscillations

2.6.1 Explain this paradox: a simple harmonic oscillator $m\ddot{x} = -kx$ is a system that oscillates in one dimension (along the x-axis). But the text says one-dimensional systems can't oscillate.

2.6.2 (No periodic solutions to $\dot{x} = f(x)$) Here's an analytic proof that periodic solutions are impossible for a vector field on a line. Suppose on the contrary that $x(t)$ is a nontrivial periodic solution, i.e., $x(t) = x(t+T)$ for some $T > 0$, and $x(t) \neq x(t+s)$ for all $0 < s < T$. Derive a contradiction by considering $\int_t^{t+T} f(x)\frac{dx}{dt}\,dt$.

2.7 Potentials

For each of the following vector fields, plot the potential function $V(x)$ and identify all the equilibrium points and their stability.

2.7.1 $\dot{x} = x(1-x)$

2.7.2 $\dot{x} = 3$

2.7.3 $\dot{x} = \sin x$

2.7.4 $\dot{x} = 2 + \sin x$

2.7.5 $\dot{x} = -\sinh x$

2.7.6 $\dot{x} = r + x - x^3$, for various values of r.

2.7.7 (Another proof that solutions to $\dot{x} = f(x)$ can't oscillate) Let $\dot{x} = f(x)$ be a vector field on the line. Use the existence of a potential function $V(x)$ to show that solutions $x(t)$ cannot oscillate.

2.8 Solving Equations on the Computer

2.8.1 (Slope field) The slope is constant along horizontal lines in Figure 2.8.2. Why should we have expected this?

2.8.2 Sketch the slope field for the following differential equations. Then "integrate" the equation manually by drawing trajectories that are everywhere parallel to the local slope.
a) $\dot{x} = x$ b) $\dot{x} = 1 - x^2$ c) $\dot{x} = 1 - 4x(1-x)$ d) $\dot{x} = \sin x$

2.8.3 (Calibrating the Euler method) The goal of this problem is to test the Euler method on the initial value problem $\dot{x} = -x$, $x(0) = 1$.
a) Solve the problem analytically. What is the exact value of $x(1)$?
b) Using the Euler method with step size $\Delta t = 1$, estimate $x(1)$ numerically—call the result $\hat{x}(1)$. Then repeat, using $\Delta t = 10^{-n}$, for $n = 1, 2, 3, 4$.
c) Plot the error $E = |\hat{x}(1) - x(1)|$ as a function of Δt. Then plot $\ln E$ vs. $\ln t$. Explain the results.

2.8.4 Redo Exercise 2.8.3, using the improved Euler method.

2.8.5 Redo Exercise 2.8.3, using the Runge–Kutta method.

2.8.6 (Analytically intractable problem) Consider the initial value problem $\dot{x} = x + e^{-x}$, $x(0) = 0$. In contrast to Exercise 2.8.3, this problem can't be solved analytically.
a) Sketch the solution $x(t)$ for $t \geq 0$.
b) Using some analytical arguments, obtain rigorous bounds on the value of x at $t = 1$. In other words, prove that $a < x(1) < b$, for a, b to be determined. By being clever, try to make a and b as close together as possible. (Hint: Bound the given vector field by approximate vector fields that can be integrated analytically.)
c) Now for the numerical part: Using the Euler method, compute x at $t = 1$, correct to three decimal places. How small does the stepsize need to be to obtain the desired accuracy? (Give the order of magnitude, not the exact number.)

d) Repeat part (b), now using the Runge–Kutta method. Compare the results for stepsizes $\Delta t = 1$, $\Delta t = 0.1$, and $\Delta t = 0.01$.

2.8.7 (Error estimate for Euler method) In this question you'll use Taylor series expansions to estimate the error in taking one step by the Euler method. The exact solution and the Euler approximation both start at $x = x_0$ when $t = t_0$. We want to compare the exact value $x(t_1) \equiv x(t_0 + \Delta t)$ with the Euler approximation $x_1 = x_0 + f(x_0)\Delta t$.

a) Expand $x(t_1) \equiv x(t_0 + \Delta t)$ as a Taylor series in Δt, through terms of $O(\Delta t^2)$. Express your answer solely in terms of x_0, Δt, and f and its derivatives at x_0.

b) Show that the local error $|x(t_1) - x_1| \sim C(\Delta t)^2$ and give an explicit expression for the constant C. (Generally one is more interested in the global error incurred after integrating over a time interval of fixed length $T = n\Delta t$. Since each step produces an $O(\Delta t)^2$ error, and we take $n = T/\Delta t = O(\Delta t^{-1})$ steps, the global error $|x(t_n) - x_n|$ is $O(\Delta t)$, as claimed in the text.)

2.8.8 (Error estimate for the improved Euler method) Use the Taylor series arguments of Exercise 2.8.7 to show that the local error for the improved Euler method is $O(\Delta t^3)$.

2.8.9 (Error estimate for Runge–Kutta) Show that the Runge–Kutta method produces a local error of size $O(\Delta t^5)$.
(Warning: This calculation involves massive amounts of algebra, but if you do it correctly, you'll be rewarded by seeing many wonderful cancellations. Teach yourself *Mathematica*, *Maple*, or some other symbolic manipulation language, and do the problem on the computer.)

3

BIFURCATIONS

3.0 Introduction

As we've seen in Chapter 2, the dynamics of vector fields on the line is very limited: all solutions either settle down to equilibrium or head out to $\pm\infty$. Given the triviality of the dynamics, what's interesting about one-dimensional systems? Answer: *Dependence on parameters.* The qualitative structure of the flow can change as parameters are varied. In particular, fixed points can be created or destroyed, or their stability can change. These qualitative changes in the dynamics are called ***bifurcations***, and the parameter values at which they occur are called ***bifurcation points***.

Bifurcations are important scientifically—they provide models of transitions and instabilities as some *control parameter* is varied. For example, consider the buckling of a beam. If a small weight is placed on top of the beam in Figure 3.0.1, the beam can support the load and remain vertical. But if the load is too heavy, the vertical position becomes unstable, and the beam may buckle.

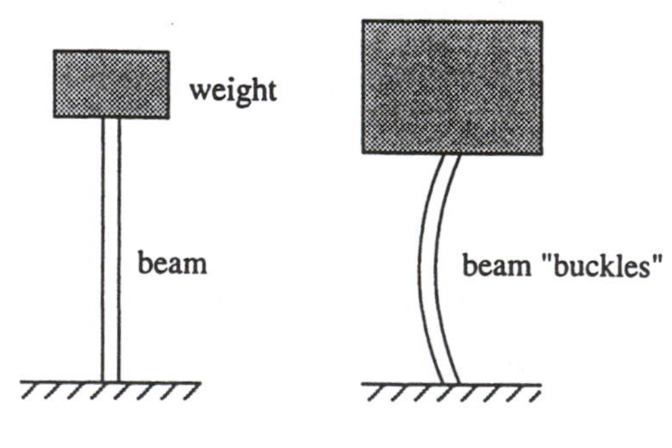

Figure 3.0.1

Here the weight plays the role of the control parameter, and the deflection of the beam from vertical plays the role of the dynamical variable x.

One of the main goals of this book is to help you develop a solid and practical understanding of bifurcations. This chapter introduces the simplest examples: bifurcations of fixed points for flows on the line. We'll use these bifurcations to model such dramatic phenomena as the onset of coherent radiation in a laser and the outbreak of an insect population. (In later chapters, when we step up to two- and three-dimensional phase spaces, we'll explore additional types of bifurcations and their scientific applications.)

We begin with the most fundamental bifurcation of all.

3.1 Saddle-Node Bifurcation

The saddle-node bifurcation is the basic mechanism by which fixed points are *created and destroyed.* As a parameter is varied, two fixed points move toward each other, collide, and mutually annihilate.

The prototypical example of a saddle-node bifurcation is given by the first-order system

$$\dot{x} = r + x^2 \qquad (1)$$

where r is a parameter, which may be positive, negative, or zero. When r is negative, there are two fixed points, one stable and one unstable (Figure 3.1.1a).

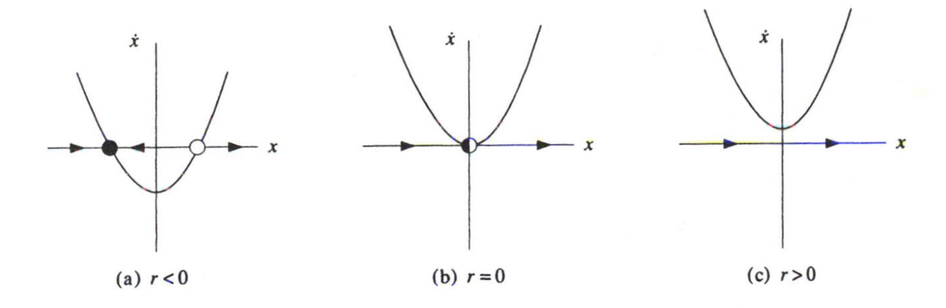

Figure 3.1.1

As r approaches 0 from below, the parabola moves up and the two fixed points move toward each other. When $r = 0$, the fixed points coalesce into a half-stable fixed point at $x^* = 0$ (Figure 3.1.1b). This type of fixed point is extremely delicate—it vanishes as soon as $r > 0$, and now there are no fixed points at all (Figure 3.1.1c).

In this example, we say that a *bifurcation* occurred at $r = 0$, since the vector fields for $r < 0$ and $r > 0$ are qualitatively different.

Graphical Conventions

There are several other ways to depict a saddle-node bifurcation. We can show a stack of vector fields for discrete values of r (Figure 3.1.2).

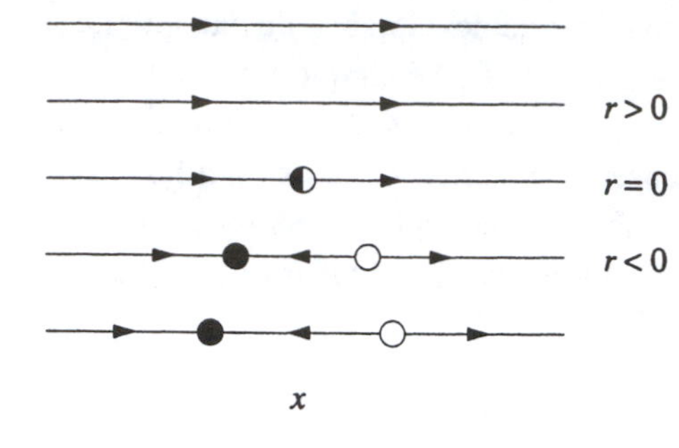

Figure 3.1.2

This representation emphasizes the dependence of the fixed points on r. In the limit of a *continuous* stack of vector fields, we have a picture like Figure 3.1.3. The curve shown is $r=-x^2$, i.e., $\dot{x}=0$, which gives the fixed points for different r. To distinguish between stable and unstable fixed points, we use a solid line for stable points and a broken line for unstable ones.

However, the most common way to depict the bifurcation is to invert the axes of Figure 3.1.3. The rationale is that r plays the role of an independent variable, and so should be plotted horizontally

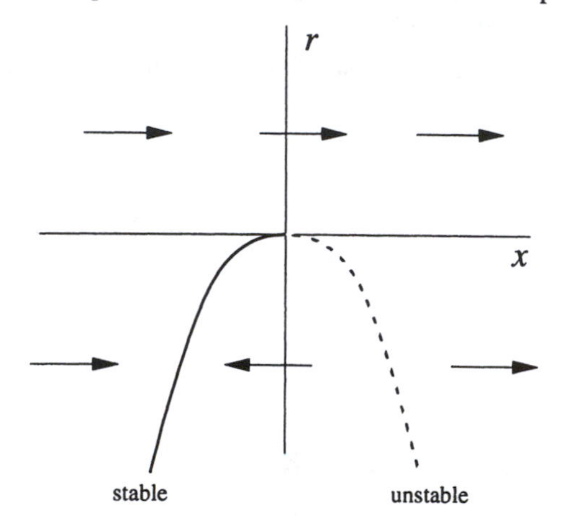

Figure 3.1.3

(Figure 3.1.4). The drawback is that now the x-axis has to be plotted vertically, which looks strange at first. Arrows are sometimes included in the picture, but not always. This picture is called the ***bifurcation diagram*** for the saddle-node bifurcation.

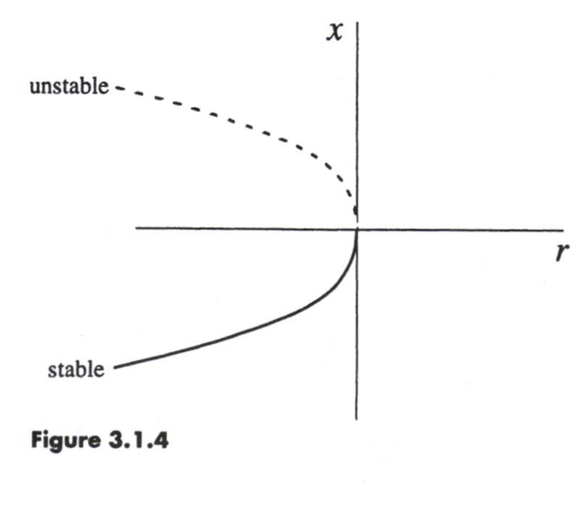

Figure 3.1.4

Terminology

Bifurcation theory is rife with conflicting terminology. The subject really hasn't settled down yet, and different people use different words for the same thing. For example, the saddle-node bifurcation is sometimes called a *fold bifurcation* (because the curve in Figure 3.1.4 has a fold in it) or a *turning-point bifurcation* (because the point $(x,r)=(0,0)$ is a "turning point.") Admittedly, the term *saddle-node* doesn't make much sense for vector fields on the line. The name derives from a completely analogous bifurcation seen in a higher-dimensional context, such as vector fields on the plane, where fixed points known as saddles and nodes can collide and annihilate (see Section 8.1).

The prize for most inventive terminology must go to Abraham and Shaw (1988), who write of a ***blue sky bifurcation***. This term comes from viewing a saddle-node bifurcation in the other direction: a pair of fixed points appears "out of the clear blue sky" as a parameter is varied. For example, the vector field

$$\dot{x} = r - x^2 \tag{2}$$

has no fixed points for $r<0$, but then one materializes when $r=0$ and splits into two when $r>0$ (Figure 3.1.5). Incidentally, this example also explains why we use the word "bifurcation": it means "splitting into two branches."

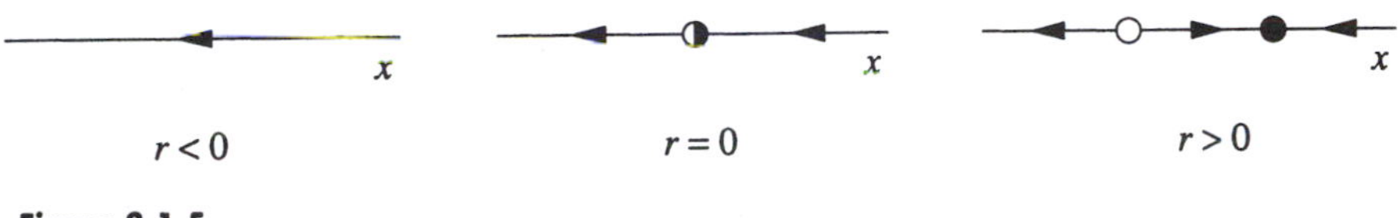

Figure 3.1.5

EXAMPLE 3.1.1:

Give a linear stability analysis of the fixed points in Figure 3.1.5.

Solution: The fixed points for $\dot{x} = f(x) = r - x^2$ are given by $x^* = \pm\sqrt{r}$. There are two fixed points for $r \geq 0$, and none for $r<0$. To determine linear stability, we compute $f'(x^*) = -2x^*$. Thus $x^* = +\sqrt{r}$ is stable, since $f'(x^*)<0$. Similarly $x^* = -\sqrt{r}$ is unstable. At the bifurcation point $r=0$, we find $f'(x^*)=0$; the linearization vanishes when the fixed points coalesce. ■

EXAMPLE 3.1.2:

Show that the first-order system $\dot{x} = r - x - e^{-x}$ undergoes a saddle-node bifurcation as r is varied, and find the value of r at the bifurcation point.

Solution: The fixed points satisfy $f(x) = r - x - e^{-x} = 0$. But now we run into a difficulty—in contrast to Example 3.1.1, we can't find the fixed points explicitly as a function of r. Instead we adopt a geometric approach. One method would be to graph the function $f(x) = r - x - e^{-x}$ for different values of r, look for its roots x^*, and then sketch the vector field on the x-axis. This method is fine, but there's an easier way. The point is that the two functions $r-x$ and e^{-x} have much more familiar graphs than their difference $r-x-e^{-x}$. So we plot $r-x$ and e^{-x} on the same picture (Figure 3.1.6a). Where the line $r-x$ intersects the curve e^{-x}, we have $r-x=e^{-x}$ and so $f(x)=0$. *Thus, intersections of the line and the curve correspond to fixed points for the system.* This picture also allows us to read off the direction of flow on the x-axis: the flow is to the right where the line lies above the curve, since $r-x>e^{-x}$ and therefore $\dot{x}>0$. Hence, the fixed point on the right is stable, and the one on the left is unstable.

Now imagine we start decreasing the parameter r. The line $r-x$ slides down and the fixed points approach each other. At some critical value $r=r_c$, the line becomes *tangent* to the curve and the fixed points coalesce in a saddle-node bifurcation (Figure 3.1.6b). For r below this critical value, the line lies below the curve and there are no fixed points (Figure 3.1.6c).

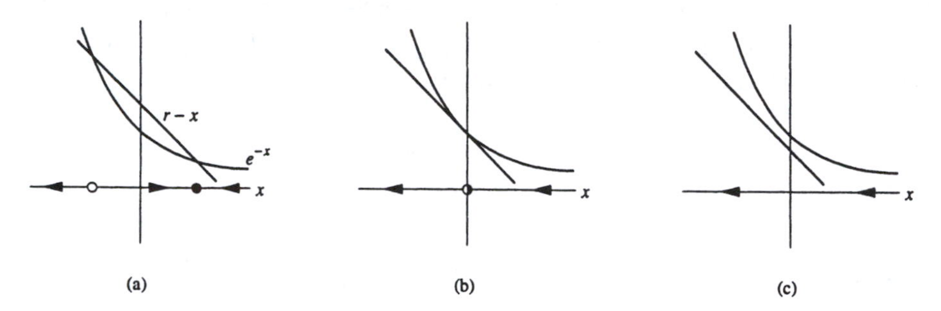

Figure 3.1.6

To find the bifurcation point r_c, we impose the condition that the graphs of $r-x$ and e^{-x} intersect *tangentially*. Thus we demand equality of the functions *and* their derivatives:

$$e^{-x}=r-x$$

and

$$\tfrac{d}{dx}e^{-x}=\tfrac{d}{dx}(r-x).$$

The second equation implies $-e^{-x}=-1$, so $x=0$. Then the first equation yields $r=1$. Hence the bifurcation point is $r_c=1$, and the bifurcation occurs at $x=0$. ■

Normal Forms

In a certain sense, the examples $\dot{x}=r-x^2$ or $\dot{x}=r+x^2$ are representative of *all* saddle-node bifurcations; that's why we called them "prototypical." The idea is that, close to a saddle-node bifurcation, the dynamics typically look like $\dot{x}=r-x^2$ or $\dot{x}=r+x^2$.

For instance, consider Example 3.1.2 near the bifurcation at $x = 0$ and $r = 1$. Using the Taylor expansion for e^{-x} about $x = 0$, we find

$$\begin{aligned}\dot{x} &= r - x - e^{-x} \\ &= r - x - \left[1 - x + \frac{x^2}{2!} + \cdots\right] \\ &= (r-1) - \frac{x^2}{2} + \cdots\end{aligned}$$

to leading order in x. This has the same algebraic form as $\dot{x} = r - x^2$, and can be made to agree exactly by appropriate rescalings of x and r.

It's easy to understand why saddle-node bifurcations typically have this algebraic form. We just ask ourselves: how can two fixed points of $\dot{x} = f(x)$ collide and disappear as a parameter r is varied? Graphically, fixed points occur where the graph of $f(x)$ intersects the x-axis. For a saddle-node bifurcation to be possible, we need two nearby roots of $f(x)$; this means $f(x)$ must look locally "bowl-shaped" or parabolic (Figure 3.1.7).

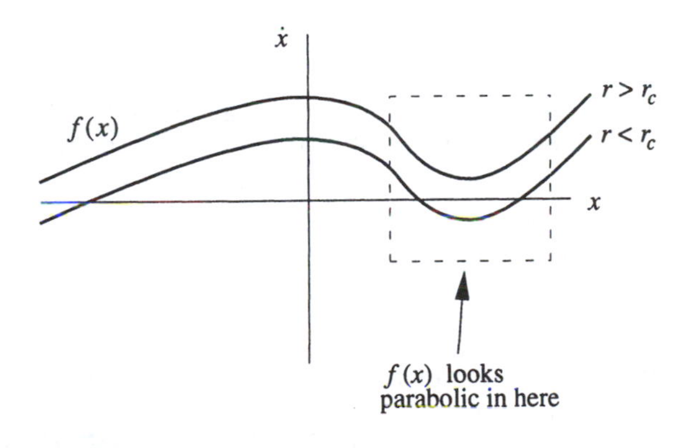

Figure 3.1.7

Now we use a microscope to zoom in on the behavior near the bifurcation. As r varies, we see a parabola intersecting the x-axis, then becoming tangent to it, and then failing to intersect. This is exactly the scenario in the prototypical Figure 3.1.1.

Here's a more algebraic version of the same argument. We regard f as a function of both x and r, and examine the behavior of $\dot{x} = f(x, r)$ near the bifurcation at $x = x^*$ and $r = r_c$. Taylor's expansion yields

$$\begin{aligned}\dot{x} &= f(x, r) \\ &= f(x^*, r_c) + (x - x^*)\left.\frac{\partial f}{\partial x}\right|_{(x^*, r_c)} + (r - r_c)\left.\frac{\partial f}{\partial r}\right|_{(x^*, r_c)} + \tfrac{1}{2}(x - x^*)^2 \left.\frac{\partial^2 f}{\partial x^2}\right|_{(x^*, r_c)} + \cdots\end{aligned}$$

where we have neglected quadratic terms in $(r - r_c)$ and cubic terms in $(x - x^*)$. Two of the terms in this equation vanish: $f(x^*, r_c) = 0$ since x^* is a fixed point, and $\partial f/\partial x|_{(x^*, r_c)} = 0$ by the tangency condition of a saddle-node bifurcation. Thus

$$\dot{x} = a(r - r_c) + b(x - x^*)^2 + \cdots \qquad (3)$$

where $a = \partial f/\partial r|_{(x^*, r_c)}$ and $b = \frac{1}{2}\partial^2 f/\partial x^2|_{(x^*, r_c)}$. Equation (3) agrees with the form of our prototypical examples. (We are assuming that $a, b \neq 0$, which is the typical case; for instance, it would be a very special situation if the second derivative $\partial^2 f/\partial x^2$ also happened to vanish at the fixed point.)

What we have been calling prototypical examples are more conventionally known as ***normal forms*** for the saddle-node bifurcation. There is much, much more to normal forms than we have indicated here. We will be seeing their importance throughout this book. For a more detailed and precise discussion, see Guckenheimer and Holmes (1983) or Wiggins (1990).

3.2 Transcritical Bifurcation

There are certain scientific situations where a fixed point must exist for all values of a parameter and can never be destroyed. For example, in the logistic equation and other simple models for the growth of a single species, there is a fixed point at zero population, regardless of the value of the growth rate. However, such a fixed point may *change its stability* as the parameter is varied. The transcritical bifurcation is the standard mechanism for such changes in stability.

The normal form for a transcritical bifurcation is

$$\dot{x} = rx - x^2. \qquad (1)$$

This looks like the logistic equation of Section 2.3, but now we allow x and r to be either positive or negative.

Figure 3.2.1 shows the vector field as r varies. Note that there is a fixed point at $x^* = 0$ for *all* values of r.

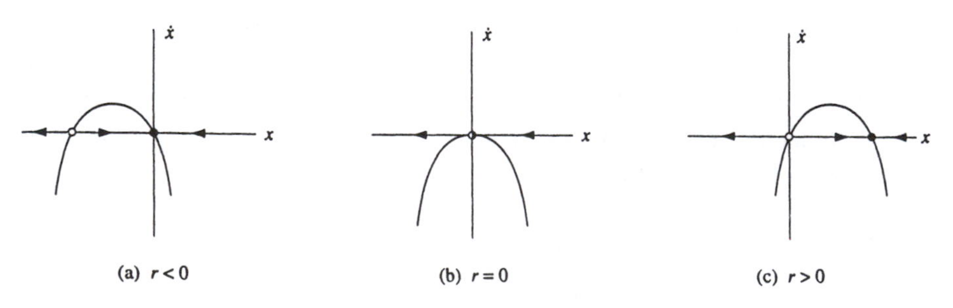

Figure 3.2.1

For $r<0$, there is an unstable fixed point at $x^*=r$ and a stable fixed point at $x^*=0$. As r increases, the unstable fixed point approaches the origin, and coalesces with it when $r=0$. Finally, when $r>0$, the origin has become unstable, and $x^*=r$ is now stable. Some people say that an ***exchange of stabilities*** has taken place between the two fixed points.

Please note the important difference between the saddle-node and transcritical bifurcations: in the transcritical case, the two fixed points don't disappear after the bifurcation—instead they just switch their stability.

Figure 3.2.2 shows the bifurcation diagram for the transcritical bifurcation. As in Figure 3.1.4, the parameter r is regarded as the independent variable, and the fixed points $x^*=0$ and $x^*=r$ are shown as dependent variables.

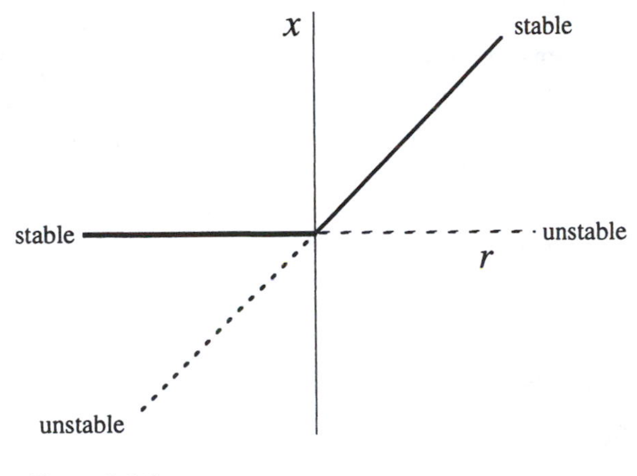

Figure 3.2.2

EXAMPLE 3.2.1:

Show that the first-order system $\dot{x}=x(1-x^2)-a(1-e^{-bx})$ undergoes a transcritical bifurcation at $x=0$ when the parameters a, b satisfy a certain equation, to be determined. (This equation defines a ***bifurcation curve*** in the (a,b) parameter space.) Then find an approximate formula for the fixed point that bifurcates from $x=0$, assuming that the parameters are close to the bifurcation curve.

Solution: Note that $x=0$ is a fixed point for all (a,b). This makes it plausible that the fixed point will bifurcate transcritically, if it bifurcates at all. For small x, we find

$$\begin{aligned} 1-e^{-bx} &= 1-\left[1-bx+\tfrac{1}{2}b^2x^2+O(x^3)\right] \\ &= bx-\tfrac{1}{2}b^2x^2+O(x^3) \end{aligned}$$

and so

$$\begin{aligned} \dot{x} &= x-a(bx-\tfrac{1}{2}b^2x^2)+O(x^3) \\ &= (1-ab)x+(\tfrac{1}{2}ab^2)x^2+O(x^3). \end{aligned}$$

Hence a transcritical bifurcation occurs when $ab = 1$; this is the equation for the bifurcation curve. The nonzero fixed point is given by the solution of $1 - ab + (\frac{1}{2}ab^2)x \approx 0$, i.e.,

$$x^* \approx \frac{2(ab-1)}{ab^2}.$$

This formula is approximately correct only if x^* is small, since our series expansions are based on the assumption of small x. Thus the formula holds only when ab is close to 1, which means that the parameters must be close to the bifurcation curve. ■

EXAMPLE 3.2.2:

Analyze the dynamics of $\dot{x} = r \ln x + x - 1$ near $x = 1$, and show that the system undergoes a transcritical bifurcation at a certain value of r. Then find new variables X and R such that the system reduces to the approximate normal form $\dot{X} \approx RX - X^2$ near the bifurcation.

Solution: First note that $x = 1$ is a fixed point for all values of r. Since we are interested in the dynamics near this fixed point, we introduce a new variable $u = x - 1$, where u is small. Then

$$\begin{aligned} \dot{u} &= \dot{x} \\ &= r \ln(1+u) + u \\ &= r\left[u - \tfrac{1}{2}u^2 + O(u^3)\right] + u \\ &\approx (r+1)u - \tfrac{1}{2}ru^2 + O(u^3). \end{aligned}$$

Hence a transcritical bifurcation occurs at $r_c = -1$.

To put this equation into normal form, we first need to get rid of the coefficient of u^2. Let $u = av$, where a will be chosen later. Then the equation for v is

$$\dot{v} = (r+1)v - (\tfrac{1}{2}ra)v^2 + O(v^3).$$

So if we choose $a = 2/r$, the equation becomes

$$\dot{v} = (r+1)v - v^2 + O(v^3).$$

Now if we let $R = r + 1$ and $X = v$, we have achieved the approximate normal form $\dot{X} \approx RX - X^2$, where cubic terms of order $O(X^3)$ have been neglected. In terms of the original variables, $X = v = u/a = \frac{1}{2}r(x-1)$. ■

To be a bit more accurate, the theory of normal forms assures us that we can find a change of variables such that the system becomes $\dot{X} = RX - X^2$, with *strict*, rather than approximate, equality. Our solution above gives an approximation to the necessary change of variables. If we wanted a better approximation, we would retain the cubic terms in the series expansions (and perhaps even higher-order terms if we're really feeling heroic) and we would have to do a more elaborate calculation to eliminate these higher-order terms. See Exercises 3.2.6 and 3.2.7 for a taste of such calculations, or see the books of Guckenheimer and Holmes (1983), Wiggins (1990), or Manneville (1990).

3.3 Laser Threshold

Now it's time to apply our mathematics to a scientific example. We analyze an extremely simplified model for a laser, following the treatment given by Haken (1983).

Physical Background

We are going to consider a particular type of laser known as a solid-state laser, which consists of a collection of special "laser-active" atoms embedded in a solid-state matrix, bounded by partially reflecting mirrors at either end. An external energy source is used to excite or "pump" the atoms out of their ground states (Figure 3.3.1).

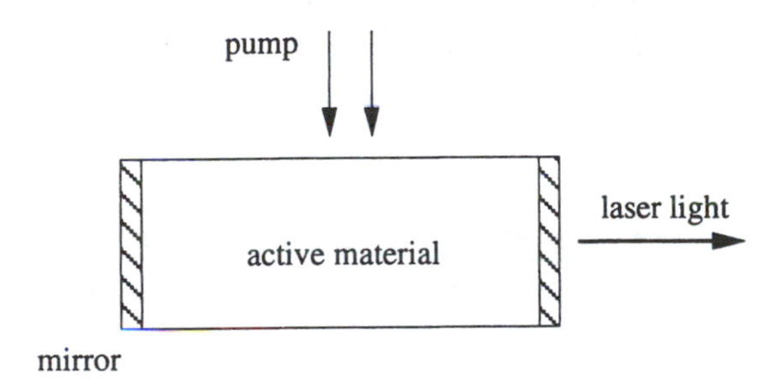

Figure 3.3.1

Each atom can be thought of as a little antenna radiating energy. When the pumping is relatively weak, the laser acts just like an ordinary *lamp*: the excited atoms oscillate independently of one another and emit randomly phased light waves.

Now suppose we increase the strength of the pumping. At first nothing different happens, but then suddenly, when the pump strength exceeds a certain threshold, the atoms begin to oscillate in phase—the lamp has turned into a ***laser***. Now the trillions of little antennas act like one giant antenna and produce a beam of radiation that is much more coherent and intense than that produced below the laser threshold.

This sudden onset of coherence is amazing, considering that the atoms are being excited completely at random by the pump! Hence the process is *self-organizing*: the coherence develops because of a cooperative interaction among the atoms themselves.

Model

A proper explanation of the laser phenomenon would require us to delve into quantum mechanics. See Milonni and Eberly (1988) for an intuitive discussion.

Instead we consider a simplified model of the essential physics (Haken 1983, p. 127). The dynamical variable is the number of photons $n(t)$ in the laser field. Its rate of change is given by

$$\begin{aligned}\dot{n} &= \text{gain} - \text{loss} \\ &= GnN - kn.\end{aligned}$$

The gain term comes from the process of *stimulated emission,* in which photons stimulate excited atoms to emit additional photons. Because this process occurs via random encounters between photons and excited atoms, it occurs at a rate proportional to n and to the number of excited atoms, denoted by $N(t)$. The parameter $G > 0$ is known as the gain coefficient. The loss term models the escape of photons through the endfaces of the laser. The parameter $k > 0$ is a rate constant; its reciprocal $\tau = 1/k$ represents the typical lifetime of a photon in the laser.

Now comes the key physical idea: after an excited atom emits a photon, it drops down to a lower energy level and is no longer excited. Thus N decreases by the emission of photons. To capture this effect, we need to write an equation relating N to n. Suppose that in the absence of laser action, the pump keeps the number of excited atoms fixed at N_0. Then the *actual* number of excited atoms will be reduced by the laser process. Specifically, we assume

$$N(t) = N_0 - \alpha n,$$

where $\alpha > 0$ is the rate at which atoms drop back to their ground states. Then

$$\begin{aligned}\dot{n} &= Gn(N_0 - \alpha n) - kn \\ &= (GN_0 - k)n - (\alpha G)n^2.\end{aligned}$$

We're finally on familiar ground—this is a first-order system for $n(t)$. Figure 3.3.2 shows the corresponding vector field for different values of the pump strength N_0. Note that only positive values of n are physically meaningful.

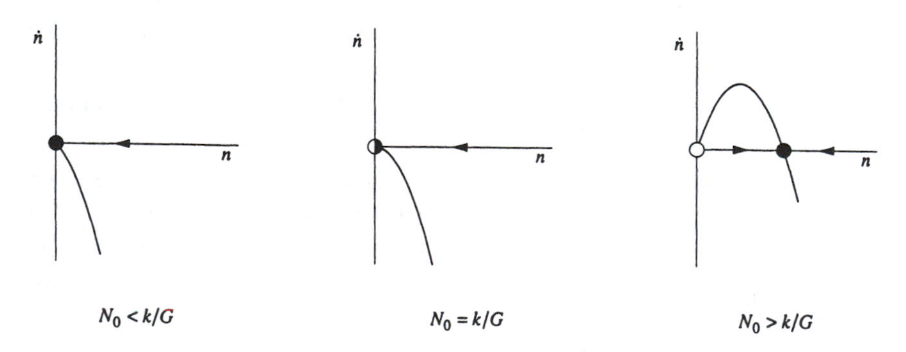

Figure 3.3.2

When $N_0 < k/G$, the fixed point at $n^* = 0$ is stable. This means that there is no stimulated emission and the laser acts like a lamp. As the pump strength N_0 is increased, the system undergoes a transcritical bifurcation when $N_0 = k/G$. For $N_0 > k/G$, the origin loses stability and a stable fixed point appears at $n^* = (GN_0 - k)/\alpha G > 0$, corresponding to spontaneous laser action. Thus $N_0 = k/G$ can be interpreted as the ***laser threshold*** in this model. Figure 3.3.3 summarizes our results.

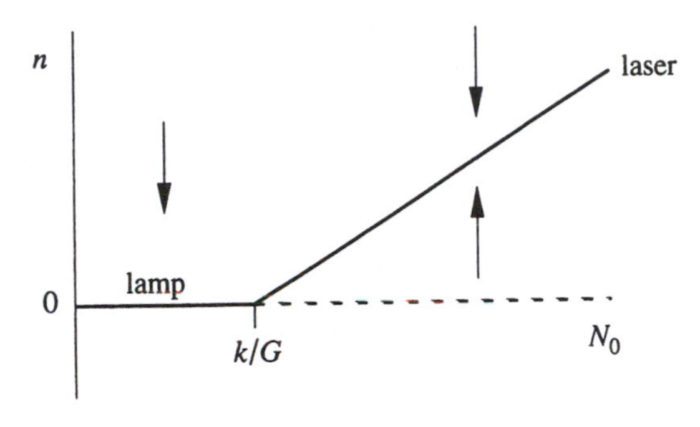

Figure 3.3.3

Although this model correctly predicts the existence of a threshold, it ignores the dynamics of the excited atoms, the existence of spontaneous emission, and several other complications. See Exercises 3.3.1 and 3.3.2 for improved models.

3.4 Pitchfork Bifurcation

We turn now to a third kind of bifurcation, the so-called pitchfork bifurcation. This bifurcation is common in physical problems that have a ***symmetry***. For example, many problems have a spatial symmetry between left and right. In such cases, fixed points tend to appear and disappear in symmetrical pairs. In the buckling example of Figure 3.0.1, the beam is stable in the vertical position if the load is small. In this case there is a stable fixed point corresponding to zero deflection. But if the load exceeds the buckling threshold, the beam may buckle to *either* the left or the right. The vertical position has gone unstable, and two new symmetrical fixed points, corresponding to left- and right-buckled configurations, have been born.

There are two very different types of pitchfork bifurcation. The simpler type is called *supercritical*, and will be discussed first.

Supercritical Pitchfork Bifurcation

The normal form of the supercritical pitchfork bifurcation is

$$\dot{x} = rx - x^3. \qquad (1)$$

Note that this equation is ***invariant*** under the change of variables $x \to -x$. That is, if we replace x by $-x$ and then cancel the resulting minus signs on both sides of the equation, we get (1) back again. This invariance is the mathematical expression of the left–right symmetry mentioned earlier. (More technically, one says that the vector field is *equivariant*, but we'll use the more familiar language.)

Figure 3.4.1 shows the vector field for different values of r.

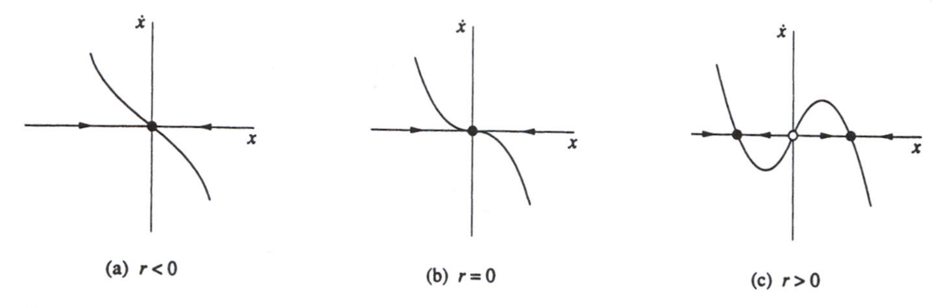

Figure 3.4.1

When $r < 0$, the origin is the only fixed point, and it is stable. When $r = 0$, the origin is still stable, but much more weakly so, since the linearization vanishes. Now solutions no longer decay exponentially fast—instead the decay is a much slower algebraic function of time (recall Exercise 2.4.9). This lethargic decay is called ***critical slowing down*** in the physics literature. Finally, when $r > 0$, the origin has become unstable. Two new stable fixed points appear on either side of the origin, symmetrically located at $x^* = \pm\sqrt{r}$.

The reason for the term "pitchfork" becomes clear when we plot the bifurcation diagram (Figure 3.4.2). Actually, pitchfork trifurcation might be a better word!

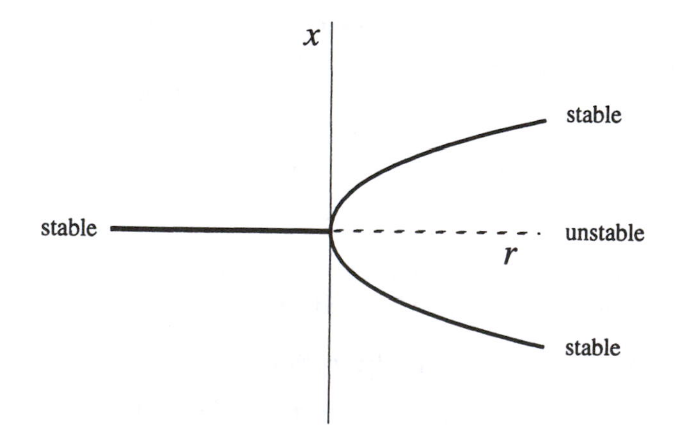

Figure 3.4.2

EXAMPLE 3.4.1:

Equations similar to $\dot{x} = -x + \beta \tanh x$ arise in statistical mechanical models of magnets and neural networks (see Exercise 3.6.7 and Palmer 1989). Show that this equation undergoes a supercritical pitchfork bifurcation as β is varied. Then give a *numerically accurate* plot of the fixed points for each β.

Solution: We use the strategy of Example 3.1.2 to find the fixed points. The graphs of $y = x$ and $y = \beta \tanh x$ are shown in Figure 3.4.3; their intersections correspond to fixed points. The key thing to realize is that as β increases, the tanh curve becomes steeper at the origin (its slope there is β). Hence for $\beta < 1$ the origin is the only fixed point. A pitchfork bifurcation occurs at $\beta = 1$, $x^* = 0$, when the tanh curve develops a slope of 1 at the origin. Finally, when $\beta > 1$, two new stable fixed points appear, and the origin becomes unstable.

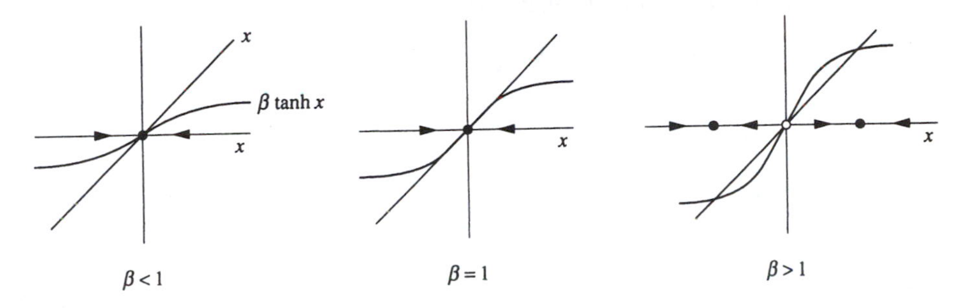

Figure 3.4.3

Now we want to compute the fixed points x^* for each β. Of course, one fixed point always occurs at $x^* = 0$; we are looking for the other, nontrivial fixed points. One approach is to solve the equation $x^* = \beta \tanh x^*$ numerically, using the Newton–Raphson method or some other root-finding scheme. (See Press et al. (1986) for a friendly and informative discussion of numerical methods.)

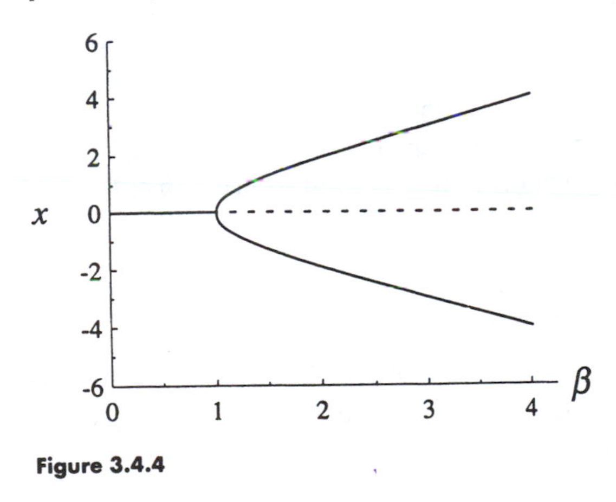

Figure 3.4.4

But there's an easier way, which comes from changing our point of view. Instead of studying the dependence of x^* on β, we think of x^* as the *independent* variable, and then compute $\beta = x^*/\tanh x^*$. This gives us a table of pairs (x^*, β). For each pair, we plot β horizontally and x^* vertically. This yields the bifurcation diagram (Figure 3.4.4).

The shortcut used here exploits the fact that $f(x,\beta) = -x + \beta \tanh x$ depends more simply on β than on x. This is frequently the case in bifurcation problems—the dependence on the control parameter is usually simpler than the dependence on x. ■

EXAMPLE 3.4.2:

Plot the potential $V(x)$ for the system $\dot{x} = rx - x^3$, for the cases $r < 0$, $r = 0$, and $r > 0$.

Solution: Recall from Section 2.7 that the potential for $\dot{x} = f(x)$ is defined by $f(x) = -dV/dx$. Hence we need to solve $-dV/dx = rx - x^3$. Integration yields $V(x) = -\frac{1}{2}rx^2 + \frac{1}{4}x^4$, where we neglect the arbitrary constant of integration. The corresponding graphs are shown in Figure 3.4.5.

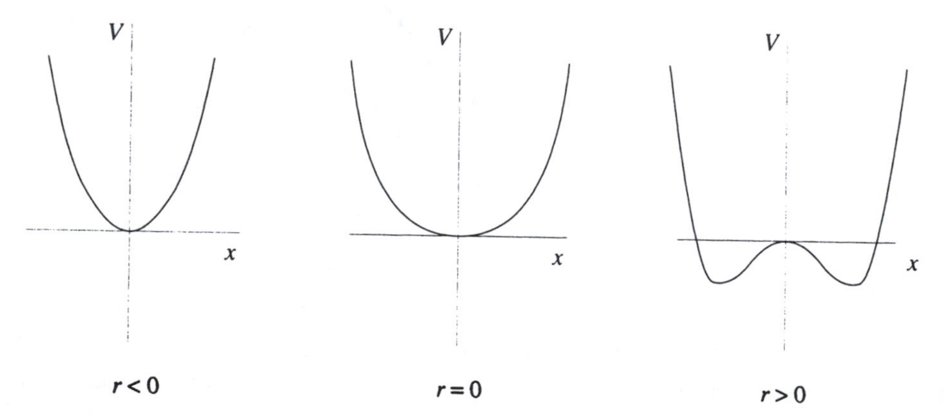

Figure 3.4.5

When $r < 0$, there is a quadratic minimum at the origin. At the bifurcation value $r = 0$, the minimum becomes a much flatter quartic. For $r > 0$, a local *maximum* appears at the origin, and a symmetric pair of minima occur to either side of it. ■

Subcritical Pitchfork Bifurcation

In the supercritical case $\dot{x} = rx - x^3$ discussed above, the cubic term is *stabilizing*: it acts as a restoring force that pulls $x(t)$ back toward $x = 0$. If instead the cubic term were *destabilizing*, as in

$$\dot{x} = rx + x^3, \tag{2}$$

then we'd have a ***subcritical*** pitchfork bifurcation. Figure 3.4.6 shows the bifurcation diagram.

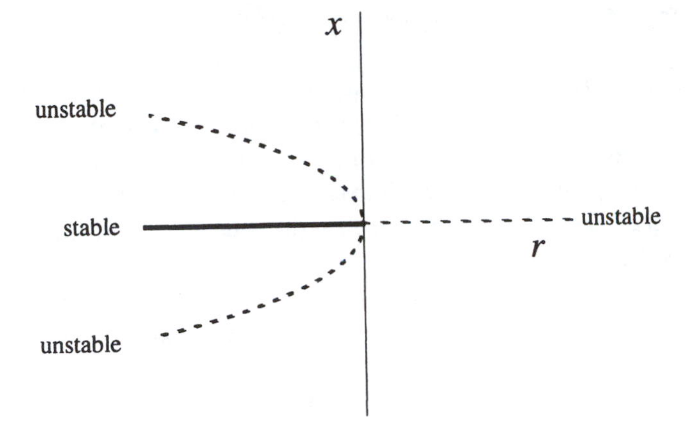

Figure 3.4.6

Compared to Figure 3.4.2, the pitchfork is inverted. The nonzero fixed points $x^* = \pm\sqrt{-r}$ are *unstable,* and exist only *below* the bifurcation ($r < 0$), which motivates the term "subcritical." More importantly, the origin is stable for $r < 0$ and unstable for $r > 0$, as in the supercritical case, but now the instability for $r > 0$ is not opposed by the cubic term—in fact the cubic term lends a helping hand in driving the trajectories out to infinity! This effect leads to *blow-up*: one can show that $x(t) \to \pm\infty$ in finite time, starting from any initial condition $x_0 \neq 0$ (Exercise 2.5.3).

In real physical systems, such an explosive instability is usually opposed by the stabilizing influence of higher-order terms. Assuming that the system is still symmetric under $x \to -x$, the first stabilizing term must be x^5. Thus the canonical example of a system with a subcritical pitchfork bifurcation is

$$\dot{x} = rx + x^3 - x^5. \tag{3}$$

There's no loss in generality in assuming that the coefficients of x^3 and x^5 are 1 (Exercise 3.5.8).

The detailed analysis of (3) is left to you (Exercises 3.4.14 and 3.4.15). But we will summarize the main results here. Figure 3.4.7 shows the bifurcation diagram for (3). For small x, the picture looks just like Figure 3.4.6: the origin is locally stable for $r < 0$, and two backward-bending branches of unstable fixed points bifurcate from the origin when $r = 0$. The new feature, due to the x^5 term, is that the unstable branches turn around and become stable at $r = r_s$, where $r_s < 0$. These stable ***large-amplitude*** branches exist for all $r > r_s$.

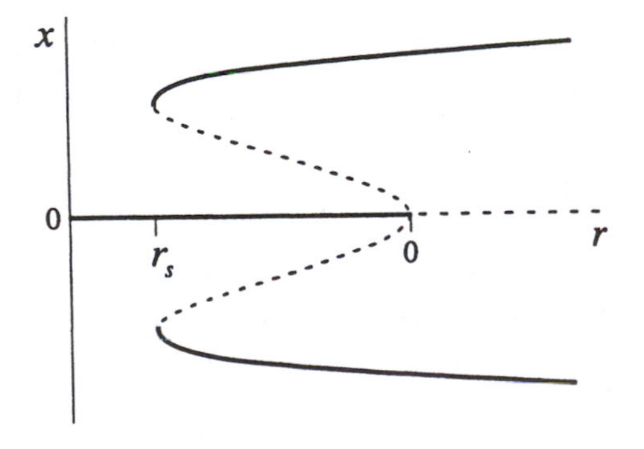

Figure 3.4.7

There are several things to note about Figure 3.4.7:

1. In the range $r_s < r < 0$, two qualitatively different stable states coexist, namely the origin and the large-amplitude fixed points. The initial condition x_0 determines which fixed point is approached as $t \to \infty$. One consequence is that the origin is stable to small perturbations, but not to large ones—in this sense the origin is *locally* stable, but not *globally* stable.
2. The existence of different stable states allows for the possibility of ***jumps*** and ***hysteresis*** as r is varied. Suppose we start the system in the state $x^* = 0$, and then slowly increase the parameter r (indicated by an arrow along the r-axis of Figure 3.4.8).

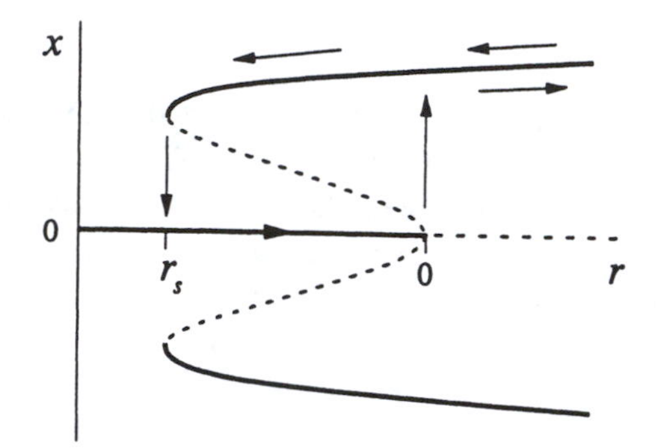

Figure 3.4.8

Then the state remains at the origin until $r = 0$, when the origin loses stability. Now the slightest nudge will cause the state to *jump* to one of the large-amplitude branches. With further increases of r, the state moves out along the large-amplitude branch. If r is now decreased, the state remains on the large-amplitude branch, even when r is decreased below 0! We have to lower r even further (down past r_s) to get the state to jump back to the origin. This lack of reversibility as a parameter is varied is called *hysteresis.*

3. The bifurcation at r_s is a saddle-node bifurcation, in which stable and unstable fixed points are born "out of the clear blue sky" as r is increased (see Section 3.1).

Terminology

As usual in bifurcation theory, there are several other names for the bifurcations discussed here. The supercritical pitchfork is sometimes called a forward bifurcation, and is closely related to a continuous or second-order phase transition in statistical mechanics. The subcritical bifurcation is sometimes called an inverted or backward bifurcation, and is related to discontinuous or first-order phase transitions. In the engineering literature, the supercritical bifurcation is sometimes called soft or safe, because the nonzero fixed points are born at small amplitude; in contrast, the subcritical bifurcation is hard or dangerous, because of the jump from zero to large amplitude.

3.5 Overdamped Bead on a Rotating Hoop

In this section we analyze a classic problem from first-year physics, the bead on a rotating hoop. This problem provides an example of a bifurcation in a mechanical system. It also illustrates the subtleties involved in replacing Newton's law, which is a second-order equation, by a simpler first-order equation.

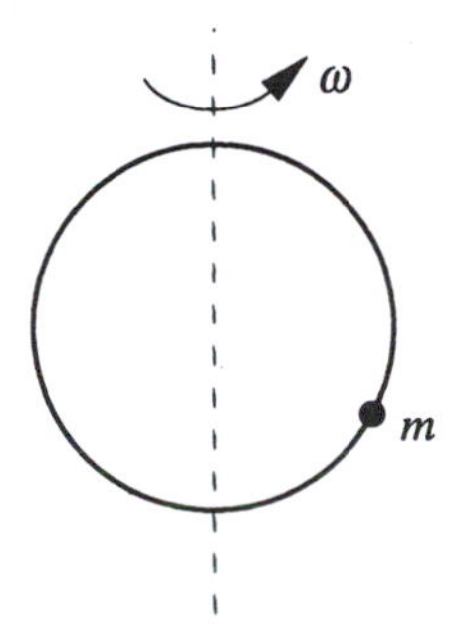

Figure 3.5.1

The mechanical system is shown in Figure 3.5.1. A bead of mass m slides along a wire hoop of radius r. The hoop is constrained to rotate at a constant angular velocity ω about its vertical axis. The problem is to analyze the motion of the bead, given that it is acted on by both gravitational and centrifugal forces. This is the usual statement of the problem, but now we want to add a new twist: suppose that there's also a frictional force on the bead that opposes its motion. To be specific, imagine that the whole system is immersed in a vat of molasses or some other very viscous fluid, and that the friction is due to viscous damping.

Let ϕ be the angle between the bead and the downward vertical direction. By convention, we restrict ϕ to the range $-\pi < \phi \le \pi$, so there's only one angle for each point on the hoop. Also, let $\rho = r\sin\phi$ denote the distance of the bead from the vertical axis. Then the coordinates are as shown in Figure 3.5.2.

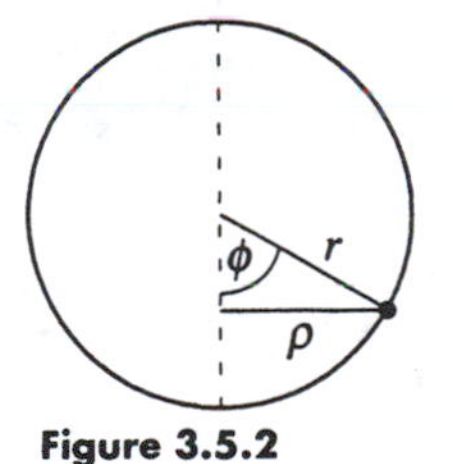

Figure 3.5.2

Now we write Newton's law for the bead. There's a downward gravitational force mg, a sideways centrifugal force $m\rho\omega^2$, and a tangential damping force $b\dot{\phi}$. (The constants g and b are taken to be positive; negative signs will be added later as needed.) The hoop is assumed to be rigid, so we only have to resolve the forces along the tangential direction, as shown in Figure 3.5.3. After substituting $\rho = r\sin\phi$ in the centrifugal term, and recalling that the tangential acceleration is $r\ddot{\phi}$, we obtain the governing equation

$$mr\ddot{\phi} = -b\dot{\phi} - mg\sin\phi + mr\omega^2\sin\phi\cos\phi. \qquad (1)$$

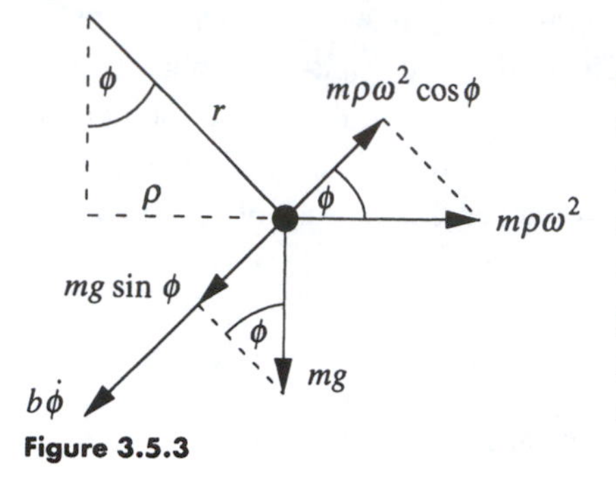

Figure 3.5.3

This is a *second-order* differential equation, since the second derivative $\ddot{\phi}$ is the highest one that appears. We are not yet equipped to analyze second-order equations, so we would like to find some conditions under which we can safely neglect the $mr\ddot{\phi}$ term. Then (1) reduces to a first-order equation, and we can apply our machinery to it.

Of course, this is a dicey business: we can't just neglect terms because we feel like it! But we will for now, and then at the end of this section we'll try to find a regime where our approximation is valid.

Analysis of the First-Order System

Our concern now is with the first-order system

$$\begin{aligned} b\dot{\phi} &= -mg\sin\phi + mr\omega^2 \sin\phi\cos\phi \\ &= mg\sin\phi\left(\frac{r\omega^2}{g}\cos\phi - 1\right). \end{aligned} \tag{2}$$

The fixed points of (2) correspond to equilibrium positions for the bead. What's your intuition about where such equilibria can occur? We would expect the bead to remain at rest if placed at the top or the bottom of the hoop. Can other fixed points occur? And what about stability? Is the bottom always stable?

Equation (2) shows that there are always fixed points where $\sin\phi = 0$, namely $\phi^* = 0$ (the bottom of the hoop) and $\phi^* = \pi$ (the top). The more interesting result is that there are two *additional* fixed points if

$$\frac{r\omega^2}{g} > 1,$$

that is, *if the hoop is spinning fast enough.* These fixed points satisfy $\phi^* = \pm\cos^{-1}\left(g/r\omega^2\right)$. To visualize them, we introduce a parameter

$$\gamma = \frac{r\omega^2}{g}$$

and solve $\cos\phi^* = 1/\gamma$ graphically. We plot $\cos\phi$ vs. ϕ, and look for intersections with the constant function $1/\gamma$, shown as a horizontal line in Figure 3.5.4. For $\gamma < 1$ there are no intersections, whereas for $\gamma > 1$ there is a symmetrical pair of in-

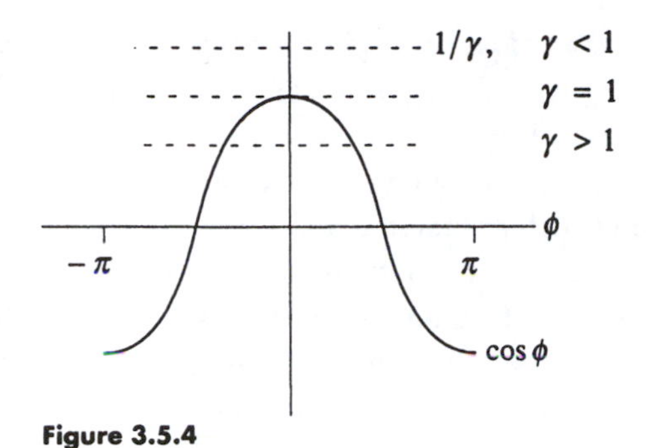

Figure 3.5.4

tersections to either side of $\phi^* = 0$. As $\gamma \to \infty$, these intersections approach $\pm\pi/2$. Figure 3.5.5 plots the fixed points on the hoop for the cases $\gamma < 1$ and $\gamma > 1$.

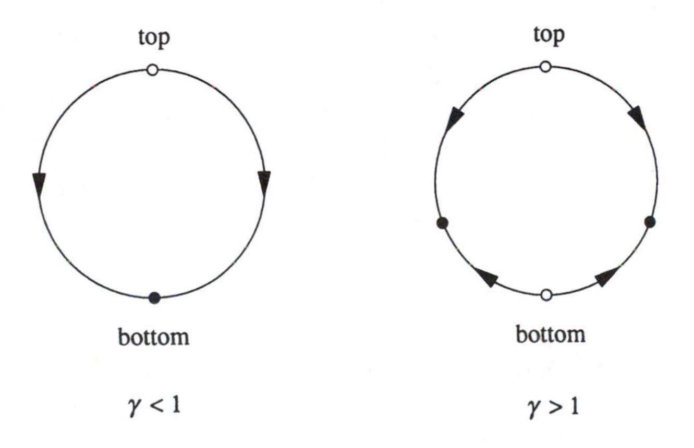

Figure 3.5.5

To summarize our results so far, let's plot *all* the fixed points as a function of the parameter γ (Figure 3.5.6). As usual, solid lines denote stable fixed points and broken lines denote unstable fixed points.

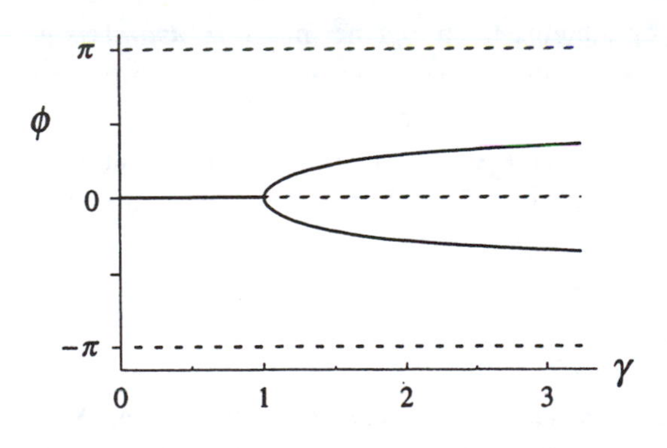

Figure 3.5.6

We now see that a ***supercritical pitchfork bifurcation*** occurs at $\gamma = 1$. It's left to you to check the stability of the fixed points, using linear stability analysis or graphical methods (Exercise 3.5.2).

Here's the physical interpretation of the results: When $\gamma < 1$, the hoop is rotating slowly and the centrifugal force is too weak to balance the force of gravity. Thus the bead slides down to the bottom and stays there. But if $\gamma > 1$, the hoop is spinning fast enough that the bottom becomes unstable. Since the centrifugal force *grows* as the bead moves farther from the bottom, any slight displacement of the bead will be *amplified.* The bead is therefore pushed up the hoop until gravity balances the centrifugal force; this balance occurs at $\phi^* = \pm\cos^{-1}\left(g/r\omega^2\right)$. Which of these two fixed points is actually selected depends on the initial disturbance. Even though the two fixed points are entirely symmetrical, an asymmetry in the initial conditions will lead to one of them being chosen—physicists sometimes refer to these as ***symmetry-broken*** solutions. In other words, the solution has less symmetry than the governing equation.

What *is* the symmetry of the governing equation? Clearly the left and right halves of the hoop are physically equivalent—this is reflected by the invariance of (1) and (2) under the change of variables $\phi \to -\phi$. As we mentioned in Section 3.4, pitchfork bifurcations are to be expected in situations where such a symmetry exists.

Dimensional Analysis and Scaling

Now we need to address the question: When is it valid to neglect the inertia term $mr\ddot{\phi}$ in (1)? At first sight the limit $m \to 0$ looks promising, but then we notice that we're throwing out the baby with the bathwater: the centrifugal and gravitational terms vanish in this limit too! So we have to be more careful.

In problems like this, it is helpful to express the equation in ***dimensionless*** form (at present, all the terms in (1) have the dimensions of force.) The advantage of a dimensionless formulation is that we know how to define *small*—it means "much less than 1." Furthermore, nondimensionalizing the equation reduces the number of parameters by lumping them together into ***dimensionless groups***. This reduction always simplifies the analysis. For an excellent introduction to dimensional analysis, see Lin and Segel (1988).

There are often several ways to nondimensionalize an equation, and the best choice might not be clear at first. Therefore we proceed in a flexible fashion. We define a dimensionless time τ by

$$\tau = \frac{t}{T}$$

where T is a ***characteristic time scale*** to be chosen later. When T is chosen correctly, the new derivatives $d\phi/d\tau$ and $d^2\phi/d\tau^2$ should be $O(1)$, i.e., of order unity. To express these new derivatives in terms of the old ones, we use the chain rule:

$$\dot{\phi} \equiv \frac{d\phi}{dt} = \frac{d\phi}{d\tau}\frac{d\tau}{dt} = \frac{1}{T}\frac{d\phi}{d\tau}$$

and similarly

$$\ddot{\phi} = \frac{1}{T^2}\frac{d^2\phi}{d\tau^2}.$$

(The easy way to remember these formulas is to formally substitute $T\tau$ for t.) Hence (1) becomes

$$\frac{mr}{T^2}\frac{d^2\phi}{d\tau^2} = -\frac{b}{T}\frac{d\phi}{d\tau} - mg\sin\phi + mr\omega^2\sin\phi\cos\phi.$$

Now since this equation is a balance of forces, we nondimensionalize it by dividing by a force mg. This yields the dimensionless equation

$$\left(\frac{r}{gT^2}\right)\frac{d^2\phi}{d\tau^2} = -\left(\frac{b}{mgT}\right)\frac{d\phi}{d\tau} - \sin\phi + \left(\frac{r\omega^2}{g}\right)\sin\phi\cos\phi. \qquad (3)$$

Each of the terms in parentheses is a dimensionless group. We recognize the group $r\omega^2/g$ in the last term—that's our old friend γ from earlier in the section.

We are interested in the regime where the left-hand side of (3) is negligible compared to all the other terms, and where all the terms on the right-hand side are of comparable size. Since the derivatives are $O(1)$ by assumption, and $\sin\phi \approx O(1)$, we see that we need

$$\frac{b}{mgT} \approx O(1), \text{ and } \frac{r}{gT^2} << 1.$$

The first of these requirements sets the time scale T: a natural choice is

$$T = \frac{b}{mg}.$$

Then the condition $r/gT^2 << 1$ becomes

$$\frac{r}{g}\left(\frac{mg}{b}\right)^2 << 1, \qquad (4)$$

or equivalently,

$$b^2 >> m^2 gr.$$

This can be interpreted as saying that the *damping is very strong*, or that the mass is very small, now in a precise sense.

The condition (4) motivates us to introduce a dimensionless group

$$\varepsilon = \frac{m^2 gr}{b^2}. \tag{5}$$

Then (3) becomes

$$\varepsilon \frac{d^2\phi}{d\tau^2} = -\frac{d\phi}{d\tau} - \sin\phi + \gamma \sin\phi \cos\phi . \tag{6}$$

As advertised, the dimensionless Equation (6) is simpler than (1): the five parameters m, g, r, ω, and b have been replaced by two dimensionless groups γ and ε.

In summary, our dimensional analysis suggests that in the ***overdamped*** limit $\varepsilon \to 0$, (6) should be well approximated by the first-order system

$$\frac{d\phi}{d\tau} = f(\phi) \tag{7}$$

where

$$\begin{aligned} f(\phi) &= -\sin\phi + \gamma \sin\phi \cos\phi \\ &= \sin\phi(\gamma \cos\phi - 1). \end{aligned}$$

A Paradox

Unfortunately, *there is something fundamentally wrong with our idea of replacing a second-order equation by a first-order equation.* The trouble is that a second-order equation requires *two* initial conditions, whereas a first-order equation has only *one*. In our case, the bead's motion is determined by its initial position and velocity. These two quantities can be chosen completely independent of each other. But that's not true for the first-order system: given the initial position, the initial velocity is dictated by the equation $d\phi/d\tau = f(\phi)$. Thus the solution to the first-order system will not, in general, be able to satisfy *both* initial conditions.

We seem to have run into a paradox. Is (7) valid in the overdamped limit or not? If it is valid, how can we satisfy the two arbitrary initial conditions demanded by (6)?

The resolution of the paradox requires us to analyze the second-order system (6). We haven't dealt with second-order systems before—that's the subject of Chapter 5. But read on if you're curious; some simple ideas are all we need to finish the problem.

Phase Plane Analysis

Throughout Chapters 2 and 3, we have exploited the idea that a first-order system $\dot{x} = f(x)$ can be regarded as a vector field on a line. By analogy, the *second-order* system (6) can be regarded as a vector field on a *plane*, the so-called ***phase plane.***

The plane is spanned by two axes, one for the angle ϕ and one for the angular velocity $d\phi/d\tau$. To simplify the notation, let

$$\Omega = \phi' \equiv d\phi/d\tau$$

where prime denotes differentiation with respect to τ. Then an initial condition for (6) corresponds to a point (ϕ_0, Ω_0) in the phase plane (Figure 3.5.7). As time evolves, the phase point $(\phi(t), \Omega(t))$ moves around in the phase plane along a ***trajectory*** determined by the solution to (6).

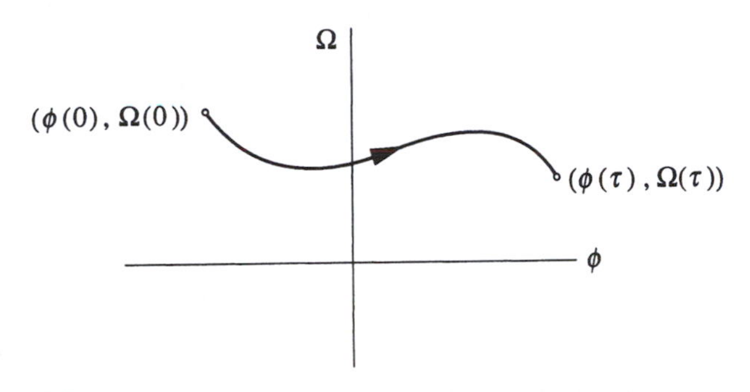

Figure 3.5.7

Our goal now is to see what those trajectories actually look like. As before, the key idea is that *the differential equation can be interpreted as a vector field on the phase space.* To convert (6) into a vector field, we first rewrite it as

$$\varepsilon\Omega' = f(\phi) - \Omega.$$

Along with the definition $\phi' = \Omega$, this yields the ***vector field***

$$\phi' = \Omega \qquad \text{(8a)}$$

$$\Omega' = \frac{1}{\varepsilon}\big(f(\phi) - \Omega\big). \qquad \text{(8b)}$$

We interpret the vector (ϕ', Ω') at the point (ϕ, Ω) as the local velocity of a phase fluid flowing steadily on the plane. Note that the velocity vector now has two components, one in the ϕ-direction and one in the Ω-direction. To visualize the trajectories, we just imagine how the phase point would move as it is carried along by the phase fluid.

In general, the pattern of trajectories would be difficult to picture, but the present case is simple because we are only interested in the limit $\varepsilon \to 0$. In this limit, *all trajectories slam straight up or down onto the curve C defined by* $f(\phi) = \Omega$, *and then slowly ooze along this curve until they reach a fixed point* (Figure 3.5.8).

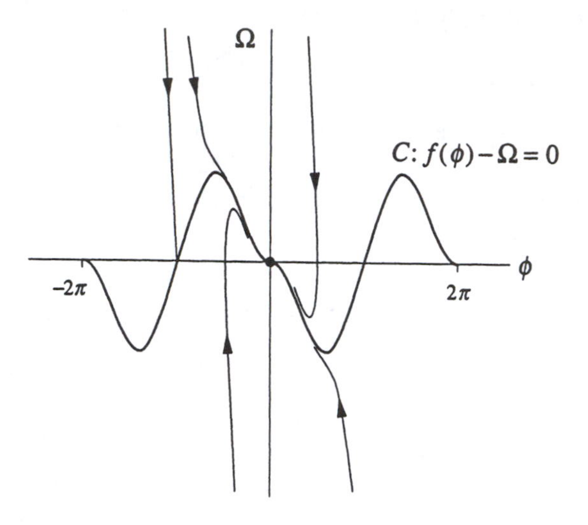

Figure 3.5.8

To arrive at this striking conclusion, let's do an order-of-magnitude calculation. Suppose that the phase point lies off the curve C. For instance, suppose (ϕ, Ω) lies an $O(1)$ distance below the curve C, i.e., $\Omega < f(\phi)$ and $f(\phi) - \Omega \approx O(1)$. Then (8b) shows that Ω' is enormously positive: $\Omega' \approx O(1/\varepsilon) >> 1$. Thus the phase point zaps like lightning up to the region where $f(\phi) - \Omega \approx O(\varepsilon)$. In the limit $\varepsilon \to 0$, this region is indistinguishable from C. Once the phase point is on C, it evolves according to $\Omega \approx f(\phi)$; that is, it approximately satisfies the first-order equation $\phi' = f(\phi)$.

Our conclusion is that a typical trajectory is made of two parts: a rapid initial ***transient***, during which the phase point zaps onto the curve where $\phi' = f(\phi)$, followed by a much slower drift along this curve.

Now we see how the paradox is resolved: The second-order system (6) *does* behave like the first-order system (7), but only after a rapid initial transient. During this transient, it is *not* correct to neglect the term $\varepsilon\, d^2\phi/d\tau^2$. The problem with our earlier approach is that we used only a single time scale $T = b/mg$; this time scale is characteristic of the slow drift process, but not of the rapid transient (Exercise 3.5.5).

A Singular Limit

The difficulty we have encountered here occurs throughout science and engineering. In some limit of interest (here, the limit of strong damping), the term containing the highest order derivative drops out of the governing equation. Then the initial conditions or boundary conditions can't be satisfied. Such a limit is often called *singular*. For example, in fluid mechanics, the limit of high Reynolds number is a singular limit; it accounts for the presence of extremely thin "boundary layers" in the flow over airplane wings. In our problem, the rapid transient played the role of a boundary layer—it is a thin layer of *time* that occurs near the boundary $t = 0$.

The branch of mathematics that deals with singular limits is called *singular perturbation theory*. See Jordan and Smith (1987) or Lin and Segel (1988) for an introduction. Another problem with a singular limit will be discussed briefly in Section 7.5.

3.6 Imperfect Bifurcations and Catastrophes

As we mentioned earlier, pitchfork bifurcations are common in problems that have a symmetry. For example, in the problem of the bead on a rotating hoop (Section 3.5), there was a perfect symmetry between the left and right sides of the hoop. But in many real-world circumstances, the symmetry is only approximate—an imperfection leads to a slight difference between left and right. We now want to see what happens when such imperfections are present.

For example, consider the system

$$\dot{x} = h + rx - x^3. \qquad (1)$$

If $h = 0$, we have the normal form for a supercritical pitchfork bifurcation, and there's a perfect symmetry between x and $-x$. But this symmetry is broken when $h \neq 0$; for this reason we refer to h as an ***imperfection parameter***.

Equation (1) is a bit harder to analyze than other bifurcation problems we've considered previously, because we have *two* independent parameters to worry about (h and r). To keep things straight, we'll think of r as fixed, and then examine the effects of varying h. The first step is to analyze the fixed points of (1). These can be found explicitly, but we'd have to invoke the messy formula for the roots of a cubic equation. It's clearer to use a graphical approach, as in Example 3.1.2. We plot the graphs of $y = rx - x^3$ and $y = -h$ on the same axes, and look for intersections (Figure 3.6.1). These intersections occur at the fixed points of (1). When $r \leq 0$, the cubic is monotonically decreasing, and so it intersects the horizontal line $y = -h$ in exactly one point (Figure 3.6.1a). The more interesting case is $r > 0$; then one, two, or three intersections are possible, depending on the value of h (Figure 3.6.1b).

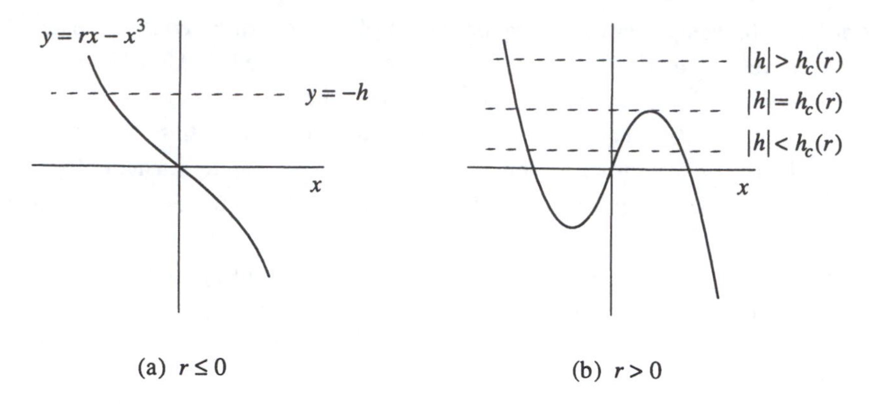

Figure 3.6.1

The critical case occurs when the horizontal line is just *tangent* to either the local minimum or maximum of the cubic; then we have a *saddle-node bifurcation*. To find the values of h at which this bifurcation occurs, note that the cubic has a local maximum when $\frac{d}{dx}(rx - x^3) = r - 3x^2 = 0$. Hence

$$x_{\text{max}} = \sqrt{\frac{r}{3}},$$

and the value of the cubic at the local maximum is

$$rx_{\text{max}} - (x_{\text{max}})^3 = \frac{2r}{3}\sqrt{\frac{r}{3}}.$$

Similarly, the value at the minimum is the negative of this quantity. Hence saddle-node bifurcations occur when $h = \pm h_c(r)$, where

$$h_c(r) = \frac{2r}{3}\sqrt{\frac{r}{3}}.$$

Equation (1) has three fixed points for $|h| < h_c(r)$ and one fixed point for $|h| > h_c(r)$.

To summarize the results so far, we plot the ***bifurcation curves*** $h = \pm h_c(r)$ in the (r, h) plane (Figure 3.6.2). Note that the two bifurcation curves meet tangentially at $(r, h) = (0, 0)$; such a point is called a ***cusp point***. We also label the regions that correspond to different numbers of fixed points. Saddle-node bifurcations occur all along the boundary of the regions, except at the cusp point, where we have a *codimension-2 bifurcation*. (This fancy terminology essentially means that we have had to tune *two* parameters, h and r, to achieve this type of bifurcation. Until now, all our bifurcations could be achieved by tuning a single parameter, and were therefore *codimension-1* bifurcations.)

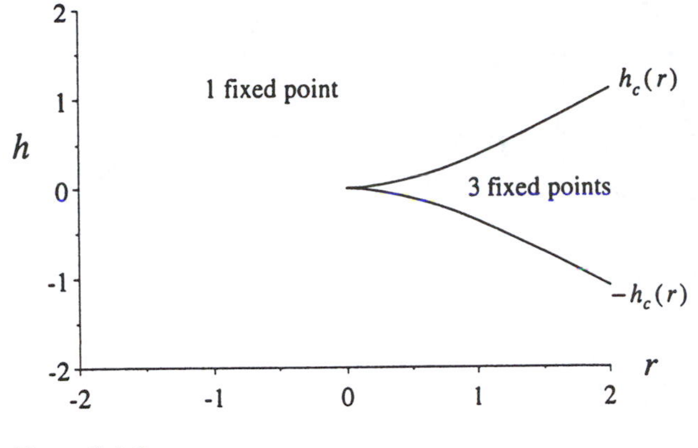

Figure 3.6.2

Pictures like Figure 3.6.2 will prove very useful in our future work. We will refer to such pictures as ***stability diagrams***. They show the different types of behavior that occur as we move around in ***parameter space*** (here, the (r,h) plane).

Now let's present our results in a more familiar way by showing the bifurcation diagram of x^* vs. r, for fixed h (Figure 3.6.3).

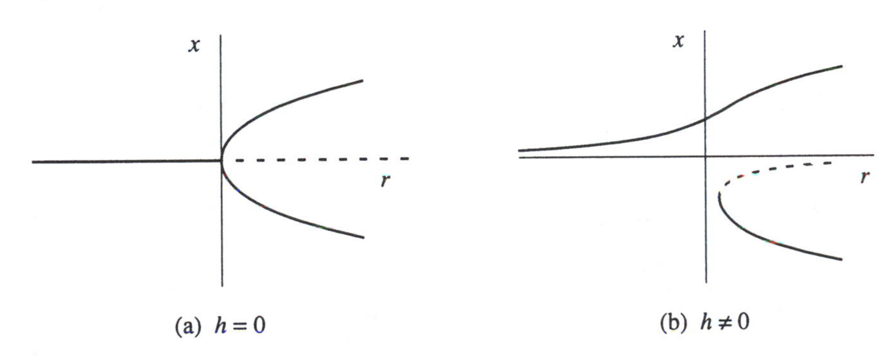

Figure 3.6.3

When $h=0$ we have the usual pitchfork diagram (Figure 3.6.3a) but when $h \neq 0$, the pitchfork disconnects into two pieces (Figure 3.6.3b). The upper piece consists entirely of stable fixed points, whereas the lower piece has both stable and unstable branches. As we increase r from negative values, there's no longer a sharp transition at $r=0$; the fixed point simply glides smoothly along the upper branch. Furthermore, the lower branch of stable points is not accessible unless we make a fairly large disturbance.

Alternatively, we could plot x^* vs. h, for fixed r (Figure 3.6.4).

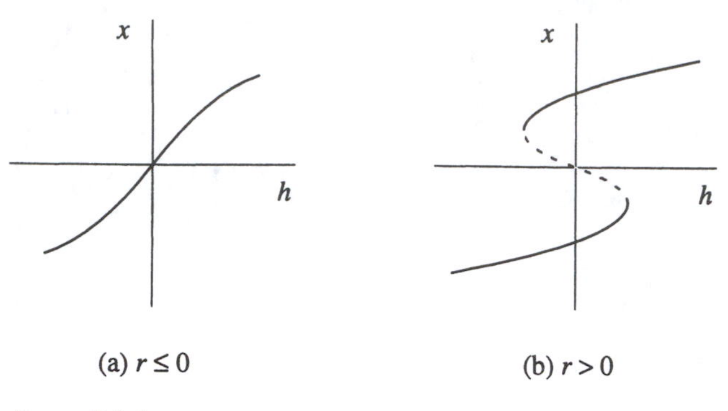

Figure 3.6.4

When $r \le 0$ there's one stable fixed point for each h (Figure 3.6.4a). However, when $r > 0$ there are three fixed points when $|h| < h_c(r)$, and one otherwise (Figure 3.6.4b). In the triple-valued region, the middle branch is unstable and the upper and lower branches are stable. Note that these graphs look like Figure 3.6.1 rotated by 90°.

There is one last way to plot the results, which may appeal to you if you like to picture things in three dimensions. This method of presentation contains all of the others as cross sections or projections. If we plot the fixed points x^* above the (r,h) plane, we get the ***cusp catastrophe*** surface shown in Figure 3.6.5. The surface folds over on itself in certain places. The projection of these folds onto the (r,h) plane yields the bifurcation curves shown in Figure 3.6.2. A cross section at fixed h yields Figure 3.6.3, and a cross section at fixed r yields Figure 3.6.4.

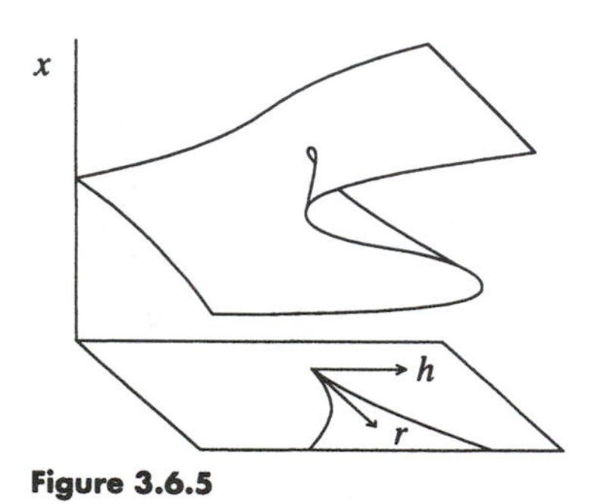

Figure 3.6.5

The term *catastrophe* is motivated by the fact that as parameters change, the state of the system can be carried over the edge of the upper surface, after which it drops discontinuously to the lower surface (Figure 3.6.6). This jump could be truly catastrophic for the equilibrium of a bridge or a building. We will see scientific examples of catastrophes in the context of insect outbreaks (Section 3.7) and in the following example from mechanics.

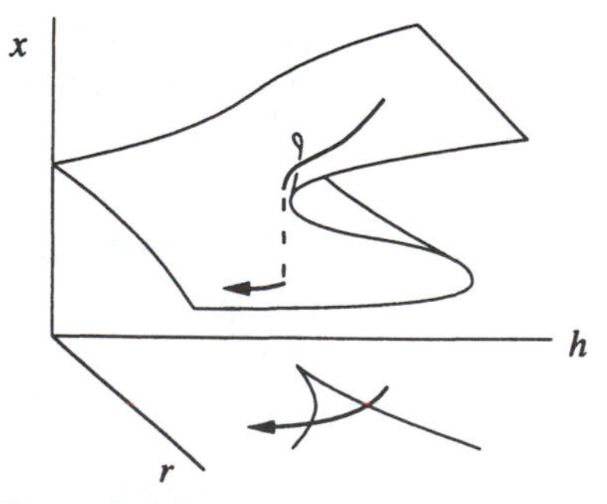

Figure 3.6.6

For more about catastrophe theory, see Zeeman (1977) or Poston and Stewart (1978). Incidentally, there was a violent controversy about this subject in the late

1970s. If you like watching fights, have a look at Zahler and Sussman (1977) and Kolata (1977).

Bead on a Tilted Wire

As a simple example of imperfect bifurcation and catastrophe, consider the following mechanical system (Figure 3.6.7).

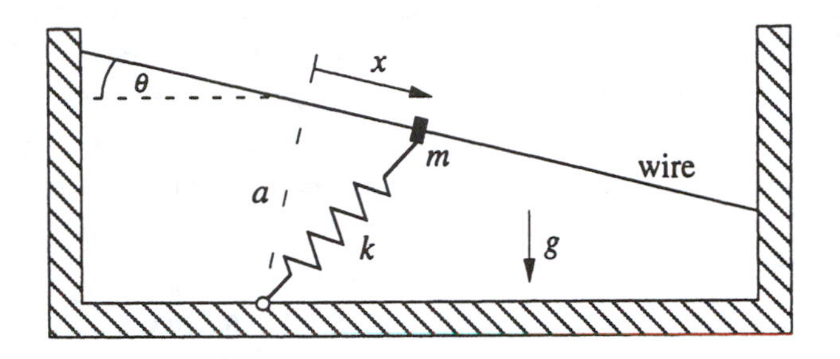

Figure 3.6.7

A bead of mass m is constrained to slide along a straight wire inclined at an angle θ with respect to the horizontal. The mass is attached to a spring of stiffness k and relaxed length L_0, and is also acted on by gravity. We choose coordinates along the wire so that $x = 0$ occurs at the point closest to the support point of the spring; let a be the distance between this support point and the wire.

In Exercises 3.5.4 and 3.6.5, you are asked to analyze the equilibrium positions of the bead. But first let's get some physical intuition. When the wire is horizontal ($\theta = 0$), there is perfect symmetry between the left and right sides of the wire, and $x = 0$ is always an equilibrium position. The stability of this equilibrium depends on the relative sizes of L_0 and a: if $L_0 < a$, the spring is in tension and so the equilibrium should be stable. But if $L_0 > a$, the spring is compressed and so we expect an *unstable* equilibrium at $x = 0$ and a pair of stable equilibria to either side of it. Exercise 3.5.4 deals with this simple case.

The problem becomes more interesting when we tilt the wire ($\theta \neq 0$). For small tilting, we expect that there are still three equilibria if $L_0 > a$. However if the tilt becomes too steep, perhaps you can see intuitively that the uphill equilibrium might suddenly disappear, causing the bead to jump catastrophically to the downhill equilibrium. You might even want to build this mechanical system and try it. Exercise 3.6.5 asks you to work through the mathematical details.

3.7 Insect Outbreak

For a biological example of bifurcation and catastrophe, we turn now to a model for the sudden outbreak of an insect called the spruce budworm. This insect is a serious pest in eastern Canada, where it attacks the leaves of the balsam fir tree. When an outbreak occurs, the budworms can defoliate and kill most of the fir trees in the forest in about four years.

Ludwig et al. (1978) proposed and analyzed an elegant model of the interaction between budworms and the forest. They simplified the problem by exploiting a separation of time scales: the budworm population evolves on a *fast* time scale (they can increase their density fivefold in a year, so they have a characteristic time scale of months), whereas the trees grow and die on a *slow* time scale (they can completely replace their foliage in about 7–10 years, and their life span in the absence of budworms is 100–150 years.) Thus, as far as the budworm dynamics are concerned, the forest variables may be treated as constants. At the end of the analysis, we will allow the forest variables to drift very slowly—this drift ultimately triggers an outbreak.

Model

The proposed model for the budworm population dynamics is

$$\dot{N} = RN\left(1 - \frac{N}{K}\right) - p(N).$$

In the absence of predators, the budworm population $N(t)$ is assumed to grow logistically with growth rate R and carrying capacity K. The carrying capacity depends on the amount of foliage left on the trees, and so it is a slowly drifting parameter; at this stage we treat it as fixed. The term $p(N)$ represents the death rate due to *predation*, chiefly by birds, and is assumed to have the shape shown in Figure 3.7.1. There is almost no predation when budworms are scarce; the birds seek food elsewhere. However, once the population exceeds a certain critical level $N = A$, the predation turns on sharply and then saturates (the birds are eating as fast as they can). Ludwig et al. (1978) assumed the specific form

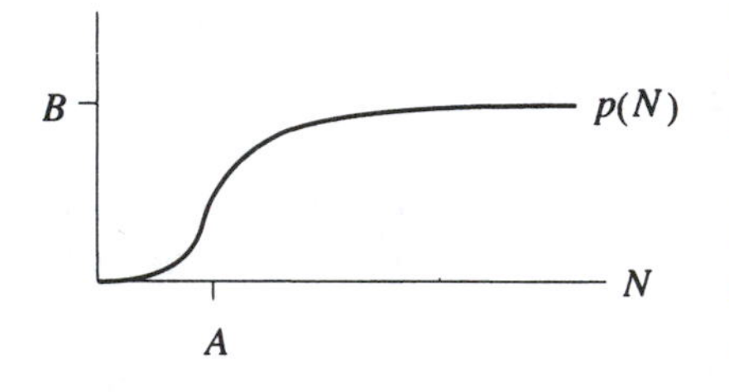

Figure 3.7.1

$$p(N) = \frac{BN^2}{A^2 + N^2}$$

where $A, B > 0$. Thus the full model is

$$\dot{N} = RN\left(1 - \frac{N}{K}\right) - \frac{BN^2}{A^2 + N^2}. \qquad (1)$$

We now have several questions to answer. What do we mean by an "outbreak" in the context of this model? The idea must be that, as parameters drift, the budworm population suddenly jumps from a low to a high level. But what do we mean by "low" and "high," and are there solutions with this character? To answer these questions, it is convenient to recast the model into a dimensionless form, as in Section 3.5.

Dimensionless Formulation

The model (1) has four parameters: R, K, A, and B. As usual, there are various ways to nondimensionalize the system. For example, both A and K have the same dimension as N, and so either N/A or N/K could serve as a dimensionless population level. It often takes some trial and error to find the best choice. In this case, our heuristic will be to scale the equation so that all the dimensionless groups are pushed into the *logistic* part of the dynamics, with none in the *predation* part. This turns out to ease the graphical analysis of the fixed points.

To get rid of the parameters in the predation term, we divide (1) by B and then let

$$x = N/A,$$

which yields

$$\frac{A}{B}\frac{dx}{dt} = \frac{R}{B}Ax\left(1-\frac{Ax}{K}\right) - \frac{x^2}{1+x^2}. \qquad (2)$$

Equation (2) suggests that we should introduce a dimensionless time τ and dimensionless groups r and k, as follows:

$$\tau = \frac{Bt}{A}, \qquad r = \frac{RA}{B}, \qquad k = \frac{K}{A}.$$

Then (2) becomes

$$\frac{dx}{d\tau} = rx\left(1-\frac{x}{k}\right) - \frac{x^2}{1+x^2}, \qquad (3)$$

which is our final dimensionless form. Here r and k are the dimensionless growth rate and carrying capacity, respectively.

Analysis of Fixed Points

Equation (3) has a fixed point at $x^* = 0$; it is *always unstable* (Exercise 3.7.1). The intuitive explanation is that the predation is extremely weak for small x, and so the budworm population grows exponentially for x near zero.

The other fixed points of (3) are given by the solutions of

$$r\left(1-\frac{x}{k}\right) = \frac{x}{1+x^2}. \qquad (4)$$

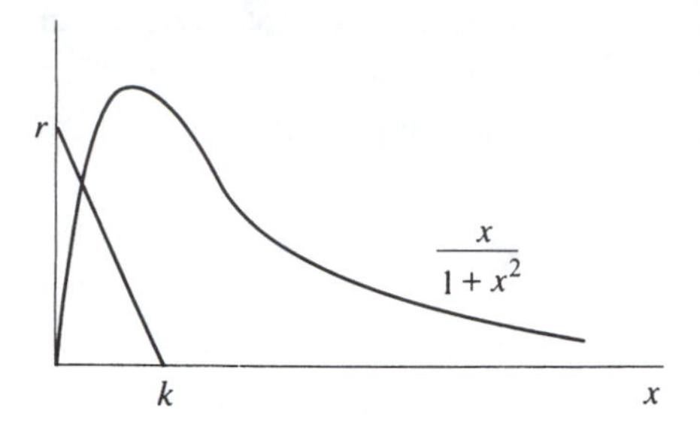

Figure 3.7.2

This equation is easy to analyze graphically—we simply graph the right- and left-hand sides of (4), and look for intersections (Figure 3.7.2). The left-hand side of (4) represents a straight line with x-intercept equal to k and a y-intercept equal to r, and the right-hand side represents a curve that is *independent of the parameters*! Hence, as we vary the parameters r and k, the line moves but the curve doesn't—this convenient property is what motivated our choice of nondimensionalization.

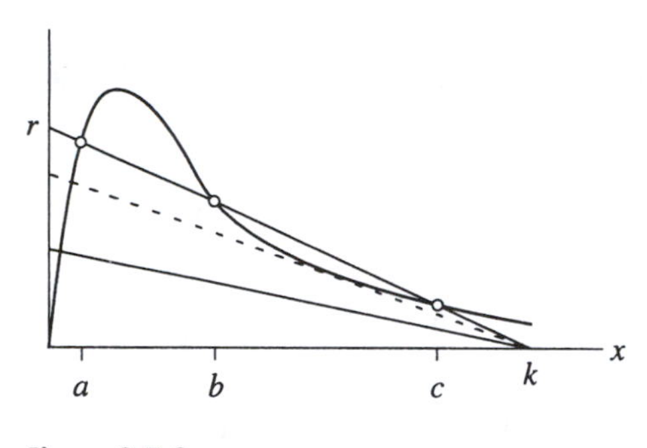

Figure 3.7.3

Figure 3.7.2 shows that if k is sufficiently small, there is exactly one intersection for any $r > 0$. However, for large k, we can have one, two, or three intersections, depending on the value of r (Figure 3.7.3). Let's suppose that there are three intersections a, b, and c. As we decrease r with k fixed, the line rotates counterclockwise about k. Then the fixed points b and c approach each other and eventually coalesce in a *saddle-node bifurcation* when the line intersects the curve *tangentially* (dashed line in Figure 3.7.3). After the bifurcation, the only remaining fixed point is a (in addition to $x^* = 0$, of course). Similarly, a and b can collide and annihilate as r is *increased.*

To determine the stability of the fixed points, we recall that $x^* = 0$ is unstable, and also observe that the stability type must alternate as we move along the x-axis. Hence a is stable, b is unstable, and c is stable. Thus, for r and k in the range corresponding to three positive fixed points, the vector field is qualitatively like that shown in Figure 3.7.4. The smaller stable fixed point a is called the ***refuge*** level of the budworm population, while the larger stable point c is the ***outbreak*** level. From the point of view of pest control, one would like to keep the population at a and away from c. The fate of the system is determined by the initial condition x_0; an outbreak occurs

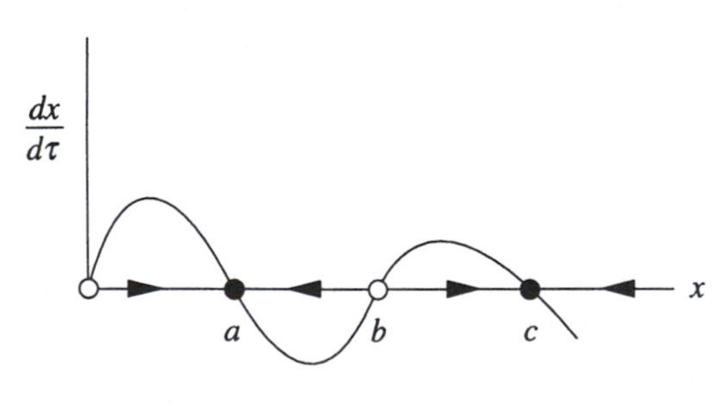

Figure 3.7.4

if and only if $x_0 > b$. In this sense the unstable equilibrium b plays the role of a ***threshold***.

An outbreak can also be triggered by a saddle-node bifurcation. If the parameters r and k drift in such a way that the fixed point a disappears, then the population will jump suddenly to the outbreak level c. The situation is made worse by the hysteresis effect—even if the parameters are restored to their values before the outbreak, the population will not drop back to the refuge level.

Calculating the Bifurcation Curves

Now we compute the curves in (k,r) space where the system undergoes saddle-node bifurcations. The calculation is somewhat harder than that in Section 3.6: we will not be able to write r explicitly as a function of k, for example. Instead, the bifurcation curves will be written in the ***parametric form*** $(k(x), r(x))$, where x runs through all positive values. (Please don't be confused by this traditional terminology—one would call x the "parameter" in these parametric equations, even though r and k are themselves parameters in a different sense.)

As discussed earlier, the condition for a saddle-node bifurcation is that the line $r(1-x/k)$ intersects the curve $x/(1+x^2)$ tangentially. Thus we require *both*

$$r\left(1-\frac{x}{k}\right)=\frac{x}{1+x^2} \tag{5}$$

and

$$\frac{d}{dx}\left[r\left(1-\frac{x}{k}\right)\right]=\frac{d}{dx}\left[\frac{x}{1+x^2}\right]. \tag{6}$$

After differentiation, (6) reduces to

$$-\frac{r}{k}=\frac{1-x^2}{(1+x^2)^2}. \tag{7}$$

We substitute this expression for r/k into (5), which allows us to express r solely in terms of x. The result is

$$r=\frac{2x^3}{(1+x^2)^2}. \tag{8}$$

Then inserting (8) into (7) yields

$$k=\frac{2x^3}{x^2-1}. \tag{9}$$

The condition $k>0$ implies that x must be restricted to the range $x>1$.

Together (8) and (9) define the bifurcation curves. For each $x>1$, we plot the corresponding point $(k(x), r(x))$ in the (k, r) plane. The resulting curves are shown in Figure 3.7.5. (Exercise 3.7.2 deals with some of the analytical properties of these curves.)

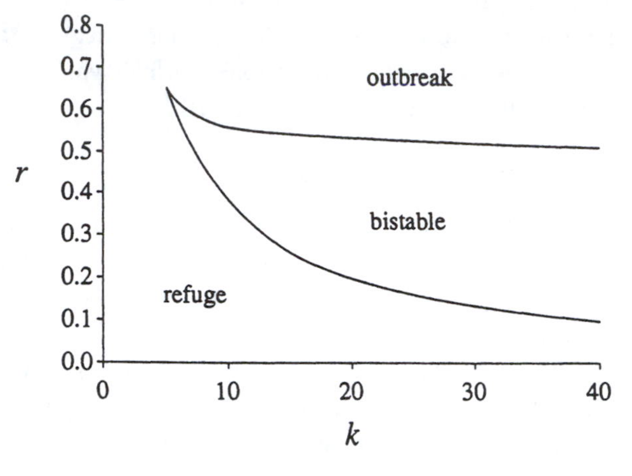

Figure 3.7.5

The different regions in Figure 3.7.5 are labeled according to the stable fixed points that exist. The refuge level a is the only stable state for low r, and the outbreak level c is the only stable state for large r. In the ***bistable*** region, both stable states exist.

The stability diagram is very similar to Figure 3.6.2. It too can be regarded as the projection of a cusp catastrophe surface, as schematically illustrated in Figure 3.7.6. You are hereby challenged to graph the surface accurately!

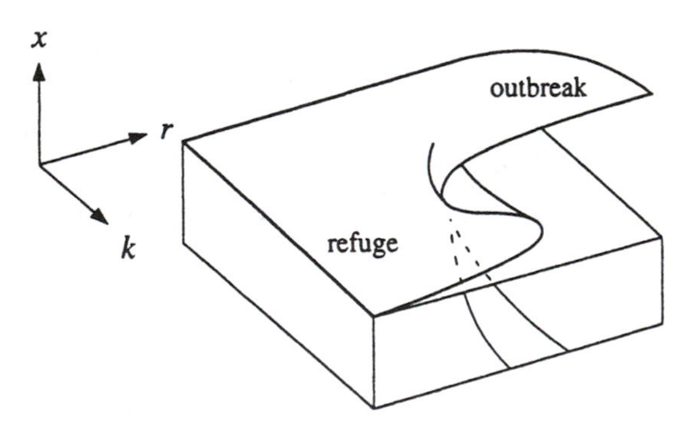

Figure 3.7.6

Comparison with Observations

Now we need to decide on biologically plausible values of the dimensionless groups $r = RA/B$ and $k = K/A$. A complication is that these parameters may drift slowly as the condition of the forest changes. According to Ludwig et al. (1978), r increases as the forest grows, while k remains fixed.

They reason as follows: let S denote the average size of the trees, interpreted as the total surface area of the branches in a stand. Then the carrying capacity K should be proportional to the available foliage, so $K = K'S$. Similarly, the half-saturation parameter A in the predation term should be proportional to S; predators such as birds search *units of foliage*, not acres of forest, and so the relevant quantity A' must have the dimensions of budworms per unit of branch area. Hence $A = A'S$ and therefore

$$r = \frac{RA'}{B}S, \qquad k = \frac{K'}{A'}. \tag{10}$$

The experimental observations suggest that for a young forest, typically $k \approx 300$ and $r < 1/2$ so the parameters lie in the bistable region. The budworm population is kept down by the birds, which find it easy to search the small number of branches per acre. However, as the forest grows, S increases and therefore the point (k, r) drifts upward in parameter space toward the outbreak region of Figure 3.7.5. Ludwig et al. (1978) estimate that $r \approx 1$ for a fully mature forest, which lies dangerously in the outbreak region. After an outbreak occurs, the fir trees die and the forest is taken over by birch trees. But they are less efficient at using nutrients and eventually the fir trees come back—this recovery takes about 50–100 years (Murray 1989).

We conclude by mentioning some of the approximations in the model presented here. The tree dynamics have been neglected; see Ludwig et al. (1978) for a discussion of this longer time-scale behavior. We've also neglected the *spatial* distribution of budworms and their possible dispersal—see Ludwig et al. (1979) and Murray (1989) for treatments of this aspect of the problem.

EXERCISES FOR CHAPTER 3

3.1 Saddle-Node Bifurcation

For each of the following exercises, sketch all the qualitatively different vector fields that occur as r is varied. Show that a saddle-node bifurcation occurs at a critical value of r, to be determined. Finally, sketch the bifurcation diagram of fixed points x^* versus r.

3.1.1 $\dot{x} = 1 + rx + x^2$ **3.1.2** $\dot{x} = r - \cosh x$

3.1.3 $\dot{x} = r + x - \ln(1 + x)$ **3.1.4** $\dot{x} = r + \frac{1}{2}x - x/(1 + x)$

3.1.5 (Unusual bifurcations) In discussing the normal form of the saddle-node bifurcation, we mentioned the assumption that $a = \partial f/\partial r|_{(x^*,r_c)} \neq 0$. To see what can happen if $\partial f/\partial r|_{(x^*,r_c)} = 0$, sketch the vector fields for the following examples, and then plot the fixed points as a function of r.

a) $\dot{x} = r^2 - x^2$
b) $\dot{x} = r^2 + x^2$

3.2 Transcritical Bifurcation

For each of the following exercises, sketch all the qualitatively different vector fields that occur as r is varied. Show that a transcritical bifurcation occurs at a critical value of r, to be determined. Finally, sketch the bifurcation diagram of fixed points x^* vs. r.

3.2.1 $\dot{x} = rx + x^2$

3.2.2 $\dot{x} = rx - \ln(1+x)$

3.2.3 $\dot{x} = x - rx(1-x)$

3.2.4 $\dot{x} = x(r - e^x)$

3.2.5 (Chemical kinetics) Consider the chemical reaction system

$$A + X \underset{k_{-1}}{\overset{k_1}{\rightleftharpoons}} 2X \qquad X + B \xrightarrow{k_2} C.$$

This is a generalization of Exercise 2.3.2; the new feature is that X is used up in the production of C.

a) Assuming that both A and B are kept at constant concentrations a and b, show that the law of mass action leads to an equation of the form $\dot{x} = c_1 x - c_2 x^2$, where x is the concentration of X, and c_1 and c_2 are constants to be determined.
b) Show that $x^* = 0$ is stable when $k_2 b > k_1 a$, and explain why this makes sense chemically.

The next two exercises concern the normal form for the transcritical bifurcation. In Example 3.2.2, we showed how to reduce the dynamics near a transcritical bifurcation to the approximate form $\dot{X} = RX - X^2 + O(X^3)$. Our goal now is to show that the $O(X^3)$ terms can always be eliminated by a suitable nonlinear change of variables; in other words, the reduction to normal form can be made *exact*, not just approximate.

3.2.6 (Eliminating the cubic term) Consider the system $\dot{X} = RX - X^2 + aX^3 + O(X^4)$, where $R \neq 0$. We want to find a new variable x such that the system transforms into $\dot{x} = Rx - x^2 + O(x^4)$. This would be a big improvement, since the cubic term has been eliminated and the error term has been bumped up to fourth order.

Let $x = X + bX^3 + O(X^4)$, where b will be chosen later to eliminate the cubic term in the differential equation for x. This is called a *near-identity transformation,* since x and X are practically equal; they differ by a tiny cubic term. (We have skipped the quadratic term X^2, because it is not needed—you should check this later.) Now we need to rewrite the system in terms of x; this calculation requires a few steps.

a) Show that the near-identity transformation can be inverted to yield $X = x + cx^3 + O(x^4)$, and solve for c.

b) Write $\dot{x} = \dot{X} + 3bX^2\dot{X} + O(X^4)$, and substitute for X and $\dot{X}$ on the right-hand side, so that everything depends only on x. Multiply the resulting series expansions and collect terms, to obtain $\dot{x} = Rx - x^2 + kx^3 + O(x^4)$, where k depends on a, b, and R.

c) Now the moment of triumph: Choose b so that $k = 0$.

d) Is is really necessary to make the assumption that $R \neq 0$? Explain.

3.2.7 (Eliminating any higher-order term) Now we generalize the method of the last exercise. Suppose we have managed to eliminate a number of higher-order terms, so that the system has been transformed into $\dot{X} = RX - X^2 + a_n X^n + O(X^{n+1})$, where $n \geq 3$. Use the near-identity transformation $x = X + b_n X^n + O(X^{n+1})$ and the previous strategy to show that the system can be rewritten as $\dot{x} = Rx - x^2 + O(x^{n+1})$ for an appropriate choice of b_n. Thus we can eliminate as many higher-order terms as we like.

3.3 Laser Threshold

3.3.1 (An improved model of a laser) In the simple laser model considered in Section 3.3, we wrote an *algebraic* equation relating N, the number of excited atoms, to n, the number of laser photons. In more realistic models, this would be replaced by a *differential* equation. For instance, Milonni and Eberly (1988) show that after certain reasonable approximations, quantum mechanics leads to the system

$$\dot{n} = GnN - kn$$
$$\dot{N} = -GnN - fN + p.$$

Here G is the gain coefficient for stimulated emission, k is the decay rate due to loss of photons by mirror transmission, scattering, etc., f is the decay rate for spontaneous emission, and p is the pump strength. All parameters are positive, except p, which can have either sign.

This two-dimensional system will be analyzed in Exercise 8.1.13. For now, let's convert it to a one-dimensional system, as follows.

a) Suppose that N relaxes much more rapidly than n. Then we may make the quasi-static approximation $\dot{N} \approx 0$. Given this approximation, express $N(t)$ in terms of $n(t)$ and derive a first-order system for n. (This procedure is often called ***adiabatic elimination***, and one says that the evolution of $N(t)$ is *slaved* to that of $n(t)$. See Haken (1983).)

b) Show that $n^* = 0$ becomes unstable for $p > p_c$, where p_c is to be determined.

c) What type of bifurcation occurs at the laser threshold p_c?

d) (Hard question) For what range of parameters is it valid to make the approximation used in (a)?

3.3.2 (Maxwell–Bloch equations) The Maxwell–Bloch equations provide an even more sophisticated model for a laser. These equations describe the dynamics of the electric field E, the mean polarization P of the atoms, and the population inversion D:

$$\dot{E} = \kappa(P - E)$$

$$\dot{P} = \gamma_1(ED - P)$$

$$\dot{D} = \gamma_2(\lambda + 1 - D - \lambda EP)$$

where κ is the decay rate in the laser cavity due to beam transmission, γ_1 and γ_2 are decay rates of the atomic polarization and population inversion, respectively, and λ is a pumping energy parameter. The parameter λ may be positive, negative, or zero; all the other parameters are positive.

These equations are similar to the Lorenz equations and can exhibit chaotic behavior (Haken 1983, Weiss and Vilaseca 1991). However, many practical lasers do not operate in the chaotic regime. In the simplest case $\gamma_1, \gamma_2 >> \kappa$; then P and D relax rapidly to steady values, and hence may be adiabatically eliminated, as follows.

a) Assuming $\dot{P} \approx 0$, $\dot{D} \approx 0$, express P and D in terms of E, and thereby derive a first-order equation for the evolution of E.

b) Find all the fixed points of the equation for E.

c) Draw the bifurcation diagram of E^* vs. λ. (Be sure to distinguish between stable and unstable branches.)

3.4 Pitchfork Bifurcation

In the following exercises, sketch all the qualitatively different vector fields that occur as r is varied. Show that a pitchfork bifurcation occurs at a critical value of r (to be determined) and classify the bifurcation as supercritical or subcritical. Finally, sketch the bifurcation diagram of x^* vs. r.

3.4.1 $\dot{x} = rx + 4x^3$ **3.4.2** $\dot{x} = rx - \sinh x$

3.4.3 $\dot{x} = rx - 4x^3$ **3.4.4** $\dot{x} = x + \dfrac{rx}{1+x^2}$

The next exercises are designed to test your ability to distinguish among the various types of bifurcations—it's easy to confuse them! In each case, find the values of r at which bifurcations occur, and classify those as saddle-node, transcritical, supercritical pitchfork, or subcritical pitchfork. Finally, sketch the bifurcation diagram of fixed points x^* vs. r.

3.4.5 $\dot{x} = r - 3x^2$ **3.4.6** $\dot{x} = rx - \dfrac{x}{1+x}$

3.4.7 $\dot{x} = 5 - re^{-x^2}$ **3.4.8** $\dot{x} = rx - \dfrac{x}{1+x^2}$

3.4.9 $\dot{x} = x + \tanh(rx)$ **3.4.10** $\dot{x} = rx + \dfrac{x^3}{1+x^2}$

3.4.11 (An interesting bifurcation diagram) Consider the system $\dot{x} = rx - \sin x$.

a) For the case $r = 0$, find and classify all the fixed points, and sketch the vector field.

b) Show that when $r > 1$, there is only one fixed point. What kind of fixed point is it?

c) As r decreases from ∞ to 0, classify *all* the bifurcations that occur.

d) For $0 < r << 1$, find an approximate formula for values of r at which bifurcations occur.

e) Now classify all the bifurcations that occur as r decreases from 0 to $-\infty$.

f) Plot the bifurcation diagram for $-\infty < r < \infty$, and indicate the stability of the various branches of fixed points.

3.4.12 ("Quadfurcation") With tongue in cheek, we pointed out that the pitchfork bifurcation could be called a "trifurcation," since three branches of fixed points appear for $r > 0$. Can you construct an example of a "quadfurcation," in which $\dot{x} = f(x,r)$ has no fixed points for $r < 0$ and four branches of fixed points for $r > 0$? Extend your results to the case of an arbitrary number of branches, if possible.

3.4.13 (Computer work on bifurcation diagrams) For the vector fields below, use a computer to obtain a quantitatively accurate plot of the values of x^* vs. r, where $0 \le r \le 3$. In each case, there's an easy way to do this, and a harder way using the Newton-Raphson method.

a) $\dot{x} = r - x - e^{-x}$ b) $\dot{x} = 1 - x - e^{-rx}$

3.4.14 (Subcritical pitchfork) Consider the system $\dot{x} = rx + x^3 - x^5$, which exhibits a subcritical pitchfork bifurcation.

a) Find algebraic expressions for all the fixed points as r varies.

b) Sketch the vector fields as r varies. Be sure to indicate all the fixed points and their stability.

c) Calculate r_s, the parameter value at which the nonzero fixed points are born in a saddle-node bifurcation.

3.4.15 (First-order phase transition) Consider the potential $V(x)$ for the system $\dot{x} = rx + x^3 - x^5$. Calculate r_c, where r_c is defined by the condition that V has three equally deep wells, i.e., the values of V at the three local minima are equal.

(Note: In equilibrium statistical mechanics, one says that a *first-order phase transition* occurs at $r = r_c$. For this value of r, there is equal probability of finding the system in the state corresponding to any of the three minima. The freezing of water into ice is the most familiar example of a first-order phase transition.)

3.4.16 (Potentials) In parts (a)–(c), let $V(x)$ be the potential, in the sense that $\dot{x} = -dV/dx$. Sketch the potential as a function of r. Be sure to show all the qualitatively different cases, including bifurcation values of r.

a) (Saddle-node) $\dot{x} = r - x^2$
b) (Transcritical) $\dot{x} = rx - x^2$
c) (Subcritical pitchfork) $\dot{x} = rx + x^3 - x^5$

3.5 Overdamped Bead on a Rotating Hoop

3.5.1 Consider the bead on the rotating hoop discussed in Section 3.5. Explain in physical terms why the bead cannot have an equilibrium position with $\phi > \pi/2$.

3.5.2 Do the linear stability analysis for all the fixed points for Equation (3.5.7), and confirm that Figure 3.5.6 is correct.

3.5.3 Show that Equation (3.5.7) reduces to $\frac{d\phi}{d\tau} = A\phi - B\phi^3 + O(\phi^5)$ near $\phi = 0$. Find A and B.

3.5.4 (Bead on a horizontal wire) A bead of mass m is constrained to slide along a straight horizontal wire. A spring of relaxed length L_0 and spring constant k is attached to the mass and to a support point a distance h from the wire (Figure 1).

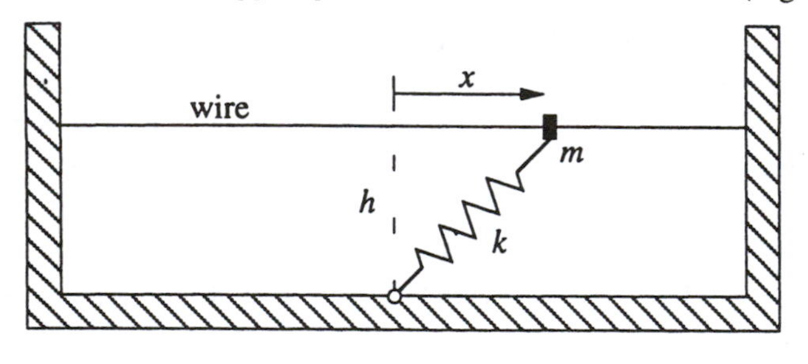

Figure 1

Finally, suppose that the motion of the bead is opposed by a viscous damping force $b\dot{x}$.

a) Write Newton's law for the motion of the bead.
b) Find all possible equilibria, i.e., fixed points, as functions of k, h, m, b, and L_0.
c) Suppose $m = 0$. Classify the stability of all the fixed points, and draw a bifurcation diagram.
d) If $m \neq 0$, how small does m have to be to be considered negligible? In what sense is it negligible?

3.5.5 (Time scale for the rapid transient) While considering the bead on the rotating hoop, we used phase plane analysis to show that the equation

$$\varepsilon \frac{d^2\phi}{d\tau^2} + \frac{d\phi}{d\tau} = f(\phi)$$

has solutions that rapidly relax to the curve where $\frac{d\phi}{d\tau} = f(\phi)$.

a) Estimate the time scale T_{fast} for this rapid transient in terms of ε, and then express T_{fast} in terms of the original dimensional quantities m, g, r, ω, and b.

b) Rescale the original differential equation, using T_{fast} as the characteristic time scale, instead of $T_{slow} = b/mg$. Which terms in the equation are negligible on this time scale?

c) Show that $T_{fast} << T_{slow}$ if $\varepsilon << 1$. (In this sense, the time scales T_{fast} and T_{slow} are *widely separated.*)

3.5.6 (A model problem about singular limits) Consider the *linear* differential equation

$$\varepsilon \ddot{x} + \dot{x} + x = 0,$$

subject to the initial conditions $x(0) = 1$, $\dot{x}(0) = 0$.

a) Solve the problem analytically for all $\varepsilon > 0$.

b) Now suppose $\varepsilon << 1$. Show that there are two widely separated time scales in the problem, and estimate them in terms of ε.

c) Graph the solution $x(t)$ for $\varepsilon << 1$, and indicate the two time scales on the graph.

d) What do you conclude about the validity of replacing $\varepsilon \ddot{x} + \dot{x} + x = 0$ with its singular limit $\dot{x} + x = 0$?

e) Give two physical analogs of this problem, one involving a mechanical system, and another involving an electrical circuit. In each case, find the dimensionless combination of parameters corresponding to ε, and state the physical meaning of the limit $\varepsilon << 1$.

3.5.7 (Nondimensionalizing the logistic equation) Consider the logistic equation $\dot{N} = rN(1 - N/K)$, with initial condition $N(0) = N_0$.

a) This system has three dimensional parameters r, K, and N_0. Find the dimensions of each of these parameters.

b) Show that the system can be rewritten in the dimensionless form

$$\frac{dx}{d\tau} = x(1-x), \qquad x(0) = x_0$$

for appropriate choices of the dimensionless variables x, x_0, and τ.

c) Find a different nondimensionalization in terms of variables u and τ, where u is chosen such that the initial condition is always $u_0 = 1$.

d) Can you think of any advantage of one nondimensionalization over the other?

3.5.8 (Nondimensionalizing the subcritical pitchfork) The first-order system $\dot{u} = au + bu^3 - cu^5$, where $b, c > 0$, has a subcritical pitchfork bifurcation at $a = 0$. Show that this equation can be rewritten as

$$\frac{dx}{d\tau} = rx + x^3 - x^5$$

where $x = u/U$, $\tau = t/T$, and U, T, and r are to be determined in terms of a, b, and c.

3.6 Imperfect Bifurcations and Catastrophes

3.6.1 (Warm-up question about imperfect bifurcation) Does Figure 3.6.3b correspond to $h > 0$ or to $h < 0$?

3.6.2 (Imperfect transcritical bifurcation) Consider the system $\dot{x} = h + rx - x^2$. When $h = 0$, this system undergoes a transcritical bifurcation at $r = 0$. Our goal is to see how the bifurcation diagram of x^* vs. r is affected by the imperfection parameter h.

a) Plot the bifurcation diagram for $\dot{x} = h + rx - x^2$, for $h < 0$, $h = 0$, and $h > 0$.

b) Sketch the regions in the (r, h) plane that correspond to qualitatively different vector fields, and identify the bifurcations that occur on the boundaries of those regions.

c) Plot the potential $V(x)$ corresponding to all the different regions in the (r, h) plane.

3.6.3 (A perturbation to the supercritical pitchfork) Consider the system $\dot{x} = rx + ax^2 - x^3$, where $-\infty < a < \infty$. When $a = 0$, we have the normal form for the supercritical pitchfork. The goal of this exercise is to study the effects of the new parameter a.

a) For each a, there is a bifurcation diagram of x^* vs. r. As a varies, these bifurcation diagrams can undergo qualitative changes. Sketch all the qualitatively different bifurcation diagrams that can be obtained by varying a.

b) Summarize your results by plotting the regions in the (r, a) plane that correspond to qualitatively different classes of vector fields. Bifurcations occur on the boundaries of these regions; identify the types of bifurcations that occur.

3.6.4 (Imperfect saddle-node) What happens if you add a small imperfection to a system that has a saddle-node bifurcation?

3.6.5 (Mechanical example of imperfect bifurcation and catastrophe) Consider the bead on a tilted wire discussed at the end of Section 3.6.

a) Show that the equilibrium positions of the bead satisfy

$$mg\sin\theta = kx\left(1 - \frac{L_0}{\sqrt{x^2 + a^2}}\right).$$

b) Show that this equilibrium equation can be written in dimensionless form as

$$1 - \frac{h}{u} = \frac{R}{\sqrt{1+u^2}}$$

for appropriate choices of R, h, and u.

c) Give a graphical analysis of the dimensionless equation for the cases $R < 1$ and $R > 1$. How many equilibria can exist in each case?

d) Let $r = R - 1$. Show that the equilibrium equation reduces to $h + ru - \frac{1}{2}u^3 \approx 0$ for small r, h, and u.

e) Find an approximate formula for the saddle-node bifurcation curves in the limit of small r, h, and u.

f) Show that the *exact* equations for the bifurcation curves can be written in parametric form as

$$h(u) = -u^3, \qquad R(u) = (1+u^2)^{3/2},$$

where $-\infty < u < \infty$. (Hint: You may want to look at Section 3.7.) Check that this result reduces to the approximate result in part (d).

g) Give a numerically accurate plot of the bifurcation curves in the (r, h) plane.

h) Interpret your results physically, in terms of the original dimensional variables.

3.6.6 (Patterns in fluids) Ahlers (1989) gives a fascinating review of experiments on one-dimensional patterns in fluid systems. In many cases, the patterns first emerge via supercritical or subcritical pitchfork bifurcations from a spatially uniform state. Near the bifurcation, the dynamics of the amplitude of the patterns are given approximately by $\tau\dot{A} = \varepsilon A - gA^3$ in the supercritical case, or $\tau\dot{A} = \varepsilon A - gA^3 - kA^5$ in the subcritical case. Here $A(t)$ is the amplitude, τ is a typical time scale, and ε is a small dimensionless parameter that measures the distance from the bifurcation. The parameter $g > 0$ in the supercritical case, whereas $g < 0$ and $k > 0$ in the subcritical case. (In this context, the equation $\tau\dot{A} = \varepsilon A - gA^3$ is often called the *Landau equation.*)

a) Dubois and Bergé (1978) studied the supercritical bifurcation that arises in Rayleigh–Bénard convection, and showed experimentally that the steady-state amplitude depended on ε according to the power law $A^* \propto \varepsilon^{\beta}$, where $\beta = 0.50 \pm 0.01$. What does the Landau equation predict?

b) The equation $\tau\dot{A} = \varepsilon A - gA^3 - kA^5$ is said to undergo a *tricritical bifurcation* when $g = 0$; this case is the borderline between supercritical and subcritical bifurcations. Find the relation between A^* and ε when $g = 0$.

c) In experiments on Taylor–Couette vortex flow, Aitta et al. (1985) were able to change the parameter g continuously from positive to negative by varying the aspect ratio of their experimental set-up. Assuming that the equation is modified to $\tau\dot{A} = h + \varepsilon A - gA^3 - kA^5$, where $h > 0$ is a slight imperfection, sketch the bifurcation diagram of A^* vs. ε in the three cases $g > 0$, $g = 0$, and $g < 0$. Then look up the actual data in Aitta et al. (1985, Figure 2) or see Ahlers (1989, Figure 15).

d) In the experiments of part (c), the amplitude $A(t)$ was found to evolve toward a steady state in the manner shown in Figure 2 (redrawn from Ahlers (1989), Figure 18). The results are for the imperfect subcritical case $g < 0$, $h \neq 0$. In the experiments, the parameter ε was switched at $t = 0$ from a negative value to a positive value ε_f. In Figure 2, ε_f increases from the bottom to the top.

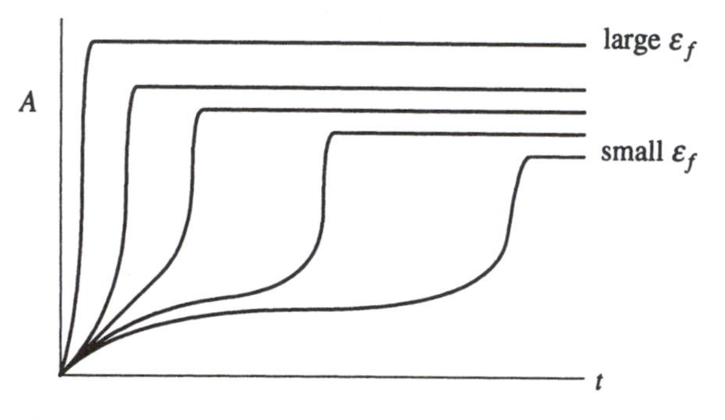

Figure 2

Explain intuitively why the curves have this strange shape. Why do the curves for large ε_f go almost straight up to their steady state, whereas the curves for small ε_f rise to a plateau before increasing sharply to their final level? (Hint: Graph $\dot{A}$ vs. A for different ε_f.)

3.6.7 (Simple model of a magnet) A magnet can be modeled as an enormous collection of electronic spins. In the simplest model, known as the ***Ising model***, the spins can point only up or down, and are assigned the values $S_i = \pm 1$, for $i = 1, \ldots, N \gg 1$. For quantum mechanical reasons, the spins like to point in the same direction as their neighbors; on the other hand, the randomizing effects of temperature tend to disrupt any such alignment.

An important macroscopic property of the magnet is its average spin or *magnetization*

$$m = \left| \frac{1}{N} \sum_{i=1}^{N} S_i \right| .$$

At high temperature the spins point in random directions and so $m \approx 0$; the material is in the *paramagnetic* state. As the temperature is lowered, m remains near zero until a critical temperature T_c is reached. Then a ***phase transition*** occurs and the material spontaneously magnetizes. Now $m > 0$; we have a *ferromagnet*.

But the symmetry between up and down spins means that there are *two* possible ferromagnetic states. This symmetry can be broken by applying an external magnetic field h, which favors either the up or down direction. Then, in an approximation called *mean-field theory*, the equation governing the equilibrium value of m is

$$h = T \tanh^{-1} m - Jnm$$

where J and n are constants; $J > 0$ is the ferromagnetic coupling strength and n is the number of neighbors of each spin (Ma 1985, p. 459).

a) Analyze the solutions m^* of $h = T \tanh^{-1} m - Jnm$, using a graphical approach.
b) For the special case $h = 0$, find the critical temperature T_c at which a phase transition occurs.

3.7 Insect Outbreak

3.7.1 (Warm-up question about insect outbreak model) Show that the fixed point $x^* = 0$ is *always unstable* for Equation (3.7.3).

3.7.2 (Bifurcation curves for insect outbreak model)
a) Using Equations (3.7.8) and (3.7.9), sketch $r(x)$ and $k(x)$ vs. x. Determine the limiting behavior of $r(x)$ and $k(x)$ as $x \to 1$ and $x \to \infty$.
b) Find the exact values of r, k, and x at the cusp point shown in Figure 3.7.5.

3.7.3 (A model of a fishery) The equation $\dot{N} = rN(1 - \frac{N}{K}) - H$ provides an extremely simple model of a fishery. In the absence of fishing, the population is assumed to grow logistically. The effects of fishing are modeled by the term $-H$, which says that fish are caught or "harvested" at a constant rate $H > 0$, independent of their population N. (This assumes that the fishermen aren't worried about fishing the population dry—they simply catch the same number of fish every day.)
a) Show that the system can be rewritten in dimensionless form as

$$\frac{dx}{d\tau} = x(1 - x) - h,$$

for suitably defined dimensionless quantities x, τ, and h.
b) Plot the vector field for different values of h.
c) Show that a bifurcation occurs at a certain value h_c, and classify this bifurcation.
d) Discuss the long-term behavior of the fish population for $h < h_c$ and $h > h_c$, and give the biological interpretation in each case.

There's something silly about this model—the population can become negative! A better model would have a fixed point at zero population for all values of H. See the next exercise for such an improvement.

3.7.4 (Improved model of a fishery) A refinement of the model in the last exercise is

$$\dot{N} = rN\left(1 - \frac{N}{K}\right) - H\frac{N}{A+N}$$

where $H > 0$ and $A > 0$. This model is more realistic in two respects: it has a fixed point at $N = 0$ for all values of the parameters, and the rate at which fish are caught decreases with N. This is plausible—when fewer fish are available, it is harder to find them and so the daily catch drops.

a) Give a biological interpretation of the parameter A; what does it measure?

b) Show that the system can be rewritten in dimensionless form as

$$\frac{dx}{d\tau} = x(1-x) - h\frac{x}{a+x},$$

for suitably defined dimensionless quantities x, τ, a, and h.

c) Show that the system can have one, two, or three fixed points, depending on the values of a and h. Classify the stability of the fixed points in each case.

d) Analyze the dynamics near $x = 0$ and show that a bifurcation occurs when $h = a$. What type of bifurcation is it?

e) Show that another bifurcation occurs when $h = \frac{1}{4}(a+1)^2$, for $a < a_c$, where a_c is to be determined. Classify this bifurcation.

f) Plot the stability diagram of the system in (a, h) parameter space. Can hysteresis occur in any of the stability regions?

3.7.5 (A biochemical switch) Zebra stripes and butterfly wing patterns are two of the most spectacular examples of biological pattern formation. Explaining the development of these patterns is one of the outstanding problems of biology; see Murray (1989) for an excellent review of our current knowledge.

As one ingredient in a model of pattern formation, Lewis et al. (1977) considered a simple example of a biochemical switch, in which a gene G is activated by a biochemical signal substance S. For example, the gene may normally be inactive but can be "switched on" to produce a pigment or other gene product when the concentration of S exceeds a certain threshold. Let $g(t)$ denote the concentration of the gene product, and assume that the concentration s_0 of S is fixed. The model is

$$\dot{g} = k_1 s_0 - k_2 g + \frac{k_3 g^2}{k_4{}^2 + g^2}$$

where the k's are positive constants. The production of g is stimulated by s_0 at a rate k_1, and by an *autocatalytic* or positive feedback process (the nonlinear term). There is also a linear degradation of g at a rate k_2.

a) Show that the system can be put in the dimensionless form

$$\frac{dx}{d\tau} = s - rx + \frac{x^2}{1+x^2}$$

where $r > 0$ and $s \geq 0$ are dimensionless groups.

b) Show that if $s = 0$, there are two positive fixed points x^* if $r < r_c$, where r_c is to be determined.

c) Assume that initially there is no gene product, i.e., $g(0) = 0$, and suppose s is slowly increased from zero (the activating signal is turned on); what happens to $g(t)$? What happens if s then goes back to zero? Does the gene turn off again?

d) Find parametric equations for the bifurcation curves in (r, s) space, and classify the bifurcations that occur.

e) Use the computer to give a quantitatively accurate plot of the stability diagram in (r, s) space.

For further discussion of this model, see Lewis et al. (1977); Edelstein–Keshet (1988), Section 7.5; or Murray (1989), Chapter 15.

3.7.6 (Model of an epidemic) In pioneering work in epidemiology, Kermack and McKendrick (1927) proposed the following simple model for the evolution of an epidemic. Suppose that the population can be divided into three classes: $x(t)$ = number of healthy people; $y(t)$ = number of sick people; $z(t)$ = number of dead people. Assume that the total population remains constant in size, except for deaths due to the epidemic. (That is, the epidemic evolves so rapidly that we can ignore the slower changes in the populations due to births, emigration, or deaths by other causes.)

Then the model is

$$\begin{aligned} \dot{x} &= -kxy \\ \dot{y} &= kxy - \ell y \\ \dot{z} &= \ell y \end{aligned}$$

where k and ℓ are positive constants. The equations are based on two assumptions:

(i) Healthy people get sick at a rate proportional to the product of x and y. This would be true if healthy and sick people encounter each other at a rate proportional to their numbers, and if there were a constant probability that each such encounter would lead to transmission of the disease.

(ii) Sick people die at a constant rate ℓ.

The goal of this exercise is to reduce the model, which is a *third-order system*, to a first-order system that can analyzed by our methods. (In Chapter 6 we will see a simpler analysis.)

a) Show that $x + y + z = N$, where N is constant.

b) Use the $\dot{x}$ and $\dot{z}$ equation to show that $x(t) = x_0 \exp(-kz(t)/\ell)$, where $x_0 = x(0)$.

c) Show that z satisfies the first-order equation $\dot{z} = \ell\left[N - z - x_0 \exp(-kz/\ell)\right]$.

d) Show that this equation can be nondimensionalized to

$$\frac{du}{d\tau} = a - bu - e^{-u}$$

by an appropriate rescaling.

e) Show that $a \geq 1$ and $b > 0$.

f) Determine the number of fixed points u^* and classify their stability.

g) Show that the maximum of $\dot{u}(t)$ occurs at the same time as the maximum of both $\dot{z}(t)$ and $y(t)$. (This time is called the ***peak*** of the epidemic, denoted t_{peak}. At this time, there are more sick people and a higher daily death rate than at any other time.)

h) Show that if $b < 1$, then $\dot{u}(t)$ is increasing at $t = 0$ and reaches its maximum at some time $t_{\text{peak}} > 0$. Thus things get worse before they get better. (The term ***epidemic*** is reserved for this case.) Show that $\dot{u}(t)$ eventually decreases to 0.

i) On the other hand, show that $t_{\text{peak}} = 0$ if $b > 1$. (Hence no epidemic occurs if $b > 1$.)

j) The condition $b = 1$ is the *threshold* condition for an epidemic to occur. Can you give a biological interpretation of this condition?

k) Kermack and McKendrick showed that their model gave a good fit to data from the Bombay plague of 1906. How would you improve the model to make it more appropriate for AIDS? Which assumptions need revising?

For an introduction to models of epidemics, see Murray (1989), Chapter 19, or Edelstein–Keshet (1988). Models of AIDS are discussed by Murray (1989) and May and Anderson (1987). An excellent review and commentary on the Kermack–McKendrick papers is given by Anderson (1991).

4

FLOWS ON THE CIRCLE

4.0 Introduction

So far we've concentrated on the equation $\dot{x} = f(x)$, which we visualized as a vector field on the line. Now it's time to consider a new kind of differential equation and its corresponding phase space. This equation,

$$\dot{\theta} = f(\theta),$$

corresponds to a ***vector field on the circle.*** Here θ is a point on the circle and $\dot{\theta}$ is the velocity vector at that point, determined by the rule $\dot{\theta} = f(\theta)$. Like the line, the circle is one-dimensional, but it has an important new property: by flowing in one direction, a particle can eventually return to its starting place (Figure 4.0.1). Thus periodic solutions become possible for the first time in this book! To put it another way, *vector fields on the circle provide the most basic model of systems that can oscillate.*

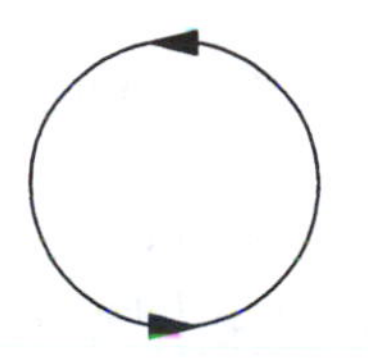

Figure 4.0.1

However, in all other respects, flows on the circle are similar to flows on the line, so this will be a short chapter. We will discuss the dynamics of some simple oscillators, and then show that these equations arise in a wide variety of applications. For example, the flashing of fireflies and the voltage oscillations of superconducting Josephson junctions have been modeled by the same equation, even though their oscillation frequencies differ by about ten orders of magnitude!

4.1 Examples and Definitions

Let's begin with some examples, and then give a more careful definition of vector fields on the circle.

EXAMPLE 4.1.1:

Sketch the vector field on the circle corresponding to $\dot{\theta} = \sin\theta$.

Solution: We assign coordinates to the circle in the usual way, with $\theta = 0$ in the direction of "east," and with θ increasing counterclockwise.

To sketch the vector field, we first find the fixed points, defined by $\dot{\theta} = 0$. These occur at $\theta^* = 0$ and $\theta^* = \pi$. To determine their stability, note that $\sin\theta > 0$ on the upper semicircle. Hence $\dot{\theta} > 0$, so the flow is counterclockwise. Similarly, the flow is clockwise on the lower semicircle, where $\dot{\theta} < 0$. Hence $\theta^* = \pi$ is stable and $\theta^* = 0$ is unstable, as shown in Figure 4.1.1.

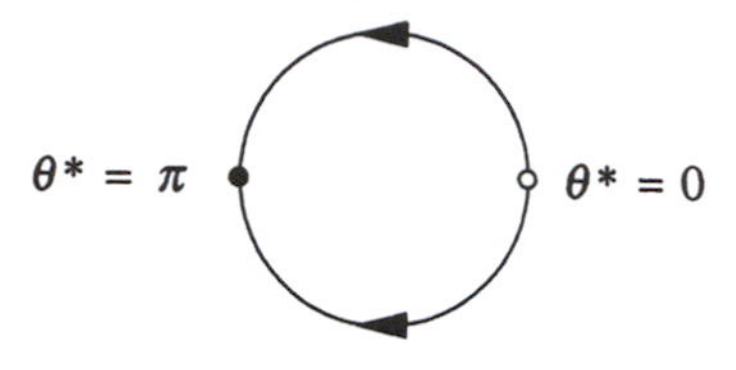

Figure 4.1.1

Actually, we've seen this example before—it's given in Section 2.1. There we regarded $\dot{x} = \sin x$ as a vector field on the *line*. Compare Figure 2.1.1 with Figure 4.1.1 and notice how much clearer it is to think of this system as a vector field on the circle. ■

EXAMPLE 4.1.2:

Explain why $\dot{\theta} = \theta$ cannot be regarded as a vector field on the circle, for θ in the range $-\infty < \theta < \infty$.

Solution: The velocity is not uniquely defined. For example, $\theta = 0$ and $\theta = 2\pi$ are two labels for the same point on the circle, but the first label implies a velocity of 0 at that point, while the second implies a velocity of 2π. ■

If we try to avoid this non-uniqueness by restricting θ to the range $-\pi < \theta \le \pi$, then the velocity vector jumps discontinuously at the point corresponding to $\theta = \pi$. Try as we might, there's no way to consider $\dot{\theta} = \theta$ as a smooth vector field on the entire circle.

Of course, there's no problem regarding $\dot{\theta} = \theta$ as a vector field on the *line*, because then $\theta = 0$ and $\theta = 2\pi$ are different points, and so there's no conflict about how to define the velocity at each of them.

Example 4.1.2 suggests how to define vector fields on the circle. Here's a geometric definition: A ***vector field on the circle*** is a rule that assigns a unique velocity vector to each point on the circle.

In practice, such vector fields arise when we have a first-order system $\dot{\theta} = f(\theta)$, where $f(\theta)$ is a real-valued, 2π-*periodic* function. That is, $f(\theta + 2\pi) = f(\theta)$ for all real θ. Moreover, we assume (as usual) that $f(\theta)$ is smooth enough to guarantee existence and uniqueness of solutions. Although this system could be regarded as a special case of a vector field on the line, it is usually clearer to think of it as a vector field on the circle (as in Example 4.1.1). This means that we don't distinguish between θ's that differ by an integer multiple of 2π. Here's where the periodicity of $f(\theta)$ becomes important—it ensures that the velocity $\dot{\theta}$ is uniquely defined at each point θ on the circle, in the sense that $\dot{\theta}$ is the same, whether we call that point θ or $\theta + 2\pi$, or $\theta + 2\pi k$ for any integer k.

4.2 Uniform Oscillator

A point on a circle is often called an *angle* or a ***phase.*** Then the simplest oscillator of all is one in which the phase θ changes uniformly:

$$\dot{\theta} = \omega$$

where ω is a constant. The solution is

$$\theta(t) = \omega t + \theta_0,$$

which corresponds to uniform motion around the circle at an angular frequency ω. This solution is ***periodic,*** in the sense that $\theta(t)$ changes by 2π, and therefore returns to the same point on the circle, after a time $T = 2\pi/\omega$. We call T the ***period*** of the oscillation.

Notice that we have said nothing about the *amplitude* of the oscillation. There really is no amplitude variable in our system. If we had an amplitude as well as a phase variable, we'd be in a *two-dimensional* phase space; this situation is more complicated and will be discussed later in the book. (Or if you prefer, you can imagine that the oscillation occurs at some *fixed* amplitude, corresponding to the radius of our circular phase space. In any case, amplitude plays no role in the dynamics.)

EXAMPLE 4.2.1:

Two joggers, Speedy and Pokey, are running at a steady pace around a circular track. It takes Speedy T_1 seconds to run once around the track, whereas it takes Pokey $T_2 > T_1$ seconds. Of course, Speedy will periodically overtake Pokey; how long does it take for Speedy to lap Pokey once, assuming that they start together?

Solution: Let $\theta_1(t)$ be Speedy's position on the track. Then $\dot{\theta}_1 = \omega_1$ where $\omega_1 = 2\pi/T_1$. This equation says that Speedy runs at a steady pace and completes a circuit every T_1 seconds. Similarly, suppose that $\dot{\theta}_2 = \omega_2 = 2\pi/T_2$ for Pokey.

The condition for Speedy to lap Pokey is that the angle between them has increased by 2π. Thus if we define the ***phase difference*** $\phi = \theta_1 - \theta_2$, we want to find how long it takes for ϕ to increase by 2π (Figure 4.2.1). By subtraction we find $\dot{\phi} = \dot{\theta}_1 - \dot{\theta}_2 = \omega_1 - \omega_2$. Thus ϕ increases by 2π after a time

Figure 4.2.1

$$T_{\text{lap}} = \frac{2\pi}{\omega_1 - \omega_2} = \left(\frac{1}{T_1} - \frac{1}{T_2} \right)^{-1}. \blacksquare$$

Example 4.2.1 illustrates an effect called the ***beat phenomenon.*** Two noninteracting oscillators with different frequencies will periodically go in and out of phase with each other. You may have heard this effect on a Sunday morning: sometimes the bells of two different churches will ring simultaneously, then slowly drift apart, and then eventually ring together again. If the oscillators *interact* (for example, if the two joggers try to stay together or the bell ringers can hear each other), then we can get more interesting effects, as we will see in Section 4.5 on the flashing rhythm of fireflies.

4.3 Nonuniform Oscillator

The equation

$$\dot{\theta} = \omega - a \sin\theta \qquad (1)$$

arises in many different branches of science and engineering. Here is a partial list:

Electronics (phase-locked loops)
Biology (oscillating neurons, firefly flashing rhythm, human sleep-wake cycle)
Condensed-matter physics (Josephson junction, charge-density waves)
Mechanics (Overdamped pendulum driven by a constant torque)

Some of these applications will be discussed later in this chapter and in the exercises.

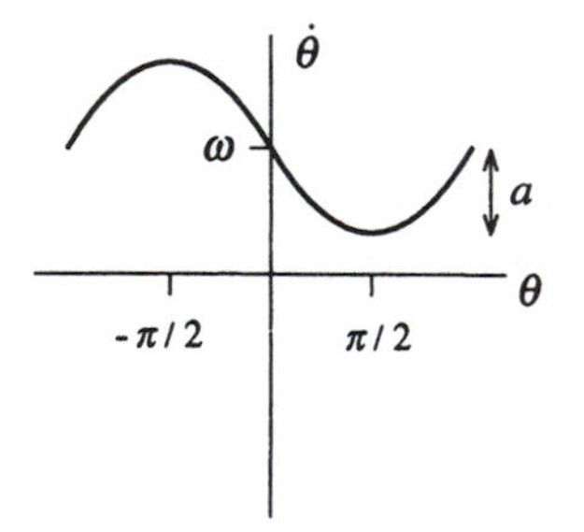

Figure 4.3.1

To analyze (1), we assume that $\omega > 0$ and $a \geq 0$ for convenience; the results for negative ω and a are similar. A typical graph of $f(\theta) = \omega - a\sin\theta$ is shown in Figure 4.3.1. Note that ω is the mean and a is the amplitude.

Vector Fields

If $a = 0$, (1) reduces to the uniform oscillator. The parameter a introduces a *nonuniformity* in the flow around the circle: the flow is fastest at $\theta = -\pi/2$ and slowest at $\theta = \pi/2$ (Figure 4.3.2a). This nonuniformity becomes more pronounced as a increases. When a is slightly less than ω, the oscillation is very jerky: the phase point $\theta(t)$ takes a long time to pass through a ***bottleneck*** near $\theta = \pi/2$, after which it zips around the rest of the circle on a much faster time scale. When $a = \omega$, the system stops oscillating altogether: a half-stable fixed point has been born in a *saddle-node bifurcation* at $\theta = \pi/2$ (Figure 4.3.2b). Finally, when $a > \omega$, the half-stable fixed point splits into a stable and unstable fixed point (Figure 4.3.2c). All trajectories are attracted to the stable fixed point as $t \to \infty$.

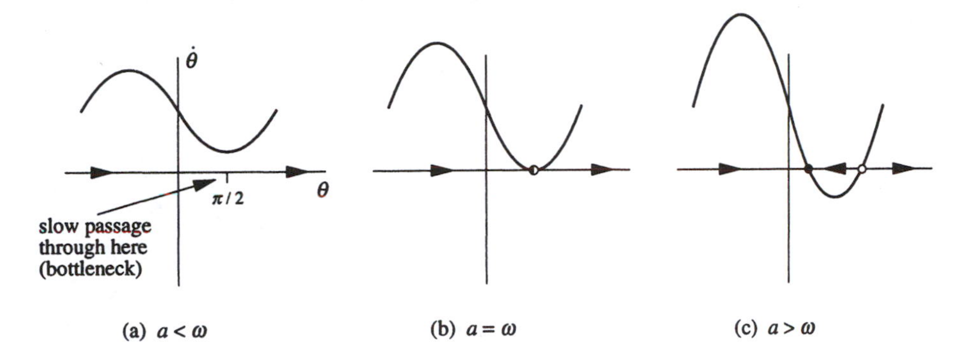

Figure 4.3.2

The same information can be shown by plotting the vector fields on the circle (Figure 4.3.3).

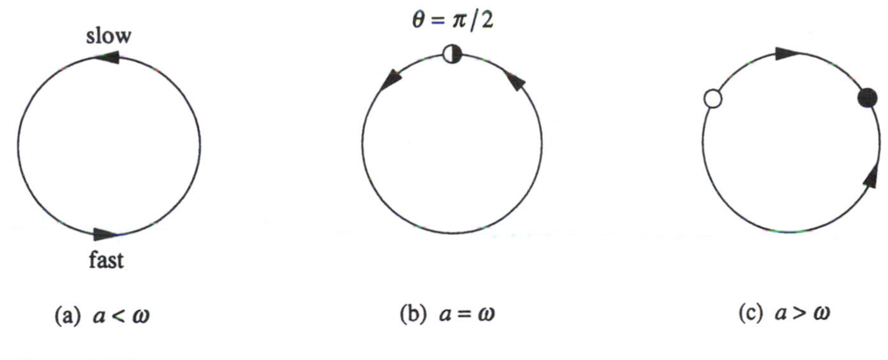

Figure 4.3.3

EXAMPLE 4.3.1:

Use linear stability analysis to classify the fixed points of (1) for $a > \omega$.

Solution: The fixed points θ^* satisfy

$$\sin\theta^* = \omega/a, \qquad \cos\theta^* = \pm\sqrt{1-(\omega/a)^2}.$$

Their linear stability is determined by

$$f'(\theta^*) = -a\cos\theta^* = \mp a\sqrt{1-(\omega/a)^2}.$$

Thus the fixed point with $\cos\theta^* > 0$ is the stable one, since $f'(\theta^*) < 0$. This agrees with Figure 4.3.2c. ■

Oscillation Period

For $a < \omega$, the period of the oscillation can be found analytically, as follows: the time required for θ to change by 2π is given by

$$\begin{aligned} T &= \int dt = \int_0^{2\pi} \frac{dt}{d\theta}\,d\theta \\ &= \int_0^{2\pi} \frac{d\theta}{\omega - a\sin\theta} \end{aligned}$$

where we have used (1) to replace $dt/d\theta$. This integral can be evaluated by complex variable methods, or by the substitution $u = \tan\frac{\theta}{2}$. (See Exercise 4.3.2 for details.) The result is

$$T = \frac{2\pi}{\sqrt{\omega^2 - a^2}}. \qquad (2)$$

Figure 4.3.4 shows the graph of T as a function of a.

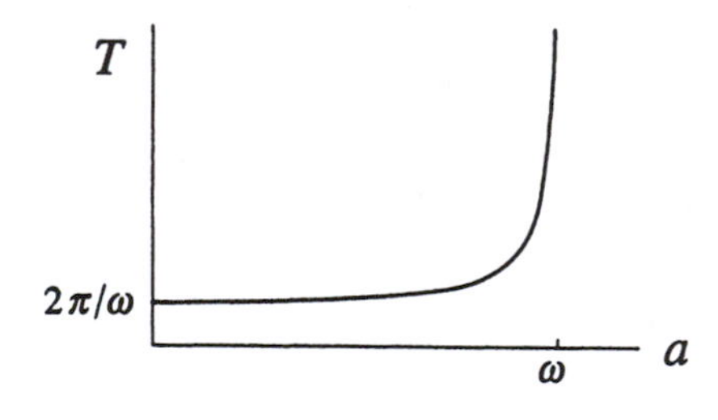

Figure 4.3.4

When $a = 0$, Equation (2) reduces to $T = 2\pi/\omega$, the familiar result for a uniform oscillator. The period increases with a and diverges as a approaches ω from below (we denote this limit by $a \to \omega^-$).

We can estimate the order of the divergence by noting that

$$\begin{aligned} \sqrt{\omega^2 - a^2} &= \sqrt{\omega + a}\sqrt{\omega - a} \\ &\approx \sqrt{2\omega}\sqrt{\omega - a} \end{aligned}$$

as $a \to \omega^-$. Hence

$$T \approx \left(\frac{\pi\sqrt{2}}{\sqrt{\omega}}\right)\frac{1}{\sqrt{\omega - a}}, \tag{3}$$

which shows that T blows up like $(a_c - a)^{-1/2}$, where $a_c = \omega$. Now let's explain the origin of this ***square-root scaling law.***

Ghosts and Bottlenecks

The square-root scaling law found above is a *very general feature of systems that are close to a saddle-node bifurcation.* Just after the fixed points collide, there is a saddle-node remnant or ***ghost*** that leads to slow passage through a bottleneck.

For example, consider $\dot{\theta} = \omega - a\sin\theta$ for decreasing values of a, starting with $a > \omega$. As a decreases, the two fixed points approach each other, collide, and disappear (this sequence was shown earlier in Figure 4.3.3, except now you have to read from right to left.) For a slightly less than ω, the fixed points near $\pi/2$ no longer exist, but they still make themselves felt through a saddle-node ghost (Figure 4.3.5).

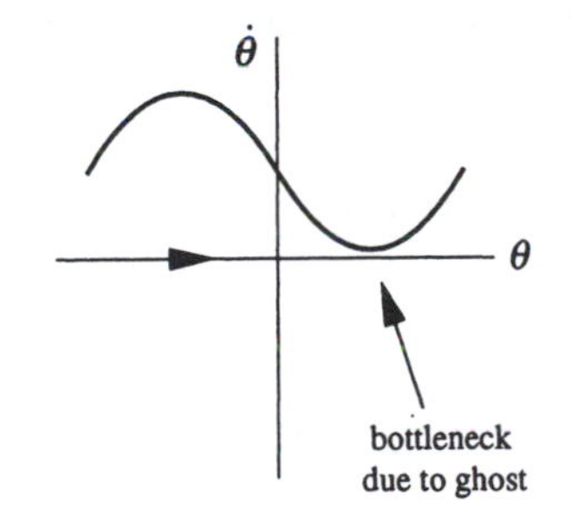

Figure 4.3.5

A graph of $\theta(t)$ would have the shape shown in Figure 4.3.6. Notice how the trajectory spends practically all its time getting through the bottleneck.

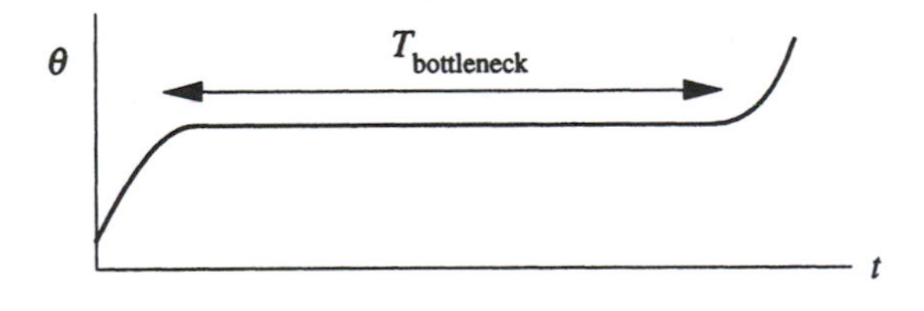

Figure 4.3.6

Now we want to derive a general scaling law for the time required to pass through a bottleneck. The only thing that matters is the behavior of $\dot{\theta}$ in the immediate vicinity of the minimum, since the time spent there dominates all other time scales in the problem. Generically, $\dot{\theta}$ looks *parabolic* near its minimum. Then the problem simplifies tremendously: the dynamics can be reduced to the normal form for a saddle-node bifurcation! By a local rescaling of space, we can rewrite the vector field as

$$\dot{x} = r + x^2$$

where r is proportional to the distance from the bifurcation, and $0 < r \ll 1$. The graph of $\dot{x}$ is shown in Figure 4.3.7.

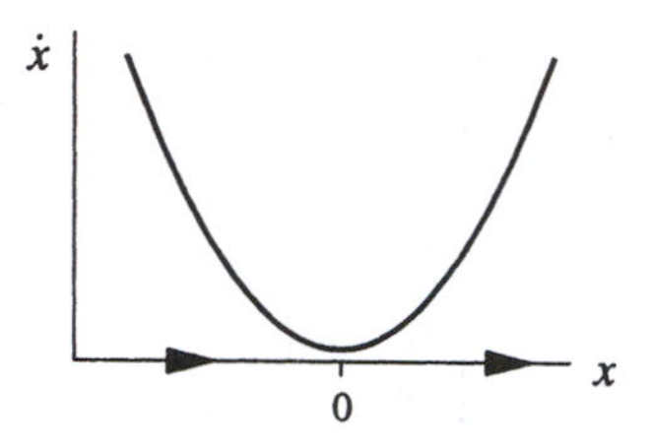

Figure 4.3.7

To estimate the time spent in the bottleneck, we calculate the time taken for x to go from $-\infty$ (all the way on one side of the bottleneck) to $+\infty$ (all the way on the other side). The result is

$$T_{\text{bottleneck}} \approx \int_{-\infty}^{\infty} \frac{dx}{r + x^2} = \frac{\pi}{\sqrt{r}}, \qquad (4)$$

which shows the generality of the square-root scaling law. (Exercise 4.3.1 reminds you how to evaluate the integral in (4).)

EXAMPLE 4.3.2:

Estimate the period of $\dot{\theta} = \omega - a \sin\theta$ in the limit $a \to \omega^-$, using the normal form method instead of the exact result.

Solution: The period will be essentially the time required to get through the bottleneck. To estimate this time, we use a Taylor expansion about $\theta = \pi/2$, where the bottleneck occurs. Let $\phi = \theta - \pi/2$, where ϕ is small. Then

$$\begin{aligned}\dot{\phi} &= \omega - a\sin(\phi + \tfrac{\pi}{2}) \\ &= \omega - a\cos\phi \\ &= \omega - a + \tfrac{1}{2}a\phi^2 + \cdots\end{aligned}$$

which is close to the desired normal form. If we let

$$x = (a/2)^{1/2}\phi, \qquad r = \omega - a$$

then $(2/a)^{1/2}\dot{x} \approx r + x^2$, to leading order in x. Separating variables yields

$$T \approx (2/a)^{1/2} \int_{-\infty}^{\infty} \frac{dx}{r + x^2} = (2/a)^{1/2} \frac{\pi}{\sqrt{r}}.$$

Now we substitute $r = \omega - a$. Furthermore, since $a \to \omega^-$, we may replace $2/a$ by $2/\omega$. Hence

$$T \approx \left(\frac{\pi\sqrt{2}}{\sqrt{\omega}}\right) \frac{1}{\sqrt{\omega - a}},$$

which agrees with (3). ■

4.4 Overdamped Pendulum

We now consider a simple mechanical example of a nonuniform oscillator: an overdamped pendulum driven by a constant torque. Let θ denote the angle between the pendulum and the downward vertical, and suppose that θ increases counterclockwise (Figure 4.4.1).

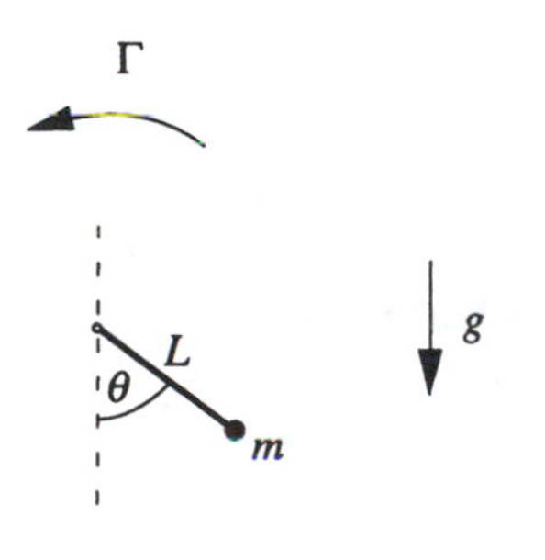

Figure 4.4.1

Then Newton's law yields

$$mL^2\ddot{\theta} + b\dot{\theta} + mgL\sin\theta = \Gamma \qquad (1)$$

where m is the mass and L is the length of the pendulum, b is a viscous damping constant, g is the acceleration due to gravity, and Γ is a constant applied torque. All of these parameters are positive. In particular, $\Gamma > 0$ implies that the applied torque drives the pendulum counterclockwise, as shown in Figure 4.4.1.

Equation (1) is a second-order system, but in the *overdamped limit* of extremely large b, it may be approximated by a first-order system (see Section 3.5 and Exercise 4.4.1). In this limit the inertia term $mL^2\ddot{\theta}$ is negligible and so (1) becomes

$$b\dot{\theta} + mgL\sin\theta = \Gamma. \qquad (2)$$

To think about this problem physically, you should imagine that the pendulum is immersed in molasses. The torque Γ enables the pendulum to plow through its viscous surroundings. Please realize that this is the *opposite* limit from the familiar frictionless case in which energy is conserved, and the pendulum swings back and forth forever. In the present case, energy is lost to damping and pumped in by the applied torque.

To analyze (2), we first nondimensionalize it. Dividing by mgL yields

$$\frac{b}{mgL}\dot{\theta} = \frac{\Gamma}{mgL} - \sin\theta.$$

Hence, if we let

$$\tau = \frac{mgL}{b}t, \qquad \gamma = \frac{\Gamma}{mgL} \tag{3}$$

then

$$\theta' = \gamma - \sin\theta \tag{4}$$

where $\theta' = d\theta/d\tau$.

The dimensionless group γ is the ratio of the applied torque to the maximum gravitational torque. If $\gamma > 1$ then the applied torque can never be balanced by the gravitational torque and *the pendulum will overturn continually.* The rotation rate is nonuniform, since gravity helps the applied torque on one side and opposes it on the other (Figure 4.4.2).

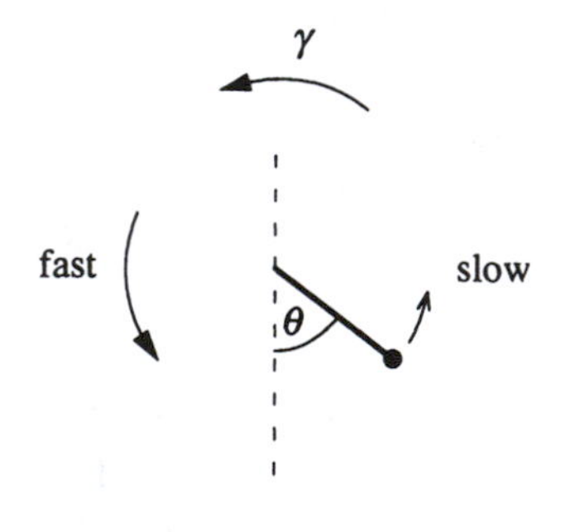

Figure 4.4.2

As $\gamma \to 1^+$, the pendulum takes longer and longer to climb past $\theta = \pi/2$ on the slow side. When $\gamma = 1$ a fixed point appears at $\theta^* = \pi/2$, and then splits into two when $\gamma < 1$ (Figure 4.4.3). On physical grounds, it's clear that the lower of the two equilibrium positions is the stable one.

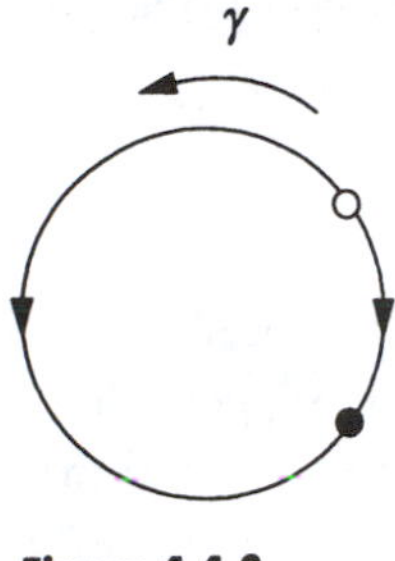

Figure 4.4.3

As γ decreases, the two fixed points move farther apart. Finally, when $\gamma = 0$, the applied torque vanishes and there is an unstable equilibrium at the top (inverted pendulum) and a stable equilibrium at the bottom.

4.5 Fireflies

Fireflies provide one of the most spectacular examples of synchronization in nature. In some parts of southeast Asia, thousands of male fireflies gather in trees at night and flash on and off in unison. Meanwhile the female fireflies cruise overhead, looking for males with a handsome light.

To really appreciate this amazing display, you have to see a movie or videotape of it. A good example is shown in David Attenborough's (1992) television series *The Trials of Life,* in the episode called "Talking to Strangers." See Buck and Buck (1976) for a beautifully written introduction to synchronous fireflies, and Buck (1988) for a more recent review. For mathematical models of synchronous fireflies, see Mirollo and Strogatz (1990) and Ermentrout (1991).

How does the synchrony occur? Certainly the fireflies don't start out synchronized; they arrive in the trees at dusk, and the synchrony builds up gradually as the night goes on. The key is that *the fireflies influence each other*: When one firefly sees the flash of another, it slows down or speeds up so as to flash more nearly in phase on the next cycle.

Hanson (1978) studied this effect experimentally, by periodically flashing a light at a firefly and watching it try to synchronize. For a range of periods close to the firefly's natural period (about 0.9 sec), the firefly was able to match its frequency to the periodic stimulus. In this case, one says that the firefly had been ***entrained*** by the stimulus. However, if the stimulus was too fast or too slow, the firefly could not keep up and entrainment was lost—then a kind of beat phenomenon occurred. But in contrast to the simple beat phenomenon of Section 4.2, the phase difference between stimulus and firefly did not increase uniformly. The phase difference increased slowly during part of the beat cycle, as the firefly struggled in vain to synchronize, and then it increased rapidly through 2π, after which the firefly tried again on the next beat cycle. This process is called *phase walk-through* or ***phase drift.***

Model

Ermentrout and Rinzel (1984) proposed a simple model of the firefly's flashing rhythm and its response to stimuli. Suppose that $\theta(t)$ is the phase of the firefly's flashing rhythm, where $\theta = 0$ corresponds to the instant when a flash is emitted. Assume that in the absence of stimuli, the firefly goes through its cycle at a frequency ω, according to $\dot{\theta} = \omega$.

Now suppose there's a periodic stimulus whose phase Θ satisfies

$$\dot{\Theta} = \Omega, \tag{1}$$

where $\Theta = 0$ corresponds to the flash of the stimulus. We model the firefly's response to this stimulus as follows: If the stimulus is ahead in the cycle, then we assume that the firefly speeds up in an attempt to synchronize. Conversely, the firefly slows down if it's flashing too early. A simple model that incorporates these assumptions is

$$\dot{\theta} = \omega + A\sin(\Theta - \theta) \tag{2}$$

where $A > 0$. For example, if Θ is ahead of θ (i.e., $0 < \Theta - \theta < \pi$) the firefly speeds up ($\dot{\theta} > \omega$). The ***resetting strength*** A measures the firefly's ability to modify its instantaneous frequency.

Analysis

To see whether entrainment can occur, we look at the dynamics of the phase difference $\phi = \Theta - \theta$. Subtracting (2) from (1) yields

$$\dot{\phi} = \dot{\Theta} - \dot{\theta} = \Omega - \omega - A\sin\phi, \tag{3}$$

which is a *nonuniform oscillator* equation for $\phi(t)$. Equation (3) can be nondimensionalized by introducing

$$\tau = At, \quad \mu = \frac{\Omega - \omega}{A}. \tag{4}$$

Then

$$\phi' = \mu - \sin\phi \tag{5}$$

where $\phi' = d\phi/d\tau$. The dimensionless group μ is a measure of the frequency difference, relative to the resetting strength. When μ is small, the frequencies are relatively close together and we expect that entrainment should be possible. This is confirmed by Figure 4.5.1, where we plot the vector fields for (5), for different values of $\mu \geq 0$. (The case $\mu < 0$ is similar.)

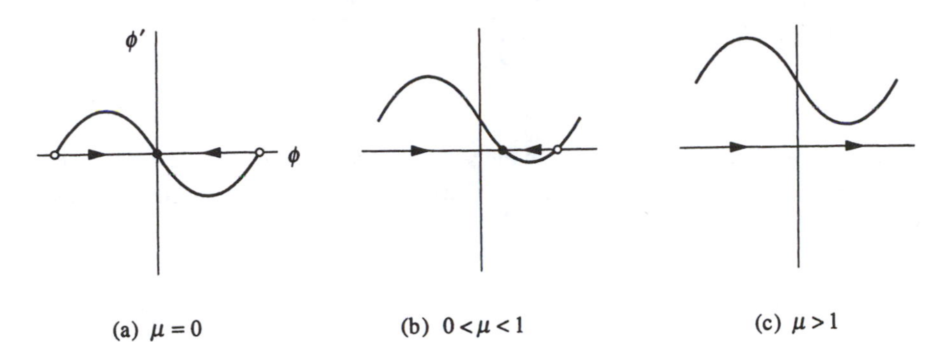

Figure 4.5.1

When $\mu = 0$, all trajectories flow toward a stable fixed point at $\phi^* = 0$ (Figure 4.5.1a). Thus the firefly eventually entrains with *zero phase difference* in the case $\Omega = \omega$. In other words, the firefly and the stimulus flash *simultaneously* if the firefly is driven at its natural frequency.

Figure 4.5.1b shows that for $0 < \mu < 1$, the curve in Figure 4.5.1a lifts up and the stable and unstable fixed points move closer together. All trajectories are still attracted to a stable fixed point, but now $\phi^* > 0$. Since the phase difference approaches a constant, one says that the firefly's rhythm is ***phase-locked*** to the stimulus.

Phase-locking means that the firefly and the stimulus run with the same instantaneous frequency, although they no longer flash in unison. The result $\phi^* > 0$ implies that the stimulus flashes *ahead* of the firefly in each cycle. This makes sense—we assumed $\mu > 0$, which means that $\Omega > \omega$; the stimulus is inherently faster than the firefly, and drives it faster than it wants to go. Thus the firefly falls behind. But it never gets lapped—it always lags in phase by a constant amount ϕ^*.

If we continue to increase μ, the stable and unstable fixed points eventually coalesce in a saddle-node bifurcation at $\mu = 1$. For $\mu > 1$ both fixed points have disappeared and now phase-locking is lost; the phase difference ϕ increases indefinitely, corresponding to *phase drift* (Figure 4.5.1c). (Of course, once ϕ reaches 2π the oscillators are in phase again.) Notice that the phases don't separate at a uniform rate, in qualitative agreement with the experiments of Hanson (1978): ϕ increases most slowly when it passes under the minimum of the sine wave in Figure 4.5.1c, at $\phi = \pi/2$, and most rapidly when it passes under the maximum at $\phi = -\pi/2$.

The model makes a number of specific and testable predictions. Entrainment is predicted to be possible only within a symmetric interval of driving frequencies, specifically $\omega - A \leq \Omega \leq \omega + A$. This interval is called the ***range of entrainment*** (Figure 4.5.2).

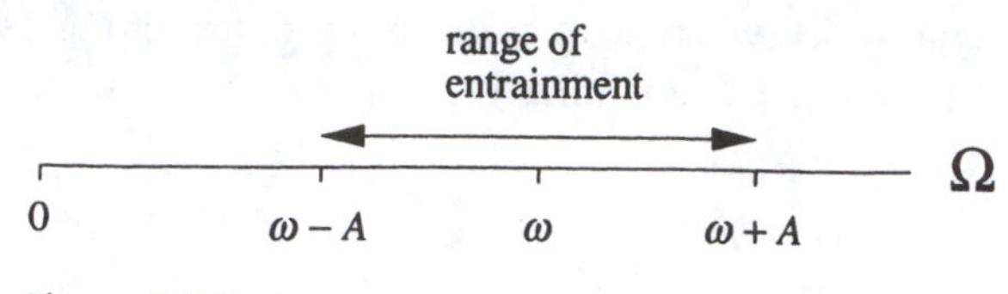

Figure 4.5.2

By measuring the range of entrainment experimentally, one can nail down the value of the parameter A. Then the model makes a rigid prediction for the phase difference during entrainment, namely

$$\sin\phi^* = \frac{\Omega-\omega}{A} \tag{6}$$

where $-\pi/2 \le \phi^* \le \pi/2$ corresponds to the *stable* fixed point of (3).

Moreover, for $\mu > 1$, the period of phase drift may be predicted as follows. The time required for ϕ to change by 2π is given by

$$T_{\text{drift}} = \int dt = \int_0^{2\pi} \frac{dt}{d\phi} d\phi$$

$$= \int_0^{2\pi} \frac{d\phi}{\Omega-\omega-A\sin\phi}\ .$$

To evaluate this integral, we invoke (2) of Section 4.3, which yields

$$T_{\text{drift}} = \frac{2\pi}{\sqrt{(\Omega-\omega)^2-A^2}}\ . \tag{7}$$

Since A and ω are presumably fixed properties of the firefly, the predictions (6) and (7) could be tested simply by varying the drive frequency Ω. Such experiments have yet to be done.

Actually, the biological reality about synchronous fireflies is more complicated. The model presented here is reasonable for certain species, such as *Pteroptyx cribellata,* which behave as if A and ω were fixed. However, the species that is best at synchronizing, *Pteroptyx malaccae,* is actually able to shift its frequency ω toward the drive frequency Ω (Hanson 1978). In this way it is able to achieve nearly zero phase difference, even when driven at periods that differ from its natural period by ±15 percent! A model of this remarkable effect has been presented by Ermentrout (1991).

4.6 Superconducting Josephson Junctions

Josephson junctions are superconducting devices that are capable of generating voltage oscillations of extraordinarily high frequency, typically 10^{10}–10^{11} cycles per second. They have great technological promise as amplifiers, voltage standards, detectors, mixers, and fast switching devices for digital circuits. Josephson junctions can detect electric potentials as small as one quadrillionth of a volt, and they have been used to detect far-infrared radiation from distant galaxies. For an introduction to Josephson junctions, as well as superconductivity more generally, see Van Duzer and Turner (1981).

Although quantum mechanics is required to explain the *origin* of the Josephson effect, we can nevertheless describe the *dynamics* of Josephson junctions in classical terms. Josephson junctions have been particularly useful for experimental studies of nonlinear dynamics, because the equation governing a single junction is the same as that for a pendulum! In this section we will study the dynamics of a single junction in the overdamped limit. In later sections we will discuss underdamped junctions, as well as arrays of enormous numbers of junctions coupled together.

Physical Background

A Josephson junction consists of two closely spaced superconductors separated by a weak connection (Figure 4.6.1). This connection may be provided by an insulator, a normal metal, a semiconductor, a weakened superconductor, or some other material that weakly couples the two superconductors. The two superconducting regions may be characterized by quantum mechanical wave functions $\psi_1 e^{i\phi_1}$ and $\psi_2 e^{i\phi_2}$ respectively. Normally a much more complicated description would be necessary because there are $\sim 10^{23}$ electrons to deal with, but in the superconducting ground state, these electrons form "Cooper pairs" that can be described by a *single* macroscopic wave function. This implies an astonishing degree of coherence among the electrons. The Cooper pairs act like a miniature version of synchronous fireflies: they all adopt the same phase, because this turns out to minimize the energy of the superconductor.

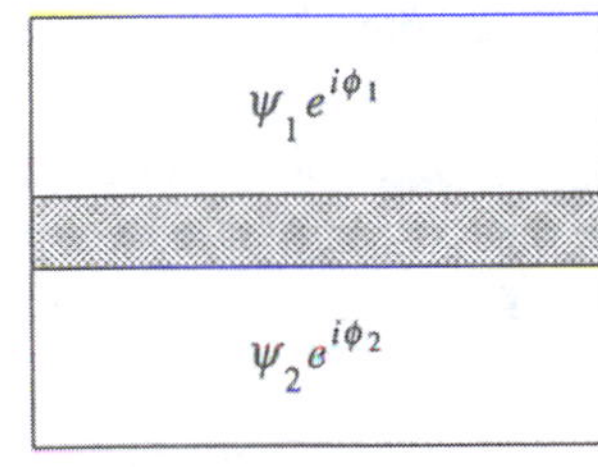

superconductor #1

weak coupling

superconductor #2

Figure 4.6.1

As a 22-year-old graduate student, Brian Josephson (1962) suggested that it should be possible for a current to pass between the two superconductors, even if there were no voltage difference between them. Although this behavior would be impossible classically, it could occur because of quantum mechanical *tunneling* of Cooper pairs across the junction. An observation of this "Josephson effect" was made by Anderson and Rowell in 1963.

Incidentally, Josephson won the Nobel Prize in 1973, after which he lost interest in mainstream physics and was rarely heard from again. See Josephson (1982) for an interview in which he reminisces about his early work and discusses his more recent interests in transcendental meditation, consciousness, language, and even psychic spoon-bending and paranormal phenomena.

The Josephson Relations

We now give a more quantitative discussion of the Josephson effect. Suppose that a Josephson junction is connected to a dc current source (Figure 4.6.2), so that a constant current $I > 0$ is driven through the junction. Using quantum mechanics, one can show that if this current is less than a certain ***critical current*** I_c, no voltage will be developed across the junction; that is, the junction acts as if it had zero resistance! However, the phases of the two superconductors will be driven apart to a constant phase difference $\phi = \phi_2 - \phi_1$, where ϕ satisfies the *Josephson current-phase relation*

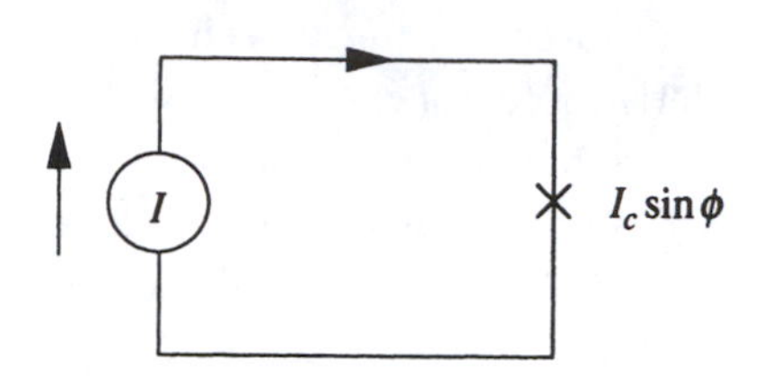

Figure 4.6.2

$$I = I_c \sin\phi. \tag{1}$$

Equation (1) implies that the phase difference increases as the ***bias current*** I increases.

When I exceeds I_c, a constant phase difference can no longer be maintained and a voltage develops across the junction. The phases on the two sides of the junction begin to slip with respect to each other, with the rate of slippage governed by the *Josephson voltage-phase relation*

$$V = \frac{\hbar}{2e}\dot{\phi}. \tag{2}$$

Here $V(t)$ is the instantaneous voltage across the junction, $\hbar$ is Planck's constant divided by 2π, and e is the charge on the electron. For an elementary derivation of the Josephson relations (1) and (2), see Feynman's argument (Feynman et al. (1965), Vol. III), also reproduced in Van Duzer and Turner (1981).

Equivalent Circuit and Pendulum Analog

The relation (1) applies only to the *supercurrent* carried by the electron pairs. In general, the total current passing through the junction will also contain contributions from a *displacement current* and an *ordinary current.* Representing the displacement current by a capacitor, and the ordinary current by a resistor, we arrive at the equivalent circuit shown in Figure 4.6.3, first analyzed by Stewart (1968) and McCumber (1968).

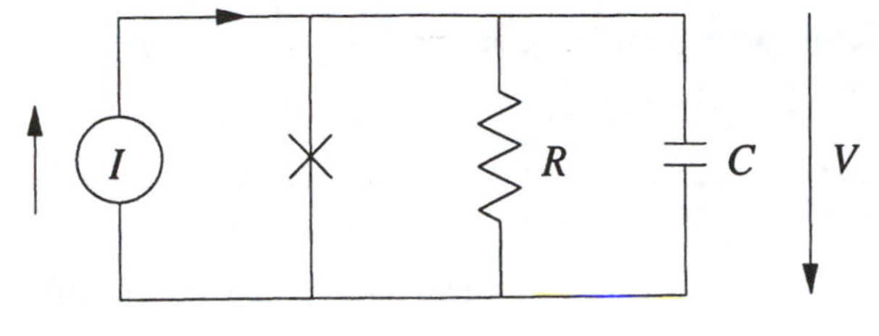

Figure 4.6.3

Now we apply Kirchhoff's voltage and current laws. For this parallel circuit, the voltage drop across each branch must be equal, and hence all the voltages are equal to V, the voltage across the junction. Hence the current through the capacitor equals $C\dot{V}$ and the current through the resistor equals V/R. The sum of these currents and the supercurrent $I_c \sin\phi$ must equal the bias current I; hence

$$C\dot{V} + \frac{V}{R} + I_c \sin\phi = I. \tag{3}$$

Equation (3) may be rewritten solely in terms of the phase difference ϕ, thanks to (2). The result is

$$\frac{\hbar C}{2e}\ddot{\phi} + \frac{\hbar}{2eR}\dot{\phi} + I_c \sin\phi = I, \tag{4}$$

which is precisely analogous to the equation governing a damped pendulum driven by a constant torque! In the notation of Section 4.4, the pendulum equation is

$$mL^2\ddot{\theta} + b\dot{\theta} + mgL\sin\theta = \Gamma.$$

Hence the analogies are as follows:

Pendulum	*Josephson junction*
Angle θ	Phase difference ϕ
Angular velocity $\dot{\theta}$	Voltage $\frac{\hbar}{2e}\dot{\phi}$
Mass m	Capacitance C
Applied torque Γ	Bias current I
Damping constant b	Conductance $1/R$
Maximum gravitational torque mgL	Critical current I_c

This mechanical analog has often proved useful in visualizing the dynamics of Josephson junctions. Sullivan and Zimmerman (1971) actually constructed such a mechanical analog, and measured the average rotation rate of the pendulum as a function of the applied torque; this is the analog of the physically important $I-V$ curve (current–voltage curve) for the Josephson junction.

Typical Parameter Values

Before analyzing (4), we mention some typical parameter values for Josephson junctions. The critical current is typically in the range $I_c \approx 1\ \mu$A - 1 mA, and a typical voltage is $I_c R \approx 1$ mV. Since $2e/h \approx 4.83 \times 10^{14}$ Hz/V, a typical frequency is on the order of 10^{11} Hz. Finally, a typical length scale for Josephson junctions is around 1 μm, but this depends on the geometry and the type of coupling used.

Dimensionless Formulation

If we divide (4) by I_c and define a dimensionless time

$$\tau = \frac{2eI_c R}{\hbar} t, \tag{5}$$

we obtain the dimensionless equation

$$\beta\phi'' + \phi' + \sin\phi = \frac{I}{I_c} \tag{6}$$

where $\phi' = d\phi/d\tau$. The dimensionless group β is defined by

$$\beta = \frac{2eI_c R^2 C}{\hbar}.$$

and is called the ***McCumber parameter.*** It may be thought of as a dimensionless capacitance. Depending on the size, the geometry, and the type of coupling used in the Josephson junction, the value of β can range from $\beta \approx 10^{-6}$ to much larger values ($\beta \approx 10^{6}$).

We are not yet prepared to analyze (6) in general. For now, let's restrict ourselves to the *overdamped limit* $\beta << 1$. Then the term $\beta\phi''$ may be neglected after a rapid initial transient, as discussed in Section 3.5, and so (6) reduces to a nonuniform oscillator:

$$\phi' = \frac{I}{I_c} - \sin\phi. \tag{7}$$

As we know from Section 4.3, the solutions of (7) tend to a stable fixed point when $I < I_c$, and vary periodically when $I > I_c$.

EXAMPLE 4.6.1:

Find the ***current–voltage curve*** analytically in the overdamped limit. In other words, find the average value of the voltage $\langle V \rangle$ as a function of the constant applied current I, assuming that all transients have decayed and the system has reached steady-state operation. Then plot $\langle V \rangle$ vs. I.

Solution: It is sufficient to find $\langle \phi' \rangle$, since $\langle V \rangle = (\hbar/2e)\langle \dot{\phi} \rangle$ from the voltage-phase relation (2), and

$$\langle \dot{\phi} \rangle = \left\langle \frac{d\phi}{dt} \right\rangle = \left\langle \frac{d\tau}{dt} \frac{d\phi}{d\tau} \right\rangle = \frac{2eI_c R}{\hbar} \langle \phi' \rangle,$$

from the definition of τ in (5); hence

$$\langle V \rangle = I_c R \langle \phi' \rangle. \tag{8}$$

There are two cases to consider. When $I \le I_c$, all solutions of (7) approach a fixed point $\phi^* = \sin^{-1}(I/I_c)$, where $-\pi/2 \le \phi^* \le \pi/2$. Thus $\phi' = 0$ in steady state, and so $\langle V \rangle = 0$ for $I \le I_c$.

When $I > I_c$, all solutions of (7) are periodic with period

$$T = \frac{2\pi}{\sqrt{(I/I_c)^2 - 1}}, \tag{9}$$

where the period is obtained from (2) of Section 4.3, and time is measured in units of τ. We compute $\langle \phi' \rangle$ by taking the average over one cycle:

$$\langle \phi' \rangle = \frac{1}{T} \int_0^T \frac{d\phi}{d\tau} d\tau = \frac{1}{T} \int_0^{2\pi} d\phi = \frac{2\pi}{T}. \tag{10}$$

Combining (8)–(10) yields

$$\langle V \rangle = I_c R \sqrt{(I/I_c)^2 - 1} \quad \text{for } I > I_c.$$

In summary, we have found

$$\langle V \rangle = \begin{cases} 0 & \text{for } I \le I_c \\ I_c R \sqrt{(I/I_c)^2 - 1} & \text{for } I > I_c. \end{cases} \tag{11}$$

The I–V curve (11) is shown in Figure 4.6.4.

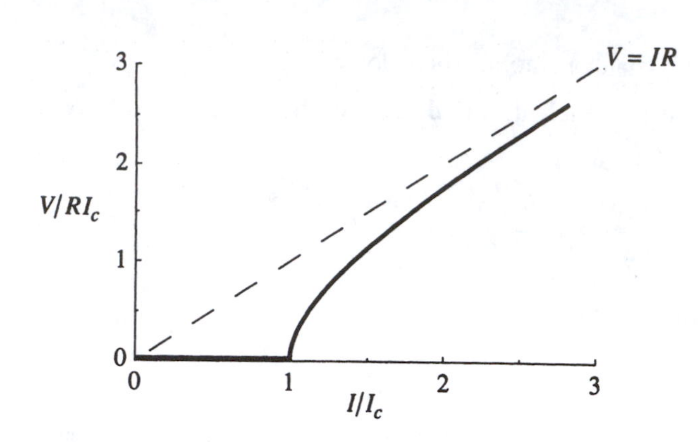

Figure 4.6.4

As I increases, the voltage remains zero until $I > I_c$; then $\langle V \rangle$ rises sharply and eventually asymptotes to the Ohmic behavior $\langle V \rangle \approx IR$ for $I >> I_c$. ■

The analysis given in Example 4.6.1 applies only to the overdamped limit $\beta << 1$. The behavior of the system becomes much more interesting if β is not negligible. In particular, the $I-V$ curve can be ***hysteretic,*** as shown in Figure 4.6.5. As the bias current is increased slowly from $I = 0$, the voltage remains at $V = 0$ until $I > I_c$. Then the voltage jumps up to a nonzero value, as shown by the upward arrow in Figure 4.6.5. The voltage increases with further increases of I. However, if we now slowly *decrease* I, the voltage doesn't drop back to zero at I_c—we have to go *below* I_c before the voltage returns to zero.

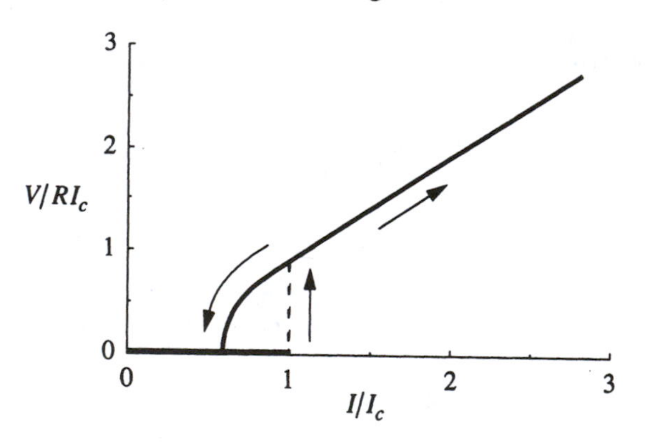

Figure 4.6.5

The hysteresis comes about because the system has ***inertia*** when $\beta \neq 0$. We can make sense of this by thinking in terms of the pendulum analog. The critical current I_c is analogous to the critical torque Γ_c needed to get the pendulum overturning. Once the pendulum has started whirling, its inertia keeps it going so that even if the torque is lowered *below* Γ_c, the rotation continues. The torque has to be lowered even further before the pendulum will fail to make it over the top.

In more mathematical terms, we'll show in Section 8.5 that this hysteresis occurs because a *stable fixed point coexists with a stable periodic solution.* We have never seen anything like *this* before! For vector fields on the line, only fixed points can exist; for vector fields on the circle, both fixed points and periodic solutions can exist, *but not simultaneously.* Here we see just one example of the new kinds of phenomena that can occur in two-dimensional systems. It's time to take the plunge.

EXERCISES FOR CHAPTER 4

4.1 Examples and Definitions

4.1.1 For which real values of a does the equation $\dot{\theta} = \sin(a\theta)$ give a well-defined vector field on the circle?

For each of the following vector fields, find and classify all the fixed points, and sketch the phase portrait on the circle.

4.1.2 $\dot{\theta} = 1 + 2\cos\theta$

4.1.3 $\dot{\theta} = \sin 2\theta$

4.1.4 $\dot{\theta} = \sin^3\theta$

4.1.5 $\dot{\theta} = \sin\theta + \cos\theta$

4.1.6 $\dot{\theta} = 3 + \cos 2\theta$

4.1.7 $\dot{\theta} = \sin k\theta$ where k is a positive integer.

4.1.8 (Potentials for vector fields on the circle)

a) Consider the vector field on the circle given by $\dot{\theta} = \cos\theta$. Show that this system has a single-valued potential $V(\theta)$, i.e., for each point on the circle, there is a well-defined value of V such that $\dot{\theta} = -dV/d\theta$. (As usual, θ and $\theta + 2\pi k$ are to be regarded as the same point on the circle, for each integer k.)

b) Now consider $\dot{\theta} = 1$. Show that there is no single-valued potential $V(\theta)$ for this vector field on the circle.

c) What's the general rule? When does $\dot{\theta} = f(\theta)$ have a single-valued potential?

4.1.9 In Exercises 2.6.2 and 2.7.7, you were asked to give two analytical proofs that periodic solutions are impossible for vector fields on the line. Review these arguments and explain why they *don't* carry over to vector fields on the circle. Specifically which parts of the argument fail?

4.2 Uniform Oscillator

4.2.1 (Church bells) The bells of two different churches are ringing. One bell rings every 3 seconds, and the other rings every 4 seconds. Assume that the bells have just rung at the same time. How long will it be until the next time they ring together? Answer the question in two ways: using common sense, and using the method of Example 4.2.1.

4.2.2 (Beats arising from linear superpositions) Graph $x(t) = \sin 8t + \sin 9t$ for $-20 < t < 20$. You should find that the amplitude of the oscillations is *modulated*—it grows and decays periodically.

a) What is the period of the amplitude modulations?

b) Solve this problem analytically, using a trigonometric identity that converts sums of sines and cosines to products of sines and cosines.

(In the old days, this beat phenomenon was used to tune musical instruments. You would strike a tuning fork at the same time as you played the desired note on the instrument. The combined sound $A_1 \sin \omega_1 t + A_2 \sin \omega_2 t$ would get louder and softer as the two vibrations went in and out of phase. Each maximum of total amplitude is called a beat. When the time between beats is long, the instrument is nearly in tune.)

4.2.3 (The clock problem) Here's an old chestnut from high school algebra: At 12:00, the hour hand and minute hand of a clock are perfectly aligned. When is the *next* time they will be aligned? (Solve the problem by the methods of this section, and also by some alternative approach of your choosing.)

4.3 Nonuniform Oscillator

4.3.1 As shown in the text, the time required to pass through a saddle-node bottleneck is approximately $T_{\text{bottleneck}} = \int_{-\infty}^{\infty} \frac{dx}{r + x^2}$. To evaluate this integral, let $x = \sqrt{r} \tan \theta$, use the identity $1 + \tan^2 \theta = \sec^2 \theta$, and change the limits of integration appropriately. Thereby show that $T_{\text{bottleneck}} = \pi / \sqrt{r}$.

4.3.2 The oscillation period for the nonuniform oscillator is given by the integral $T = \int_{-\pi}^{\pi} \frac{d\theta}{\omega - a \sin \theta}$, where $\omega > a > 0$. Evaluate this integral as follows.

a) Let $u = \tan \frac{\theta}{2}$. Solve for θ and then express $d\theta$ in terms of u and du.

b) Show that $\sin \theta = 2u/(1 + u^2)$. (Hint: Draw a right triangle with base 1 and height u. Then $\frac{\theta}{2}$ is the angle opposite the side of length u, since $u = \tan \frac{\theta}{2}$ by definition. Finally, invoke the half-angle formula $\sin \theta = 2 \sin \frac{\theta}{2} \cos \frac{\theta}{2}$.)

c) Show that $u \to \pm\infty$ as $\theta \to \pm\pi$, and use that fact to rewrite the limits of integration.

d) Express T as an integral with respect to u.

e) Finally, complete the square in the denominator of the integrand of (d), and reduce the integral to the one studied in Exercise 4.3.1, for a suitable choice of x and r.

For each of the following questions, draw the phase portrait as function of the control parameter μ. Classify the bifurcations that occur as μ varies, and find all the bifurcation values of μ.

4.3.3 $\dot{\theta} = \mu \sin\theta - \sin 2\theta$

4.3.4 $\dot{\theta} = \dfrac{\sin\theta}{\mu + \cos\theta}$

4.3.5 $\dot{\theta} = \mu + \cos\theta + \cos 2\theta$

4.3.6 $\dot{\theta} = \mu + \sin\theta + \cos 2\theta$

4.3.7 $\dot{\theta} = \dfrac{\sin\theta}{\mu + \sin\theta}$

4.3.8 $\dot{\theta} = \dfrac{\sin 2\theta}{1 + \mu \sin\theta}$

4.3.9 (Alternative derivation of scaling law) For systems close to a saddle-node bifurcation, the scaling law $T_{\text{bottleneck}} \sim O(r^{-1/2})$ can also be derived as follows.

a) Suppose that x has a characteristic scale $O(r^a)$, where a is unknown for now. Then $x = r^a u$, where $u \sim O(1)$. Similarly, suppose $t = r^b \tau$, with $\tau \sim O(1)$. Show that $\dot{x} = r + x^2$ is thereby transformed to $r^{a-b} \dfrac{du}{d\tau} = r + r^{2a} u^2$.

b) Assume that all terms in the equation have the same order with respect to r, and thereby derive $a = \frac{1}{2}$, $b = -\frac{1}{2}$.

4.3.10 (Nongeneric scaling laws) In deriving the square-root scaling law for the time spent passing through a bottleneck, we assumed that $\dot{x}$ had a quadratic minimum. This is the generic case, but what if the minimum were of higher order? Suppose that the bottleneck is governed by $\dot{x} = r + x^{2n}$, where $n > 1$ is an integer. Using the method of Exercise 4.3.9, show that $T_{\text{bottleneck}} \approx c r^b$, and determine b and c.

(It's acceptable to leave c in the form of a definite integral. If you know complex variables and residue theory, you should be able to evaluate c exactly by integrating around the boundary of the pie-slice $\left\{ z = re^{i\theta} : 0 \le \theta \le \pi/n,\ 0 \le r \le R \right\}$ and letting $R \to \infty$.)

4.4 Overdamped Pendulum

4.4.1 (Validity of overdamped limit) Find the conditions under which it is valid to approximate the equation $mL^2\ddot{\theta} + b\dot{\theta} + mgL\sin\theta = \Gamma$ by its overdamped limit $b\dot{\theta} + mgL\sin\theta = \Gamma$.

4.4.2 (Understanding $\sin\theta(t)$) By imagining the rotational motion of an overdamped pendulum, sketch $\sin\theta(t)$ vs. t for a typical solution of $\theta' = \gamma - \sin\theta$. How does the shape of the waveform depend on γ? Make a series of graphs for different γ, including the limiting cases $\gamma \approx 1$ and $\gamma >> 1$. For the pendulum, what physical quantity is proportional to $\sin\theta(t)$?

4.4.3 (Understanding $\dot{\theta}(t)$) Redo Exercise 4.4.2, but now for $\dot{\theta}(t)$ instead of $\sin\theta(t)$.

4.4.4 (Torsional spring) Suppose that our overdamped pendulum is connected to a torsional spring. As the pendulum rotates, the spring winds up and generates an opposing torque $-k\theta$. Then the equation of motion becomes $b\dot{\theta} + mgL\sin\theta = \Gamma - k\theta$.

a) Does this equation give a well-defined vector field on the circle?
b) Nondimensionalize the equation.
c) What does the pendulum do in the long run?
d) Show that many bifurcations occur as k is varied from 0 to ∞. What kind of bifurcations are they?

4.5 Fireflies

4.5.1 (Triangle wave) In the firefly model, the sinusoidal form of the firefly's response function was chosen somewhat arbitrarily. Consider the alternative model $\dot{\Theta} = \Omega$, $\dot{\theta} = \omega + Af(\Theta - \theta)$, where f is given now by a triangle wave, not a sine wave. Specifically, let

$$f(\phi) = \begin{cases} \phi, & -\frac{\pi}{2} \le \phi \le \frac{\pi}{2} \\ \pi - \phi, & \frac{\pi}{2} \le \phi \le \frac{3\pi}{2} \end{cases}$$

on the interval $-\frac{\pi}{2} \le \phi \le \frac{3\pi}{2}$, and extend f periodically outside this interval.

a) Graph $f(\phi)$.
b) Find the range of entrainment.
c) Assuming that the firefly is phase-locked to the stimulus, find a formula for the phase difference ϕ^*.
d) Find a formula for T_{drift}.

4.5.2 (General response function) Redo as much of the previous exercise as possible, assuming only that $f(\phi)$ is a smooth, 2π-periodic function with a single maximum and minimum on the interval $-\pi \le \phi \le \pi$.

4.5.3 (Excitable systems) Suppose you stimulate a neuron by injecting it with a pulse of current. If the stimulus is small, nothing dramatic happens: the neuron increases its membrane potential slightly, and then relaxes back to its resting potential. However, if the stimulus exceeds a certain threshold, the neuron will "fire" and produce a large voltage spike before returning to rest. Surprisingly, the size of the spike doesn't depend much on the size of the stimulus—anything above threshold will elicit essentially the same response.

Similar phenomena are found in other types of cells and even in some chemical reactions (Winfree 1980, Rinzel and Ermentrout 1989, Murray 1989). These systems are called ***excitable***. The term is hard to define precisely, but roughly speaking, an excitable system is characterized by two properties: (1) it has a unique, globally attracting rest state, and (2) a large enough stimulus can send the system on a long excursion through phase space before it returns to the resting state.

This exercise deals with the simplest caricature of an excitable system. Let $\dot{\theta} = \mu + \sin\theta$, where μ is slightly less than 1.

a) Show that the system satisfies the two properties mentioned above. What object plays the role of the "rest state"? And the "threshold"?

b) Let $V(t) = \cos\theta(t)$. Sketch $V(t)$ for various initial conditions. (Here V is analogous to the neuron's membrane potential, and the initial conditions correspond to different perturbations from the rest state.)

4.6 Superconducting Josephson Junctions

4.6.1 (Current and voltage oscillations) Consider a Josephson junction in the overdamped limit $\beta = 0$.

a) Sketch the supercurrent $I_c \sin\phi(t)$ as a function of t, assuming first that I/I_c is slightly greater than 1, and then assuming that $I/I_c >> 1$. (Hint: In each case, visualize the flow on the circle, as given by Equation (4.6.7).)

b) Sketch the instantaneous voltage $V(t)$ for the two cases considered in (a).

4.6.2 (Computer work) Check your qualitative solution to Exercise 4.6.1 by integrating Equation (4.6.7) numerically, and plotting the graphs of $I_c \sin\phi(t)$ and $V(t)$.

4.6.3 (Washboard potential) Here's another way to visualize the dynamics of an overdamped Josephson junction. As in Section 2.7, imagine a particle sliding down a suitable potential.

a) Find the potential function corresponding to Equation (4.6.7). Show that it is *not* a single-valued function on the circle.

b) Graph the potential as a function of ϕ, for various values of I/I_c. Here ϕ is to be regarded as a real number, not an angle.

c) What is the effect of increasing I?

The potential in (b) is often called the "washboard potential" (Van Duzer and Turner 1981, p. 179) because its shape is reminiscent of a tilted, corrugated washboard.

4.6.4 (Resistively loaded array) ***Arrays*** of coupled Josephson junctions raise many fascinating questions. Their dynamics are not yet understood in detail. The questions are technologically important because arrays can produce much greater power output than a single junction, and also because arrays provide a reasonable model of the (still mysterious) high-temperature superconductors. For an introduction to some of the dynamical questions of current interest, see Tsang et al. (1991) and Strogatz and Mirollo (1993).

Figure 1 shows an array of two identical overdamped Josephson junctions. The junctions are in series with each other, and in parallel with a resistive "load" R.

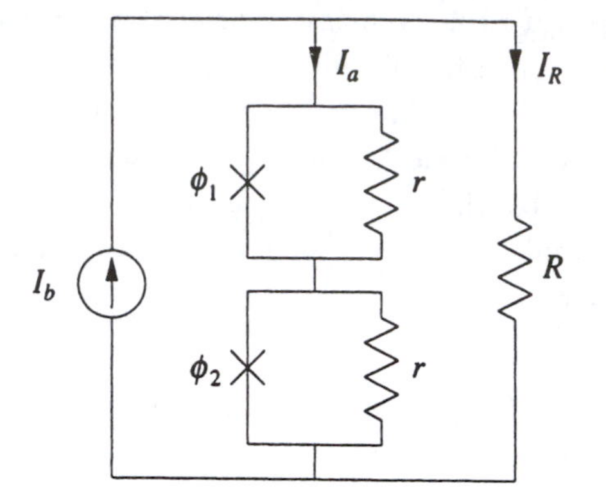

Figure 1

The goal of this exercise is to derive the governing equations for this circuit. In particular, we want to find differential equations for ϕ_1 and ϕ_2.

a) Write an equation relating the dc bias current I_b to the current I_a flowing through the array and the current I_R flowing through the load resistor.

b) Let V_1 and V_2 denote the voltages across the first and second Josephson junctions. Show that $I_a = I_c \sin\phi_1 + V_1/r$ and $I_a = I_c \sin\phi_2 + V_2/r$.

c) Let $k = 1,2$. Express V_k in terms of $\dot{\phi}_k$.

d) Using the results above, along with Kirchhoff's voltage law, show that

$$I_b = I_c \sin\phi_k + \frac{\hbar}{2er}\dot{\phi}_k + \frac{\hbar}{2eR}(\dot{\phi}_1 + \dot{\phi}_2) \text{ for } k = 1,2.$$

e) The equations in part (d) can be written in more standard form as equations for $\dot{\phi}_k$, as follows. Add the equations for $k = 1,2$, and use the result to eliminate the term $(\dot{\phi}_1 + \dot{\phi}_2)$. Show that the resulting equations take the form

$$\dot{\phi}_k = \Omega + a\sin\phi_k + K\sum_{j=1}^{2}\sin\phi_j,$$

and write down explicit expressions for the parameters Ω, a, K.

4.6.5 (N junctions, resistive load) Generalize Exercise 4.6.4 as follows. Instead of the two Josephson junctions in Figure 1, consider an array of N junctions in series. As before, assume the array is in parallel with a resistive load R, and that the junctions are identical, overdamped, and driven by a constant bias current I_b. Show that the governing equations can be written in dimensionless form as

$$\frac{d\phi_k}{d\tau} = \Omega + a\sin\phi_k + \tfrac{1}{N}\sum_{j=1}^{N}\sin\phi_j \ , \text{ for } k = 1,\ldots,N,$$

and write down explicit expressions for the dimensionless groups Ω and a and the dimensionless time τ. (See Example 8.7.4 and Tsang et al. (1991) for further discussion.)

4.6.6 (N junctions, RLC load) Generalize Exercise 4.6.4 to the case where there are N junctions in series, and where the load is a resistor R in series with a capacitor C and an inductor L. Write differential equations for ϕ_k and for Q, where Q is the charge on the load capacitor. (See Strogatz and Mirollo 1993.)

PART TWO

TWO-DIMENSIONAL FLOWS

5

LINEAR SYSTEMS

5.0 Introduction

As we've seen, in one-dimensional phase spaces the flow is extremely confined—all trajectories are forced to move monotonically or remain constant. In higher-dimensional phase spaces, trajectories have much more room to maneuver, and so a wider range of dynamical behavior becomes possible. Rather than attack all this complexity at once, we begin with the simplest class of higher-dimensional systems, namely *linear systems in two dimensions*. These systems are interesting in their own right, and, as we'll see later, they also play an important role in the classification of fixed points of *nonlinear* systems. We begin with some definitions and examples.

5.1 Definitions and Examples

A ***two-dimensional linear system*** is a system of the form

$$\dot{x} = ax + by$$
$$\dot{y} = cx + dy$$

where a, b, c, d are parameters. If we use boldface to denote vectors, this system can be written more compactly in matrix form as

$$\dot{\mathbf{x}} = A\mathbf{x},$$

where

$$A = \begin{pmatrix} a & b \\ c & d \end{pmatrix} \text{ and } \mathbf{x} = \begin{pmatrix} x \\ y \end{pmatrix}.$$

Such a system is ***linear*** in the sense that if $\mathbf{x}_1$ and $\mathbf{x}_2$ are solutions, then so is any linear combination $c_1\mathbf{x}_1 + c_2\mathbf{x}_2$. Notice that $\dot{\mathbf{x}} = \mathbf{0}$ when $\mathbf{x} = \mathbf{0}$, so $\mathbf{x}^* = \mathbf{0}$ is always a fixed point for any choice of A.

The solutions of $\dot{\mathbf{x}} = A\mathbf{x}$ can be visualized as trajectories moving on the (x, y) plane, in this context called the ***phase plane***. Our first example presents the phase plane analysis of a familiar system.

EXAMPLE 5.1.1:

As discussed in elementary physics courses, the vibrations of a mass hanging from a linear spring are governed by the linear differential equation

$$m\ddot{x} + kx = 0 \tag{1}$$

where m is the mass, k is the spring constant, and x is the displacement of the mass from equilibrium (Figure 5.1.1). Give a phase plane analysis of this ***simple harmonic oscillator***.

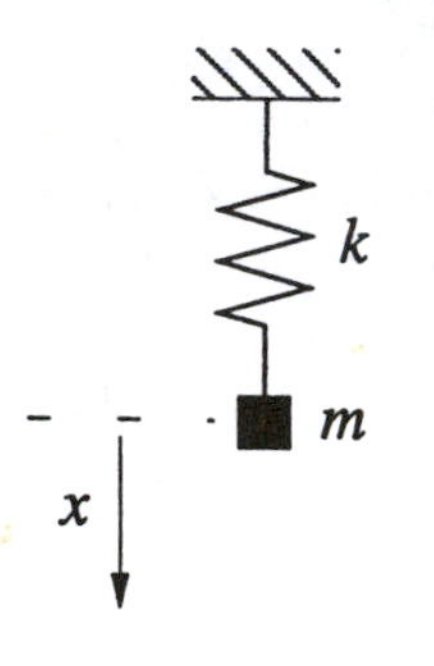

Figure 5.1.1

Solution: As you probably recall, it's easy to solve (1) analytically in terms of sines and cosines. But that's precisely what makes linear equations so special! For the *nonlinear* equations of ultimate interest to us, it's usually impossible to find an analytical solution. We want to develop methods for deducing the behavior of equations like (1) *without actually solving them.*

The motion in the phase plane is determined by a vector field that comes from the differential equation (1). To find this vector field, we note that the ***state*** of the system is characterized by its current position x and velocity v; if we know the values of *both* x and v, then (1) uniquely determines the future states of the system. Therefore we rewrite (1) in terms of x and v, as follows:

$$\dot{x} = v \tag{2a}$$
$$\dot{v} = -\tfrac{k}{m}x. \tag{2b}$$

Equation (2a) is just the definition of velocity, and (2b) is the differential equation (1) rewritten in terms of v. To simplify the notation, let $\omega^2 = k/m$. Then (2) becomes

$$\dot{x} = v \tag{3a}$$
$$\dot{v} = -\omega^2 x. \tag{3b}$$

The system (3) assigns a vector $(\dot{x}, \dot{v}) = (v, -\omega^2 x)$ at each point (x, v), and therefore represents a ***vector field*** on the phase plane.

For example, let's see what the vector field looks like when we're on the x-axis. Then $v = 0$ and so $(\dot{x}, \dot{v}) = (0, -\omega^2 x)$. Hence the vectors point vertically downward for positive x and vertically upward for negative x (Figure 5.1.2). As x gets larger in magnitude, the vectors $(0, -\omega^2 x)$ get longer. Similarly, on the v-axis, the vector field is $(\dot{x}, \dot{v}) = (v, 0)$, which points to the right when $v > 0$ and to the left when $v < 0$. As we move around in phase space, the vectors change direction as shown in Figure 5.1.2.

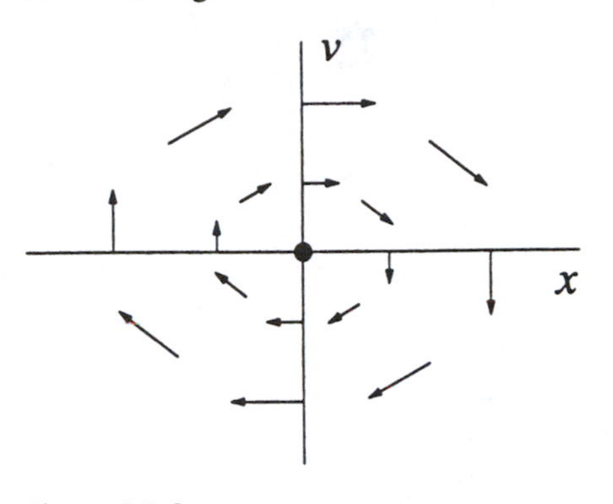

Figure 5.1.2

Just as in Chapter 2, it is helpful to visualize the vector field in terms of the motion of an imaginary fluid. In the present case, we imagine that a fluid is flowing steadily on the phase plane with a local velocity given by $(\dot{x}, \dot{v}) = (v, -\omega^2 x)$. Then, to find the trajectory starting at (x_0, v_0), we place an imaginary particle or ***phase point*** at (x_0, v_0) and watch how it is carried around by the flow.

The flow in Figure 5.1.2 swirls about the origin. The origin is special, like the eye of a hurricane: a phase point placed there would remain motionless, because $(\dot{x}, \dot{v}) = (0, 0)$ when $(x, v) = (0, 0)$; hence the origin is a ***fixed point***. But a phase point starting anywhere else would circulate around the origin and eventually return to its starting point. Such trajectories form ***closed orbits***, as shown in Figure 5.1.3. Figure 5.1.3 is called the ***phase portrait*** of the system—it shows the overall picture of trajectories in phase space.

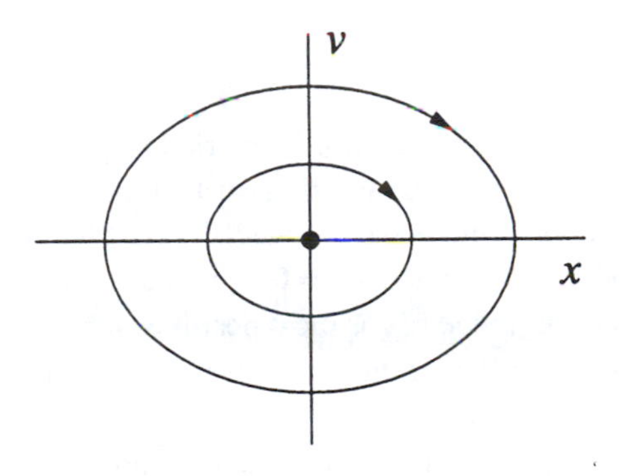

Figure 5.1.3

What do fixed points and closed orbits have to do with the original problem of a mass on a spring? The answers are beautifully simple. The fixed point $(x, v) = (0, 0)$ corresponds to static equilibrium of the system: the mass is at rest at its equilibrium position and will remain there forever, since the spring is relaxed. The closed orbits have a more interesting interpretation: they correspond to periodic motions, i.e., oscillations of the mass. To see this, just look at some points on a closed orbit (Figure 5.1.4). When the displacement x is most negative, the velocity v is zero; this corresponds to one extreme of the oscillation, where the spring is most compressed (Figure 5.1.4).

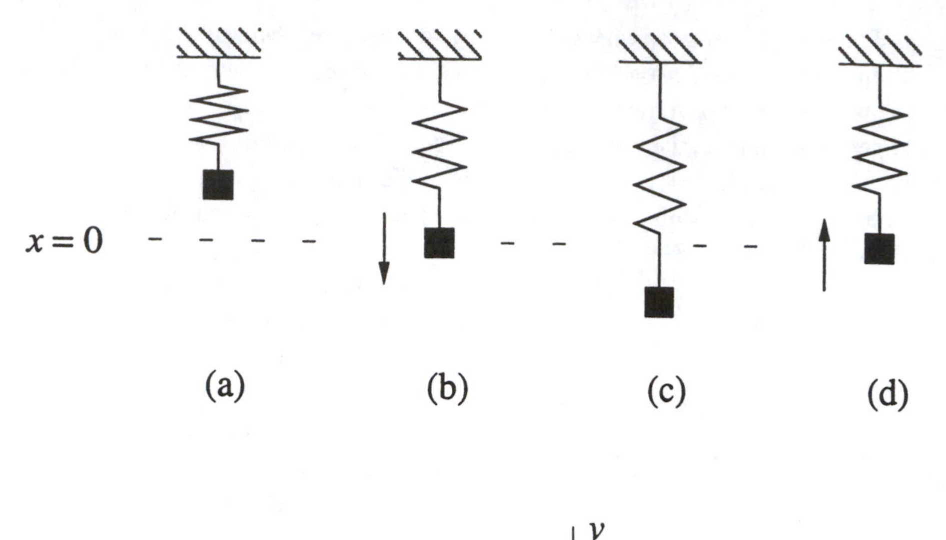

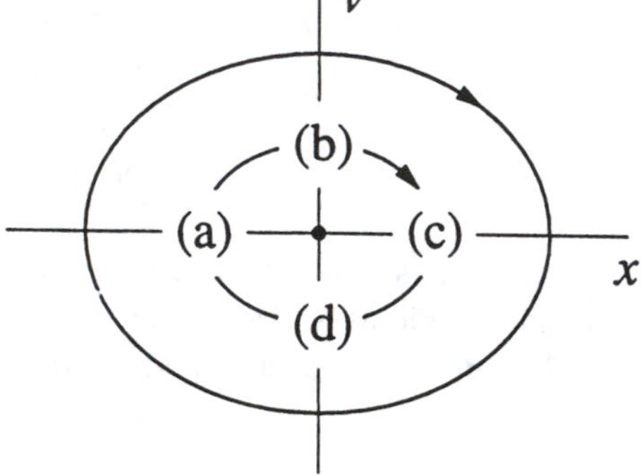

Figure 5.1.4

In the next instant as the phase point flows along the orbit, it is carried to points where x has increased and v is now positive; the mass is being pushed back toward its equilibrium position. But by the time the mass has reached $x = 0$, it has a large positive velocity (Figure 5.1.4b) and so it overshoots $x = 0$. The mass eventually comes to rest at the other end of its swing, where x is most positive and v is zero again (Figure 5.1.4c). Then the mass gets pulled up again and eventually completes the cycle (Figure 5.1.4d).

The shape of the closed orbits also has an interesting physical interpretation. The orbits in Figures 5.1.3 and 5.1.4 are actually *ellipses* given by the equation $\omega^2 x^2 + v^2 = C$, where $C \geq 0$ is a constant. In Exercise 5.1.1, you are asked to derive this geometric result, and to show that it is equivalent to conservation of energy. ■

EXAMPLE 5.1.2:

Solve the linear system $\dot{\mathbf{x}} = A\mathbf{x}$, where $A = \begin{pmatrix} a & 0 \\ 0 & -1 \end{pmatrix}$. Graph the phase portrait as a varies from $-\infty$ to $+\infty$, showing the qualitatively different cases.

Solution: The system is

$$\begin{pmatrix} \dot{x} \\ \dot{y} \end{pmatrix} = \begin{pmatrix} a & 0 \\ 0 & -1 \end{pmatrix} \begin{pmatrix} x \\ y \end{pmatrix}.$$

Matrix multiplication yields

$$\dot{x} = ax$$
$$\dot{y} = -y$$

which shows that the two equations are ***uncoupled***; there's no x in the y-equation and vice versa. In this simple case, each equation may be solved separately. The solution is

$$x(t) = x_0 e^{at} \qquad (1a)$$
$$y(t) = y_0 e^{-t}. \qquad (1b)$$

The phase portraits for different values of a are shown in Figure 5.1.5. In each case, $y(t)$ decays exponentially. When $a < 0$, $x(t)$ also decays exponentially and so all trajectories approach the origin as $t \to \infty$. However, the direction of approach depends on the size of a compared to -1.

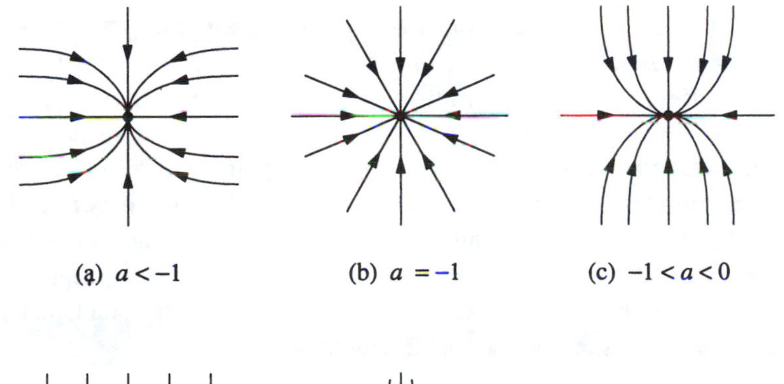

(a) $a < -1$ (b) $a = -1$ (c) $-1 < a < 0$

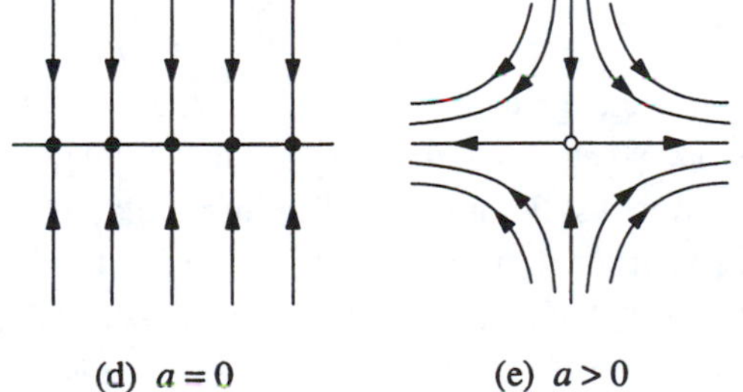

(d) $a = 0$ (e) $a > 0$

Figure 5.1.5

In Figure 5.1.5a, we have $a < -1$, which implies that $x(t)$ decays more rapidly than $y(t)$. The trajectories approach the origin tangent to the *slower* direction (here, the y-direction). The intuitive explanation is that when a is very negative, the trajectory slams horizontally onto the y-axis, because the decay of $x(t)$ is almost instantaneous. Then the trajectory dawdles along the y-axis toward the origin, and so the approach is tangent to the y-axis. On the other hand, if we look *backwards* along a trajectory ($t \to -\infty$), then the trajectories all become parallel to the faster decaying direction (here, the x-direction). These conclusions are easily proved by looking at the slope $dy/dx = \dot{y}/\dot{x}$ along the trajectories; see Exercise 5.1.2. In Figure 5.1.5a, the fixed point $\mathbf{x}^* = \mathbf{0}$ is called a ***stable node***.

Figure 5.1.5b shows the case $a = -1$. Equation (1) shows that $y(t)/x(t) = y_0/x_0 =$ constant, and so all trajectories are straight lines through the origin. This is a very special case—it occurs because the decay rates in the two directions are precisely equal. In this case, $\mathbf{x}^*$ is called a symmetrical node or ***star***.

When $-1 < a < 0$, we again have a node, but now the trajectories approach $\mathbf{x}^*$ along the x-direction, which is the more slowly decaying direction for this range of a (Figure 5.1.5c).

Something dramatic happens when $a = 0$ (Figure 5.1.5d). Now (1a) becomes $x(t) \equiv x_0$ and so there's an entire ***line of fixed points*** along the x-axis. All trajectories approach these fixed points along vertical lines.

Finally when $a > 0$ (Figure 5.1.5e), $\mathbf{x}^*$ becomes unstable, due to the exponential growth in the x-direction. Most trajectories veer away from $\mathbf{x}^*$ and head out to infinity. An exception occurs if the trajectory starts on the y-axis; then it walks a tightrope to the origin. In forward time, the trajectories are asymptotic to the x-axis; in backward time, to the y-axis. Here $\mathbf{x}^* = \mathbf{0}$ is called a ***saddle point***. The y-axis is called the ***stable manifold*** of the saddle point $\mathbf{x}^*$, defined as the set of initial conditions $\mathbf{x}_0$ such that $\mathbf{x}(t) \to \mathbf{x}^*$ as $t \to \infty$. Likewise, the ***unstable manifold*** of $\mathbf{x}^*$ is the set of initial conditions such that $\mathbf{x}(t) \to \mathbf{x}^*$ as $t \to -\infty$. Here the unstable manifold is the x-axis. Note that a typical trajectory asymptotically approaches the unstable manifold as $t \to \infty$, and approaches the stable manifold as $t \to -\infty$. This sounds backwards, but it's right! ■

Stability Language

It's useful to introduce some language that allows us to discuss the stability of different types of fixed points. This language will be especially useful when we analyze fixed points of *nonlinear* systems. For now we'll be informal; precise definitions of the different types of stability will be given in Exercise 5.1.10.

We say that $\mathbf{x}^* = \mathbf{0}$ is an ***attracting*** fixed point in Figures 5.1.5a–c; all trajectories that start near $\mathbf{x}^*$ approach it as $t \to \infty$. That is, $\mathbf{x}(t) \to \mathbf{x}^*$ as $t \to \infty$. In fact $\mathbf{x}^*$ attracts *all* trajectories in the phase plane, so it could be called ***globally attracting***.

There's a completely different notion of stability which relates to the behavior of trajectories for *all* time, not just as $t \to \infty$. We say that a fixed point $\mathbf{x}^*$ is ***Liapunov stable*** if all trajectories that start sufficiently close to $\mathbf{x}^*$ remain close to it for all time. In Figures 5.1.5a–d, the origin is Liapunov stable.

Figure 5.1.5d shows that a fixed point can be Liapunov stable but not attracting. This situation comes up often enough that there is a special name for it. When a fixed point is Liapunov stable but not attracting, it is called ***neutrally stable***. Nearby trajectories are neither attracted to nor repelled from a neutrally stable point. As a second example, the equilibrium point of the simple harmonic oscillator (Figure 5.1.3) is neutrally stable. Neutral stability is commonly encountered in mechanical systems in the absence of friction. Conversely, it's possible for a fixed point to be attracting but not Liapunov stable; thus, neither notion of stability implies the other. An example is given by the following vector field on the circle: $\dot{\theta} = 1 - \cos\theta$ (Figure 5.1.6). Here $\theta^* = 0$ attracts all trajectories as $t \to \infty$, but it is not Liapunov stable; there are trajectories that start infinitesimally close to θ^* but go on a very large excursion before returning to θ^*.

Figure 5.1.6

However, in practice the two types of stability often occur together. If a fixed point is *both* Liapunov stable and attracting, we'll call it ***stable***, or sometimes ***asymptotically stable***.

Finally, $\mathbf{x}^*$ is ***unstable*** in Figure 5.1.5e, because it is neither attracting nor Liapunov stable.

A graphical convention: we'll use open dots to denote unstable fixed points, and solid black dots to denote Liapunov stable fixed points. This convention is consistent with that used in previous chapters.

5.2 Classification of Linear Systems

The examples in the last section had the special feature that two of the entries in the matrix A were zero. Now we want to study the general case of an arbitrary 2×2 matrix, with the aim of classifying all the possible phase portraits that can occur.

Example 5.1.2 provides a clue about how to proceed. Recall that the x and y axes played a crucial geometric role. They determined the direction of the trajectories as $t \to \pm\infty$. They also contained special ***straight-line trajectories***: a trajectory starting on one of the coordinate axes stayed on that axis forever, and exhibited simple exponential growth or decay along it.

For the general case, we would like to find the analog of these straight-line trajectories. That is, we seek trajectories of the form

$$\mathbf{x}(t) = e^{\lambda t}\mathbf{v}, \qquad (2)$$

where $\mathbf{v} \neq \mathbf{0}$ is some fixed vector to be determined, and λ is a growth rate, also to be determined. If such solutions exist, they correspond to exponential motion along the line spanned by the vector $\mathbf{v}$.

To find the conditions on $\mathbf{v}$ and λ, we substitute $\mathbf{x}(t) = e^{\lambda t}\mathbf{v}$ into $\dot{\mathbf{x}} = A\mathbf{x}$, and obtain $\lambda e^{\lambda t}\mathbf{v} = e^{\lambda t}A\mathbf{v}$. Canceling the nonzero scalar factor $e^{\lambda t}$ yields

$$A\mathbf{v} = \lambda \mathbf{v}, \tag{3}$$

which says that the desired straight line solutions exist if $\mathbf{v}$ is an ***eigenvector*** of A with corresponding ***eigenvalue*** λ. In this case we call the solution (2) an ***eigensolution***.

Let's recall how to find eigenvalues and eigenvectors. (If your memory needs more refreshing, see any text on linear algebra.) In general, the eigenvalues of a matrix A are given by the ***characteristic equation*** $\det(A - \lambda I) = 0$, where I is the identity matrix. For a 2×2 matrix

$$A = \begin{pmatrix} a & b \\ c & d \end{pmatrix},$$

the characteristic equation becomes

$$\det\begin{pmatrix} a-\lambda & b \\ c & d-\lambda \end{pmatrix} = 0.$$

Expanding the determinant yields

$$\lambda^2 - \tau\lambda + \Delta = 0 \tag{4}$$

where

$$\tau = \text{trace}(A) = a + d,$$
$$\Delta = \det(A) = ad - bc.$$

Then

$$\lambda_1 = \frac{\tau + \sqrt{\tau^2 - 4\Delta}}{2}, \qquad \lambda_2 = \frac{\tau - \sqrt{\tau^2 - 4\Delta}}{2} \tag{5}$$

are the solutions of the quadratic equation (4). In other words, the eigenvalues depend only on the trace and determinant of the matrix A.

The typical situation is for the eigenvalues to be distinct: $\lambda_1 \neq \lambda_2$. In this case, a theorem of linear algebra states that the corresponding eigenvectors $\mathbf{v}_1$ and $\mathbf{v}_2$ are linearly independent, and hence span the entire plane (Figure 5.2.1). In particular, any initial condition $\mathbf{x}_0$ can be written as a linear combination of eigenvectors, say $\mathbf{x}_0 = c_1\mathbf{v}_1 + c_2\mathbf{v}_2$.

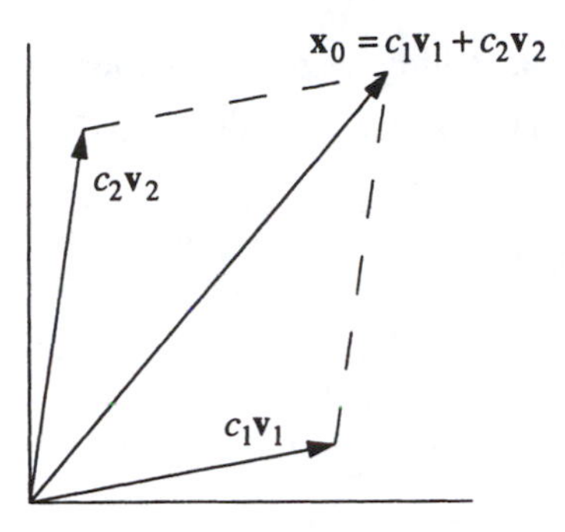

Figure 5.2.1

This observation allows us to write down the general solution for $\mathbf{x}(t)$—it is simply

$$\mathbf{x}(t) = c_1 e^{\lambda_1 t}\mathbf{v}_1 + c_2 e^{\lambda_2 t}\mathbf{v}_2 \,. \tag{6}$$

Why is this the general solution? First of all, it is a linear combination of solutions to $\dot{\mathbf{x}} = A\mathbf{x}$, and hence is itself a solution. Second, it satisfies the initial condition $\mathbf{x}(0) = \mathbf{x}_0$, and so by the existence and uniqueness theorem, it is the *only* solution. (See Section 6.2 for a general statement of the existence and uniqueness theorem.)

EXAMPLE 5.2.1:

Solve the initial value problem $\dot{x} = x + y$, $\dot{y} = 4x - 2y$, subject to the initial condition $(x_0, y_0) = (2, -3)$.

Solution: The corresponding matrix equation is

$$\begin{pmatrix} \dot{x} \\ \dot{y} \end{pmatrix} = \begin{pmatrix} 1 & 1 \\ 4 & -2 \end{pmatrix} \begin{pmatrix} x \\ y \end{pmatrix}.$$

First we find the eigenvalues of the matrix A. The matrix has $\tau = -1$ and $\Delta = -6$, so the characteristic equation is $\lambda^2 + \lambda - 6 = 0$. Hence

$$\lambda_1 = 2, \quad \lambda_2 = -3.$$

Next we find the eigenvectors. Given an eigenvalue λ, the corresponding eigenvector $\mathbf{v} = (v_1, v_2)$ satisfies

$$\begin{pmatrix} 1-\lambda & 1 \\ 4 & -2-\lambda \end{pmatrix} \begin{pmatrix} v_1 \\ v_2 \end{pmatrix} = \begin{pmatrix} 0 \\ 0 \end{pmatrix}.$$

For $\lambda_1 = 2$, this yields $\begin{pmatrix} -1 & 1 \\ 4 & -4 \end{pmatrix} \begin{pmatrix} v_1 \\ v_2 \end{pmatrix} = \begin{pmatrix} 0 \\ 0 \end{pmatrix}$, which has a nontrivial solution

$(v_1, v_2) = (1,1)$, or any scalar multiple thereof. (Of course, any multiple of an eigenvector is always an eigenvector; we try to pick the simplest multiple, but any one will do.) Similarly, for $\lambda_2 = -3$, the eigenvector equation becomes $\begin{pmatrix} 4 & 1 \\ 4 & 1 \end{pmatrix}\begin{pmatrix} v_1 \\ v_2 \end{pmatrix} = \begin{pmatrix} 0 \\ 0 \end{pmatrix}$, which has a nontrivial solution $(v_1, v_2) = (1, -4)$. In summary,

$$\mathbf{v}_1 = \begin{pmatrix} 1 \\ 1 \end{pmatrix}, \qquad \mathbf{v}_2 = \begin{pmatrix} 1 \\ -4 \end{pmatrix}.$$

Next we write the general solution as a linear combination of eigensolutions. From (6), the general solution is

$$\mathbf{x}(t) = c_1 \begin{pmatrix} 1 \\ 1 \end{pmatrix} e^{2t} + c_2 \begin{pmatrix} 1 \\ -4 \end{pmatrix} e^{-3t}. \tag{7}$$

Finally, we compute c_1 and c_2 to satisfy the initial condition $(x_0, y_0) = (2, -3)$. At $t = 0$, (7) becomes

$$\begin{pmatrix} 2 \\ -3 \end{pmatrix} = c_1 \begin{pmatrix} 1 \\ 1 \end{pmatrix} + c_2 \begin{pmatrix} 1 \\ -4 \end{pmatrix},$$

which is equivalent to the algebraic system

$$\begin{aligned} 2 &= c_1 + c_2, \\ -3 &= c_1 - 4c_2. \end{aligned}$$

The solution is $c_1 = 1$, $c_2 = 1$. Substituting back into (7) yields

$$\begin{aligned} x(t) &= e^{2t} + e^{-3t}, \\ y(t) &= e^{2t} - 4e^{-3t} \end{aligned}$$

for the solution to the initial value problem. ■

Whew! Fortunately we don't need to go through all this to draw the phase portrait of a linear system. All we need to know are the eigenvectors and eigenvalues.

EXAMPLE 5.2.2:

Draw the phase portrait for the system of Example 5.2.1.

Solution: The system has eigenvalues $\lambda_1 = 2$, $\lambda_2 = -3$. Hence the first eigensolution grows exponentially, and the second eigensolution decays. This means the origin is a *saddle point*. Its stable manifold is the line spanned by the eigenvector $\mathbf{v}_2 = (1, -4)$, corresponding to the decaying eigensolution. Similarly, the unstable manifold is the line spanned by $\mathbf{v}_1 = (1,1)$. As with all saddle points, a typical trajectory approaches the unstable manifold as $t \to \infty$, and the stable manifold as $t \to -\infty$. Figure 5.2.2 shows the phase portrait. ■

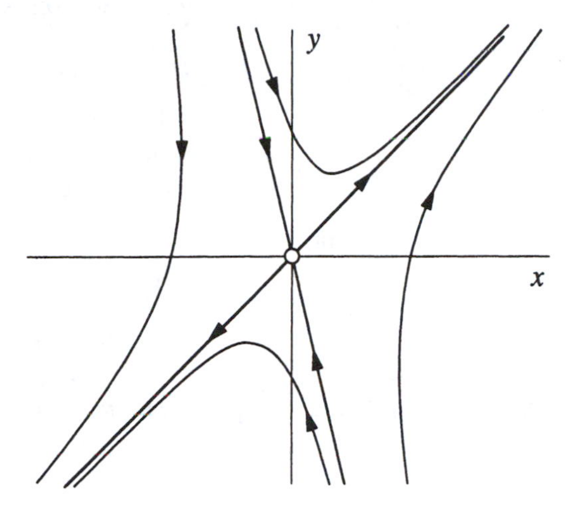

Figure 5.2.2

EXAMPLE 5.2.3:

Sketch a typical phase portrait for the case $\lambda_2 < \lambda_1 < 0$.

Solution: First suppose $\lambda_2 < \lambda_1 < 0$. Then both eigensolutions decay exponentially. The fixed point is a stable node, as in Figures 5.1.5a and 5.1.5c, except now the eigenvectors are not mutually perpendicular, in general. Trajectories typically approach the origin tangent to the ***slow eigendirection***, defined as the direction spanned by the eigenvector with the smaller $|\lambda|$. In backwards time ($t \to -\infty$), the trajectories become parallel to the fast eigendirection. Figure 5.2.3 shows the phase portrait. (If we reverse all the arrows in Figure 5.2.3, we obtain a typical phase portrait for an ***unstable node***.) ■

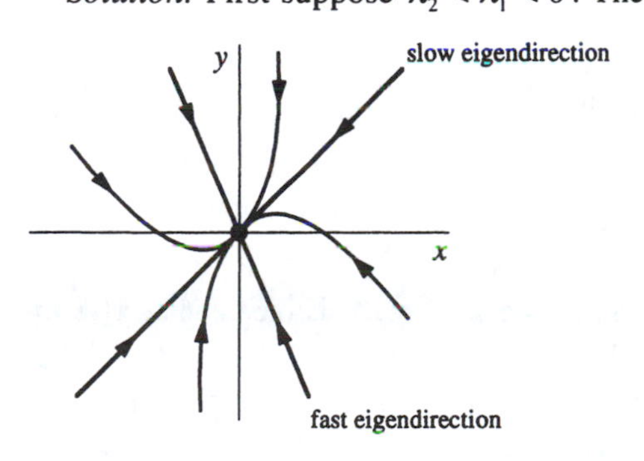

Figure 5.2.3

EXAMPLE 5.2.4:

What happens if the eigenvalues are *complex* numbers?

Solution: If the eigenvalues are complex, the fixed point is either a ***center*** (Figure 5.2.4a) or a ***spiral*** (Figure 5.2.4b). We've already seen an example of a center in the simple harmonic oscillator of Section 5.1; the origin is surrounded by a family of closed orbits. Note that centers are *neutrally stable*, since nearby trajectories are neither attracted to nor repelled from the fixed point. A spiral would occur if the harmonic oscillator were lightly damped. Then the trajectory would just fail to close, because the oscillator loses a bit of energy on each cycle.

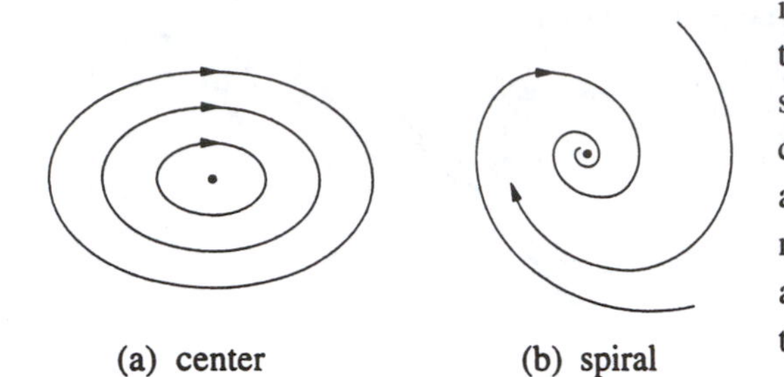

Figure 5.2.4

To justify these statements, recall that the eigenvalues are $\lambda_{1,2} = \frac{1}{2}\left(\tau \pm \sqrt{\tau^2 - 4\Delta}\right)$. Thus complex eigenvalues occur when

$$\tau^2 - 4\Delta < 0.$$

To simplify the notation, let's write the eigenvalues as

$$\lambda_{1,2} = \alpha \pm i\omega$$

where

$$\alpha = \tau/2, \qquad \omega = \tfrac{1}{2}\sqrt{4\Delta - \tau^2}\,.$$

By assumption, $\omega \neq 0$. Then the eigenvalues are distinct and so the general solution is still given by

$$\mathbf{x}(t) = c_1 e^{\lambda_1 t}\mathbf{v}_1 + c_2 e^{\lambda_2 t}\mathbf{v}_2\,.$$

But now the c's and $\mathbf{v}$'s are *complex*, since the λ's are. This means that $\mathbf{x}(t)$ involves linear combinations of $e^{(\alpha \pm i\omega)t}$. By Euler's formula, $e^{i\omega t} = \cos\omega t + i\sin\omega t$. Hence $\mathbf{x}(t)$ is a combination of terms involving $e^{\alpha t}\cos\omega t$ and $e^{\alpha t}\sin\omega t$. Such terms represent exponentially *decaying oscillations* if $\alpha = \text{Re}(\lambda) < 0$ and *growing oscillations* if $\alpha > 0$. The corresponding fixed points are ***stable*** and ***unstable spirals***, respectively. Figure 5.2.4b shows the stable case.

If the eigenvalues are pure imaginary ($\alpha = 0$), then all the solutions are periodic with period $T = 2\pi/\omega$. The oscillations have fixed amplitude and the fixed point is a center.

For both centers and spirals, it's easy to determine whether the rotation is clockwise or counterclockwise; just compute a few vectors in the vector field and the sense of rotation should be obvious. ■

EXAMPLE 5.2.5:

In our analysis of the general case, we have been assuming that the eigenvalues are distinct. What happens if the eigenvalues are *equal*?

Solution: Suppose $\lambda_1 = \lambda_2 = \lambda$. There are two possibilities: either there are two independent eigenvectors corresponding to λ, or there's only one.

If there are two independent eigenvectors, then they span the plane and so *every vector is an eigenvector with this same eigenvalue* λ. To see this, write an arbitrary vector $\mathbf{x}_0$ as a linear combination of the two eigenvectors: $\mathbf{x}_0 = c_1\mathbf{v}_1 + c_2\mathbf{v}_2$. Then

$$A\mathbf{x}_0 = A(c_1\mathbf{v}_1 + c_2\mathbf{v}_2) = c_1\lambda\mathbf{v}_1 + c_2\lambda\mathbf{v}_2 = \lambda\mathbf{x}_0$$

so $\mathbf{x}_0$ is also an eigenvector with eigenvalue λ. Since multiplication by A simply stretches every vector by a factor λ, the matrix must be a multiple of the identity:

$$A = \begin{pmatrix} \lambda & 0 \\ 0 & \lambda \end{pmatrix}.$$

Then if $\lambda \neq 0$, all trajectories are straight lines through the origin ($\mathbf{x}(t) = e^{\lambda t}\mathbf{x}_0$) and the fixed point is a ***star node*** (Figure 5.2.5).

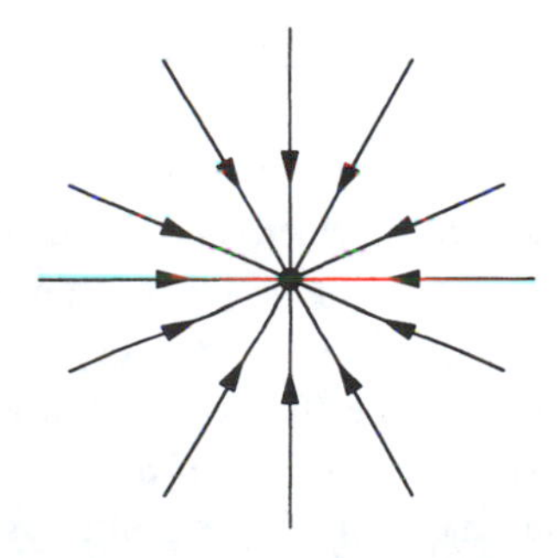

Figure 5.2.5

On the other hand, if $\lambda = 0$, the whole plane is filled with fixed points! (No surprise—the system is $\dot{\mathbf{x}} = \mathbf{0}$.)

The other possibility is that there's only one eigenvector (more accurately, the eigenspace corresponding to λ is one-dimensional.) For example, any matrix of the form $A = \begin{pmatrix} \lambda & b \\ 0 & \lambda \end{pmatrix}$, with $b \neq 0$ has only a one-dimensional eigenspace (Exercise 5.2.11).

When there's only one eigendirection, the fixed point is a ***degenerate node***. A

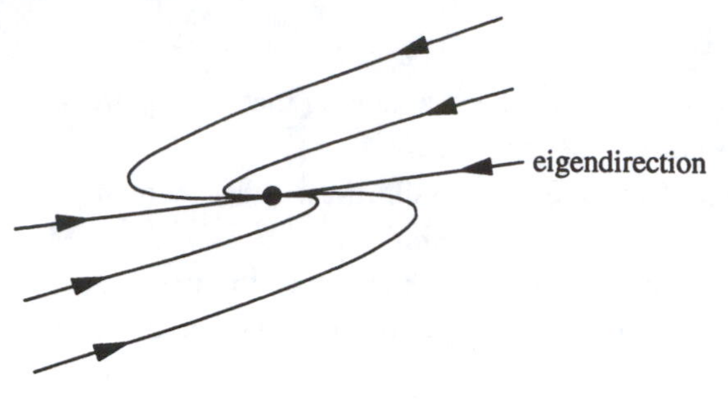

Figure 5.2.6

typical phase portrait is shown in Figure 5.2.6. As $t \to +\infty$ and also as $t \to -\infty$, all trajectories become parallel to the one available eigendirection.

A good way to think about the degenerate node is to imagine that it has been created by deforming an ordinary node. The ordinary node has two independent eigendirections; all trajectories are parallel to the slow eigendirection as $t \to \infty$, and to the fast eigendirection as $t \to -\infty$ (Figure 5.2.7a).

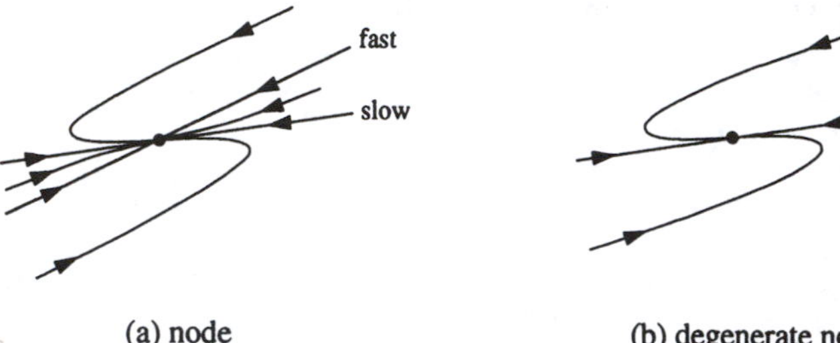

Figure 5.2.7

Now suppose we start changing the parameters of the system in such a way that the two eigendirections are scissored together. Then some of the trajectories will get squashed in the collapsing region between the two eigendirections, while the surviving trajectories get pulled around to form the degenerate node (Figure 5.2.7b).

Another way to get intuition about this case is to realize that the degenerate node is on the *borderline between a spiral and a node*. The trajectories are trying to wind around in a spiral, but they don't quite make it. ■

Classification of Fixed Points

By now you're probably tired of all the examples and ready for a simple classification scheme. Happily, there is one. We can show the type and stability of all the different fixed points on a single diagram (Figure 5.2.8).

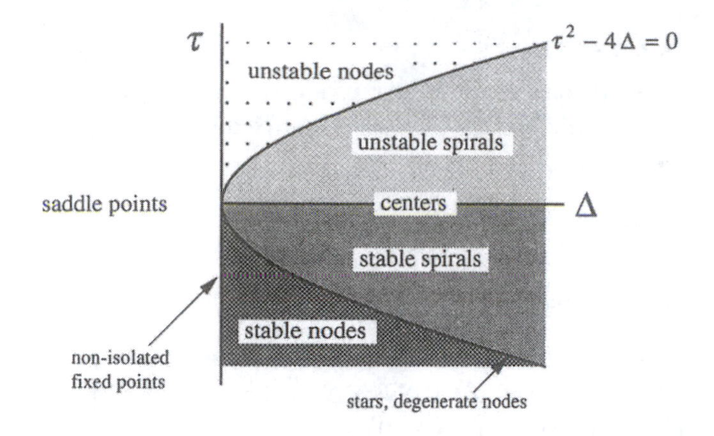

Figure 5.2.8

The axes are the trace τ and the determinant Δ of the matrix A. All of the information in the diagram is implied by the following formulas:

$$\lambda_{1,2} = \tfrac{1}{2}\left(\tau \pm \sqrt{\tau^2 - 4\Delta}\right), \qquad \Delta = \lambda_1 \lambda_2, \qquad \tau = \lambda_1 + \lambda_2.$$

The first equation is just (5). The second and third can be obtained by writing the characteristic equation in the form $(\lambda - \lambda_1)(\lambda - \lambda_2) = \lambda^2 - \tau\lambda + \Delta = 0$.

To arrive at Figure 5.2.8, we make the following observations:

If $\Delta < 0$, the eigenvalues are real and have opposite signs; hence the fixed point is a *saddle point.*

If $\Delta > 0$, the eigenvalues are either real with the same sign (*nodes*), or complex conjugate (*spirals* and *centers*). Nodes satisfy $\tau^2 - 4\Delta > 0$ and spirals satisfy $\tau^2 - 4\Delta < 0$. The parabola $\tau^2 - 4\Delta = 0$ is the borderline between nodes and spirals; star nodes and degenerate nodes live on this parabola. The stability of the nodes and spirals is determined by τ. When $\tau < 0$, both eigenvalues have negative real parts, so the fixed point is stable. Unstable spirals and nodes have $\tau > 0$. Neutrally stable centers live on the borderline $\tau = 0$, where the eigenvalues are purely imaginary.

If $\Delta = 0$, at least one of the eigenvalues is zero. Then the origin is not an isolated fixed point. There is either a whole line of fixed points, as in Figure 5.1.5d, or a plane of fixed points, if $A = 0$.

Figure 5.2.8 shows that saddle points, nodes, and spirals are the major types of fixed points; they occur in large open regions of the (Δ, τ) plane. Centers, stars, degenerate nodes, and non-isolated fixed points are ***borderline cases*** that occur along curves in the (Δ, τ) plane. Of these borderline cases, centers are by far the most important. They occur very commonly in frictionless mechanical systems where energy is conserved.

EXAMPLE 5.2.6:

Classify the fixed point $\mathbf{x}^* = \mathbf{0}$ for the system $\dot{\mathbf{x}} = A\mathbf{x}$, where $A = \begin{pmatrix} 1 & 2 \\ 3 & 4 \end{pmatrix}$.

Solution: The matrix has $\Delta = -2$; hence the fixed point is a saddle point. ■

EXAMPLE 5.2.7:

Redo Example 5.2.6 for $A = \begin{pmatrix} 2 & 1 \\ 3 & 4 \end{pmatrix}$.

Solution: Now $\Delta = 5$ and $\tau = 6$. Since $\Delta > 0$ and $\tau^2 - 4\Delta = 16 > 0$, the fixed point is a node. It is unstable, since $\tau > 0$. ■

5.3 Love Affairs

To arouse your interest in the classification of linear systems, we now discuss a simple model for the dynamics of love affairs (Strogatz 1988). The following story illustrates the idea.

Romeo is in love with Juliet, but in our version of this story, Juliet is a fickle lover. The more Romeo loves her, the more Juliet wants to run away and hide. But when Romeo gets discouraged and backs off, Juliet begins to find him strangely attractive. Romeo, on the other hand, tends to echo her: he warms up when she loves him, and grows cold when she hates him.

Let

$R(t)$ = Romeo's love/hate for Juliet at time t

$J(t)$ = Juliet's love/hate for Romeo at time t.

Positive values of R, J signify love, negative values signify hate. Then a model for their star-crossed romance is

$$\dot{R} = aJ$$
$$\dot{J} = -bR$$

where the parameters a and b are positive, to be consistent with the story.

The sad outcome of their affair is, of course, a neverending cycle of love and hate; the governing system has a center at $(R, J) = (0,0)$. At least they manage to achieve simultaneous love one-quarter of the time (Figure 5.3.1).

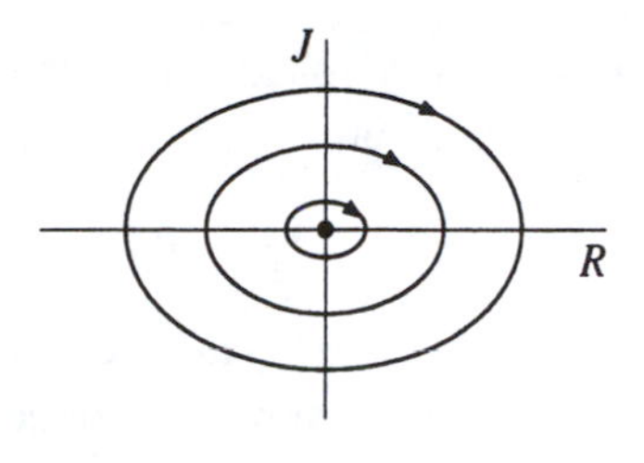

Figure 5.3.1

Now consider the forecast for lovers governed by the general linear system

$$\dot{R} = aR + bJ$$
$$\dot{J} = cR + dJ$$

where the parameters a, b, c, d may have either sign. A choice of signs specifies the romantic styles. As named by one of my students, the choice $a > 0$, $b > 0$ means that Romeo is an "eager beaver"—he gets excited by Juliet's love for him, and is further spurred on by his own affectionate feelings for her. It's entertaining to name the other three romantic styles, and to predict the outcomes for the various pairings. For example, can a "cautious lover" $(a < 0,\ b > 0)$ find true love with an eager beaver? These and other pressing questions will be considered in the exercises.

EXAMPLE 5.3.1:

What happens when two identically cautious lovers get together?

Solution: The system is

$$\dot{R} = aR + bJ$$
$$\dot{J} = bR + aJ$$

with $a < 0$, $b > 0$. Here a is a measure of cautiousness (they each try to avoid throwing themselves at the other) and b is a measure of responsiveness (they both get excited by the other's advances). We might suspect that the outcome depends on the relative size of a and b. Let's see what happens.

The corresponding matrix is

$$A = \begin{pmatrix} a & b \\ b & a \end{pmatrix}$$

which has

$$\tau = 2a < 0, \qquad \Delta = a^2 - b^2, \qquad \tau^2 - 4\Delta = 4b^2 > 0.$$

Hence the fixed point $(R,J)=(0,0)$ is a saddle point if $a^2<b^2$ and a stable node if $a^2>b^2$. The eigenvalues and corresponding eigenvectors are

$$\lambda_1=a+b, \qquad \mathbf{v}_1=(1,1), \qquad \lambda_2=a-b, \qquad \mathbf{v}_2=(1,-1).$$

Since $a+b>a-b$, the eigenvector $(1,1)$ spans the unstable manifold when the origin is a saddle point, and it spans the slow eigendirection when the origin is a stable node. Figure 5.3.2 shows the phase portrait for the two cases.

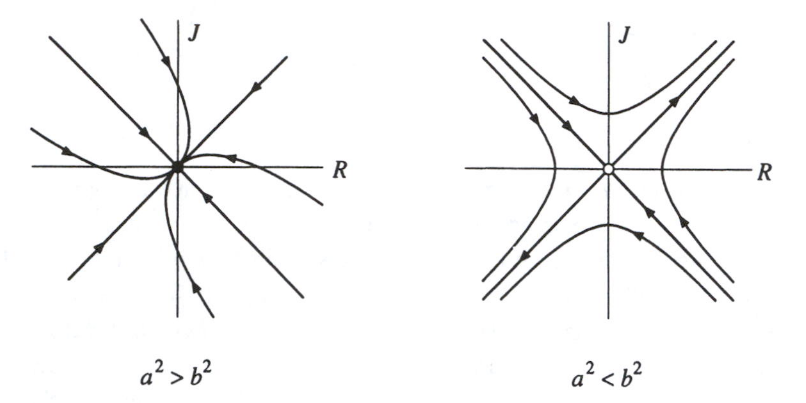

Figure 5.3.2

If $a^2>b^2$, the relationship always fizzles out to mutual indifference. The lesson seems to be that excessive caution can lead to apathy.

If $a^2<b^2$, the lovers are more daring, or perhaps more sensitive to each other. Now the relationship is explosive. Depending on their feelings initially, their relationship either becomes a love fest or a war. In either case, all trajectories approach the line $R=J$, so their feelings are eventually mutual. ■

EXERCISES FOR CHAPTER 5

5.1 Definitions and Examples

5.1.1 (Ellipses and energy conservation for the harmonic oscillator) Consider the harmonic oscillator $\dot{x}=v$, $\dot{v}=-\omega^2 x$.

a) Show that the orbits are given by ellipses $\omega^2x^2+v^2=C$, where C is any non-negative constant. (Hint: Divide the $\dot{x}$ equation by the $\dot{v}$ equation, separate the v's from the x's, and integrate the resulting separable equation.)

b) Show that this condition is equivalent to conservation of energy.

5.1.2 Consider the system $\dot{x} = ax$, $\dot{y} = -y$, where $a < -1$. Show that all trajectories become parallel to the y-direction as $t \to \infty$, and parallel to the x-direction as $t \to -\infty$.

(Hint: Examine the slope $dy/dx = \dot{y}/\dot{x}$.)

Write the following systems in matrix form.

5.1.3 $\dot{x} = -y,\ \dot{y} = -x$

5.1.4 $\dot{x} = 3x - 2y,\ \dot{y} = 2y - x$

5.1.5 $\dot{x} = 0,\ \dot{y} = x + y$

5.1.6 $\dot{x} = x,\ \dot{y} = 5x + y$

Sketch the vector field for the following systems. Indicate the length and direction of the vectors with reasonable accuracy. Sketch some typical trajectories.

5.1.7 $\dot{x} = x,\ \dot{y} = x + y$

5.1.8 $\dot{x} = -2y,\ \dot{y} = x$

5.1.9 Consider the system $\dot{x} = -y,\ \dot{y} = -x$.

a) Sketch the vector field.
b) Show that the trajectories of the system are hyperbolas of the form $x^2 - y^2 = C$. (Hint: Show that the governing equations imply $x\dot{x} - y\dot{y} = 0$ and then integrate both sides.)
c) The origin is a saddle point; find equations for its stable and unstable manifolds.
d) The system can be decoupled and solved as follows. Introduce new variables u and v, where $u = x + y$, $v = x - y$. Then rewrite the system in terms of u and v. Solve for $u(t)$ and $v(t)$, starting from an arbitrary initial condition (u_0, v_0).
e) What are the equations for the stable and unstable manifolds in terms of u and v?
f) Finally, using the answer to (d), write the general solution for $x(t)$ and $y(t)$, starting from an initial condition (x_0, y_0).

5.1.10 (Attracting and Liapunov stable) Here are the official definitions of the various types of stability. Consider a fixed point $\mathbf{x}^*$ of a system $\dot{\mathbf{x}} = \mathbf{f}(\mathbf{x})$.

We say that $\mathbf{x}^*$ is ***attracting*** if there is a $\delta > 0$ such that $\lim_{t\to\infty} \mathbf{x}(t) = \mathbf{x}^*$ whenever $\|\mathbf{x}(0) - \mathbf{x}^*\| < \delta$. In other words, any trajectory that starts within a distance δ of $\mathbf{x}^*$ is guaranteed to converge to $\mathbf{x}^*$ *eventually*. As shown schematically in Figure 1, trajectories that start nearby are allowed to stray from $\mathbf{x}^*$ in the short run, but they must approach $\mathbf{x}^*$ in the long run.

In contrast, Liapunov stability requires that nearby trajectories remain close for *all* time. We say that $\mathbf{x}^*$ is ***Liapunov stable*** if for each $\varepsilon > 0$, there is a $\delta > 0$ such that $\|\mathbf{x}(t) - \mathbf{x}^*\| < \varepsilon$ whenever $t \geq 0$ and $\|\mathbf{x}(0) - \mathbf{x}^*\| < \delta$. Thus, trajectories that start within δ of $\mathbf{x}^*$ remain within ε of $\mathbf{x}^*$ for all positive time (Figure 1).

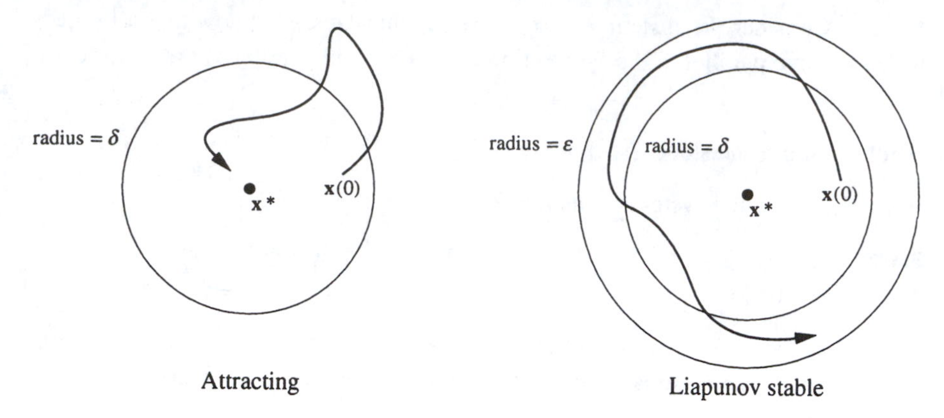

Figure 1

Finally, $\mathbf{x}^*$ is ***asymptotically stable*** if it is both attracting and Liapunov stable.

For each of the following systems, decide whether the origin is attracting, Liapunov stable, asymptotically stable, or none of the above.

a) $\dot{x} = y$, $\dot{y} = -4x$.

b) $\dot{x} = 2y$, $\dot{y} = x$

c) $\dot{x} = 0$, $\dot{y} = x$

d) $\dot{x} = 0$, $\dot{y} = -y$

e) $\dot{x} = -x$, $\dot{y} = -5y$

f) $\dot{x} = x$, $\dot{y} = y$

5.1.11 (Stability proofs) Prove that your answers to 5.1.10 are correct, using the definitions of the different types of stability. (You must produce a suitable δ to prove that the origin is attracting, or a suitable $\delta(\varepsilon)$ to prove Liapunov stability.)

5.1.12 (Closed orbits from symmetry arguments) Give a simple proof that orbits are closed for the simple harmonic oscillator $\dot{x} = v$, $\dot{v} = -x$, using *only* the symmetry properties of the vector field. (Hint: Consider a trajectory that starts on the v-axis at $(0, -v_0)$, and suppose that the trajectory intersects the x-axis at $(x, 0)$. Then use symmetry arguments to find the subsequent intersections with the v-axis and x-axis.)

5.1.13 Why do you think a "saddle point" is called by that name? What's the connection to real saddles (the kind used on horses)?

5.2 Classification of Linear Systems

5.2.1 Consider the system $\dot{x} = 4x - y$, $\dot{y} = 2x + y$.

a) Write the system as $\dot{\mathbf{x}} = A\mathbf{x}$. Show that the characteristic polynomial is $\lambda^2 - 5\lambda + 6$, and find the eigenvalues and eigenvectors of A.

b) Find the general solution of the system.

c) Classify the fixed point at the origin.

d) Solve the system subject to the initial condition $(x_0, y_0) = (3, 4)$.

5.2.2 (Complex eigenvalues) This exercise leads you through the solution of a linear system where the eigenvalues are complex. The system is $\dot{x} = x - y$, $\dot{y} = x + y$.

a) Find A and show that it has eigenvalues $\lambda_1 = 1 + i$, $\lambda_2 = 1 - i$, with eigenvectors $\mathbf{v}_1 = (i, 1)$, $\mathbf{v}_2 = (-i, 1)$. (Note that the eigenvalues are complex conjugates, and so are the eigenvectors—this is always the case for real A with complex eigenvalues.)
b) The general solution is $\mathbf{x}(t) = c_1 e^{\lambda_1 t}\mathbf{v}_1 + c_2 e^{\lambda_2 t}\mathbf{v}_2$. So in one sense we're done! But this way of writing $\mathbf{x}(t)$ involves complex coefficients and looks unfamiliar. Express $\mathbf{x}(t)$ purely in terms of real-valued functions. (Hint: Use $e^{i\omega t} = \cos\omega t + i \sin\omega t$ to rewrite $\mathbf{x}(t)$ in terms of sines and cosines, and then separate the terms that have a prefactor of i from those that don't.)

Plot the phase portrait and classify the fixed point of the following linear systems. If the eigenvectors are real, indicate them in your sketch.

5.2.3 $\dot{x} = y$, $\dot{y} = -2x - 3y$

5.2.4 $\dot{x} = 5x + 10y$, $\dot{y} = -x - y$

5.2.5 $\dot{x} = 3x - 4y$, $\dot{y} = x - y$

5.2.6 $\dot{x} = -3x + 2y$, $\dot{y} = x - 2y$

5.2.7 $\dot{x} = 5x + 2y$, $\dot{y} = -17x - 5y$

5.2.8 $\dot{x} = -3x + 4y$, $\dot{y} = -2x + 3y$

5.2.9 $\dot{x} = 4x - 3y$, $\dot{y} = 8x - 6y$

5.2.10 $\dot{x} = y$, $\dot{y} = -x - 2y$.

5.2.11 Show that any matrix of the form $A = \begin{pmatrix} \lambda & b \\ 0 & \lambda \end{pmatrix}$, with $b \neq 0$, has only a one-dimensional eigenspace corresponding to the eigenvalue λ. Then solve the system $\dot{\mathbf{x}} = A\mathbf{x}$ and sketch the phase portrait.

5.2.12 (*LRC* circuit) Consider the circuit equation $L\ddot{I} + R\dot{I} + I/C = 0$, where $L, C > 0$ and $R \geq 0$.

a) Rewrite the equation as a two-dimensional linear system.
b) Show that the origin is asymptotically stable if $R > 0$ and neutrally stable if $R = 0$.
c) Classify the fixed point at the origin, depending on whether $R^2C - 4L$ is positive, negative, or zero, and sketch the phase portrait in all three cases.

5.2.13 (Damped harmonic oscillator) The motion of a damped harmonic oscillator is described by $m\ddot{x} + b\dot{x} + kx = 0$, where $b > 0$ is the damping constant.

a) Rewrite the equation as a two-dimensional linear system.
b) Classify the fixed point at the origin and sketch the phase portrait. Be sure to show all the different cases that can occur, depending on the relative sizes of the parameters.
c) How do your results relate to the standard notions of overdamped, critically damped, and underdamped vibrations?

5.2.14 (A project about random systems) Suppose we pick a linear system at random; what's the probability that the origin will be, say, an unstable spiral? To be more specific, consider the system $\dot{\mathbf{x}} = A\mathbf{x}$, where $A = \begin{pmatrix} a & b \\ c & d \end{pmatrix}$. Suppose we pick the entries a, b, c, d independently and at random from a uniform distribution on the interval $[-1,1]$. Find the probabilities of all the different kinds of fixed points.

To check your answers (or if you hit an analytical roadblock), try the *Monte Carlo method.* Generate millions of random matrices on the computer and have the machine count the relative frequency of saddles, unstable spirals, etc.

Are the answers the same if you use a normal distribution instead of a uniform distribution?

5.3 Love Affairs

5.3.1 (Name-calling) Suggest names for the four romantic styles, determined by the signs of a and b in $\dot{R} = aR + bJ$.

5.3.2 Consider the affair described by $\dot{R} = J$, $\dot{J} = -R + J$.

a) Characterize the romantic styles of Romeo and Juliet.

b) Classify the fixed point at the origin. What does this imply for the affair?

c) Sketch $R(t)$ and $J(t)$ as functions of t, assuming $R(0) = 1$, $J(0) = 0$.

In each of the following problems, predict the course of the love affair, depending on the signs and relative sizes of a and b.

5.3.3 (Out of touch with their own feelings) Suppose Romeo and Juliet react to each other, but not to themselves: $\dot{R} = aJ$, $\dot{J} = bR$. What happens?

5.3.4 (Fire and water) Do opposites attract? Analyze $\dot{R} = aR + bJ$, $\dot{J} = -bR - aJ$.

5.3.5 (Peas in a pod) If Romeo and Juliet are romantic clones ($\dot{R} = aR + bJ$, $\dot{J} = bR + aJ$), should they expect boredom or bliss?

5.3.6 (Romeo the robot) Nothing could ever change the way Romeo feels about Juliet: $\dot{R} = 0$, $\dot{J} = aR + bJ$. Does Juliet end up loving him or hating him?

6

PHASE PLANE

6.0 Introduction

This chapter begins our study of two-dimensional *nonlinear* systems. First we consider some of their general properties. Then we classify the kinds of fixed points that can arise, building on our knowledge of linear systems (Chapter 5). The theory is further developed through a series of examples from biology (competition between two species) and physics (conservative systems, reversible systems, and the pendulum). The chapter concludes with a discussion of index theory, a topological method that provides global information about the phase portrait.

This chapter is mainly about fixed points. The next two chapters will discuss closed orbits and bifurcations in two-dimensional systems.

6.1 Phase Portraits

The general form of a vector field on the phase plane is

$$\dot{x}_1 = f_1(x_1, x_2)$$
$$\dot{x}_2 = f_2(x_1, x_2)$$

where f_1 and f_2 are given functions. This system can be written more compactly in vector notation as

$$\dot{\mathbf{x}} = \mathbf{f}(\mathbf{x})$$

where $\mathbf{x} = (x_1, x_2)$ and $\mathbf{f}(\mathbf{x}) = (f_1(\mathbf{x}), f_2(\mathbf{x}))$. Here $\mathbf{x}$ represents a point in the phase plane, and $\dot{\mathbf{x}}$ is the velocity vector at that point. By flowing along the vector field, a phase point traces out a solution $\mathbf{x}(t)$, corresponding to a trajectory winding through the phase plane (Figure 6.1.1).

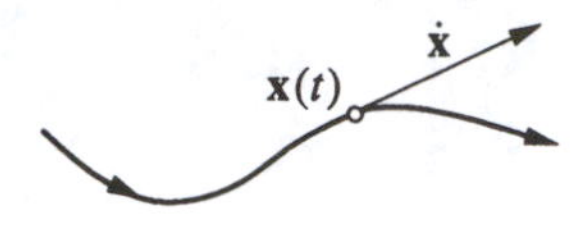

Figure 6.1.1

Furthermore, the entire phase plane is filled with trajectories, since each point can play the role of an initial condition.

For nonlinear systems, there's typically no hope of finding the trajectories analytically. Even when explicit formulas are available, they are often too complicated to provide much insight. Instead we will try to determine the *qualitative* behavior of the solutions. Our goal is to find the system's phase portrait directly from the properties of $\mathbf{f}(\mathbf{x})$. An enormous variety of phase portraits is possible; one example is shown in Figure 6.1.2.

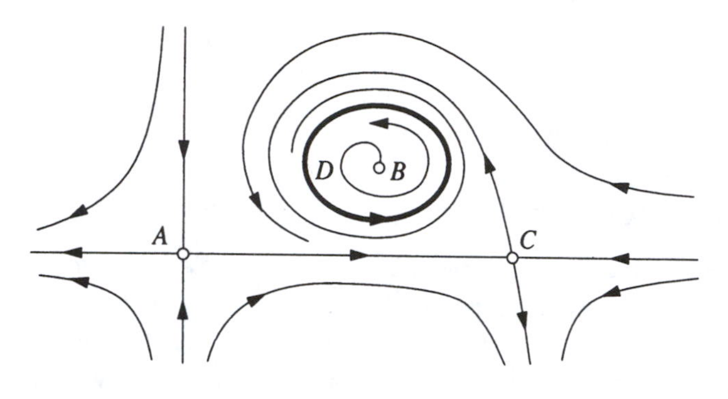

Figure 6.1.2

Some of the most salient features of any phase portrait are:

1. The ***fixed points***, like A, B, and C in Figure 6.1.2. Fixed points satisfy $\mathbf{f}(\mathbf{x}^*) = \mathbf{0}$, and correspond to steady states or equilibria of the system.
2. The ***closed orbits***, like D in Figure 6.1.2. These correspond to periodic solutions, i.e., solutions for which $\mathbf{x}(t+T) = \mathbf{x}(t)$ for all t, for some $T > 0$.
3. The arrangement of trajectories near the fixed points and closed orbits. For example, the flow pattern near A and C is similar, and different from that near B.
4. The stability or instability of the fixed points and closed orbits. Here, the fixed points A, B, and C are unstable, because nearby trajectories tend to move away from them, whereas the closed orbit D is stable.

Numerical Computation of Phase Portraits

Sometimes we are also interested in *quantitative* aspects of the phase portrait. Fortunately, numerical integration of $\dot{\mathbf{x}} = \mathbf{f}(\mathbf{x})$ is not much harder than that of $\dot{x} = f(x)$. The numerical methods of Section 2.8 still work, as long as we replace the numbers x and $f(x)$ by the vectors $\mathbf{x}$ and $\mathbf{f}(\mathbf{x})$. We will always use the Runge-Kutta method, which in vector form is

$$\mathbf{x}_{n+1} = \mathbf{x}_n + \tfrac{1}{6}(\mathbf{k}_1 + 2\mathbf{k}_2 + 2\mathbf{k}_3 + \mathbf{k}_4)$$

where

$$\begin{aligned}
\mathbf{k}_1 &= \mathbf{f}(\mathbf{x}_n)\Delta t \\
\mathbf{k}_2 &= \mathbf{f}(\mathbf{x}_n + \tfrac{1}{2}\mathbf{k}_1)\Delta t \\
\mathbf{k}_3 &= \mathbf{f}(\mathbf{x}_n + \tfrac{1}{2}\mathbf{k}_2)\Delta t \\
\mathbf{k}_4 &= \mathbf{f}(\mathbf{x}_n + \mathbf{k}_3)\Delta t.
\end{aligned}$$

A stepsize $\Delta t = 0.1$ usually provides sufficient accuracy for our purposes.

When plotting the phase portrait, it often helps to see a grid of representative vectors in the vector field. Unfortunately, the arrowheads and different lengths of the vectors tend to clutter such pictures. A plot of the ***direction field*** is clearer: short line segments are used to indicate the local direction of flow.

EXAMPLE 6.1.1:

Consider the system $\dot{x} = x + e^{-y}$, $\dot{y} = -y$. First use qualitative arguments to obtain information about the phase portrait. Then, using a computer, plot the direction field. Finally, use the Runge–Kutta method to compute several trajectories, and plot them on the phase plane.

Solution: First we find the fixed points by solving $\dot{x} = 0$, $\dot{y} = 0$ simultaneously. The only solution is $(x^*, y^*) = (-1, 0)$. To determine its stability, note that $y(t) \to 0$ as $t \to \infty$, since the solution to $\dot{y} = -y$ is $y(t) = y_0 e^{-t}$. Hence $e^{-y} \to 1$ and so in the long run, the equation for x becomes $\dot{x} \approx x + 1$; this has exponentially growing solutions, which suggests that the fixed point is unstable. In fact, if we restrict our attention to initial conditions on the x-axis, then $y_0 = 0$ and so $y(t) = 0$ for all time. Hence the flow on the x-axis is governed *strictly* by $\dot{x} = x + 1$. Therefore the fixed point is unstable.

To sketch the phase portrait, it is helpful to plot the ***nullclines***, defined as the curves where either $\dot{x} = 0$ or $\dot{y} = 0$. The nullclines indicate where the flow is purely horizontal or vertical (Figure 6.1.3). For example, the flow is horizontal where $\dot{y} = 0$, and since $\dot{y} = -y$, this occurs on the line $y = 0$. Along this line, the flow is to the right where $\dot{x} = x + 1 > 0$, that is, where $x > -1$.

Similarly, the flow is vertical where $\dot{x} = x + e^{-y} = 0$, which occurs on the curve shown in Figure 6.1.3. On the upper part of the curve where $y > 0$, the flow is downward, since $\dot{y} < 0$.

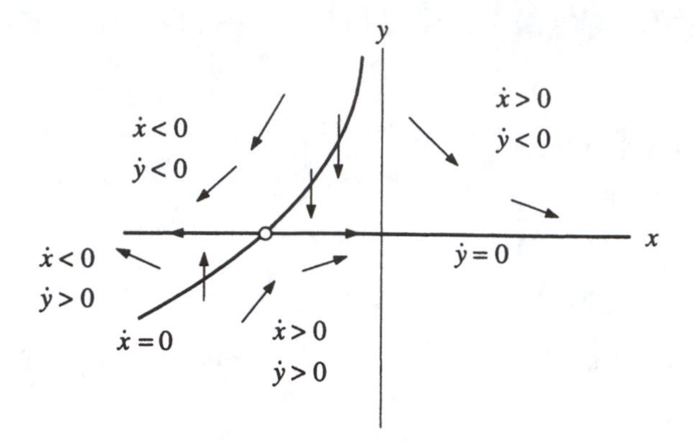

Figure 6.1.3

The nullclines also partition the plane into regions where $\dot{x}$ and $\dot{y}$ have various signs. Some of the typical vectors are sketched above in Figure 6.1.3. Even with the limited information obtained so far, Figure 6.1.3 gives a good sense of the overall flow pattern.

Now we use the computer to finish the problem. The direction field is indicated by the line segments in Figure 6.1.4, and several trajectories are shown. Note how the trajectories always follow the local slope.

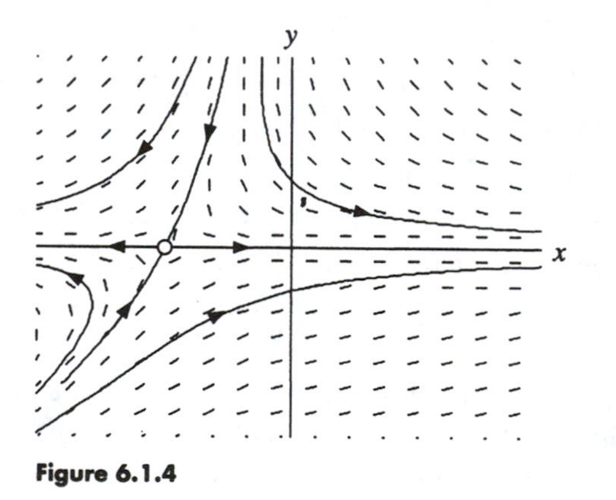

Figure 6.1.4

The fixed point is now seen to be a nonlinear version of a saddle point. ■

6.2 Existence, Uniqueness, and Topological Consequences

We have been a bit optimistic so far—at this stage, we have no guarantee that the general nonlinear system $\dot{\mathbf{x}} = \mathbf{f}(\mathbf{x})$ even *has* solutions! Fortunately the existence and uniqueness theorem given in Section 2.5 can be generalized to two-dimensional systems. We state the result for n-dimensional systems, since no extra effort is involved:

Existence and Uniqueness Theorem: Consider the initial value problem $\dot{\mathbf{x}} = \mathbf{f}(\mathbf{x})$, $\mathbf{x}(0) = \mathbf{x}_0$. Suppose that $\mathbf{f}$ is continuous and that all its partial derivatives $\partial f_i / \partial x_j$, $i, j = 1, \ldots, n$, are continuous for $\mathbf{x}$ in some open connected set $D \subset \mathbf{R}^n$. Then for $\mathbf{x}_0 \in D$, the initial value problem has a solution $\mathbf{x}(t)$ on some time interval $(-\tau, \tau)$ about $t = 0$, and the solution is unique.

In other words, existence and uniqueness of solutions are guaranteed if $\mathbf{f}$ is continuously differentiable. The proof of the theorem is similar to that for the case $n = 1$, and can be found in most texts on differential equations. Stronger versions of the theorem are available, but this one suffices for most applications.

From now on, we'll assume that all our vector fields are smooth enough to ensure the existence and uniqueness of solutions, starting from any point in phase space.

The existence and uniqueness theorem has an important corollary: *different trajectories never intersect*. If two trajectories *did* intersect, then there would be two solutions starting from the same point (the crossing point), and this would violate the uniqueness part of the theorem. In more intuitive language, a trajectory can't move in two directions at once.

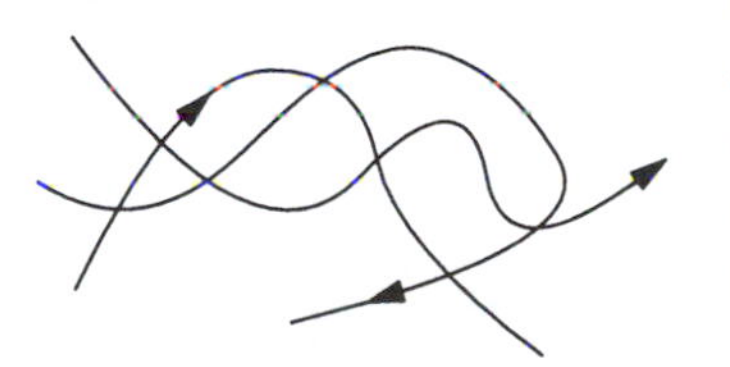

Figure 6.2.1

Because trajectories can't intersect, phase portraits always have a well-groomed look to them. Otherwise they might degenerate into a snarl of criss-crossed curves (Figure 6.2.1). The existence and uniqueness theorem prevents this from happening.

In two-dimensional phase spaces (as opposed to higher-dimensional phase spaces), these results have especially strong topological consequences. For example, suppose there is a closed orbit C in the phase plane. Then any trajectory starting inside C is trapped in there forever (Figure 6.2.2).

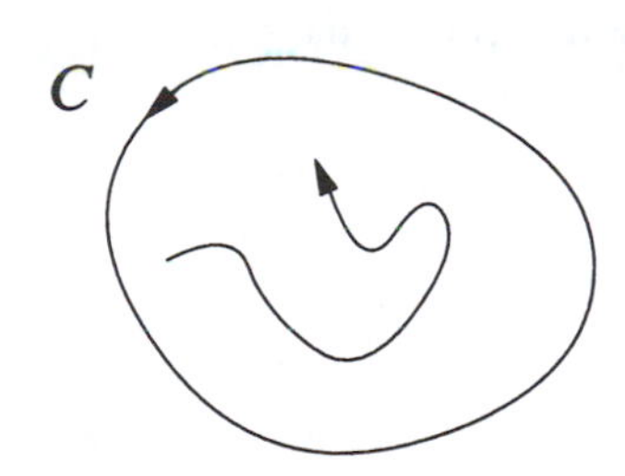

Figure 6.2.2

What is the fate of such a bounded trajectory? If there are fixed points inside C, then of course the trajectory might eventually approach one of them. But what if there *aren't* any fixed points? Your intuition may tell you that the trajectory can't meander around forever—if so, you're right. For vector fields on the plane, the ***Poincaré–Bendixson theorem*** states that if a trajectory is confined to a closed, bounded region and there are no fixed points in the region, then the trajectory must eventually approach a closed orbit. We'll discuss this important theorem in Section 7.3.

But that part of our story comes later. First we must become better acquainted with fixed points.

6.3 Fixed Points and Linearization

In this section we extend the ***linearization*** technique developed earlier for one-dimensional systems (Section 2.4). The hope is that we can approximate the phase portrait near a fixed point by that of a corresponding linear system.

Linearized System

Consider the system

$$\dot{x} = f(x,y)$$
$$\dot{y} = g(x,y)$$

and suppose that (x^*, y^*) is a fixed point, i.e.,

$$f(x^*, y^*) = 0, \qquad g(x^*, y^*) = 0.$$

Let

$$u = x - x^*, \qquad v = y - y^*$$

denote the components of a small disturbance from the fixed point. To see whether the disturbance grows or decays, we need to derive differential equations for u and v. Let's do the u-equation first:

$$\begin{aligned}
\dot{u} &= \dot{x} && \text{(since } x^* \text{ is a constant)} \\
&= f(x^* + u, y^* + v) && \text{(by substitution)} \\
&= f(x^*, y^*) + u\frac{\partial f}{\partial x} + v\frac{\partial f}{\partial y} + O(u^2, v^2, uv) && \text{(Taylor series expansion)} \\
&= u\frac{\partial f}{\partial x} + v\frac{\partial f}{\partial y} + O(u^2, v^2, uv) && \text{(since } f(x^*, y^*) = 0\text{)}.
\end{aligned}$$

To simplify the notation, we have written $\partial f/\partial x$ and $\partial f/\partial y$, but please remember that these partial derivatives are to be evaluated *at the fixed point* (x^*, y^*); thus they are *numbers*, not functions. Also, the shorthand notation $O(u^2, v^2, uv)$ denotes ***quadratic terms*** in u and v. Since u and v are small, these quadratic terms are *extremely* small.

Similarly we find

$$\dot{v} = u\frac{\partial g}{\partial x} + v\frac{\partial g}{\partial y} + O(u^2, v^2, uv).$$

Hence the disturbance (u, v) evolves according to

$$\begin{pmatrix} \dot{u} \\ \dot{v} \end{pmatrix} = \begin{pmatrix} \frac{\partial f}{\partial x} & \frac{\partial f}{\partial y} \\ \frac{\partial g}{\partial x} & \frac{\partial g}{\partial y} \end{pmatrix} \begin{pmatrix} u \\ v \end{pmatrix} + \text{quadratic terms}. \tag{1}$$

The matrix

$$A = \begin{pmatrix} \frac{\partial f}{\partial x} & \frac{\partial f}{\partial y} \\ \frac{\partial g}{\partial x} & \frac{\partial g}{\partial y} \end{pmatrix}_{(x^*, y^*)}$$

is called the ***Jacobian matrix*** at the fixed point (x^*, y^*). It is the multivariable analog of the derivative $f'(x^*)$ seen in Section 2.4.

Now since the quadratic terms in (1) are tiny, it's tempting to neglect them altogether. If we do that, we obtain the ***linearized system***

$$\begin{pmatrix} \dot{u} \\ \dot{v} \end{pmatrix} = \begin{pmatrix} \frac{\partial f}{\partial x} & \frac{\partial f}{\partial y} \\ \frac{\partial g}{\partial x} & \frac{\partial g}{\partial y} \end{pmatrix} \begin{pmatrix} u \\ v \end{pmatrix} \tag{2}$$

whose dynamics can be analyzed by the methods of Section 5.2.

The Effect of Small Nonlinear Terms

Is it really safe to neglect the quadratic terms in (1)? In other words, does the linearized system give a qualitatively correct picture of the phase portrait near (x^*, y^*)? The answer is *yes, as long as the fixed point for the linearized system is not one of the borderline cases* discussed in Section 5.2. In other words, if the linearized system predicts a saddle, node, or a spiral, then the fixed point *really is* a saddle, node, or spiral for the original nonlinear system. See Andronov et al. (1973) for a proof of this result, and Example 6.3.1 for a concrete illustration.

The borderline cases (centers, degenerate nodes, stars, or non-isolated fixed points) are much more delicate. They can be altered by small nonlinear terms, as we'll see in Example 6.3.2 and in Exercise 6.3.11.

EXAMPLE 6.3.1:

Find all the fixed points of the system $\dot{x} = -x + x^3$, $\dot{y} = -2y$, and use linearization to classify them. Then check your conclusions by deriving the phase portrait for the full nonlinear system.

Solution: Fixed points occur where $\dot{x} = 0$ and $\dot{y} = 0$ simultaneously. Hence we need $x = 0$ or $x = \pm 1$, and $y = 0$. Thus, there are three fixed points: (0,0), (1,0), and (−1,0). The Jacobian matrix at a general point (x, y) is

$$A = \begin{pmatrix} \frac{\partial \dot{x}}{\partial x} & \frac{\partial \dot{x}}{\partial y} \\ \frac{\partial \dot{y}}{\partial x} & \frac{\partial \dot{y}}{\partial y} \end{pmatrix} = \begin{pmatrix} -1+3x^2 & 0 \\ 0 & -2 \end{pmatrix}.$$

Next we evaluate A at the fixed points. At $(0,0)$, we find $A = \begin{pmatrix} -1 & 0 \\ 0 & -2 \end{pmatrix}$, so $(0,0)$ is a stable node. At $(\pm 1,0)$, $A = \begin{pmatrix} 2 & 0 \\ 0 & -2 \end{pmatrix}$, so both $(1,0)$ and $(-1,0)$ are saddle points.

Now because stable nodes and saddle points are not borderline cases, we can be certain that the fixed points for the full nonlinear system have been predicted correctly.

This conclusion can be checked explicitly for the nonlinear system, since the x and y equations are *uncoupled*; the system is essentially two independent first-order systems at right angles to each other. In the y-direction, all trajectories decay exponentially to $y = 0$. In the x-direction, the trajectories are attracted to $x = 0$ and repelled from $x = \pm 1$. The vertical lines $x = 0$ and $x = \pm 1$ are *invariant*, because $\dot{x} = 0$ on them; hence any trajectory that starts on these lines stays on them forever. Similarly, $y = 0$ is an invariant horizontal line. As a final observation, we note that the phase portrait must be symmetric in both the x and y axes, since the equations are invariant under the transformations $x \to -x$ and $y \to -y$. Putting all this information together, we arrive at the phase portrait shown in Figure 6.3.1.

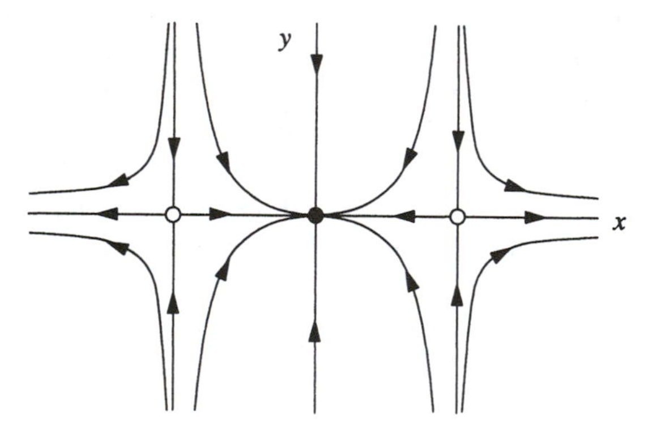

Figure 6.3.1

This picture confirms that $(0,0)$ is a stable node, and $(\pm 1,0)$ are saddles, as expected from the linearization. ■

The next example shows that small nonlinear terms can change a center into a spiral.

EXAMPLE 6.3.2:

Consider the system

$$\dot{x} = -y + ax(x^2 + y^2)$$
$$\dot{y} = x + ay(x^2 + y^2)$$

where a is a parameter. Show that the linearized system *incorrectly* predicts that the origin is a center for all values of a, whereas in fact the origin is a stable spiral if $a < 0$ and an unstable spiral if $a > 0$.

Solution: To obtain the linearization about $(x^*, y^*) = (0,0)$, we can either compute the Jacobian matrix directly from the definition, or we can take the following shortcut. For any system with a fixed point at the origin, x and y represent deviations from the fixed point, since $u = x - x^* = x$ and $v = y - y^* = y$; hence we can linearize by simply omitting nonlinear terms in x and y. Thus the linearized system is $\dot{x} = -y$, $\dot{y} = x$. The Jacobian is

$$A = \begin{pmatrix} 0 & -1 \\ 1 & 0 \end{pmatrix}$$

which has $\tau = 0$, $\Delta = 1 > 0$, so the origin is always a center, according to the linearization.

To analyze the nonlinear system, we change variables to ***polar coordinates***. Let $x = r\cos\theta$, $y = r\sin\theta$. To derive a differential equation for r, we note $x^2 + y^2 = r^2$, so $x\dot{x} + y\dot{y} = r\dot{r}$. Substituting for $\dot{x}$ and $\dot{y}$ yields

$$\begin{aligned} r\dot{r} &= x\left(-y + ax(x^2 + y^2)\right) + y\left(x + ay(x^2 + y^2)\right) \\ &= a(x^2 + y^2)^2 \\ &= ar^4. \end{aligned}$$

Hence $\dot{r} = ar^3$. In Exercise 6.3.12, you are asked to derive the following differential equation for θ:

$$\dot{\theta} = \frac{x\dot{y} - y\dot{x}}{r^2}.$$

After substituting for $\dot{x}$ and $\dot{y}$ we find $\dot{\theta} = 1$. Thus in polar coordinates the original system becomes

$$\dot{r} = ar^3$$
$$\dot{\theta} = 1.$$

The system is easy to analyze in this form, because the radial and angular motions are independent. All trajectories rotate about the origin with constant angular velocity $\dot{\theta} = 1$.

The radial motion depends on a, as shown in Figure 6.3.2.

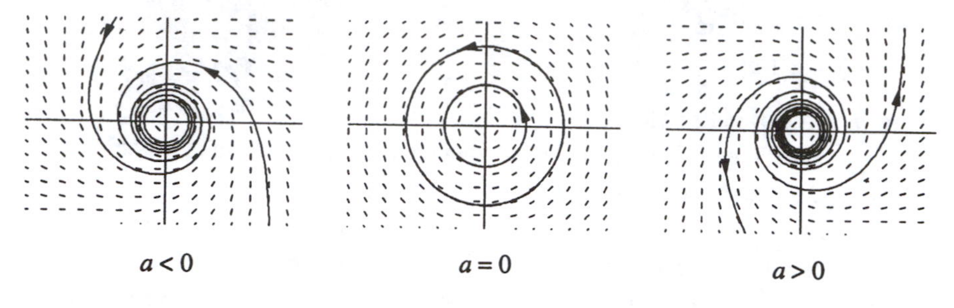

Figure 6.3.2

If $a<0$, then $r(t) \to 0$ monotonically as $t \to \infty$. In this case, the origin is a stable spiral. (However, note that the decay is extremely slow, as suggested by the computer-generated trajectories shown in Figure 6.3.2.) If $a=0$, then $r(t) = r_0$ for all t and the origin is a center. Finally, if $a>0$, then $r(t) \to \infty$ monotonically and the origin is an unstable spiral.

We can see now why centers are so delicate: all trajectories are required to close *perfectly* after one cycle. The slightest miss converts the center into a spiral. ■

Similarly, stars and degenerate nodes can be altered by small nonlinearities, but unlike centers, *their stability doesn't change.* For example, a stable star may be changed into a stable spiral (Exercise 6.3.11) but not into an unstable spiral. This is plausible, given the classification of linear systems in Figure 5.2.8: stars and degenerate nodes live squarely in the stable or unstable region, whereas centers live on the razor's edge between stability and instability.

If we're only interested in *stability*, and not in the detailed geometry of the trajectories, then we can classify fixed points more coarsely as follows:

Robust cases:

Repellers (also called *sources*): both eigenvalues have positive real part.

Attractors (also called *sinks*): both eigenvalues have negative real part.

Saddles: one eigenvalue is positive and one is negative.

Marginal cases:

Centers: both eigenvalues are pure imaginary.

Higher-order and non-isolated fixed points: at least one eigenvalue is zero.

Thus, from the point of view of stability, the marginal cases are those where at least one eigenvalue satisfies $\mathrm{Re}(\lambda) = 0$.

Hyperbolic Fixed Points, Topological Equivalence, and Structural Stability

If $\text{Re}(\lambda) \neq 0$ for both eigenvalues, the fixed point is often called ***hyperbolic***. (This is an unfortunate name—it sounds like it should mean "saddle point"—but it has become standard.) Hyperbolic fixed points are sturdy; their stability type is unaffected by small nonlinear terms. Nonhyperbolic fixed points are the fragile ones.

We've already seen a simple instance of hyperbolicity in the context of vector fields on the line. In Section 2.4 we saw that the stability of a fixed point was accurately predicted by the linearization, *as long as* $f'(x^*) \neq 0$. This condition is the exact analog of $\text{Re}(\lambda) \neq 0$.

These ideas also generalize neatly to higher-order systems. A fixed point of an nth-order system is *hyperbolic* if all the eigenvalues of the linearization lie off the imaginary axis, i.e., $\text{Re}(\lambda_i) \neq 0$ for $i = 1, \ldots, n$. The important *Hartman–Grobman theorem* states that the local phase portrait near a hyperbolic fixed point is "topologically equivalent" to the phase portrait of the linearization; in particular, the stability type of the fixed point is faithfully captured by the linearization. Here ***topologically equivalent*** means that there is a ***homeomorphism*** (a continuous deformation with a continuous inverse) that maps one local phase portrait onto the other, such that trajectories map onto trajectories and the sense of time (the direction of the arrows) is preserved.

Intuitively, two phase portraits are topologically equivalent if one is a distorted version of the other. Bending and warping are allowed, but not ripping, so closed orbits must remain closed, trajectories connecting saddle points must not be broken, etc.

Hyperbolic fixed points also illustrate the important general notion of structural stability. A phase portrait is ***structurally stable*** if its topology cannot be changed by an arbitrarily small perturbation to the vector field. For instance, the phase portrait of a saddle point is structurally stable, but that of a center is not: an arbitrarily small amount of damping converts the center to a spiral.

6.4 Rabbits versus Sheep

In the next few sections we'll consider some simple examples of phase plane analysis. We begin with the classic ***Lotka–Volterra model of competition*** between two species, here imagined to be rabbits and sheep. Suppose that both species are competing for the same food supply (grass) and the amount available is limited. Furthermore, ignore all other complications, like predators, seasonal effects, and other sources of food. Then there are two main effects we should consider:

1. Each species would grow to its carrying capacity in the absence of the other. This can be modeled by assuming logistic growth for each species (recall Section 2.3). Rabbits have a legendary ability to reproduce, so perhaps we should assign them a higher intrinsic growth rate.

2. When rabbits and sheep encounter each other, trouble starts. Sometimes the rabbit gets to eat, but more usually the sheep nudges the rabbit aside and starts nibbling (on the grass, that is). We'll assume that these conflicts occur at a rate proportional to the size of each population. (If there were twice as many sheep, the odds of a rabbit encountering a sheep would be twice as great.) Furthermore, we assume that the conflicts reduce the growth rate for each species, but the effect is more severe for the rabbits.

A specific model that incorporates these assumptions is

$$\dot{x} = x(3 - x - 2y)$$
$$\dot{y} = y(2 - x - y)$$

where

$$x(t) = \text{population of rabbits,}$$
$$y(t) = \text{population of sheep}$$

and $x, y \geq 0$. The coefficients have been chosen to reflect this scenario, but are otherwise arbitrary. In the exercises, you'll be asked to study what happens if the coefficients are changed.

To find the fixed points for the system, we solve $\dot{x} = 0$ and $\dot{y} = 0$ simultaneously. Four fixed points are obtained: (0,0), (0,2), (3,0), and (1,1). To classify them, we compute the Jacobian:

$$A = \begin{pmatrix} \frac{\partial \dot{x}}{\partial x} & \frac{\partial \dot{x}}{\partial y} \\ \frac{\partial \dot{y}}{\partial x} & \frac{\partial \dot{y}}{\partial y} \end{pmatrix} = \begin{pmatrix} 3 - 2x - 2y & -2x \\ -y & 2 - x - 2y \end{pmatrix}.$$

Now consider the four fixed points in turn:

$$(0,0): \text{ Then } A = \begin{pmatrix} 3 & 0 \\ 0 & 2 \end{pmatrix}.$$

The eigenvalues are $\lambda = 3, 2$ so (0,0) is an *unstable node.* Trajectories leave the origin parallel to the eigenvector for $\lambda = 2$, i.e. tangential to $\mathbf{v} = (0,1)$, which spans the y-axis. (Recall the general rule: at a node, trajectories are tangential to the slow eigendirection, which is the eigendirection with the smallest $|\lambda|$.) Thus, the phase portrait near (0,0) looks like Figure 6.4.1.

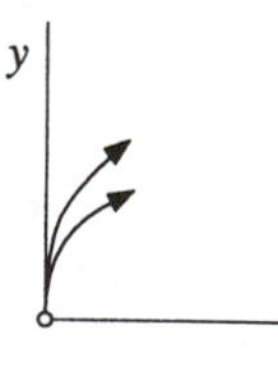

Figure 6.4.1

$$(0,2): \text{ Then } A = \begin{pmatrix} -1 & 0 \\ -2 & -2 \end{pmatrix}.$$

This matrix has eigenvalues $\lambda = -1, -2$, as can be seen from inspection, since the matrix is triangular. Hence the fixed point is a *stable node.* Trajectories approach along the eigendirection associated with $\lambda = -1$; you can check that this direction is spanned by $\mathbf{v} = (1,-2)$. Figure 6.4.2 shows the phase portrait near the fixed point $(0,2)$.

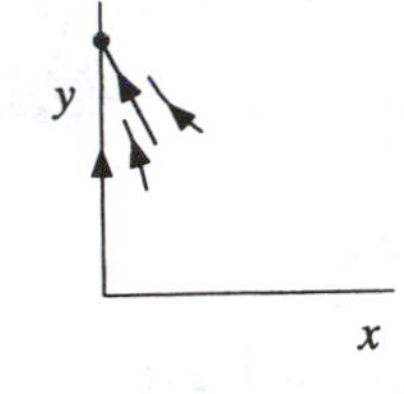

Figure 6.4.2

$(3,0)$: Then $A = \begin{pmatrix} -3 & -6 \\ 0 & -1 \end{pmatrix}$ and $\lambda = -3, -1$.

This is also a *stable node.* The trajectories approach along the slow eigendirection spanned by $\mathbf{v} = (3,-1)$, as shown in Figure 6.4.3.

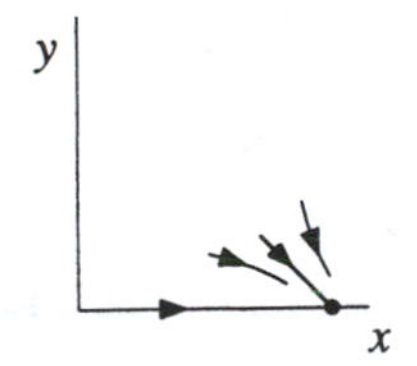

Figure 6.4.3

$(1,1)$: Then $A = \begin{pmatrix} -1 & -2 \\ -1 & -1 \end{pmatrix}$, which has $\tau = -2$, $\Delta = -1$, and $\lambda = -1 \pm \sqrt{2}$. Hence this is a *saddle point.* As you can check, the phase portrait near $(1,1)$ is as shown in Figure 6.4.4.

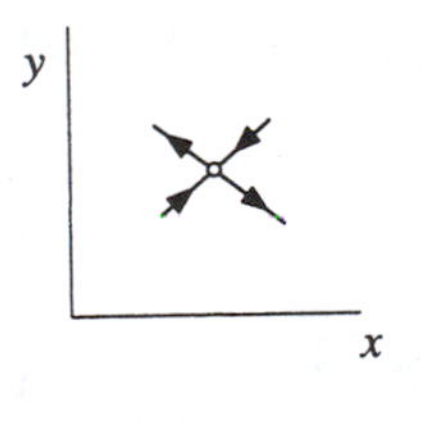

Figure 6.4.4

Combining Figures 6.4.1–6.4.4, we get Figure 6.4.5, which already conveys a good sense of the entire phase portrait. Furthermore, notice that the x and y axes contain straight-line trajectories, since $\dot{x} = 0$ when $x = 0$, and $\dot{y} = 0$ when $y = 0$.

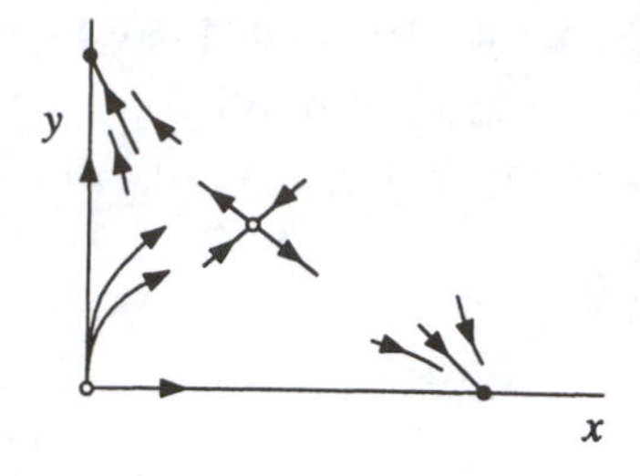

Figure 6.4.5

Now we use common sense to fill in the rest of the phase portrait (Figure 6.4.6). For example, some of the trajectories starting near the origin must go to the stable node on the x-axis, while others must go to the stable node on the y-axis. In between, there must be a special trajectory that can't decide which way to turn, and so it dives into the saddle point. This trajectory is part of the ***stable manifold*** of the saddle, drawn with a heavy line in Figure 6.4.6.

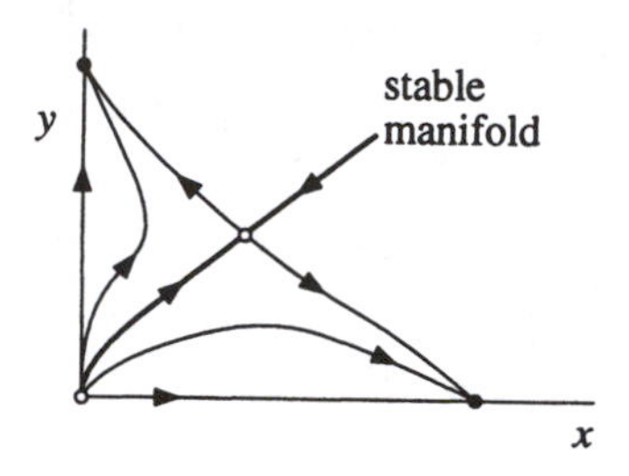

Figure 6.4.6

The other branch of the stable manifold consists of a trajectory coming in "from infinity." A computer-generated phase portrait (Figure 6.4.7) confirms our sketch.

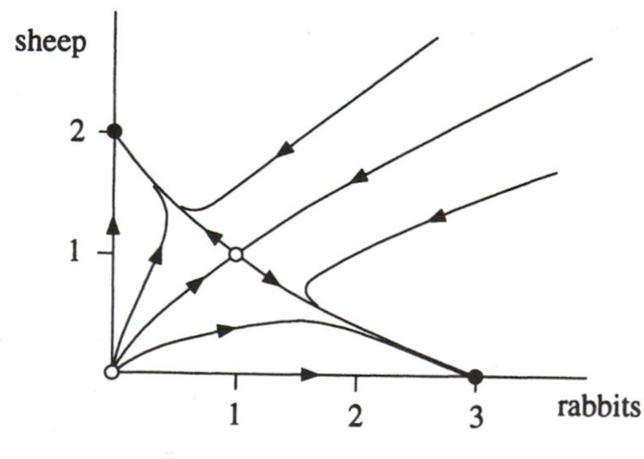

Figure 6.4.7

The phase portrait has an interesting biological interpretation. It shows that one species generally drives the other to extinction. Trajectories starting below the stable manifold lead to eventual extinction of the sheep, while those starting above lead to eventual extinction of the rabbits. This dichotomy occurs in other models of competition and has led biologists to formulate the *principle of competitive exclusion*, which states that two species competing for the same limited resource typically cannot coexist. See Pianka (1981) for a biological discussion, and

Pielou (1969), Edelstein–Keshet (1988), or Murray (1989) for additional references and analysis.

Our example also illustrates some general mathematical concepts. Given an attracting fixed point $\mathbf{x}^*$, we define its ***basin of attraction*** to be the set of initial conditions $\mathbf{x}_0$ such that $\mathbf{x}(t) \to \mathbf{x}^*$ as $t \to \infty$. For instance, the basin of attraction for the node at (3,0) consists of all the points lying below the stable manifold of the saddle. This basin is shown as the shaded region in Figure 6.4.8.

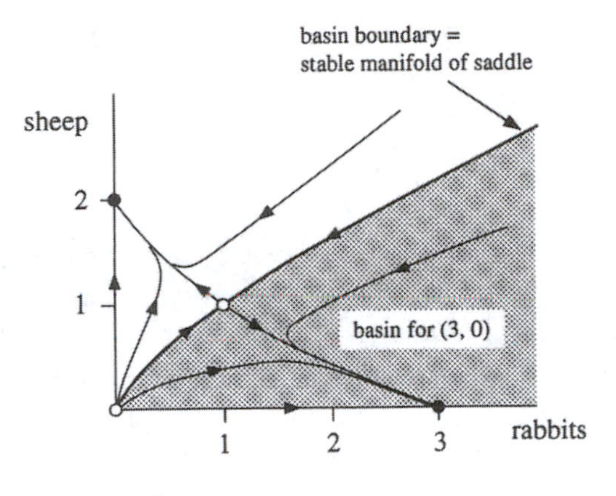

Figure 6.4.8

Because the stable manifold separates the basins for the two nodes, it is called the ***basin boundary***. For the same reason, the two trajectories that comprise the stable manifold are traditionally called ***separatrices***. Basins and their boundaries are important because they partition the phase space into regions of different long-term behavior.

6.5 Conservative Systems

Newton's law $F = ma$ is the source of many important second-order systems. For example, consider a particle of mass m moving along the x-axis, subject to a nonlinear force $F(x)$. Then the equation of motion is

$$m\ddot{x} = F(x).$$

Notice that we are assuming that F is independent of both $\dot{x}$ and t; hence there is no damping or friction of any kind, and there is no time-dependent driving force.

Under these assumptions, we can show that *energy is conserved*, as follows. Let $V(x)$ denote the ***potential energy***, defined by $F(x) = -dV/dx$. Then

$$m\ddot{x} + \frac{dV}{dx} = 0. \tag{1}$$

Now comes a trick worth remembering: multiply both sides by $\dot{x}$ and notice that the left-hand side becomes an exact time-derivative!

$$m\dot{x}\ddot{x} + \frac{dV}{dx}\dot{x} = 0 \Rightarrow \frac{d}{dt}\left[\tfrac{1}{2}m\dot{x}^2 + V(x)\right] = 0$$

where we've used the chain rule

$$\frac{d}{dt}V(x(t)) = \frac{dV}{dx}\frac{dx}{dt}$$

in reverse. Hence, for a given solution $x(t)$, the total ***energy***

$$E = \tfrac{1}{2}m\dot{x}^2 + V(x)$$

is constant as a function of time. The energy is often called a conserved quantity, a constant of motion, or a first integral. Systems for which a conserved quantity exists are called ***conservative systems***.

Let's be a bit more general and precise. Given a system $\dot{\mathbf{x}} = \mathbf{f}(\mathbf{x})$, a ***conserved quantity*** is a real-valued continuous function $E(\mathbf{x})$ that is constant on trajectories, i.e. $dE/dt = 0$. To avoid trivial examples, we also require that $E(\mathbf{x})$ be nonconstant on every open set. Otherwise a constant function like $E(\mathbf{x}) \equiv 0$ would qualify as a conserved quantity for every system, and so *every* system would be conservative! Our caveat rules out this silliness.

The first example points out a basic fact about conservative systems.

EXAMPLE 6.5.1:

Show that *a conservative system cannot have any attracting fixed points.*

Solution: Suppose $\mathbf{x}^*$ were an attracting fixed point. Then all points in its basin of attraction would have to be at the same energy $E(\mathbf{x}^*)$ (because energy is constant on trajectories and all trajectories in the basin flow to $\mathbf{x}^*$). Hence $E(\mathbf{x})$ must be a *constant function* for $\mathbf{x}$ in the basin. But this contradicts our definition of a conservative system, in which we required that $E(\mathbf{x})$ be nonconstant on all open sets. ■

If attracting fixed points can't occur, then what kind of fixed points *can* occur? One generally finds saddles and centers, as in the next example.

EXAMPLE 6.5.2:

Consider a particle of mass $m = 1$ moving in a double-well potential $V(x) = -\tfrac{1}{2}x^2 + \tfrac{1}{4}x^4$. Find and classify all the equilibrium points for the system. Then plot the phase portrait and interpret the results physically.

Solution: The force is $-dV/dx = x - x^3$, so the equation of motion is

$$\ddot{x} = x - x^3.$$

This can be rewritten as the vector field

$$\dot{x} = y$$
$$\dot{y} = x - x^3$$

where y represents the particle's velocity. Equilibrium points occur where $(\dot{x}, \dot{y}) = (0,0)$. Hence the equilibria are $(x^*, y^*) = (0,0)$ and $(\pm 1, 0)$. To classify these fixed points we compute the Jacobian:

$$A = \begin{pmatrix} 0 & 1 \\ 1 - 3x^2 & 0 \end{pmatrix}.$$

At (0,0), we have $\Delta = -1$, so the origin is a saddle point. But when $(x^*, y^*) = (\pm 1, 0)$, we find $\tau = 0$, $\Delta = 2$; hence these equilibria are predicted to be centers.

At this point you should be hearing warning bells—in Section 6.3 we saw that small nonlinear terms can easily destroy a center predicted by the linear approximation. But that's not the case here, because of energy conservation. The trajectories are closed curves defined by the ***contours*** of constant energy, i.e.,

$$E = \tfrac{1}{2}y^2 - \tfrac{1}{2}x^2 + \tfrac{1}{4}x^4 = \text{constant}.$$

Figure 6.5.1 shows the trajectories corresponding to different values of E. To decide which way the arrows point along the trajectories, we simply compute the vector $(\dot{x}, \dot{y})$ at a few convenient locations. For example, $\dot{x} > 0$ and $\dot{y} = 0$ on the positive y-axis, so the motion is to the right. The orientation of neighboring trajectories follows by continuity.

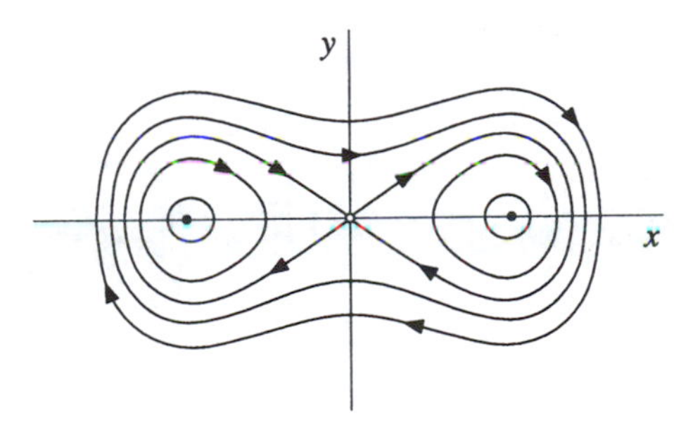

Figure 6.5.1

As expected, the system has a saddle point at $(0,0)$ and centers at $(1,0)$ and $(-1,0)$. Each of the neutrally stable centers is surrounded by a family of small closed orbits. There are also large closed orbits that encircle all three fixed points.

Thus solutions of the system are typically *periodic*, except for the equilibrium solutions and two very special trajectories: these are the trajectories that appear to start and end at the origin. More precisely, these trajectories approach the origin as $t \to \pm\infty$. Trajectories that start and end at the same fixed point are called ***homoclinic orbits***. They are common in conservative systems, but are rare otherwise. Notice that a homoclinic orbit does *not* correspond to a periodic solution, because the trajectory takes forever trying to reach the fixed point.

Finally, let's connect the phase portrait to the motion of an undamped particle in a double-well potential (Figure 6.5.2).

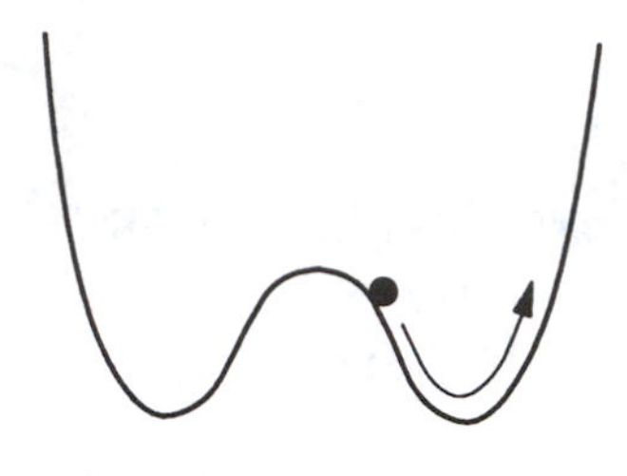

Figure 6.5.2

The neutrally stable equilibria correspond to the particle at rest at the bottom of one of the wells, and the small closed orbits represent small oscillations about these equilibria. The large orbits represent more energetic oscillations that repeatedly take the particle back and forth over the hump. Do you see what the saddle point and the homoclinic orbits mean physically? ■

EXAMPLE 6.5.3:

Sketch the graph of the energy function $E(x, y)$ for Example 6.5.2.

Solution: The graph of $E(x, y)$ is shown in Figure 6.5.3. The energy E is plotted above each point (x, y) of the phase plane. The resulting surface is often called the ***energy surface*** for the system.

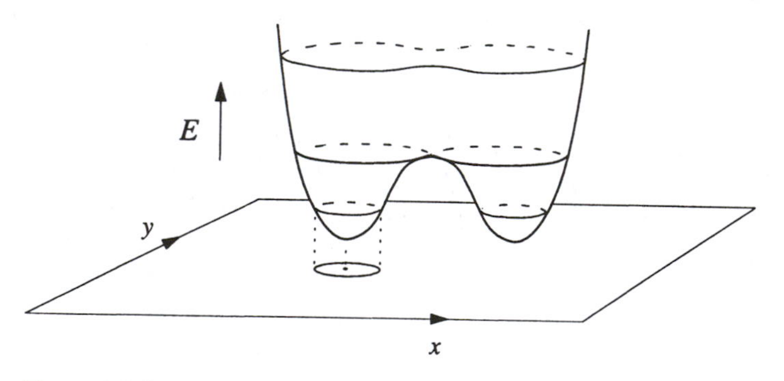

Figure 6.5.3

Figure 6.5.3 shows that the local minima of E project down to centers in the phase plane. Contours of slightly higher energy correspond to the small orbits surrounding the centers. The saddle point and its homoclinic orbits lie at even higher energy, and the large orbits that encircle all three fixed points are the most energetic of all.

It's sometimes helpful to think of the flow as occurring on the energy surface itself, rather than in the phase plane. But notice—the trajectories must maintain a constant height E, so they would run *around* the surface, not down it. ∎

Nonlinear Centers

Centers are ordinarily very delicate but, as the examples above suggest, they are much more robust when the system is conservative. We now present a theorem about nonlinear centers in second-order conservative systems.

The theorem says that centers occur at the local minima of the energy function. This is physically plausible—one expects neutrally stable equilibria and small oscillations to occur at the bottom of *any* potential well, no matter what its shape.

Theorem 6.5.1: (Nonlinear centers for conservative systems) Consider the system $\dot{\mathbf{x}} = \mathbf{f}(\mathbf{x})$, where $\mathbf{x} = (x, y) \in \mathbf{R}^2$, and $\mathbf{f}$ is continuously differentiable. Suppose there exists a conserved quantity $E(\mathbf{x})$ and suppose that $\mathbf{x}^*$ is an isolated fixed point (i.e., there are no other fixed points in a small neighborhood surrounding $\mathbf{x}^*$). If $\mathbf{x}^*$ is a local minimum of E, then all trajectories sufficiently close to $\mathbf{x}^*$ are closed.

Ideas behind the proof: Since E is constant on trajectories, each trajectory is contained in some contour of E. Near a local maximum or minimum, the contours are *closed*. (We won't prove this, but Figure 6.5.3 should make it seem obvious.) The only remaining question is whether the trajectory actually goes all the way around the contour or whether it stops at a fixed point on the contour. But because we're assuming that $\mathbf{x}^*$ is an *isolated* fixed point, there cannot be any fixed points on contours sufficiently close to $\mathbf{x}^*$. Hence all trajectories in a sufficiently small neighborhood of $\mathbf{x}^*$ are closed orbits, and therefore $\mathbf{x}^*$ is a center. ∎

Two remarks about this result:

1. The theorem is valid for local *maxima* of E also. Just replace the function E by $-E$, and maxima get converted to minima; then Theorem 6.5.1 applies.
2. We need to assume that $\mathbf{x}^*$ is isolated. Otherwise there are counterexamples due to fixed points on the energy contour—see Exercise 6.5.12.

Another theorem about nonlinear centers will be presented in the next section.

6.6 Reversible Systems

Many mechanical systems have ***time-reversal symmetry***. This means that their dynamics look the same whether time runs forward or backward. For example, if you were watching a movie of an undamped pendulum swinging back and forth, you wouldn't see any physical absurdities if the movie were run backward.

In fact, any mechanical system of the form $m\ddot{x} = F(x)$ is symmetric under time reversal. If we make the change of variables $t \to -t$, the second derivative $\ddot{x}$ stays the same and so the equation is unchanged. Of course, the velocity $\dot{x}$ would be reversed. Let's see what this means in the phase plane. The equivalent system is

$$\begin{aligned} \dot{x} &= y \\ \dot{y} &= \tfrac{1}{m} F(x) \end{aligned}$$

where y is the velocity. If we make the change of variables $t \to -t$ and $y \to -y$, both equations stay the same. Hence if $(x(t),\ y(t))$ is a solution, then so is $(x(-t),\ -y(-t))$. Therefore every trajectory has a twin: they differ only by time-reversal and a reflection in the x-axis (Figure 6.6.1).

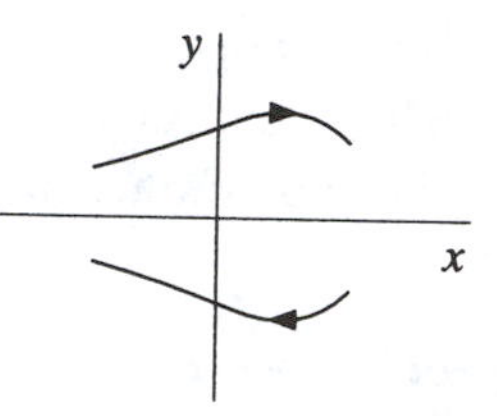

Figure 6.6.1

The trajectory above the x-axis looks just like the one below the x-axis, except the arrows are reversed.

More generally, let's define a ***reversible system*** to be *any* second-order system that is invariant under $t \to -t$ and $y \to -y$. For example, any system of the form

$$\begin{aligned} \dot{x} &= f(x, y) \\ \dot{y} &= g(x, y), \end{aligned}$$

where f is *odd* in y and g is *even* in y (i.e., $f(x,-y) = -f(x,y)$ and $g(x,-y) = g(x,y)$) is reversible.

Reversible systems are different from conservative systems, but they have many of the same properties. For instance, the next theorem shows that centers are robust in reversible systems as well.

Theorem 6.6.1: (Nonlinear centers for reversible systems) Suppose the origin $\mathbf{x}^* = \mathbf{0}$ is a linear center for the continuously differentiable system

$$\begin{aligned} \dot{x} &= f(x, y) \\ \dot{y} &= g(x, y), \end{aligned}$$

and suppose that the system is reversible. Then sufficiently close to the origin, all trajectories are closed curves.

Ideas behind the proof: Consider a trajectory that starts on the positive x-axis near the origin (Figure 6.6.2). Sufficiently near the origin, the flow swirls around the origin, thanks to the dominant influence of the linear center, and so the trajectory eventually intersects the *negative* x-axis. (This is the step where our proof lacks rigor, but the claim should seem plausible.)

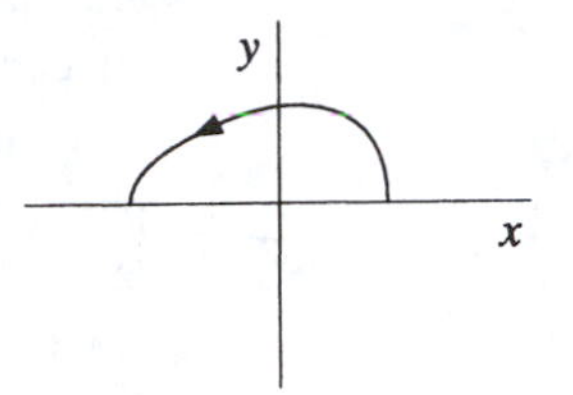

Figure 6.6.2

Now we use reversibility. By reflecting the trajectory across the x-axis, and changing the sign of t, we obtain a twin trajectory with the same endpoints but with its arrow reversed (Figure 6.6.3).

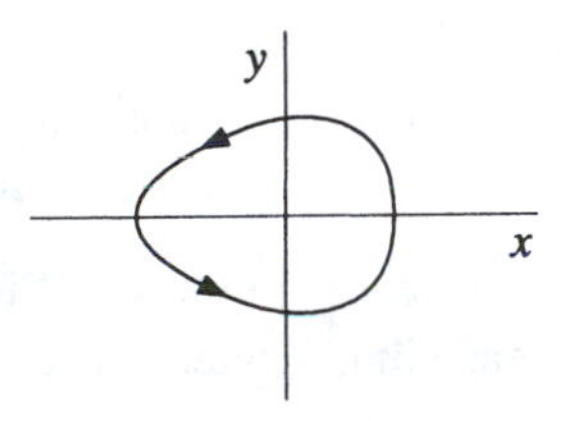

Figure 6.6.3

Together the two trajectories form a closed orbit, as desired. Hence all trajectories sufficiently close to the origin are closed. ■

EXAMPLE 6.6.1:

Show that the system

$$\dot{x} = y - y^3$$
$$\dot{y} = -x - y^2$$

has a nonlinear center at the origin, and plot the phase portrait.

Solution: We'll show that the hypotheses of the theorem are satisfied. The Jacobian at the origin is

$$A = \begin{pmatrix} 0 & 1 \\ -1 & 0 \end{pmatrix}.$$

This has $\tau = 0$, $\Delta > 0$, so the origin is a linear center. Furthermore, the system is reversible, since the equations are invariant under the transformation $t \to -t$, $y \to -y$. By Theorem 6.6.1, the origin is a *nonlinear* center.

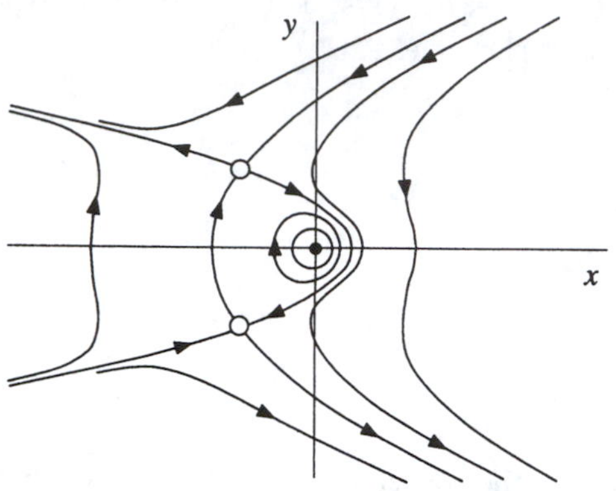

Figure 6.6.4

The other fixed points of the system are $(-1,1)$ and $(-1,-1)$. They are saddle points, as is easily checked by computing the linearization. A computer-generated phase portrait is shown in Figure 6.6.4. It looks like some exotic sea creature, perhaps a manta ray. The reversibility symmetry is apparent. The trajectories above the x-axis have twins below the x-axis, with arrows reversed.

Notice that the twin saddle points are joined by a pair of trajectories. They are called ***heteroclinic trajectories*** or ***saddle connections***. Like homoclinic orbits, heteroclinic trajectories are much more common in reversible or conservative systems than in other types of systems. ■

Although we have relied on the computer to plot Figure 6.6.4, it can be sketched on the basis of qualitative reasoning alone. For example, the existence of the heteroclinic trajectories can be deduced rigorously using reversibility arguments (Exercise 6.6.6). The next example illustrates the spirit of such arguments.

EXAMPLE 6.6.2:

Using reversibility arguments alone, show that the system

$$\dot{x} = y$$
$$\dot{y} = x - x^2$$

has a homoclinic orbit in the half-plane $x \geq 0$.

Solution: Consider the unstable manifold of the saddle point at the origin. This manifold leaves the origin along the vector $(1,1)$, since this is the unstable eigendirection for the linearization. Hence, close to the origin, part of the unstable manifold lies in the first quadrant $x, y > 0$. Now imagine a phase point with coordinates $(x(t), y(t))$ moving along the unstable manifold, starting from x, y small and positive. At first, $x(t)$ must increase since $\dot{x} = y > 0$. Also, $y(t)$ increases initially, since $\dot{y} = x - x^2 > 0$ for small x. Thus the phase point moves up and to the right. Its horizontal velocity is continually increasing, so at some time it must cross the vertical line $x=1$. Then $\dot{y}<0$ so $y(t)$ decreases, eventually reaching $y=0$. Figure 6.6.5 shows the situation.

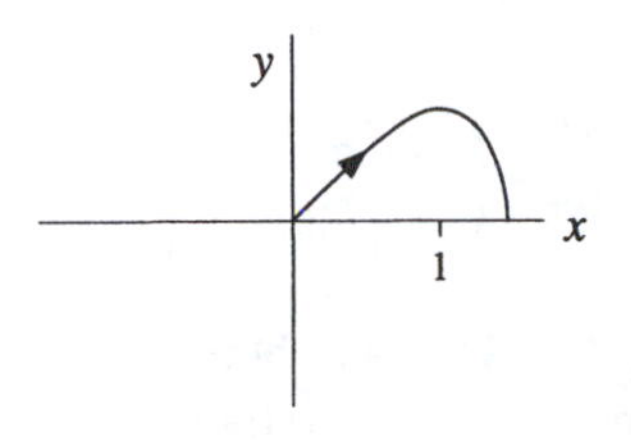

Figure 6.6.5

Now, *by reversibility*, there must be a twin trajectory with the same endpoints but with arrow reversed (Figure 6.6.6). Together the two trajectories form the desired homoclinic orbit. ■

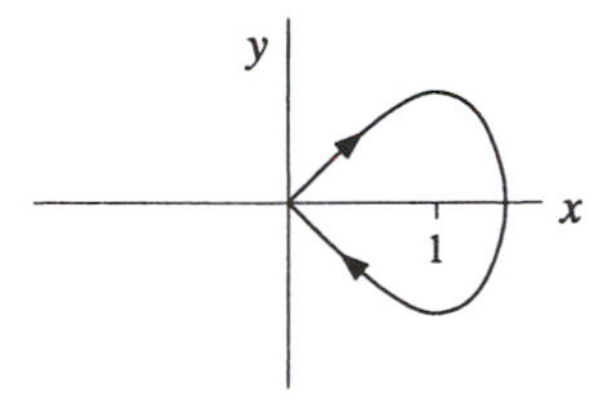

Figure 6.6.6

There is a more general definition of reversibility which extends nicely to higher-order systems. Consider any mapping $R(\mathbf{x})$ of the phase space to itself that satisfies $R^2(\mathbf{x})=\mathbf{x}$. In other words, if the mapping is applied *twice*, all points go back to where they started. In our two-dimensional examples, a reflection about the x-axis (or any axis through the origin) has this property. Then the system $\dot{\mathbf{x}}=\mathbf{f}(\mathbf{x})$ is ***reversible*** if it is invariant under the change of variables $t\to -t$, $\mathbf{x}\to R(\mathbf{x})$.

Our next example illustrates this more general notion of reversibility, and also highlights the main difference between reversible and conservative systems.

EXAMPLE 6.6.3:

Show that the system

$$\dot{x} = -2\cos x - \cos y$$
$$\dot{y} = -2\cos y - \cos x$$

is reversible, but *not* conservative. Then plot the phase portrait.

Solution: The system is invariant under the change of variables $t\to -t$, $x\to -x$, and $y\to -y$. Hence the system is reversible, with $R(x,y)=(-x,-y)$ in the preceding notation.

To show that the system is not conservative, it suffices to show that it has an attracting fixed point. (Recall that a conservative system can never have an attracting fixed point—see Example 6.5.1.)

The fixed points satisfy $2\cos x=-\cos y$ and $2\cos y=-\cos x$. Solving these equations simultaneously yields $\cos x^* = \cos y^* = 0$. Hence there are four fixed points, given by $(x^*, y^*) = (\pm\frac{\pi}{2}, \pm\frac{\pi}{2})$.

We claim that $(x^*, y^*) = (-\frac{\pi}{2}, -\frac{\pi}{2})$ is an attracting fixed point. The Jacobian there is

$$A = \begin{pmatrix} 2\sin x^* & \sin y^* \\ \sin x^* & 2\sin y^* \end{pmatrix} = \begin{pmatrix} -2 & -1 \\ -1 & -2 \end{pmatrix},$$

which has $\tau = -4$, $\Delta = 3$, $\tau^2 - 4\Delta = 4$. Therefore the fixed point is a stable node. This shows that the system is not conservative.

The other three fixed points can be shown to be an unstable node and two saddles. A computer-generated phase portrait is shown in Figure 6.6.7.

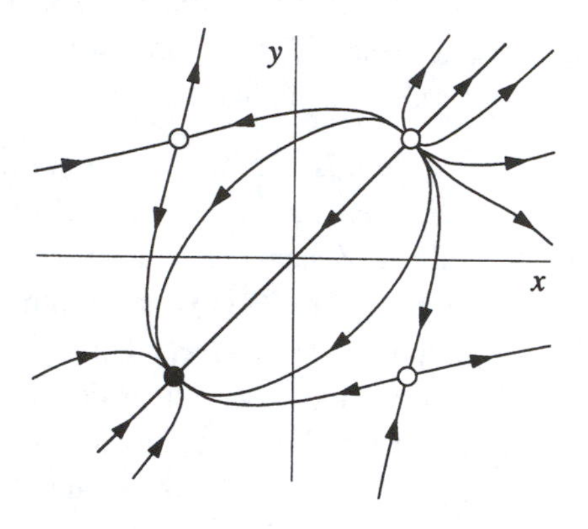

Figure 6.6.7

To see the reversibility symmetry, compare the dynamics at any two points (x, y) and $R(x, y) = (-x, -y)$. The trajectories look the same, but the arrows are reversed. In particular, the stable node at $(-\frac{\pi}{2}, -\frac{\pi}{2})$ is the twin of the unstable node at $(\frac{\pi}{2}, \frac{\pi}{2})$. ■

The system in Example 6.6.3 is closely related to a model of two superconducting Josephson junctions coupled through a resistive load (Tsang et al. 1991). For further discussion, see Exercise 6.6.9 and Example 8.7.4. Reversible, nonconservative systems also arise in the context of lasers (Politi et al. 1986) and fluid flows (Stone, Nadim, and Strogatz 1991 and Exercise 6.6.8).

6.7 Pendulum

Do you remember the first nonlinear system you ever studied in school? It was probably the pendulum. But in elementary courses, the pendulum's essential nonlinearity is sidestepped by the small-angle approximation $\sin\theta \approx \theta$. Enough of that! In this section we use phase plane methods to analyze the pendulum, even in the dreaded large-angle regime where the pendulum whirls over the top.

In the absence of damping and external driving, the motion of a pendulum is governed by

$$\frac{d^2\theta}{dt^2}+\frac{g}{L}\sin\theta=0 \tag{1}$$

where θ is the angle from the downward vertical, g is the acceleration due to gravity, and L is the length of the pendulum (Figure 6.7.1).

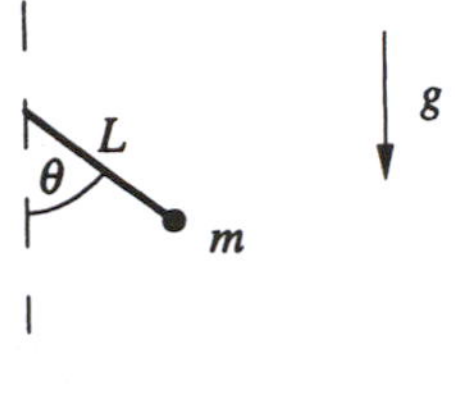

Figure 6.7.1

We nondimensionalize (1) by introducing a frequency $\omega=\sqrt{g/L}$ and a dimensionless time $\tau=\omega t$. Then the equation becomes

$$\ddot{\theta}+\sin\theta=0 \tag{2}$$

where the overdot denotes differentiation with respect to τ. The corresponding system in the phase plane is

$$\dot{\theta}=v \tag{3a}$$
$$\dot{v}=-\sin\theta \tag{3b}$$

where v is the (dimensionless) angular velocity.

The fixed points are $(\theta^*, v^*)=(k\pi,\ 0)$, where k is any integer. There's no physical difference between angles that differ by 2π, so we'll concentrate on the two fixed points $(0,0)$ and $(\pi,0)$. At $(0,0)$, the Jacobian is

$$A=\begin{pmatrix}0 & 1\\ -1 & 0\end{pmatrix}$$

so the origin is a linear center.

In fact, the origin is a *nonlinear* center, for two reasons. First, the system (3) is *reversible*: the equations are invariant under the transformation $\tau\to-\tau$, $v\to-v$. Then Theorem 6.6.1 implies that the origin is a nonlinear center.

Second, the system is also *conservative*. Multiplying (2) by $\dot{\theta}$ and integrating yields

$$\dot{\theta}(\ddot{\theta}+\sin\theta)=0 \ \Rightarrow\ \tfrac{1}{2}\dot{\theta}^2-\cos\theta=\text{constant}.$$

The energy function

$$E(\theta, v) = \tfrac{1}{2}v^2 - \cos\theta \tag{4}$$

has a local minimum at (0,0), since $E \approx \frac{1}{2}(v^2 + \theta^2) - 1$ for small (θ, v). Hence Theorem 6.5.1 provides a second proof that the origin is a nonlinear center. (This argument also shows that the closed orbits are approximately *circular*, with $\theta^2 + v^2 \approx 2(E+1)$.)

Now that we've beaten the origin to death, consider the fixed point at $(\pi, 0)$. The Jacobian is

$$A = \begin{pmatrix} 0 & 1 \\ 1 & 0 \end{pmatrix}.$$

The characteristic equation is $\lambda^2 - 1 = 0$. Therefore $\lambda_1 = -1$, $\lambda_2 = 1$; the fixed point is a saddle. The corresponding eigenvectors are $\mathbf{v}_1 = (1, -1)$ and $\mathbf{v}_2 = (1, 1)$.

The phase portrait near the fixed points can be sketched from the information obtained so far (Figure 6.7.2).

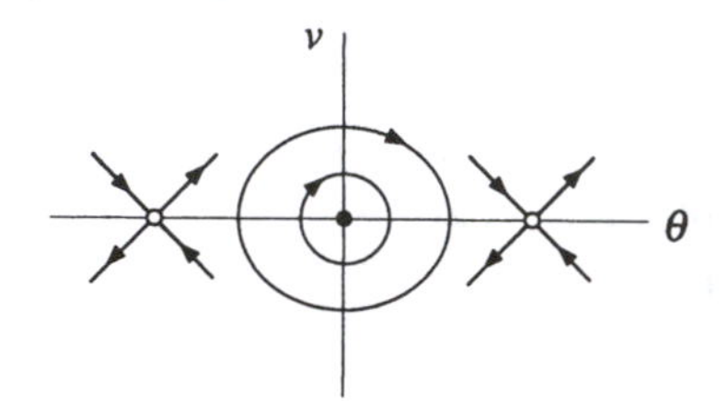

Figure 6.7.2

To fill in the picture, we include the energy contours $E = \frac{1}{2}v^2 - \cos\theta$ for different values of E. The resulting phase portrait is shown in Figure 6.7.3. The picture is periodic in the θ-direction, as we'd expect.

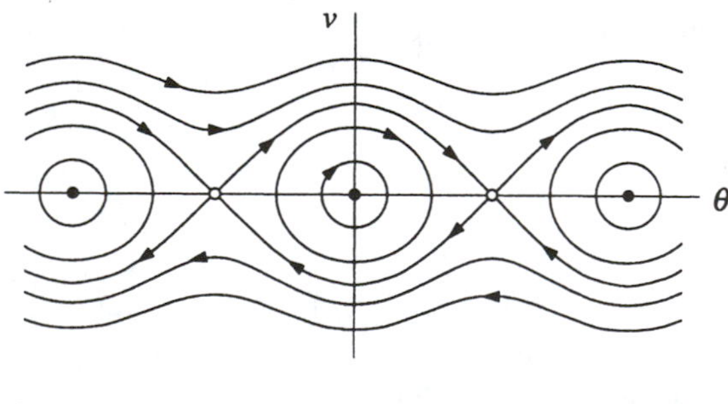

Figure 6.7.3

Now for the physical interpretation. The center corresponds to a state of neutrally stable equilibrium, with the pendulum at rest and hanging straight down. This is the lowest possible energy state ($E = -1$). The small orbits surrounding the center represent small oscillations about equilibrium, traditionally called ***librations***. As E increases, the orbits grow. The critical case is $E = 1$, corresponding to the heteroclinic trajectories joining the saddles in Figure 6.7.3. The saddles represent an *inverted* pendulum at rest;

hence the heteroclinic trajectories represent delicate motions in which the pendulum slows to a halt precisely as it approaches the inverted position. For $E > 1$, the pendulum whirls repeatedly over the top. These ***rotations*** should also be regarded as periodic solutions, since $\theta = -\pi$ and $\theta = +\pi$ are the same physical position.

Cylindrical Phase Space

The phase portrait for the pendulum is more illuminating when wrapped onto the surface of a cylinder (Figure 6.7.4). In fact, a cylinder is the *natural* phase space for the pendulum, because it incorporates the fundamental geometric difference between v and θ: the angular velocity v is a real number, whereas θ is an *angle*.

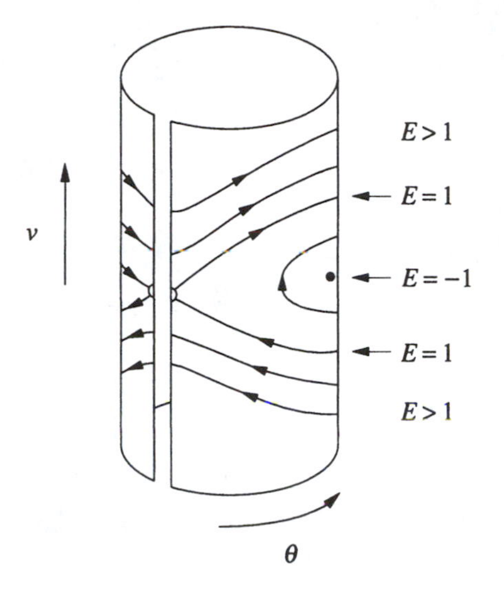

Figure 6.7.4

There are several advantages to the cylindrical representation. Now the periodic whirling motions *look* periodic—they are the closed orbits that encircle the cylinder for $E > 1$. Also, it becomes obvious that the saddle points in Figure 6.7.3 are all the same physical state (an inverted pendulum at rest). The heteroclinic trajectories of Figure 6.7.3 become homoclinic orbits on the cylinder.

There is an obvious symmetry between the top and bottom half of Figure 6.7.4. For example, both homoclinic orbits have the same energy and shape. To highlight this symmetry, it is interesting (if a bit mind-boggling at first) to plot the *energy* vertically instead of the angular velocity v (Figure 6.7.5). Then the orbits on the cylinder remain at constant height, while the cylinder gets bent into a ***U-tube***. The two arms of the tube are distinguished by the sense of rotation of the pendulum, either clockwise or counterclock-

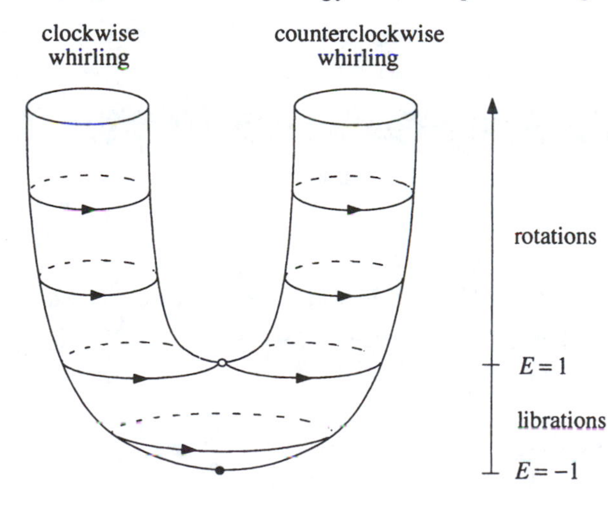

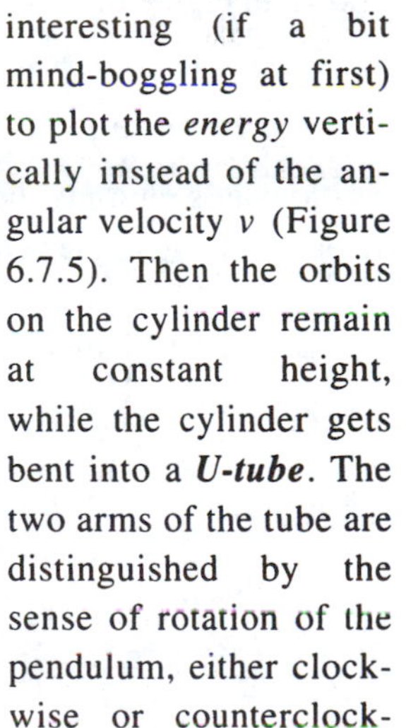

Figure 6.7.5

wise. At low energies, this distinction no longer exists; the pendulum oscillates to and fro. The homoclinic orbits lie at $E = 1$, the borderline between rotations and librations.

At first you might think that the trajectories are drawn incorrectly on one of the arms of the U-tube. It might seem that the arrows for clockwise and counterclockwise motions should go in *opposite* directions. But if you think about the coordinate system shown in Figure 6.7.6, you'll see that the picture is correct.

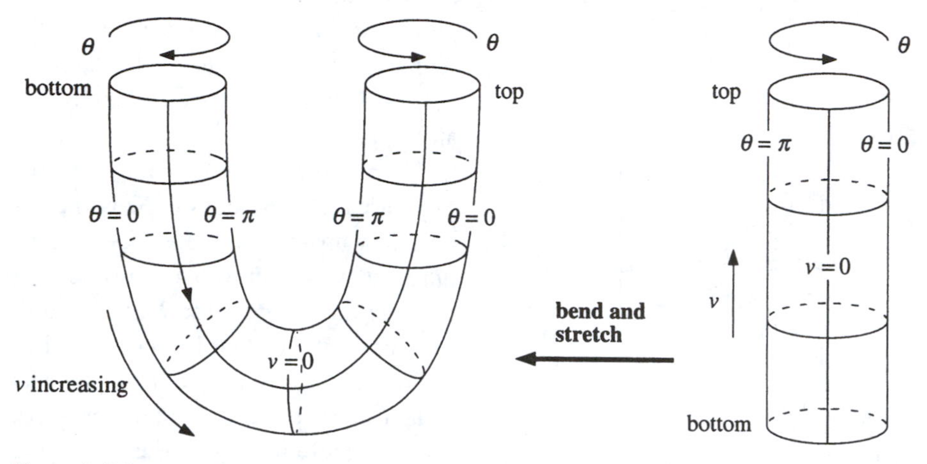

Figure 6.7.6

The point is that the direction of increasing θ has reversed when the bottom of the cylinder is bent around to form the U-tube. (Please understand that Figure 6.7.6 shows the coordinate system, not the actual trajectories; the trajectories were shown in Figure 6.7.5.)

Damping

Now let's return to the phase plane, and suppose that we add a small amount of linear damping to the pendulum. The governing equation becomes

$$\ddot{\theta} + b\dot{\theta} + \sin\theta = 0$$

where $b > 0$ is the damping strength. Then centers become stable spirals while saddles remain saddles. A computer-generated phase portrait is shown in Figure 6.7.7.

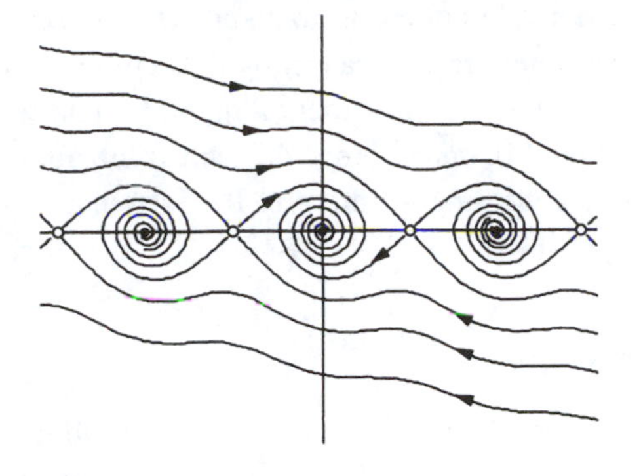

Figure 6.7.7

The picture on the U-tube is clearer. *All trajectories continually lose altitude,* except for the fixed points (Figure 6.7.8).

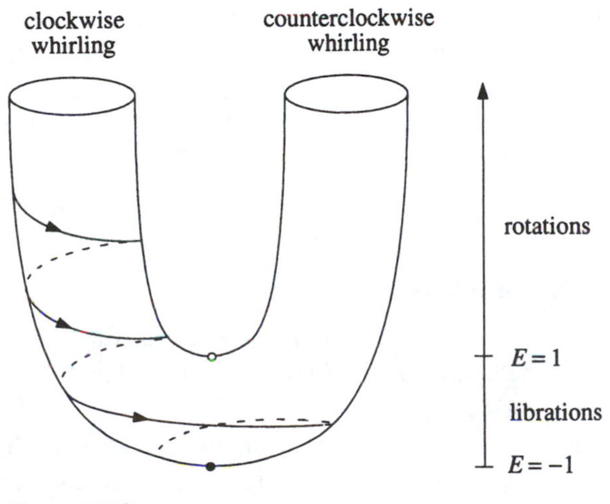

Figure 6.7.8

We can see this explicitly by computing the change in energy along a trajectory:

$$\frac{dE}{d\tau} = \frac{d}{d\tau}\left(\tfrac{1}{2}\dot{\theta}^2 - \cos\theta\right) = \dot{\theta}\left(\ddot{\theta} + \sin\theta\right) = -b\dot{\theta}^2 \le 0.$$

Hence E decreases monotonically along trajectories, except at fixed points where $\dot{\theta} \equiv 0$.

The trajectory shown in Figure 6.7.8 has the following physical interpretation: the pendulum is initially whirling clockwise. As it loses energy, it has a harder time rotating over the top. The corresponding trajectory spirals down the arm of the U-tube until $E < 1$; then the pendulum doesn't have enough energy to whirl, and so it settles down into a small oscillation about the bottom. Eventually the motion damps out and the pendulum comes to rest at its stable equilibrium.

This example shows how far we can go with pictures—without invoking any difficult formulas, we were able to extract all the important features of the pendulum's dynamics. It would be much more difficult to obtain these results analytically, and much more confusing to interpret the formulas, even if we *could* find them.

6.8 Index Theory

In Section 6.3 we learned how to linearize a system about a fixed point. Linearization is a prime example of a *local* method: it gives us a detailed microscopic view of the trajectories near a fixed point, but it can't tell us what happens to the trajectories after they leave that tiny neighborhood. Furthermore, if the vector field starts with quadratic or higher-order terms, the linearization tells us nothing.

In this section we discuss index theory, a method that provides *global* information about the phase portrait. It enables us to answer such questions as: Must a closed trajectory always encircle a fixed point? If so, what types of fixed points are permitted? What types of fixed points can coalesce in bifurcations? The method also yields information about the trajectories near higher-order fixed points. Finally, we can sometimes use index arguments to rule out the possibility of closed orbits in certain parts of the phase plane.

The Index of a Closed Curve

The index of a closed curve C is an integer that measures the winding of the vector field on C. The index also provides information about any fixed points that might happen to lie inside the curve, as we'll see.

This idea may remind you of a concept in electrostatics. In that subject, one often introduces a hypothetical closed surface (a "Gaussian surface") to probe a configuration of electric charges. By studying the behavior of the electric field on the surface, one can determine the total amount of charge *inside* the surface. Amazingly, the behavior *on* the surface tells us what's happening far away *inside* the surface! In the present context, the electric field is analogous to our vector field, the Gaussian surface is analogous to the curve C, and the total charge is analogous to the index.

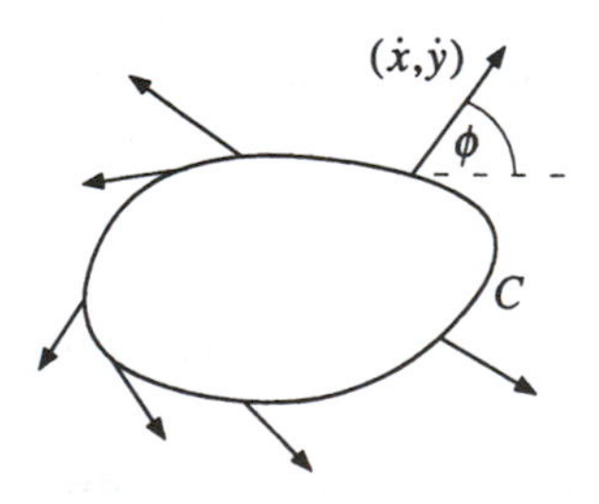

Figure 6.8.1

Now let's make these notions precise. Suppose that $\dot{\mathbf{x}} = \mathbf{f}(\mathbf{x})$ is a smooth vector field on the phase plane. Consider a closed curve C (Figure 6.8.1). This curve is *not* necessarily a trajectory—it's simply a loop that we're putting in the phase plane to probe the behavior of the vector field. We also assume that C is a "simple closed curve" (i.e., it doesn't intersect itself) and that it doesn't pass through any fixed points of the system. Then at each point $\mathbf{x}$ on C, the vector field $\dot{\mathbf{x}} = (\dot{x}, \dot{y})$ makes a well-defined angle

$$\phi = \tan^{-1}(\dot{y}/\dot{x})$$

with the positive x-axis (Figure 6.8.1).

As $\mathbf{x}$ moves counterclockwise around C, the angle ϕ changes *continuously* since the vector field is smooth. Also, when $\mathbf{x}$ returns to its starting place, ϕ returns to its original direction. Hence, over one circuit, ϕ has changed by an *integer* multiple of 2π. Let $[\phi]_C$ be the net change in ϕ over one circuit. Then the ***index of the closed curve*** C with respect to the vector field $\mathbf{f}$ is defined as

$$I_C = \tfrac{1}{2\pi}[\phi]_C .$$

Thus, I_C is the net number of counterclockwise revolutions made by the vector field as $\mathbf{x}$ moves once counterclockwise around C.

To compute the index, we do not need to know the vector field everywhere; we only need to know it along C. The first two examples illustrate this point.

EXAMPLE 6.8.1:

Given that the vector field varies along C as shown in Figure 6.8.2, find I_C.

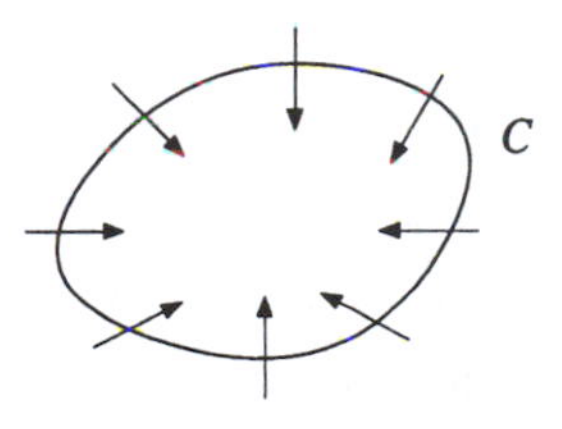

Figure 6.8.2

Solution: As we traverse C once counterclockwise, the vectors rotate through one full turn in the same sense. Hence $I_C = +1$.

If you have trouble visualizing this, here's a foolproof method. Number the vectors in counterclockwise order, starting anywhere on C (Figure 6.8.3a). Then transport these vectors *(without rotation!)* such that their tails lie at a common origin (Figure 6.8.3b). The index equals the net number of counterclockwise revolutions made by the numbered vectors.

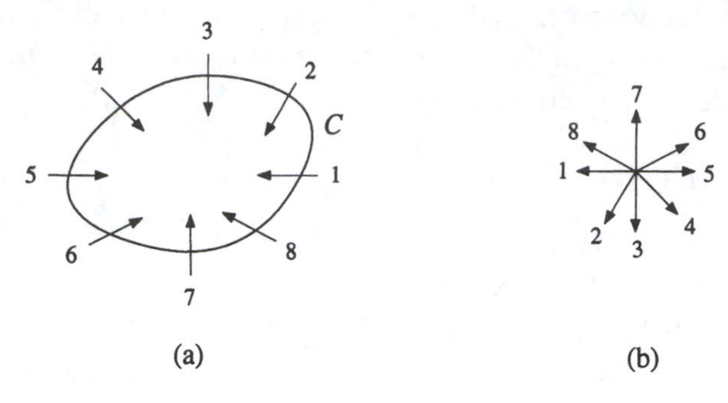

Figure 6.8.3

As Figure 6.8.3b shows, the vectors rotate once counterclockwise as we go in increasing order from vector #1 to vector #8. Hence $I_C = +1$. ∎

EXAMPLE 6.8.2:

Given the vector field on the closed curve shown in Figure 6.8.4a, compute I_C.

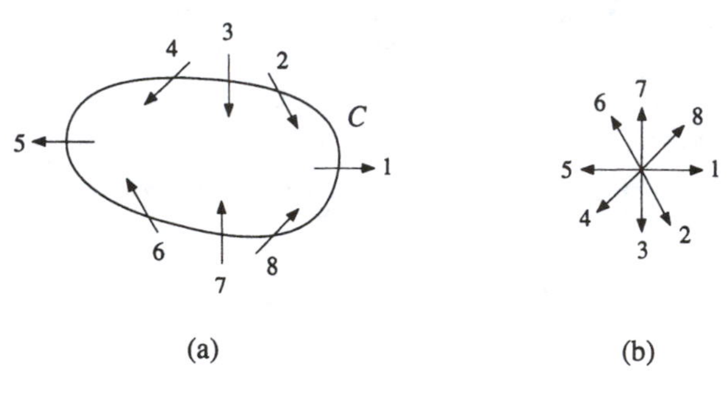

Figure 6.8.4

Solution: We use the same construction as in Example 6.8.1. As we make one circuit around C, the vectors rotate through one full turn, but now in the *opposite* sense. In other words, the vectors on C rotate *clockwise* as we go around C counterclockwise. This is clear from Figure 6.8.4b; the vectors rotate clockwise as we go in increasing order from vector #1 to vector #8. Therefore $I_C = -1$. ∎

In many cases, we are given equations for the vector field, rather than a picture of it. Then we have to draw the picture ourselves, and repeat the steps above. Sometimes this can be confusing, as in the next example.

EXAMPLE 6.8.3:

Given the vector field $\dot{x} = x^2 y$, $\dot{y} = x^2 - y^2$, find I_C, where C is the unit circle $x^2 + y^2 = 1$.

Solution: To get a clear picture of the vector field, it is sufficient to consider a few conveniently chosen points on C. For instance, at $(x, y) = (1,0)$, the vector is $(\dot{x}, \dot{y}) = (x^2 y,\ x^2 - y^2) = (0,1)$. This vector is labeled #1 in Figure 6.8.5a. Now we move counterclockwise around C, computing vectors as we go. At $(x, y) = \frac{1}{\sqrt{2}}(1,1)$, we have $(\dot{x}, \dot{y}) = (x, y) = \frac{1}{2\sqrt{2}}(1,0)$, labeled #2. The remaining vectors are found similarly. Notice that different points on the circle may be associated with the same vector; for example, vector #3 and #7 are both $(0,-1)$.

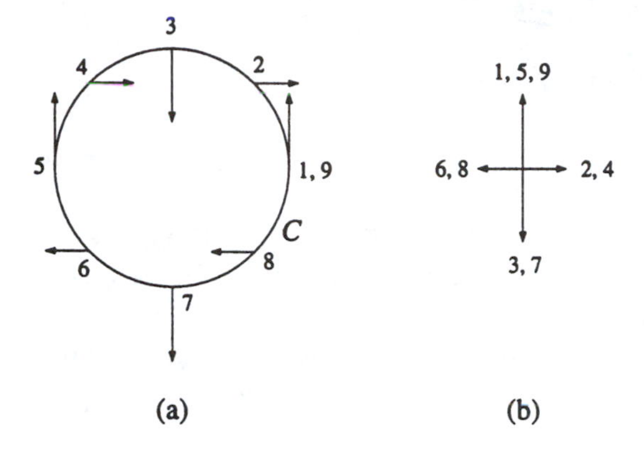

Figure 6.8.5

Now we translate the vectors over to Figure 6.8.5b. As we move from #1 to #9 in order, the vectors rotate 180° clockwise between #1 and #3, then swing back 360° counterclockwise between #3 and #7, and finally rotate 180° clockwise again between #7 and #9 as we complete the circuit of C. Thus $[\phi]_C = -\pi + 2\pi - \pi = 0$ and therefore $I_C = 0$. ■

We plotted nine vectors in this example, but you may want to plot more to see the variation of the vector field in finer detail.

Properties of the Index

Now we list some of the most important properties of the index.

1. Suppose that C can be continuously deformed into C' without passing through a fixed point. Then $I_C = I_{C'}$.

   This property has an elegant proof: Our assumptions imply that as we deform C into C', the index I_C varies *continuously*. But I_C is an integer—hence it can't change without jumping! (To put it more formally, if an integer-valued function is continuous, it must be *constant.*)

   As you think about this argument, try to see where we used the assumption that the intermediate curves don't pass through any fixed points.
2. If C doesn't enclose any fixed points, then $I_C = 0$.

   Proof: By property (1), we can shrink C to a tiny circle without changing the index. But ϕ is essentially constant on such a circle, because all the vectors point in nearly the same direction, thanks to the assumed smoothness of the vector field (Figure 6.8.6). Hence $[\phi]_C = 0$ and therefore $I_C = 0$.

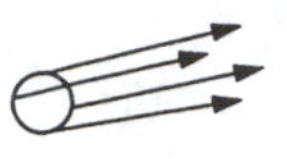

Figure 6.8.6

3. If we reverse all the arrows in the vector field by changing $t \to -t$, the index is unchanged.

   Proof: All angles change from ϕ to $\phi + \pi$. Hence $[\phi]_C$ stays the same.
4. Suppose that the closed curve C is actually a *trajectory* for the system, i.e., C is a closed orbit. Then $I_C = +1$.

   We won't prove this, but it should be clear from geometric intuition (Figure 6.8.7).

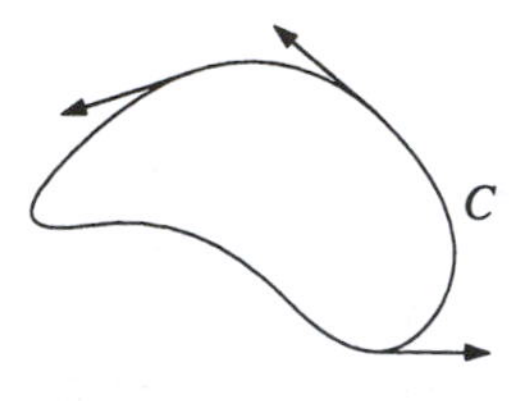

Figure 6.8.7

Notice that the vector field is everywhere tangent to C, because C is a trajectory. Hence, as $\mathbf{x}$ winds around C once, the tangent vector also rotates once in the same sense.

Index of a Point

The properties above are useful in several ways. Perhaps most importantly, they allow us to define the index of a fixed point, as follows.

Suppose $\mathbf{x}^*$ is an isolated fixed point. Then the ***index*** I of $\mathbf{x}^*$ is defined as I_C, where C is *any* closed curve that encloses $\mathbf{x}^*$ and no other fixed points. By property (1) above, I_C is independent of C and is therefore a property of $\mathbf{x}^*$ alone. Therefore we may drop the subscript C and use the notation I for the index of a point.

EXAMPLE 6.8.4:

Find the index of a stable node, an unstable node, and a saddle point.

Solution: The vector field near a stable node looks like the vector field of Example 6.8.1. Hence $I = +1$. The index is also $+1$ for an unstable node, because the only difference is that all the arrows are reversed; by property (3), this doesn't change the index! (This observation shows that *the index is not related to stability,* per se.) Finally, $I = -1$ for a saddle point, because the vector field resembles that discussed in Example 6.8.2. ■

In Exercise 6.8.1, you are asked to show that spirals, centers, degenerate nodes and stars all have $I = +1$. Thus, a saddle point is truly a different animal from all the other familiar types of isolated fixed points.

The index of a curve is related in a beautifully simple way to the indices of the fixed points inside it. This is the content of the following theorem.

Theorem 6.8.1: If a closed curve C surrounds n isolated fixed points $\mathbf{x}_1^*, \ldots, \mathbf{x}_n^*$, then

$$I_C = I_1 + I_2 + \ldots + I_n$$

where I_k is the index of $\mathbf{x}_k^*$, for $k = 1, \ldots, n$.

Ideas behind the proof: The argument is a familiar one, and comes up in multivariable calculus, complex variables, electrostatics, and various other subjects. We think of C as a balloon and suck most of the air out it, being careful not to hit any of the fixed points. The result of this deformation is a new closed curve Γ, consisting of n small circles $\gamma_1, \ldots, \gamma_n$ about the fixed points, and two-way bridges connecting these circles (Figure 6.8.8). Note that $I_\Gamma = I_C$, by property (1), since we didn't cross any fixed points during the deformation. Now let's compute I_Γ by considering $[\phi]_\Gamma$. There are contributions to $[\phi]_\Gamma$ from the small circles and from the two-way bridges. The key point is that *the contributions from the bridges cancel out*: as we move around Γ, each bridge is traversed once in one direction, and later in the opposite direction. Thus we only need to consider the contributions from the small circles. On γ_k, the angle ϕ changes by $[\phi]_{\gamma_k} = 2\pi I_k$, by definition of I_k. Hence

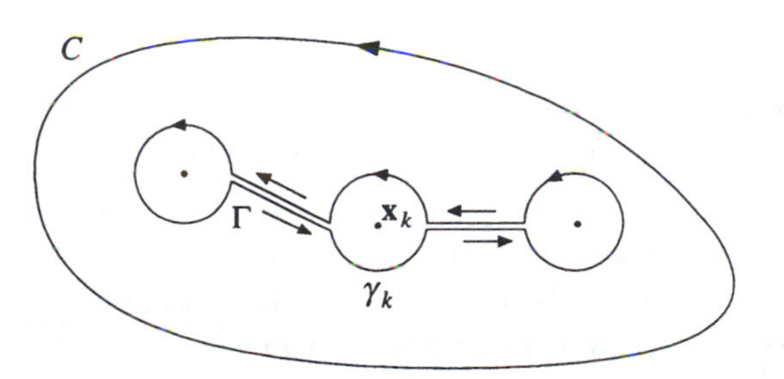

Figure 6.8.8

$$I_\Gamma = \tfrac{1}{2\pi}[\phi]_\Gamma = \tfrac{1}{2\pi}\sum_{k=1}^{n}[\phi]_{\gamma_k} = \sum_{k=1}^{n} I_k$$

and since $I_\Gamma = I_C$, we're done. ■

This theorem is reminiscent of Gauss's law in electrostatics, namely that the electric flux through a surface is proportional to the total charge enclosed. See Exercise 6.8.12 for a further exploration of this analogy between index and charge.

Theorem 6.8.2: Any closed orbit in the phase plane must enclose fixed points whose indices sum to +1.

Proof: Let C denote the closed orbit. From property (4) above, $I_C = +1$. Then Theorem 6.8.1 implies $\sum_{k=1}^{n} I_k = +1$. ∎

Theorem 6.8.2 has many practical consequences. For instance, it implies that there is always at least one fixed point inside any closed orbit in the phase plane (as you may have noticed on your own). If there is *only* one fixed point inside, it cannot be a saddle point. Furthermore, Theorem 6.8.2 can sometimes be used to rule out the possible occurrence of closed trajectories, as seen in the following examples.

EXAMPLE 6.8.5:

Show that closed orbits are impossible for the "rabbit vs. sheep" system

$$\dot{x} = x(3 - x - 2y)$$
$$\dot{y} = y(2 - x - y)$$

studied in Section 6.4. Here $x, y \geq 0$.

Solution: As shown previously, the system has four fixed points: $(0,0)$ = unstable node; $(0,2)$ and $(3,0)$ = stable nodes; and $(1,1)$ = saddle point. The index at each of these points is shown in Figure 6.8.9. Now suppose that the system had a closed trajectory. Where could it lie? There are three qualitatively different locations, indicated by the dotted curves C_1, C_2, C_3. They can be ruled out as follows: orbits like C_1 are impossible because they don't enclose any fixed points, and orbits like C_2 violate the requirement that the indices inside must sum to $+1$. But what is wrong with orbits like C_3, which satisfy the index requirement? The trouble is that such orbits always cross the x-axis or the y-axis, and these axes contain straight-line trajectories. Hence C_3 violates the rule that trajectories can't cross (recall Section 6.2). ∎

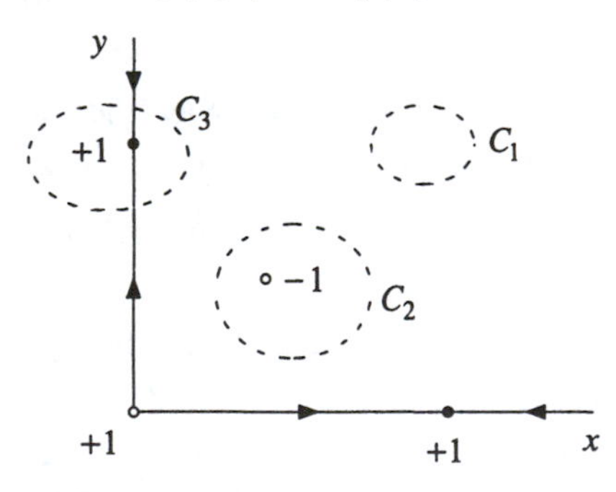

Figure 6.8.9

EXAMPLE 6.8.6:

Show that the system $\dot{x} = xe^{-x}$, $\dot{y} = 1 + x + y^2$ has no closed orbits.

Solution: This system has no fixed points: if $\dot{x} = 0$, then $x = 0$ and so $\dot{y} = 1 + y^2 \neq 0$. By Theorem 6.8.2, closed orbits cannot exist. ■

EXERCISES FOR CHAPTER 6

6.1 Phase Portraits

For each of the following systems, find the fixed points. Then sketch the nullclines, the vector field, and a plausible phase portrait.

6.1.1 $\dot{x} = x - y$, $\dot{y} = 1 - e^x$

6.1.2 $\dot{x} = x - x^3$, $\dot{y} = -y$

6.1.3 $\dot{x} = x(x - y)$, $\dot{y} = y(2x - y)$

6.1.4 $\dot{x} = y$, $\dot{y} = x(1 + y) - 1$

6.1.5 $\dot{x} = x(2 - x - y)$, $\dot{y} = x - y$

6.1.6 $\dot{x} = x^2 - y$, $\dot{y} = x - y$

6.1.7 (Nullcline vs. stable manifold) There's a confusing aspect of Example 6.1.1. The nullcline $\dot{x} = 0$ in Figure 6.1.3 has a similar shape and location as the stable manifold of the saddle, shown in Figure 6.1.4. But they're not the same curve! To clarify the relation between the two curves, sketch both of them on the same phase portrait.

(Computer work) Plot computer-generated phase portraits of the following systems. As always, you may write your own computer programs or use any ready-made software, e.g., *MacMath* (Hubbard and West 1992).

6.1.8 (van der Pol oscillator) $\dot{x} = y$, $\dot{y} = -x + y(1 - x^2)$

6.1.9 (Dipole fixed point) $\dot{x} = 2xy$, $\dot{y} = y^2 - x^2$

6.1.10 (Two-eyed monster) $\dot{x} = y + y^2$, $\dot{y} = -\frac{1}{2}x + \frac{1}{5}y - xy + \frac{6}{5}y^2$ (from Borrelli and Coleman 1987, p. 385.)

6.1.11 (Parrot) $\dot{x} = y + y^2$, $\dot{y} = -x + \frac{1}{5}y - xy + \frac{6}{5}y^2$ (from Borrelli and Coleman 1987, p. 384.)

6.1.12 (Saddle connections) A certain system is known to have exactly two fixed points, both of which are saddles. Sketch phase portraits in which

a) there is a single trajectory that connects the saddles;

b) there is no trajectory that connects the saddles.

6.1.13 Draw a phase portrait that has exactly three closed orbits and one fixed point.

6.1.14 (Series approximation for the stable manifold of a saddle point) Recall the system $\dot{x} = x + e^{-y}$, $\dot{y} = -y$ from Example 6.1.1. We showed that this system has one fixed point, a saddle at $(-1,0)$. Its unstable manifold is the x-axis, but its stable manifold is a curve that is harder to find. The goal of this exercise is to approximate this unknown curve.

a) Let (x,y) be a point on the stable manifold, and assume that (x,y) is close to $(-1,0)$. Introduce a new variable $u = x+1$, and write the stable manifold as $y = a_1 u + a_2 u^2 + O(u^3)$. To determine the coefficients, derive two expressions for dy/du and equate them.

b) Check that your analytical result produces a curve with the same shape as the stable manifold shown in Figure 6.1.4.

6.2 Existence, Uniqueness, and Topological Consequences

6.2.1 We claimed that different trajectories can never intersect. But in many phase portraits, different trajectories appear to intersect at a fixed point. Is there a contradiction here?

6.2.2 Consider the system $\dot{x} = y$, $\dot{y} = -x + (1 - x^2 - y^2)y$.

a) Let D be the open disk $x^2 + y^2 < 4$. Verify that the system satisfies the hypotheses of the existence and uniqueness theorem throughout the domain D.

b) By substitution, show that $x(t) = \sin t$, $y(t) = \cos t$ is an exact solution of the system.

c) Now consider a different solution, in this case starting from the initial condition $x(0) = \frac{1}{2}$, $y(0) = 0$. Without doing any calculations, explain why this solution *must* satisfy $x(t)^2 + y(t)^2 < 1$ for all $t < \infty$.

6.3 Fixed Points and Linearization

For each of the following systems, find the fixed points, classify them, sketch the neighboring trajectories, and try to fill in the rest of the phase portrait.

6.3.1 $\dot{x} = x - y$, $\dot{y} = x^2 - 4$

6.3.2 $\dot{x} = \sin y$, $\dot{y} = x - x^3$

6.3.3 $\dot{x} = 1 + y - e^{-x}$, $\dot{y} = x^3 - y$

6.3.4 $\dot{x} = y + x - x^3$, $\dot{y} = -y$

6.3.5 $\dot{x} = \sin y$, $\dot{y} = \cos x$

6.3.6 $\dot{x} = xy - 1$, $\dot{y} = x - y^3$

6.3.7 For each of the nonlinear systems above, plot a computer-generated phase portrait and compare to your approximate sketch.

6.3.8 (Gravitational equilibrium) A particle moves along a line joining two stationary masses, m_1 and m_2, which are separated by a fixed distance a. Let x denote the distance of the particle from m_1.

a) Show that $\ddot{x} = \dfrac{Gm_2}{(x-a)^2} - \dfrac{Gm_1}{x^2}$, where G is the gravitational constant.

b) Find the particle's equilibrium position. Is it stable or unstable?

6.3.9 Consider the system $\dot{x} = y^3 - 4x$, $\dot{y} = y^3 - y - 3x$.

a) Find all the fixed points and classify them.

b) Show that the line $x = y$ is invariant, i.e., any trajectory that starts on it stays on it.

c) Show that $|x(t) - y(t)| \to 0$ as $t \to \infty$ for all other trajectories. (Hint: Form a differential equation for $x - y$.)

d) Sketch the phase portrait.

e) If you have access to a computer, plot an accurate phase portrait on the square domain $-20 \le x, y \le 20$. (To avoid numerical instability, you'll need to use a fairly small step size, because of the strong cubic nonlinearity.) Notice the trajectories seem to approach a certain curve as $t \to -\infty$; can you explain this behavior intuitively, and perhaps find an approximate equation for this curve?

6.3.10 (Dealing with a fixed point for which linearization is inconclusive) The goal of this exercise is to sketch the phase portrait for $\dot{x} = xy$, $\dot{y} = x^2 - y$.

a) Show that the linearization predicts that the origin is a non-isolated fixed point.

b) Show that the origin is in fact an isolated fixed point.

c) Is the origin repelling, attracting, a saddle, or what? Sketch the vector field along the nullclines and at other points in the phase plane. Use this information to sketch the phase portrait.

d) Plot a computer-generated phase portrait to check your answer to (c).

(Note: This problem can also be solved by a method called *center manifold theory*, as explained in Wiggins (1990) and Guckenheimer and Holmes (1983).)

6.3.11 (Nonlinear terms can change a star into a spiral) Here's another example that shows that borderline fixed points are sensitive to nonlinear terms. Consider the system in polar coordinates given by $\dot{r} = -r$, $\dot{\theta} = 1/\ln r$.

a) Find $r(t)$ and $\theta(t)$ explicitly, given an initial condition (r_0, θ_0).

b) Show that $r(t) \to 0$ and $|\theta(t)| \to \infty$ as $t \to \infty$. Therefore the origin is a stable spiral for the nonlinear system.

c) Write the system in x, y coordinates.

d) Show that the linearized system about the origin is $\dot{x} = -x$, $\dot{y} = -y$. Thus the origin is a stable star for the linearized system.

6.3.12 (Polar coordinates) Using the identity $\theta = \tan^{-1}(y/x)$, show that $\dot{\theta} = (x\dot{y} - y\dot{x})/r^2$.

6.3.13 (Another linear center that's actually a nonlinear spiral) Consider the system $\dot{x} = -y - x^3$, $\dot{y} = x$. Show that the origin is a spiral, although the linearization predicts a center.

6.3.14 Classify the fixed point at the origin for the system $\dot{x} = -y + ax^3$, $\dot{y} = x + ay^3$, for all real values of the parameter a.

6.3.15 Consider the system $\dot{r} = r(1-r^2)$, $\dot{\theta} = 1 - \cos\theta$, where r, θ represent polar coordinates. Sketch the phase portrait and thereby show that the fixed point $r^* = 1$, $\theta^* = 0$ is attracting but not Liapunov stable.

6.3.16 (Saddle switching and structural stability) Consider the system $\dot{x} = a + x^2 - xy$, $\dot{y} = y^2 - x^2 - 1$, where a is a parameter.

a) Sketch the phase portrait for $a = 0$. Show that there is a trajectory connecting two saddle points. (Such a trajectory is called a *saddle connection*.)
b) With the aid of a computer if necessary, sketch the phase portrait for $a < 0$ and $a > 0$.

Notice that for $a \neq 0$, the phase portrait has a different topological character: the saddles are no longer connected by a trajectory. The point of this exercise is that the phase portrait in (a) is *not structurally stable*, since its topology can be changed by an arbitrarily small perturbation a.

6.3.17 (Nasty fixed point) The system $\dot{x} = xy - x^2y + y^3$, $\dot{y} = y^2 + x^3 - xy^2$ has a nasty higher-order fixed point at the origin. Using polar coordinates or otherwise, sketch the phase portrait.

6.4 Rabbits versus Sheep

Consider the following "rabbits vs. sheep" problems, where $x, y \geq 0$. Find the fixed points, investigate their stability, draw the nullclines, and sketch plausible phase portraits. Indicate the basins of attraction of any stable fixed points.

6.4.1 $\dot{x} = x(3 - x - y)$, $\dot{y} = y(2 - x - y)$

6.4.2 $\dot{x} = x(3 - 2x - y)$, $\dot{y} = y(2 - x - y)$

6.4.3 $\dot{x} = x(3 - 2x - 2y)$, $\dot{y} = y(2 - x - y)$

The next three exercises deal with competition models of increasing complexity. We assume $N_1, N_2 \geq 0$ in all cases.

6.4.4 The simplest model is $\dot{N}_1 = r_1N_1 - b_1N_1N_2$, $\dot{N}_2 = r_2N_2 - b_2N_1N_2$.

a) In what way is this model less realistic than the one considered in the text?
b) Show that by suitable rescalings of N_1, N_2, and t, the model can be nondimensionalized to $x' = x(1-y)$, $y' = y(\rho - x)$. Find a formula for the dimensionless group ρ.
c) Sketch the nullclines and vector field for the system in (b).
d) Draw the phase portrait, and comment on the biological implications.
e) Show that (almost) all trajectories are curves of the form $\rho \ln x - x = \ln y - y + C$. (Hint: Derive a differential equation for dx/dy, and separate the variables.) Which trajectories are not of the stated form?

6.4.5 Now suppose that species #1 has a finite carrying capacity K_1. Thus

$$\dot{N}_1 = r_1 N_1 (1 - N_1/K_1) - b_1 N_1 N_2$$
$$\dot{N}_2 = r_2 N_2 - b_2 N_1 N_2 .$$

Nondimensionalize the model and analyze it. Show that there are two qualitatively different kinds of phase portrait, depending on the size of K_1. (Hint: Draw the nullclines.) Describe the long-term behavior in each case.

6.4.6 Finally, suppose that both species have finite carrying capacities:

$$\dot{N}_1 = r_1 N_1 (1 - N_1/K_1) - b_1 N_1 N_2$$
$$\dot{N}_2 = r_2 N_2 (1 - N_2/K_2) - b_2 N_1 N_2 .$$

a) Nondimensionalize the model. How many dimensionless groups are needed?
b) Show that there are four qualitatively different phase portraits, as far as long-term behavior is concerned.
c) Find conditions under which the two species can stably coexist. Explain the biological meaning of these conditions. (Hint: The carrying capacities reflect the competition *within* a species, whereas the b's reflect the competition *between* species.)

6.4.7 (Two-mode laser) According to Haken (1983, p. 129), a two-mode laser produces two different kinds of photons with numbers n_1 and n_2. By analogy with the simple laser model discussed in Section 3.3, the rate equations are

$$\dot{n}_1 = G_1 N n_1 - k_1 n_1$$
$$\dot{n}_2 = G_2 N n_2 - k_2 n_2$$

where $N(t) = N_0 - \alpha_1 n_1 - \alpha_2 n_2$ is the number of excited atoms. The parameters $G_1, G_2, k_1, k_2, \alpha_1, \alpha_2, N_0$ are all positive.
a) Discuss the stability of the fixed point $n_1{}^* = n_2{}^* = 0$.
b) Find and classify any other fixed points that may exist.
c) Depending on the values of the various parameters, how many qualitatively different phase portraits can occur? For each case, what does the model predict about the long-term behavior of the laser?

6.5 Conservative Systems

6.5.1 Consider the system $\ddot{x} = x^3 - x$.
a) Find all the equilibrium points and classify them.
b) Find a conserved quantity.
c) Sketch the phase portrait.

6.5.2 Consider the system $\ddot{x} = x - x^2$.
a) Find and classify the equilibrium points.

b) Sketch the phase portrait.
c) Find an equation for the homoclinic orbit that separates closed and nonclosed trajectories.

6.5.3 Find a conserved quantity for the system $\ddot{x} = a - e^x$, and sketch the phase portrait for $a < 0$, $a = 0$, and $a > 0$.

6.5.4 Sketch the phase portrait for the system $\ddot{x} = ax - x^2$ for $a < 0$, $a = 0$, and $a > 0$.

6.5.5 Investigate the stability of the equilibrium points of the system $\ddot{x} = (x-a)(x^2-a)$ for all real values of the parameter a. (Hints: It might help to graph the right-hand side. An alternative is to rewrite the equation as $\ddot{x} = -V'(x)$ for a suitable potential energy function V and then use your intuition about particles moving in potentials.)

6.5.6 (Epidemic model revisited) In Exercise 3.7.6, you analyzed the Kermack–McKendrick model of an epidemic by reducing it to a certain first-order system. In this problem you'll see how much easier the analysis becomes in the phase plane. As before, let $x(t) \geq 0$ denote the size of the healthy population and $y(t) \geq 0$ denote the size of the sick population. Then the model is

$$\dot{x} = -kxy, \qquad \dot{y} = kxy - \ell y$$

where $k, \ell > 0$. (The equation for $z(t)$, the number of deaths, plays no role in the x, y dynamics so we omit it.)

a) Find and classify all the fixed points.
b) Sketch the nullclines and the vector field.
c) Find a conserved quantity for the system. (Hint: Form a differential equation for dy/dx. Separate the variables and integrate both sides.)
d) Plot the phase portrait. What happens as $t \to \infty$?
e) Let (x_0, y_0) be the initial condition. An *epidemic* is said to occur if $y(t)$ increases initially. Under what condition does an epidemic occur?

6.5.7 (General relativity and planetary orbits) The relativistic equation for the orbit of a planet around the sun is

$$\frac{d^2u}{d\theta^2} + u = \alpha + \varepsilon u^2$$

where $u = 1/r$ and r, θ are the polar coordinates of the planet in its plane of motion. The parameter α is positive and can be found explicitly from classical Newtonian mechanics; the term εu^2 is Einstein's correction. Here ε is a very small positive parameter.

a) Rewrite the equation as a system in the (u, v) phase plane, where $v = du/d\theta$.

b) Find all the equilibrium points of the system.

c) Show that one of the equilibria is a center in the (u, v) phase plane, according to the linearization. Is it a *nonlinear* center?

d) Show that the equilibrium point found in (c) corresponds to a circular planetary orbit.

Hamiltonian systems are fundamental to classical mechanics; they provide an equivalent but more geometric version of Newton's laws. They are also central to celestial mechanics and plasma physics, where dissipation can sometimes be neglected on the time scales of interest. The theory of Hamiltonian systems is deep and beautiful, but perhaps too specialized and subtle for a first course on nonlinear dynamics. See Arnold (1978), Lichtenberg and Lieberman (1992), Tabor (1989), or Hénon (1983) for introductions.

Here's the simplest instance of a Hamiltonian system. Let $H(p,q)$ be a smooth, real-valued function of two variables. The variable q is the "generalized coordinate" and p is the "conjugate momentum." (In some physical settings, H could also depend explicitly on time t, but we'll ignore that possibility.) Then a system of the form

$$\dot{q} = \partial H / \partial p, \qquad \dot{p} = -\partial H / \partial q$$

is called a ***Hamiltonian system*** and the function H is called the ***Hamiltonian***. The equations for $\dot{q}$ and $\dot{p}$ are called Hamilton's equations.

The next three exercises concern Hamiltonian systems.

6.5.8 (Harmonic oscillator) For a simple harmonic oscillator of mass m, spring constant k, displacement x, and momentum p, the Hamiltonian is $H = \frac{p^2}{2m} + \frac{kx^2}{2}$. Write out Hamilton's equations explicitly. Show that one equation gives the usual definition of momentum and the other is equivalent to $F = ma$. Verify that H is the total energy.

6.5.9 Show that for any Hamiltonian system, $H(x, p)$ is a conserved quantity. (Hint: Show $\dot{H} = 0$ by applying the chain rule and invoking Hamilton's equations.) Hence the trajectories lie on the contour curves $H(x, p) = C$.

6.5.10 (Inverse-square law) A particle moves in a plane under the influence of an inverse-square force. It is governed by the Hamiltonian $H(p,r) = \frac{p^2}{2} + \frac{h^2}{2r^2} - \frac{k}{r}$ where $r > 0$ is the distance from the origin and p is the radial momentum. The parameters h and k are the angular momentum and the force constant, respectively.

a) Suppose $k > 0$, corresponding to an attractive force like gravity. Sketch the phase portrait in the (r, p) plane. (Hint: Graph the "effective potential" $V(r) = h^2/2r^2 - k/r$ and then look for intersections with horizontal lines of height E. Use this information to sketch the contour curves $H(p, r) = E$ for various positive and negative values of E.)

b) Show that the trajectories are closed if $-k^2/2h^2 < E < 0$, in which case the particle is "captured" by the force. What happens if $E > 0$? What about $E = 0$?

c) If $k < 0$ (as in electric repulsion), show that there are no periodic orbits.

6.5.11 (Basins for damped double-well oscillator) Suppose we add a small amount of damping to the double-well oscillator of Example 6.5.2. The new system is $\dot{x} = y$, $\dot{y} = -by + x - x^3$, where $0 < b << 1$. Sketch the basin of attraction for the stable fixed point $(x^*, y^*) = (1, 0)$. Make the picture large enough so that the global structure of the basin is clearly indicated.

6.5.12 (Why we need to assume *isolated* minima in Theorem 6.5.1) Consider the system $\dot{x} = xy$, $\dot{y} = -x^2$.

a) Show that $E = x^2 + y^2$ is conserved.

b) Show that the origin is a fixed point, but not an isolated fixed point.

c) Since E has a local minimum at the origin, one might have thought that the origin has to be a center. But that would be a misuse of Theorem 6.5.1; the theorem does not apply here because the origin is *not* an isolated fixed point. Show that in fact the origin is not surrounded by closed orbits, and sketch the actual phase portrait.

6.5.13 (Nonlinear centers)

a) Show that the Duffing equation $\ddot{x} + x + \varepsilon x^3 = 0$ has a nonlinear center at the origin for all $\varepsilon > 0$.

b) If $\varepsilon < 0$, show that all trajectories near the origin are closed. What about trajectories that are far from the origin?

6.5.14 (Glider) Consider a glider flying at speed v at an angle θ to the horizontal. Its motion is governed approximately by the dimensionless equations

$$\dot{v} = -\sin\theta - Dv^2$$

$$v\dot{\theta} = -\cos\theta + v^2$$

where the trigonometric terms represent the effects of gravity and the v^2 terms represent the effects of drag and lift.

a) Suppose there is no drag ($D = 0$). Show that $v^3 - 3v\cos\theta$ is a conserved quantity. Sketch the phase portrait in this case. Interpret your results physically—what does the flight path of the glider look like?

b) Investigate the case of positive drag ($D > 0$).

In the next four exercises, we return to the problem of a bead on a rotating hoop, discussed in Section 3.5. Recall that the bead's motion is governed by

$$mr\ddot{\phi} = -b\dot{\phi} - mg\sin\phi + mr\omega^2 \sin\phi\cos\phi .$$

Previously, we could only treat the overdamped limit. The next four exercises deal with the dynamics more generally.

6.5.15 (Frictionless bead) Consider the undamped case $b = 0$.

a) Show that the equation can be nondimensionalized to $\phi'' = \sin\phi(\cos\phi - \gamma^{-1})$, where $\gamma = r\omega^2/g$ as before, and prime denotes differentiation with respect to dimensionless time $\tau = \omega t$.
b) Draw all the qualitatively different phase portraits as γ varies.
c) What do the phase portraits imply about the physical motion of the bead?

6.5.16 (Small oscillations of the bead) Return to the original dimensional variables. Show that when $b = 0$ and ω is sufficiently large, the system has a symmetric pair of stable equilibria. Find the approximate frequency of small oscillations about these equilibria. (Please express your answer with respect to t, not τ.)

6.5.17 (A puzzling constant of motion for the bead) Find a conserved quantity when $b = 0$. You might think that it's essentially the bead's total energy, but it isn't! Show explicitly that the bead's kinetic plus potential energy is *not* conserved. Does this make sense physically? Can you find a physical interpretation for the conserved quantity? (Hint: Think about reference frames and moving constraints.)

6.5.18 (General case for the bead) Finally, allow the damping b to be arbitrary. Define an appropriate dimensionless version of b, and plot all the qualitatively different phase portraits that occur as b and γ vary.

6.5.19 (Rabbits vs. foxes) The model $\dot{R} = aR - bRF$, $\dot{F} = -cF + dRF$ is the ***Lotka–Volterra predator-prey model***. Here $R(t)$ is the number of rabbits, $F(t)$ is the number of foxes, and $a, b, c, d > 0$ are parameters.

a) Discuss the biological meaning of each of the terms in the model. Comment on any unrealistic assumptions.
b) Show that the model can be recast in dimensionless form as $x' = x(1 - y)$, $y' = \mu y(x - 1)$.
c) Find a conserved quantity in terms of the dimensionless variables.
d) Show that the model predicts *cycles* in the populations of both species, for almost all initial conditions.

This model is popular with many textbook writers because it's simple, but some are beguiled into taking it too seriously. Mathematical biologists dismiss the Lotka–Volterra model because it is not structurally stable, and because real predator-prey cycles typically have a characteristic amplitude. In other words, realistic models should predict a *single* closed orbit, or perhaps finitely many, but not a continuous family of neutrally stable cycles. See the discussions in May (1972), Edelstein–Keshet (1988), or Murray (1989).

6.6 Reversible Systems

Show that each of the following systems is reversible, and sketch the phase portrait.

6.6.1 $\dot{x} = y(1-x^2)$, $\dot{y} = 1-y^2$

6.6.2 $\dot{x} = y$, $\dot{y} = x\cos y$

6.6.3 (Wallpaper) Consider the system $\dot{x} = \sin y$, $\dot{y} = \sin x$.

a) Show that the system is reversible.
b) Find and classify all the fixed points.
c) Show that the lines $y = \pm x$ are invariant (any trajectory that starts on them stays on them forever).
d) Sketch the phase portrait.

6.6.4 (Computer explorations) For each of the following reversible systems, try to sketch the phase portrait by hand. Then use a computer to check your sketch. If the computer reveals patterns you hadn't anticipated, try to explain them.

a) $\ddot{x} + (\dot{x})^2 + x = 3$ b) $\dot{x} = y - y^3$, $\dot{y} = x\cos y$ c) $\dot{x} = \sin y$, $\dot{y} = y^2 - x$

6.6.5 Consider equations of the form $\ddot{x} + f(\dot{x}) + g(x) = 0$, where f is an even function, and both f and g are smooth.

a) Show that the equation is invariant under the pure time-reversal symmetry $t \to -t$.
b) Show that the equilibrium points cannot be stable nodes or spirals.

6.6.6 (Manta ray) Use qualitative arguments to deduce the "manta ray" phase portrait of Example 6.6.1.

a) Plot the nullclines $\dot{x} = 0$ and $\dot{y} = 0$.
b) Find the sign of $\dot{x}$, $\dot{y}$ in different regions of the plane.
c) Calculate the eigenvalues and eigenvectors of the saddle points at $(-1, \pm 1)$.
d) Consider the unstable manifold of $(-1,-1)$. By making an argument about the signs of $\dot{x}$, $\dot{y}$, prove that this unstable manifold intersects the negative x-axis. Then use reversibility to prove the existence of a heteroclinic trajectory connecting $(-1,-1)$ to $(-1,1)$.
e) Using similar arguments, prove that another heteroclinic trajectory exists, and sketch several other trajectories to fill in the phase portrait.

6.6.7 (Oscillator with both positive and negative damping) Show that the system $\ddot{x} + x\dot{x} + x = 0$ is reversible and plot the phase portrait.

6.6.8 (Reversible system on a cylinder) While studying chaotic streamlines inside a drop immersed in a steady Stokes flow, Stone et al. (1991) encountered the system

$$\dot{x} = \frac{\sqrt{2}}{4} x(x-1)\sin\phi, \qquad \dot{\phi} = \frac{1}{2}\left[\beta - \frac{1}{\sqrt{2}}\cos\phi - \frac{1}{8\sqrt{2}} x\cos\phi\right]$$

where $0 \le x \le 1$ and $-\pi \le \phi < \pi$.

Since the system is 2π-periodic in ϕ, it may be considered as a vector field on a *cylinder.* (See Section 6.7 for another vector field on a cylinder.) The x-axis runs along the cylinder, and the ϕ-axis wraps around it. Note that the cylindrical phase space is finite, with edges given by the circles $x = 0$ and $x = 1$.

a) Show that the system is reversible.
b) Verify that for $\frac{9}{8\sqrt{2}} > \beta > \frac{1}{\sqrt{2}}$, the system has three fixed points on the cylinder, one of which is a saddle. Show that this saddle is connected to itself by a homoclinic orbit that winds around the waist of the cylinder. Using reversibility, prove that there is a *band of closed orbits* sandwiched between the circle $x = 0$ and the homoclinic orbit. Sketch the phase portrait on the cylinder, and check your results by numerical integration.
c) Show that as $\beta \to \frac{1}{\sqrt{2}}$ from above, the saddle point moves toward the circle $x = 0$, and the homoclinic orbit tightens like a noose. Show that all the closed orbits disappear when $\beta = \frac{1}{\sqrt{2}}$.
d) For $0 < \beta < \frac{1}{\sqrt{2}}$, show that there are two saddle points on the edge $x = 0$. Plot the phase portrait on the cylinder.

6.6.9 (Josephson junction array) As discussed in Exercises 4.6.4 and 4.6.5, the equations

$$\frac{d\phi_k}{d\tau} = \Omega + a\sin\phi_k + \frac{1}{N}\sum_{j=1}^{N}\sin\phi_j, \quad \text{for } k = 1, 2,$$

arise as the dimensionless circuit equations for a resistively loaded array of Josephson junctions.

a) Let $\theta_k = \phi_k - \frac{\pi}{2}$, and show that the resulting system for θ_k is reversible.
b) Show that there are four fixed points (mod 2π) when $|\Omega/(a+1)| < 1$, and none when $|\Omega/(a+1)| > 1$.
c) Using the computer, explore the various phase portraits that occur for $a = 1$, as Ω varies over the interval $0 \le \Omega \le 3$.

For more about this system, see Tsang et al. (1991).

6.6.10 Is the origin a nonlinear center for the system $\dot{x} = -y - x^2$, $\dot{y} = x$?

6.6.11 (Rotational dynamics and a phase portrait on a sphere) The rotational dynamics of an object in a shear flow are governed by

$$\dot{\theta} = \cot\phi \cos\theta, \qquad \dot{\phi} = (\cos^2\phi + A\sin^2\phi)\sin\theta,$$

where θ and ϕ are spherical coordinates that describe the orientation of the object. Our convention here is that $-\pi < \theta \le \pi$ is the "longitude," i.e., the angle around the z-axis, and $-\frac{\pi}{2} \le \phi \le \frac{\pi}{2}$ is the "latitude," i.e., the angle measured northward from the equator. The parameter A depends on the shape of the object.

a) Show that the equations are reversible in two ways: under $t \to -t$, $\theta \to -\theta$ and under $t \to -t$, $\phi \to -\phi$.

b) Investigate the phase portraits when A is positive, zero, and negative. You may sketch the phase portraits as Mercator projections (treating θ and ϕ as rectangular coordinates), but it's better to visualize the motion on the sphere, if you can.

c) Relate your results to the tumbling motion of an object in a shear flow. What happens to the orientation of the object as $t \to \infty$?

6.7 Pendulum

6.7.1 (Damped pendulum) Find and classify the fixed points of $\ddot{\theta} + b\dot{\theta} + \sin\theta = 0$ for all $b > 0$, and plot the phase portraits for the qualitatively different cases.

6.7.2 (Pendulum driven by constant torque) The equation $\ddot{\theta} + \sin\theta = \gamma$ describes the dynamics of an undamped pendulum driven by a constant torque, or an undamped Josephson junction driven by a constant bias current.

a) Find all the equilibrium points and classify them as γ varies.

b) Sketch the nullclines and the vector field.

c) Is the system conservative? If so, find a conserved quantity. Is the system reversible?

d) Sketch the phase portrait on the plane as γ varies.

e) Find the approximate frequency of small oscillations about any centers in the phase portrait.

6.7.3 (Nonlinear damping) Analyze $\ddot{\theta} + (1 + a\cos\theta)\dot{\theta} + \sin\theta = 0$, for all $a \ge 0$.

6.7.4 (Period of the pendulum) Suppose a pendulum governed by $\ddot{\theta} + \sin\theta = 0$ is swinging with an amplitude α. Using some tricky manipulations, we are going to derive a formula for $T(\alpha)$, the period of the pendulum.

a) Using conservation of energy, show that $\dot{\theta}^2 = 2(\cos\theta - \cos\alpha)$ and hence that

$$T = 4\int_0^{\alpha} \frac{d\theta}{\left[2(\cos\theta - \cos\alpha)\right]^{1/2}}.$$

b) Using the half-angle formula, show that $T = 4\int_0^{\alpha} \frac{d\theta}{\left[4(\sin^2\frac{1}{2}\alpha - \sin^2\frac{1}{2}\theta)\right]^{1/2}}$.

c) The formulas in parts (a) and (b) have the disadvantage that α appears in both the integrand and the upper limit of integration. To remove the α-dependence from the limits of integration, we introduce a new angle ϕ that runs from 0 to $\frac{\pi}{2}$ when θ runs from 0 to α. Specifically, let $(\sin\frac{1}{2}\alpha)\sin\phi = \sin\frac{1}{2}\theta$. Using this substitution, rewrite (b) as an integral with respect to ϕ. Thereby derive the exact result

$$T = 4\int_0^{\pi/2} \frac{d\phi}{\cos\frac{1}{2}\theta} = 4K(\sin^2\tfrac{1}{2}\alpha),$$

where the *complete elliptic integral of the first kind* is defined as

$$K(m) = \int_0^{\pi/2} \frac{d\phi}{(1-m\sin^2\phi)^{1/2}}, \quad \text{for } 0 \le m < 1.$$

d) By expanding the elliptic integral using the binomial series and integrating term-by-term, show that

$$T(\alpha) = 2\pi\left[1 + \tfrac{1}{16}\alpha^2 + O(\alpha^4)\right] \text{ for } \alpha \ll 1.$$

Note that larger swings take longer.

6.7.5 (Numerical solution for the period) Redo Exercise 6.7.4 using either numerical integration of the differential equation, or numerical evaluation of the elliptic integral. Specifically, compute the period $T(\alpha)$, where α runs from 0 to 180° in steps of 10°.

6.8 Index Theory

6.8.1 Show that each of the following fixed points has an index equal to +1.
a) stable spiral b) unstable spiral c) center d) star e) degenerate node

(Unusual fixed points) For each of the following systems, locate the fixed points and calculate the index. (Hint: Draw a small closed curve C around the fixed point and examine the variation of the vector field on C.)

6.8.2 $\dot{x} = x^2$, $\dot{y} = y$

6.8.3 $\dot{x} = y - x$, $\dot{y} = x^2$

6.8.4 $\dot{x} = y^3$, $\dot{y} = x$

6.8.5 $\dot{x} = xy$, $\dot{y} = x + y$

6.8.6 A closed orbit in the phase plane encircles S saddles, N nodes, F spirals, and C centers, all of the usual type. Show that $N + F + C = 1 + S$.

6.8.7 (Ruling out closed orbits) Use index theory to show that the system $\dot{x} = x(4 - y - x^2)$, $\dot{y} = y(x-1)$ has no closed orbits.

6.8.8 A smooth vector field on the phase plane is known to have exactly three closed orbits. Two of the cycles, say C_1 and C_2, lie inside the third cycle C_3. However, C_1 does not lie inside C_2, nor vice-versa.

a) Sketch the arrangement of the three cycles.

b) Show that there must be at least one fixed point in the region bounded by C_1, C_2, C_3.

6.8.9 A smooth vector field on the phase plane is known to have exactly two closed trajectories, one of which lies inside the other. The inner cycle runs clockwise, and the outer one runs counterclockwise. True or False: There must be at least one fixed point in the region between the cycles. If true, prove it. If false, provide a simple counterexample.

6.8.10 (Open-ended question for the topologically minded) Does Theorem 6.8.2 hold for surfaces other than the plane? Check its validity for various types of closed orbits on a torus, cylinder, and sphere.

6.8.11 (Complex vector fields) Let $z = x + iy$. Explore the complex vector fields $\dot{z} = z^k$ and $\dot{z} = (\bar{z})^k$, where $k > 0$ is an integer and $\bar{z} = x - iy$ is the complex conjugate of z.

a) Write the vector fields in both Cartesian and polar coordinates, for the cases $k = 1, 2, 3$.
b) Show that the origin is the only fixed point, and compute its index.
c) Generalize your results to arbitrary integer $k > 0$.

6.8.12 ("Matter and antimatter") There's an intriguing analogy between bifurcations of fixed points and collisions of particles and anti-particles. Let's explore this in the context of index theory. For example, a two-dimensional version of the saddle-node bifurcation is given by $\dot{x} = a + x^2$, $\dot{y} = -y$, where a is a parameter.

a) Find and classify all the fixed points as a varies from $-\infty$ to $+\infty$.
b) Show that the sum of the indices of all the fixed points is conserved as a varies.
c) State and prove a generalization of this result, for systems of the form $\dot{\mathbf{x}} = \mathbf{f}(\mathbf{x}, a)$, where $\mathbf{x} \in \mathbf{R}^2$ and a is a parameter.

6.8.13 (Integral formula for the index of a curve) Consider a smooth vector field $\dot{x} = f(x, y)$, $\dot{y} = g(x, y)$ on the plane, and let C be a simple closed curve that does not pass through any fixed points. As usual, let $\phi = \tan^{-1}(\dot{y}/\dot{x})$ as in Figure 6.8.1.

a) Show that $d\phi = (f\,dg - g\,df)/(f^2 + g^2)$.
b) Derive the integral formula

$$I_C = \frac{1}{2\pi} \oint_C \frac{f\,dg - g\,df}{f^2 + g^2}.$$

6.8.14 Consider the family of linear systems $\dot{x} = x\cos\alpha - y\sin\alpha$, $\dot{y} = x\sin\alpha + y\cos\alpha$, where α is a parameter that runs over the range $0 \le \alpha \le \pi$. Let C be a simple closed curve that does not pass through the origin.

a) Classify the fixed point at the origin as a function of α.

b) Using the integral derived in Exercise 6.8.13, show that I_C is *independent* of α.

c) Let C be a circle centered at the origin. Compute I_C explicitly by evaluating the integral for any convenient choice of α.

7

LIMIT CYCLES

7.0 Introduction

A ***limit cycle*** is an isolated closed trajectory. *Isolated* means that neighboring trajectories are not closed; they spiral either toward or away from the limit cycle (Figure 7.0.1).

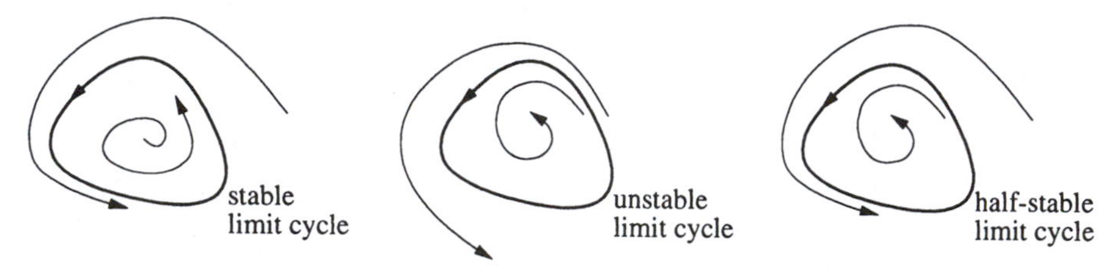

Figure 7.0.1

If all neighboring trajectories approach the limit cycle, we say the limit cycle is *stable* or *attracting*. Otherwise the limit cycle is *unstable*, or in exceptional cases, *half-stable*.

Stable limit cycles are very important scientifically—they model systems that exhibit self-sustained oscillations. In other words, these systems oscillate even in the absence of external periodic forcing. Of the countless examples that could be given, we mention only a few: the beating of a heart; the periodic firing of a pacemaker neuron; daily rhythms in human body temperature and hormone secretion; chemical reactions that oscillate spontaneously; and dangerous self-excited vibrations in bridges and airplane wings. In each case, there is a standard oscillation of some preferred period, waveform, and amplitude. If the system is perturbed slightly, it always returns to the standard cycle.

Limit cycles are inherently nonlinear phenomena; they can't occur in linear systems. Of course, a linear system $\dot{\mathbf{x}} = A\mathbf{x}$ can have closed orbits, but they won't be *isolated*; if $\mathbf{x}(t)$ is a periodic solution, then so is $c\mathbf{x}(t)$ for any constant $c \neq 0$. Hence $\mathbf{x}(t)$ is surrounded by a one-parameter family of closed orbits (Figure 7.0.2). Consequently, the amplitude of a linear oscillation is set entirely by its initial conditions; any slight disturbance to the amplitude will persist forever. In contrast, limit cycle oscillations are determined by the structure of the system itself.

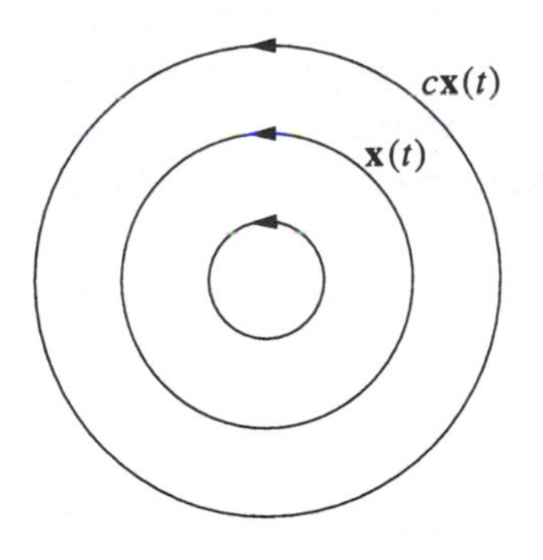

Figure 7.0.2

The next section presents two examples of systems with limit cycles. In the first case, the limit cycle is obvious by inspection, but normally it's difficult to tell whether a given system has a limit cycle, or indeed any closed orbits, from the governing equations alone. Sections 7.2–7.4 present some techniques for ruling out closed orbits or for proving their existence. The remainder of the chapter discusses analytical methods for approximating the shape and period of a closed orbit and for studying its stability.

7.1 Examples

It's straightforward to construct examples of limit cycles if we use polar coordinates.

EXAMPLE 7.1.1: A SIMPLE LIMIT CYCLE

Consider the system

$$\dot{r} = r(1 - r^2), \qquad \dot{\theta} = 1 \tag{1}$$

where $r \geq 0$. The radial and angular dynamics are uncoupled and so can be analyzed separately. Treating $\dot{r} = r(1 - r^2)$ as a vector field on the line, we see that $r^* = 0$ is an unstable fixed point and $r^* = 1$ is stable (Figure 7.1.1).

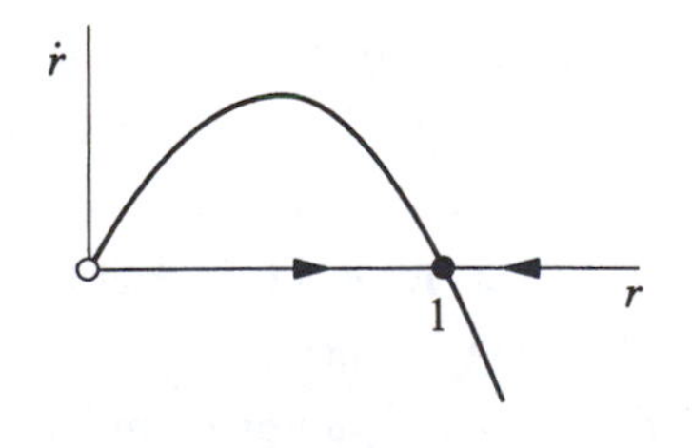

Figure 7.1.1

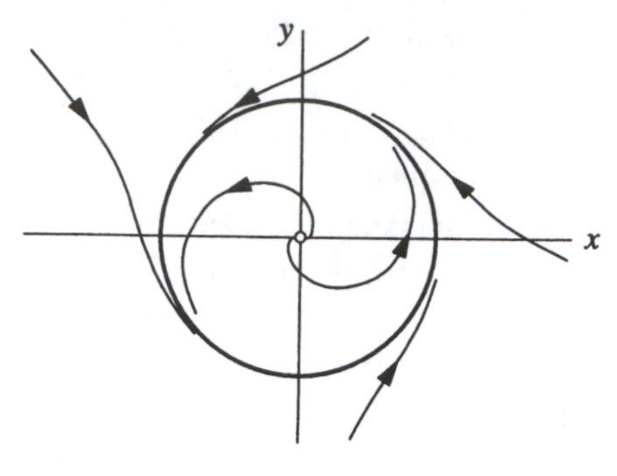

Figure 7.1.2

Hence, back in the phase plane, all trajectories (except $r^* = 0$) approach the unit circle $r^* = 1$ monotonically. Since the motion in the θ-direction is simply rotation at constant angular velocity, we see that all trajectories spiral asymptotically toward a limit cycle at $r = 1$ (Figure 7.1.2).

It is also instructive to plot solutions as functions of t. For instance, in Figure 7.1.3 we plot $x(t) = r(t)\cos\theta(t)$ for a trajectory starting outside the limit cycle. As expected, the solution settles down to a sinusoidal oscillation of constant amplitude, corresponding to the limit cycle solution $x(t) = \cos(t + \theta_0)$ of (1). ∎

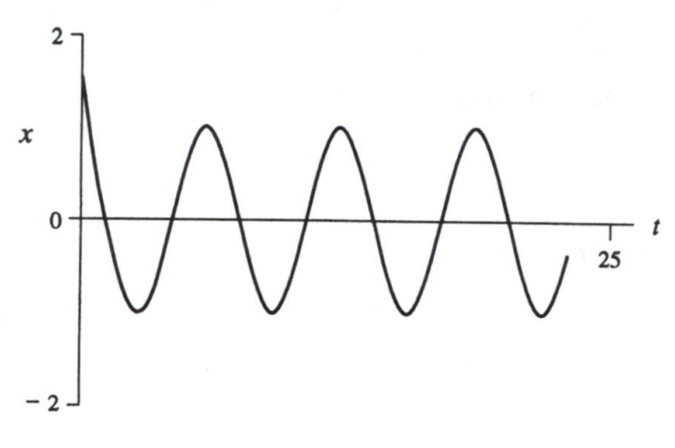

Figure 7.1.3

EXAMPLE 7.1.2: VAN DER POL OSCILLATOR

A less transparent example, but one that played a central role in the development of nonlinear dynamics, is given by the ***van der Pol equation***

$$\ddot{x} + \mu(x^2 - 1)\dot{x} + x = 0 \tag{2}$$

where $\mu \geq 0$ is a parameter. Historically, this equation arose in connection with the nonlinear electrical circuits used in the first radios (see Exercise 7.1.6 for the circuit). Equation (2) looks like a simple harmonic oscillator, but with a ***nonlinear damping*** term $\mu(x^2 - 1)\dot{x}$. This term acts like ordinary positive damping for $|x| > 1$, but like *negative* damping for $|x| < 1$. In other words, it causes large-amplitude oscillations to decay, but it pumps them back up if they become too small.

As you might guess, the system eventually settles into a self-sustained oscillation where the energy dissipated over one cycle balances the energy pumped in. This idea can be made rigorous, and with quite a bit of work, one can prove that *the van der Pol equation has a unique, stable limit cycle for each* $\mu > 0$. This result follows from a more general theorem discussed in Section 7.4.

To give a concrete illustration, suppose we numerically integrate (2) for $\mu = 1.5$, starting from $(x, \dot{x}) = (0.5, 0)$ at $t = 0$. Figure 7.1.4 plots the solution in the phase plane and Figure 7.1.5 shows the graph of $x(t)$. Now, in contrast to Example 7.1.1, the limit cycle is not a circle and the stable waveform is not a sine wave. ■

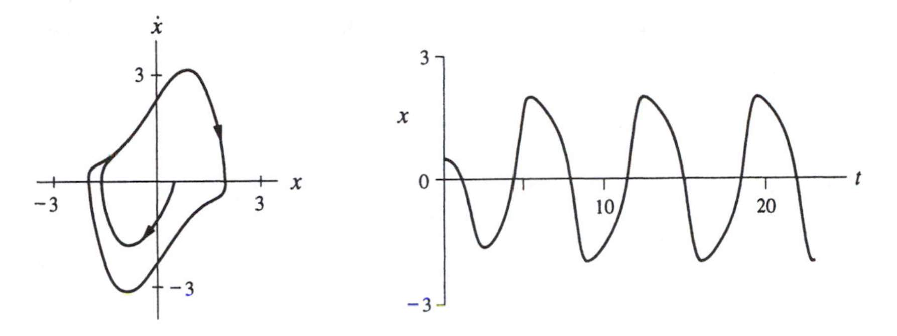

Figure 7.1.4

Figure 7.1.5

7.2 Ruling Out Closed Orbits

Suppose we have a strong suspicion, based on numerical evidence or otherwise, that a particular system has no periodic solutions. How could we prove this? In the last chapter we mentioned one method, based on index theory (see Examples 6.8.5 and 6.8.6). Now we present three other ways of ruling out closed orbits. They are of limited applicability, but they're worth knowing about, in case you get lucky.

Gradient Systems

Suppose the system can be written in the form $\dot{\mathbf{x}} = -\nabla V$, for some continuously differentiable, single-valued scalar function $V(\mathbf{x})$. Such a system is called a ***gradient system*** with ***potential function*** V.

Theorem 7.2.1: Closed orbits are impossible in gradient systems.

Proof: Suppose there were a closed orbit. We obtain a contradiction by considering the change in V after one circuit. On the one hand, $\Delta V = 0$ since V is single-valued. But on the other hand,

$$\begin{aligned}\Delta V &= \int_0^T \frac{dV}{dt}dt \\ &= \int_0^T (\nabla V \cdot \dot{\mathbf{x}})dt \\ &= -\int_0^T \|\dot{\mathbf{x}}\|^2 dt \\ &< 0\end{aligned}$$

(unless $\dot{\mathbf{x}} \equiv \mathbf{0}$, in which case the trajectory is a fixed point, not a closed orbit). This contradiction shows that closed orbits can't exist in gradient systems. ■

The trouble with Theorem 7.2.1 is that most two-dimensional systems are *not* gradient systems. (Although, curiously, all vector fields *on the line* are gradient systems; this gives another explanation for the absence of oscillations noted in Sections 2.6 and 2.7.)

EXAMPLE 7.2.1:

Show that there are no closed orbits for the system $\dot{x} = \sin y$, $\dot{y} = x\cos y$.

Solution: The system is a gradient system with potential function $V(x, y) = -x\sin y$, since $\dot{x} = -\partial V/\partial x$ and $\dot{y} = -\partial V/\partial y$. By Theorem 7.2.1, there are no closed orbits. ■

How can you tell whether a system is a gradient system? And if it is, how do you find its potential function V? See Exercises 7.2.5 and 7.2.6.

Even if the system is not a gradient system, similar techniques may still work, as in the following example. We examine the change in an energy-like function after one circuit around the putative closed orbit, and derive a contradiction.

EXAMPLE 7.2.2:

Show that the nonlinearly damped oscillator $\ddot{x} + (\dot{x})^3 + x = 0$ has no periodic solutions.

Solution: Suppose that there were a periodic solution $x(t)$ of period T. Consider the energy function $E(x, \dot{x}) = \frac{1}{2}(x^2 + \dot{x}^2)$. After one cycle, x and $\dot{x}$ return to their starting values, and therefore $\Delta E = 0$ around any closed orbit.

On the other hand, $\Delta E = \int_0^T \dot{E}\, dt$. If we can show this integral is nonzero, we've reached a contradiction. Note that $\dot{E} = \dot{x}(x + \ddot{x}) = \dot{x}(-\dot{x}^3) = -\dot{x}^4 \le 0$. Therefore $\Delta E = -\int_0^T (\dot{x})^4 dt \le 0$, with equality only if $\dot{x} \equiv 0$. But $\dot{x} \equiv 0$ would mean the trajectory is a fixed point, contrary to the original assumption that it's a closed orbit. Thus ΔE is *strictly* negative, which contradicts $\Delta E = 0$. Hence there are no periodic solutions. ■

Liapunov Functions

Even for systems that have nothing to do with mechanics, it is occasionally possible to construct an energy-like function that decreases along trajectories. Such a function is called a Liapunov function. If a Liapunov function exists, then closed orbits are forbidden, by the same reasoning as in Example 7.2.2.

To be more precise, consider a system $\dot{\mathbf{x}} = \mathbf{f}(\mathbf{x})$ with a fixed point at $\mathbf{x}^*$. Suppose that we can find a ***Liapunov function***, i.e., a continuously differentiable, real-valued function $V(\mathbf{x})$ with the following properties:

1. $V(\mathbf{x}) > 0$ for all $\mathbf{x} \neq \mathbf{x}^*$, and $V(\mathbf{x}^*) = 0$. (We say that V is *positive definite*.)
2. $\dot{V} < 0$ for all $\mathbf{x} \neq \mathbf{x}^*$. (All trajectories flow "downhill" toward $\mathbf{x}^*$.)

Then $\mathbf{x}^*$ is globally asymptotically stable: for all initial conditions, $\mathbf{x}(t) \to \mathbf{x}^*$ as $t \to \infty$. In particular the system has no closed orbits. (For a proof, see Jordan and Smith 1987.)

The intuition is that all trajectories move monotonically down the graph of $V(\mathbf{x})$ toward $\mathbf{x}^*$ (Figure 7.2.1).

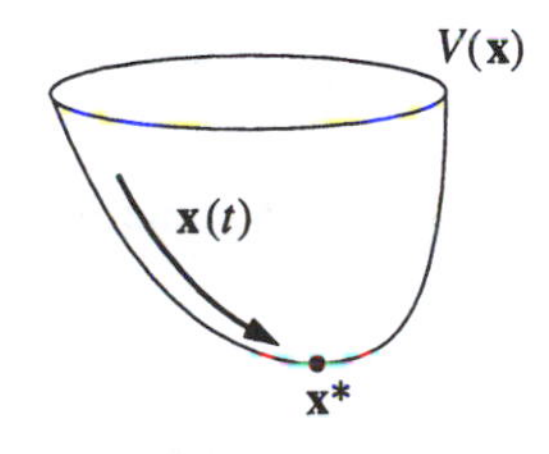

Figure 7.2.1

The solutions can't get stuck anywhere else because if they did, V would stop changing, but by assumption, $\dot{V} < 0$ everywhere except at $\mathbf{x}^*$.

Unfortunately, there is no systematic way to construct Liapunov functions. Divine inspiration is usually required, although sometimes one can work backwards. Sums of squares occasionally work, as in the following example.

EXAMPLE 7.2.3:

By constructing a Liapunov function, show that the system $\dot{x} = -x + 4y$, $\dot{y} = -x - y^3$ has no closed orbits.

Solution: Consider $V(x, y) = x^2 + ay^2$, where a is a parameter to be chosen later. Then $\dot{V} = 2x\dot{x} + 2ay\dot{y} = 2x(-x + 4y) + 2ay(-x - y^3) = -2x^2 + (8 - 2a)xy - 2ay^4$. If we choose $a = 4$, the xy term disappears and $\dot{V} = -2x^2 - 8y^4$. By inspection, $V > 0$ and $\dot{V} < 0$ for all $(x, y) \neq (0,0)$. Hence $V = x^2 + 4y^2$ is a Liapunov function and so there are no closed orbits. In fact, all trajectories approach the origin as $t \to \infty$. ■

Dulac's Criterion

The third method for ruling out closed orbits is based on Green's theorem, and is known as Dulac's criterion.

Dulac's Criterion: Let $\dot{\mathbf{x}} = \mathbf{f}(\mathbf{x})$ be a continuously differentiable vector field defined on a simply connected subset R of the plane. If there exists a continuously differentiable, real-valued function $g(\mathbf{x})$ such that $\nabla \cdot (g\dot{\mathbf{x}})$ has one sign throughout R, then there are no closed orbits lying entirely in R.

Proof: Suppose there were a closed orbit C lying entirely in the region R. Let A denote the region inside C (Figure 7.2.2). Then Green's theorem yields

$$\iint_A \nabla \cdot (g\dot{\mathbf{x}})\, dA = \oint_C g\dot{\mathbf{x}} \cdot \mathbf{n}\, d\ell$$

where $\mathbf{n}$ is the outward normal and $d\ell$ is the element of arc length along C. Look first at the double integral on the left: it must be *nonzero,* since $\nabla \cdot (g\dot{\mathbf{x}})$ has one sign in R. On the other hand, the line integral on the right equals *zero* since $\dot{\mathbf{x}} \cdot \mathbf{n} = 0$ everywhere, by the assumption that C is a trajectory (the tangent vector $\dot{\mathbf{x}}$ is orthogonal to $\mathbf{n}$). This contradiction implies that no such C can exist. ■

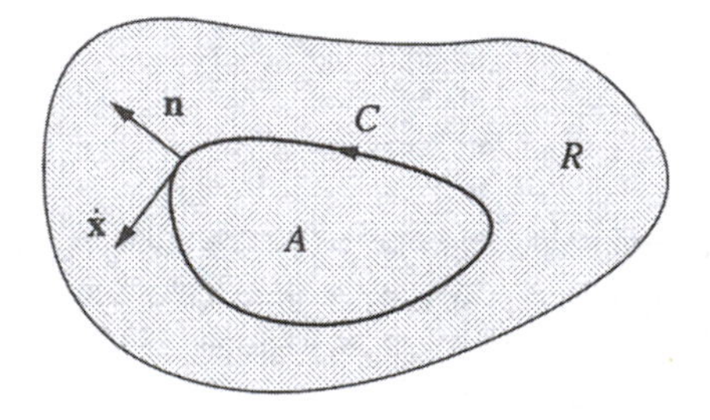

Figure 7.2.2

Dulac's criterion suffers from the same drawback as Liapunov's method: there is no algorithm for finding $g(\mathbf{x})$. Candidates that occasionally work are $g = 1$, $1/x^a y^b$, e^{ax}, and e^{ay}.

EXAMPLE 7.2.4:

Show that the system $\dot{x} = x(2 - x - y)$, $\dot{y} = y(4x - x^2 - 3)$ has no closed orbits in the positive quadrant $x, y > 0$.

Solution: A hunch tells us to pick $g = 1/xy$. Then

$$\begin{aligned}
\nabla \cdot (g\dot{\mathbf{x}}) &= \frac{\partial}{\partial x}(g\dot{x}) + \frac{\partial}{\partial y}(g\dot{y}) \\
&= \frac{\partial}{\partial x}\left(\frac{2-x-y}{y}\right) + \frac{\partial}{\partial y}\left(\frac{4x-x^2-3}{x}\right) \\
&= -1/y \\
&< 0.
\end{aligned}$$

Since the region $x, y > 0$ is simply connected and g and $\mathbf{f}$ satisfy the required smoothness conditions, Dulac's criterion implies there are no closed orbits in the positive quadrant. ■

EXAMPLE 7.2.5:

Show that the system $\dot{x} = y$, $\dot{y} = -x - y + x^2 + y^2$ has no closed orbits.

Solution: Let $g = e^{-2x}$. Then $\nabla \cdot (g\dot{\mathbf{x}}) = -2e^{-2x}y + e^{-2x}(-1+2y) = -e^{-2x} < 0$. By Dulac's criterion, there are no closed orbits. ■

7.3 Poincaré–Bendixson Theorem

Now that we know how to rule out closed orbits, we turn to the opposite task: finding methods to *establish that closed orbits exist* in particular systems. The following theorem is one of the few results in this direction. It is also one of the key theoretical results in nonlinear dynamics, because it implies that chaos can't occur in the phase plane, as discussed briefly at the end of this section.

Poincaré–Bendixson Theorem: Suppose that:

(1) R is a closed, bounded subset of the plane;
(2) $\dot{\mathbf{x}} = \mathbf{f}(\mathbf{x})$ is a continuously differentiable vector field on an open set containing R;
(3) R does not contain any fixed points; and
(4) There exists a trajectory C that is "confined" in R, in the sense that it starts in R and stays in R for all future time (Figure 7.3.1).

Then either C is a closed orbit, or it spirals toward a closed orbit as $t \to \infty$. In either case, *R contains a closed orbit* (shown as a heavy curve in Figure 7.3.1).

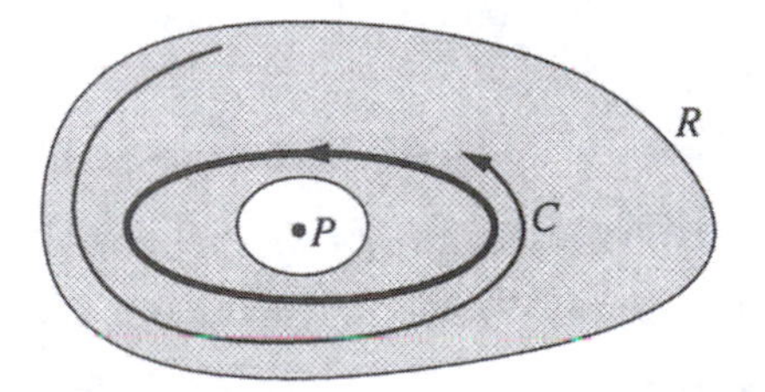

Figure 7.3.1

The proof of this theorem is subtle, and requires some advanced ideas from topology. For details, see Perko (1991), Coddington and Levinson (1955), Hurewicz (1958), or Cesari (1963).

In Figure 7.3.1, we have drawn R as a ring-shaped region because any closed orbit must encircle a fixed point (P in Figure 7.3.1) and no fixed points are allowed in R.

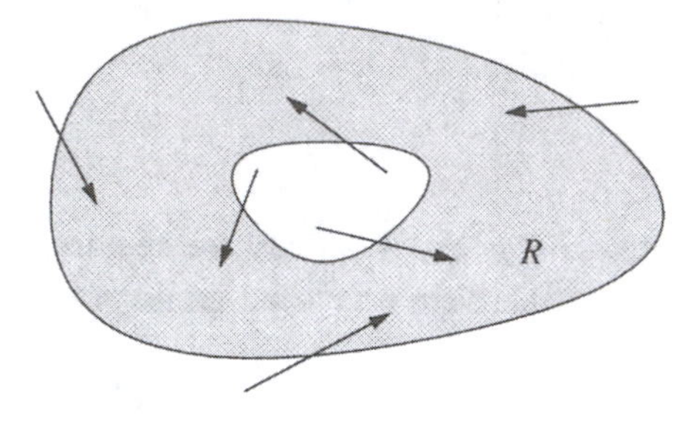

Figure 7.3.2

When applying the Poincaré–Bendixson theorem, it's easy to satisfy conditions (1)–(3); condition (4) is the tough one. How can we be sure that a confined trajectory C exists? The standard trick is to construct a ***trapping region*** R, i.e., a closed connected set such that the vector field points "inward" everywhere on the boundary of R (Figure 7.3.2). Then *all* trajectories in R are confined. If we can also arrange that there are no fixed points in R, then the Poincaré–Bendixson theorem ensures that R contains a closed orbit.

The Poincaré–Bendixson theorem can be difficult to apply in practice. One convenient case occurs when the system has a simple representation in polar coordinates, as in the following example.

EXAMPLE 7.3.1:

Consider the system

$$\begin{aligned}\dot{r} &= r(1-r^2)+\mu r\cos\theta \\ \dot{\theta} &= 1. \end{aligned} \qquad (1)$$

When $\mu = 0$, there's a stable limit cycle at $r = 1$, as discussed in Example 7.1.1. Show that a closed orbit still exists for $\mu > 0$, as long as μ is sufficiently small.

Solution: We seek two concentric circles with radii $r_{\min}$ and $r_{\max}$, such that $\dot{r} < 0$ on the outer circle and $\dot{r} > 0$ on the inner circle. Then the annulus $0 < r_{\min} \le r \le r_{\max}$ will be our desired trapping region. Note that there are no fixed points in the annulus since $\dot{\theta} > 0$; hence if $r_{\min}$ and $r_{\max}$ can be found, the Poincaré–Bendixson theorem will imply the existence of a closed orbit.

To find $r_{\min}$, we require $\dot{r} = r(1-r^2)+\mu r\cos\theta > 0$ for all θ. Since $\cos\theta \ge -1$, a sufficient condition for $r_{\min}$ is $1 - r^2 - \mu > 0$. Hence any $r_{\min} < \sqrt{1-\mu}$ will work, as long as $\mu < 1$ so that the square root makes sense. We should choose $r_{\min}$ as large as possible, to hem in the limit cycle as tightly as we can. For instance, we could pick $r_{\min} = 0.999\sqrt{1-\mu}$. (Even $r_{\min} = \sqrt{1-\mu}$ works, but more careful reasoning is required.) By a similar argument, the flow is inward on the outer circle if $r_{\max} = 1.001\sqrt{1+\mu}$.

Therefore a closed orbit exists for all $\mu < 1$, and it lies somewhere in the annulus $0.999\sqrt{1-\mu} < r < 1.001\sqrt{1+\mu}$. ■

The estimates used in Example 7.3.1 are conservative. In fact, the closed orbit can exist even if $\mu \geq 1$. Figure 7.3,3 shows a computer-generated phase portrait of (1) for $\mu = 1$. In Exercise 7.3.8, you're asked to explore what happens for larger μ, and in particular, whether there's a critical μ beyond which the closed orbit disappears. It's also possible to obtain some analytical insight about the closed orbit for small μ (Exercise 7.3.9).

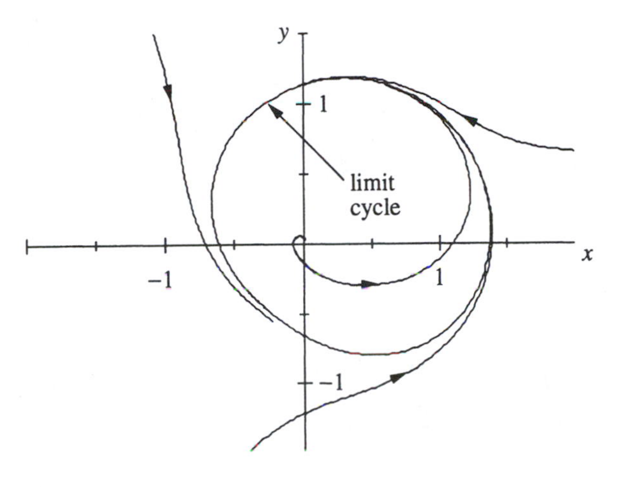

Figure 7.3.3

When polar coordinates are inconvenient, we may still be able to find an appropriate trapping region by examining the system's nullclines, as in the next example.

EXAMPLE 7.3.2:

In the fundamental biochemical process called ***glycolysis***, living cells obtain energy by breaking down sugar. In intact yeast cells as well as in yeast or muscle extracts, glycolysis can proceed in an *oscillatory* fashion, with the concentrations of various intermediates waxing and waning with a period of several minutes. For reviews, see Chance et al. (1973) or Goldbeter (1980).

A simple model of these oscillations has been proposed by Sel'kov (1968). In dimensionless form, the equations are

$$\dot{x} = -x + ay + x^2 y$$
$$\dot{y} = b - ay - x^2 y$$

where x and y are the concentrations of ADP (adenosine diphosphate) and F6P (fructose-6-phosphate), and $a, b > 0$ are kinetic parameters. Construct a trapping region for this system.

Solution: First we find the nullclines. The first equation shows that $\dot{x} = 0$ on the curve $y = x/(a + x^2)$ and the second equation shows that $\dot{y} = 0$ on the curve $y = b/(a + x^2)$. These nullclines are sketched in Figure 7.3.4, along with some representative vectors.

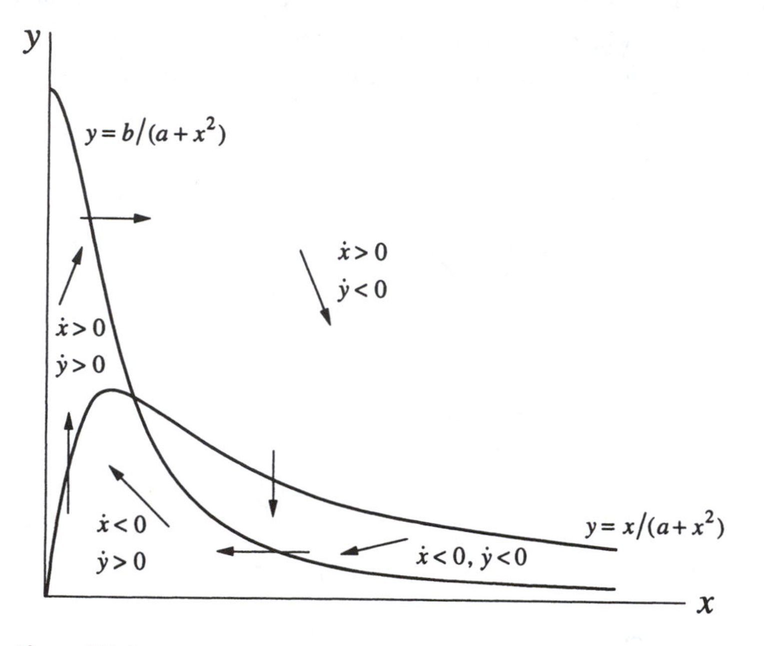

Figure 7.3.4

How did we know how to sketch these vectors? By definition, the arrows are vertical on the $\dot{x} = 0$ nullcline, and horizontal on the $\dot{y} = 0$ nullcline. The direction of flow is determined by the signs of $\dot{x}$ and $\dot{y}$. For instance, in the region above both nullclines, the governing equations imply $\dot{x} > 0$ and $\dot{y} < 0$, so the arrows point down and to the right, as shown in Figure 7.3.4.

Now consider the region bounded by the dashed line shown in Figure 7.3.5. *We claim that it's a trapping region.* To verify this, we have to show that all the vectors on the boundary point into the box. On the horizontal and vertical sides, there's no problem: the claim follows from Figure 7.3.4. The tricky part of the construction is the diagonal line of slope -1 extending from the point $(b, b/a)$ to the nullcline $y = x/(a + x^2)$. Where did this come from?

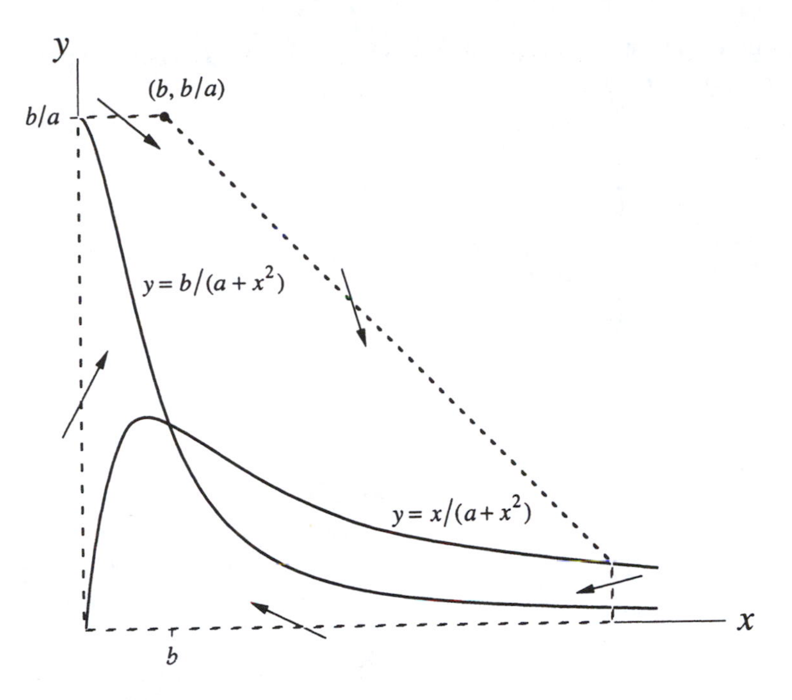

Figure 7.3.5

To get the right intuition, consider $\dot{x}$ and $\dot{y}$ in the limit of very large x. Then $\dot{x} \approx x^2 y$ and $\dot{y} \approx -x^2 y$, so $\dot{y}/\dot{x} = dy/dx \approx -1$ along trajectories. Hence the vector field at large x is roughly parallel to the diagonal line. This suggests that in a more precise calculation, we should compare the sizes of $\dot{x}$ and $-\dot{y}$, for some sufficiently large x.

In particular, consider $\dot{x} - (-\dot{y})$. We find

$$\begin{aligned}\dot{x} - (-\dot{y}) &= -x + ay + x^2 y + (b - ay - x^2 y) \\ &= b - x.\end{aligned}$$

Hence

$$-\dot{y} > \dot{x} \text{ if } x > b.$$

This inequality implies that the vector field points inward on the diagonal line in Figure 7.3.5, because dy/dx is more negative than -1, and therefore the vectors are steeper than the diagonal line. Thus the region is a trapping region, as claimed. ■

Can we conclude that there is a closed orbit inside the trapping region? No! There is a fixed point in the region (at the intersection of the nullclines), and so the conditions of the Poincaré–Bendixson theorem are not satisfied. But if this fixed point is a *repeller*, then we *can* prove the existence of a closed orbit by considering the modified "punctured" region shown in Figure 7.3.6. (The hole is infinitesimal, but drawn larger for clarity.)

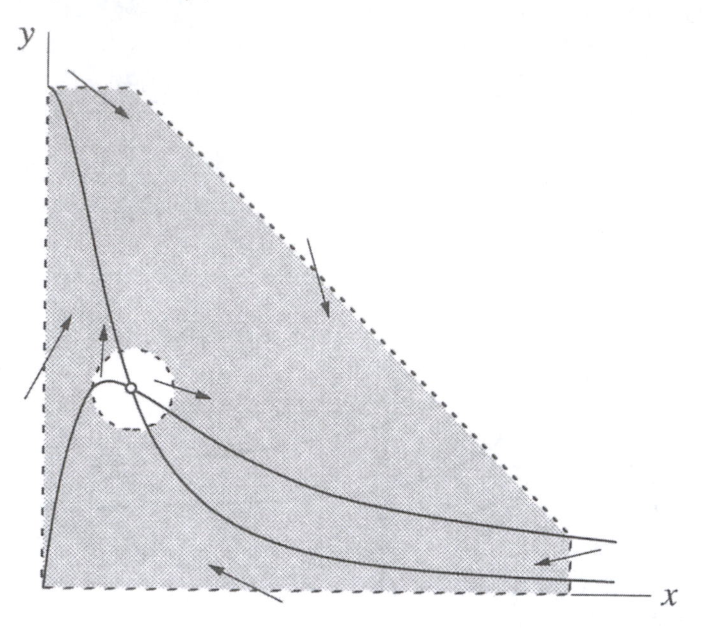

Figure 7.3.6

The repeller drives all neighboring trajectories into the shaded region, and since this region *is* free of fixed points, the Poincaré–Bendixson theorem applies.

Now we find conditions under which the fixed point is a repeller.

EXAMPLE 7.3.3:

Once again, consider the glycolytic oscillator $\dot{x} = -x + ay + x^2y$, $\dot{y} = b - ay - x^2y$ of Example 7.3.2. Prove that a closed orbit exists if a and b satisfy an appropriate condition, to be determined. (As before, $a, b > 0$.)

Solution: By the argument above, it suffices to find conditions under which the fixed point is a repeller, i.e., an unstable node or spiral. In general, the Jacobian is

$$A = \begin{pmatrix} -1 + 2xy & a + x^2 \\ -2xy & -(a + x^2) \end{pmatrix}.$$

After some algebra, we find that at the fixed point

$$x^* = b, \qquad y^* = \frac{b}{a + b^2},$$

the Jacobian has determinant $\Delta = a + b^2 > 0$ and trace

$$\tau = -\frac{b^4 + (2a-1)b^2 + (a+a^2)}{a+b^2}.$$

Hence the fixed point is unstable for $\tau > 0$, and stable for $\tau < 0$. The dividing line $\tau = 0$ occurs when

$$b^2 = \tfrac{1}{2}\left(1 - 2a \pm \sqrt{1-8a}\right).$$

This defines a curve in (a,b) space, as shown in Figure 7.3.7.

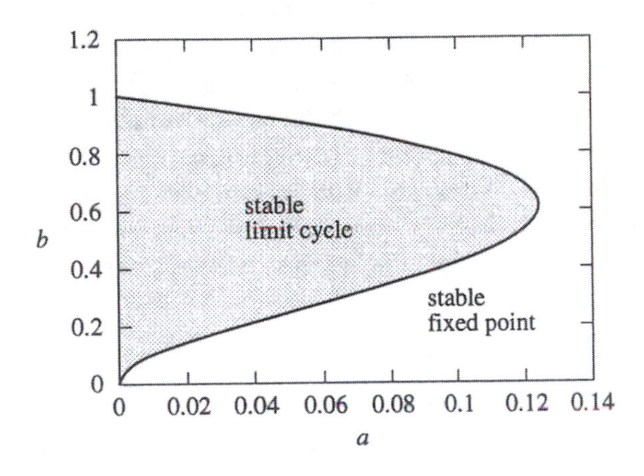

Figure 7.3.7

For parameters in the region corresponding to $\tau > 0$, we are guaranteed that the system has a closed orbit—numerical integration shows that it is actually a stable limit cycle. Figure 7.3.8 shows a computer-generated phase portrait for the typical case $a = 0.08$, $b = 0.6$. ■

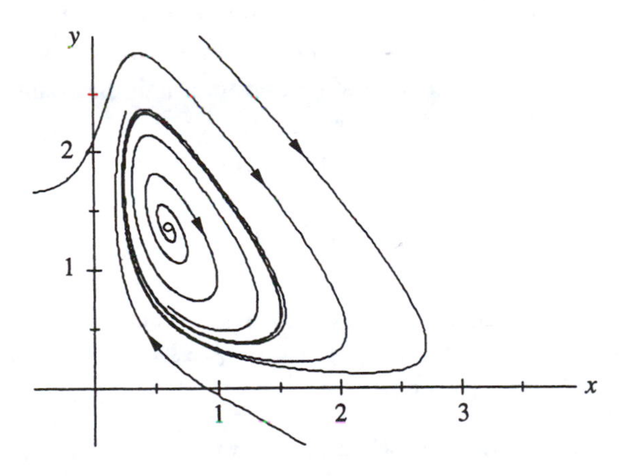

Figure 7.3.8

No Chaos in the Phase Plane

The Poincaré–Bendixson theorem is one of the central results of nonlinear dynamics. It says that the dynamical possibilities in the phase plane are very limited: if a trajectory is confined to a closed, bounded region that contains no fixed points, then the trajectory must eventually approach a closed orbit. Nothing more complicated is possible.

This result depends crucially on the two-dimensionality of the plane. In higher-dimensional systems ($n \geq 3$), the Poincaré–Bendixson theorem no longer applies, and something radically new can happen: trajectories may wander around forever in a bounded region without settling down to a fixed point or a closed orbit. In some cases, the trajectories are attracted to a complex geometric object called a *strange attractor*, a fractal set on which the motion is aperiodic and sensitive to tiny changes in the initial conditions. This sensitivity makes the motion unpredictable in the long run. We are now face to face with *chaos*. We'll discuss this fascinating topic soon enough, but for now you should appreciate that the Poincaré–Bendixson theorem implies that chaos can never occur in the phase plane.

7.4 Liénard Systems

In the early days of nonlinear dynamics, say from about 1920 to 1950, there was a great deal of research on nonlinear oscillations. The work was initially motivated by the development of radio and vacuum tube technology, and later it took on a mathematical life of its own. It was found that many oscillating circuits could be modeled by second-order differential equations of the form

$$\ddot{x} + f(x)\dot{x} + g(x) = 0, \qquad (1)$$

now known as ***Liénard's equation***. This equation is a generalization of the van der Pol oscillator $\ddot{x} + \mu(x^2 - 1)\dot{x} + x = 0$ mentioned in Section 7.1. It can also be interpreted mechanically as the equation of motion for a unit mass subject to a nonlinear damping force $-f(x)\dot{x}$ and a nonlinear restoring force $-g(x)$.

Liénard's equation is equivalent to the system

$$\begin{aligned} \dot{x} &= y \\ \dot{y} &= -g(x) - f(x)y. \end{aligned} \qquad (2)$$

The following theorem states that this system has a unique, stable limit cycle under appropriate hypotheses on f and g. For a proof, see Jordan and Smith (1987), Grimshaw (1990), or Perko (1991).

Liénard's Theorem: Suppose that $f(x)$ and $g(x)$ satisfy the following conditions:

(1) $f(x)$ and $g(x)$ are continuously differentiable for all x;

(2) $g(-x) = -g(x)$ for all x (i.e., $g(x)$ is an *odd* function);

(3) $g(x) > 0$ for $x > 0$;

(4) $f(-x) = f(x)$ for all x (i.e., $f(x)$ is an *even* function);

(5) The odd function $F(x) = \int_0^x f(u)\,du$ has exactly one positive zero at $x = a$, is negative for $0 < x < a$, is positive and nondecreasing for $x > a$, and $F(x) \to \infty$ as $x \to \infty$.

Then the system (2) has a unique, stable limit cycle surrounding the origin in the phase plane.

This result should seem plausible. The assumptions on $g(x)$ mean that the restoring force acts like an ordinary spring, and tends to reduce any displacement, whereas the assumptions on $f(x)$ imply that the damping is negative at small $|x|$ and positive at large $|x|$. Since small oscillations are pumped up and large oscillations are damped down, it is not surprising that the system tends to settle into a self-sustained oscillation of some intermediate amplitude.

EXAMPLE 7.4.1:

Show that the van der Pol equation has a unique, stable limit cycle.

Solution: The van der Pol equation $\ddot{x} + \mu(x^2 - 1)\dot{x} + x = 0$ has $f(x) = \mu(x^2 - 1)$ and $g(x) = x$, so conditions (1)–(4) of Liénard's theorem are clearly satisfied. To check condition (5), notice that

$$F(x) = \mu\left(\tfrac{1}{3}x^3 - x\right) = \tfrac{1}{3}\mu x(x^2 - 3).$$

Hence condition (5) is satisfied for $a = \sqrt{3}$. Thus the van der Pol equation has a unique, stable limit cycle. ■

There are several other classical results about the existence of periodic solutions for Liénard's equation and its relatives. See Stoker (1950), Minorsky (1962), Andronov et al. (1973), and Jordan and Smith (1987).

7.5 Relaxation Oscillations

It's time to change gears. So far in this chapter, we have focused on a qualitative question: Given a particular two-dimensional system, does it have any periodic solutions? Now we ask a quantitative question: Given that a closed orbit exists, what can we say about its shape and period? In general, such problems can't be solved exactly, but we can still obtain useful approximations if some parameter is large or small.

We begin by considering the van der Pol equation

$$\ddot{x} + \mu(x^2 - 1)\dot{x} + x = 0$$

for $\mu \gg 1$. In this *strongly nonlinear* limit, we'll see that the limit cycle consists of an extremely slow buildup followed by a sudden discharge, followed by another slow buildup, and so on. Oscillations of this type are often called ***relaxation oscillations***, because the "stress" accumulated during the slow buildup is "relaxed" during the sudden discharge. Relaxation oscillations occur in many other scientific contexts, from the stick-slip oscillations of a bowed violin string to the periodic firing of nerve cells driven by a constant current (Edelstein–Keshet 1988, Murray 1989, Rinzel and Ermentrout 1989).

EXAMPLE 7.5.1:

Give a phase plane analysis of the van der Pol equation for $\mu \gg 1$.

Solution: It proves convenient to introduce different phase plane variables from the usual "$\dot{x} = y$, $\dot{y} = \ldots$". To motivate the new variables, notice that

$$\ddot{x} + \mu\dot{x}(x^2 - 1) = \frac{d}{dt}\left(\dot{x} + \mu\left[\tfrac{1}{3}x^3 - x\right]\right).$$

So if we let

$$F(x) = \tfrac{1}{3}x^3 - x, \qquad w = \dot{x} + \mu F(x), \tag{1}$$

the van der Pol equation implies that

$$\dot{w} = \ddot{x} + \mu\dot{x}(x^2 - 1) = -x. \tag{2}$$

Hence the van der Pol equation is equivalent to (1), (2), which may be rewritten as

$$\begin{aligned} \dot{x} &= w - \mu F(x) \\ \dot{w} &= -x. \end{aligned} \tag{3}$$

One further change of variables is helpful. If we let

$$y = \frac{w}{\mu}$$

then (3) becomes

$$\begin{aligned} \dot{x} &= \mu\left[\,y - F(x)\,\right] \\ \dot{y} &= -\tfrac{1}{\mu}x. \end{aligned} \tag{4}$$

Now consider a typical trajectory in the (x, y) phase plane. The nullclines are the key to understanding the motion. We claim that all trajectories behave like that shown in Figure 7.5.1; starting from any point except the origin, the trajectory zaps horizontally onto the ***cubic nullcline*** $y = F(x)$. Then it crawls down the nullcline until it comes to the knee (point B in Figure 7.5.1), after which it zaps over to the other branch of the cubic at C. This is followed by another crawl along the cubic

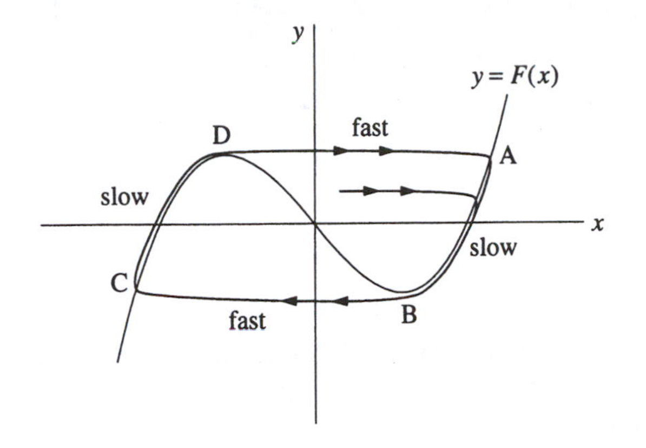

Figure 7.5.1

until the trajectory reaches the next jumping-off point at D, and the motion continues periodically after that.

To justify this picture, suppose that the initial condition is not too close to the cubic nullcline, i.e., suppose $y - F(x) \sim O(1)$. Then (4) implies $|\dot{x}| \sim O(\mu) \gg 1$ whereas $|\dot{y}| \sim O(\mu^{-1}) \ll 1$; hence the velocity is enormous in the horizontal direction and tiny in the vertical direction, so trajectories move practically horizontally. If the initial condition is *above* the nullcline, then $y - F(x) > 0$ and therefore $\dot{x} > 0$; thus the trajectory moves sideways *toward* the nullcline. However, once the trajectory gets so close that $y - F(x) \sim O(\mu^{-2})$, then $\dot{x}$ and $\dot{y}$ become comparable, both being $O(\mu^{-1})$. What happens then? The trajectory crosses the nullcline vertically, as shown in Figure 7.5.1, and then moves slowly along the backside of the branch, with a velocity of size $O(\mu^{-1})$, until it reaches the knee and can jump sideways again. ■

This analysis shows that the limit cycle has two ***widely separated time scales***: the crawls require $\Delta t \sim O(\mu)$ and the jumps require $\Delta t \sim O(\mu^{-1})$. Both time scales are apparent in the waveform of $x(t)$ shown in Figure 7.5.2, obtained by numerical integration of the van der Pol equation for $\mu = 10$ and initial condition $(x_0, y_0) = (2, 0)$.

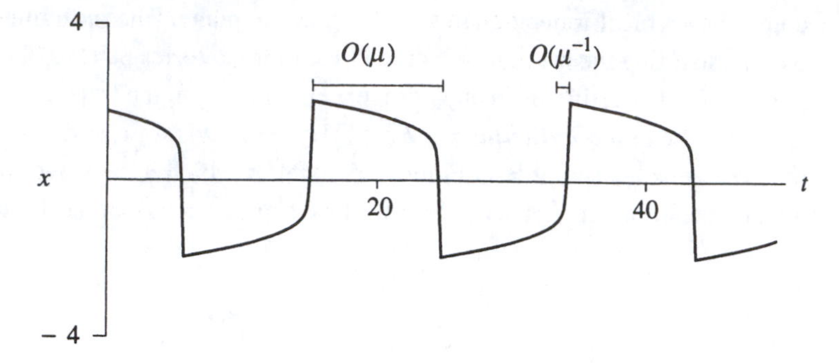

Figure 7.5.2

EXAMPLE 7.5.2:

Estimate the period of the limit cycle for the van der Pol equation for $\mu >> 1$.

Solution: The period T is essentially the time required to travel along the two ***slow branches***, since the time spent in the jumps is negligible for large μ. By symmetry, the time spent on each branch is the same. Hence $T \approx 2\int_{t_A}^{t_B} dt$. To derive an expression for dt, note that on the slow branches, $y \approx F(x)$ and thus

$$\frac{dy}{dt} \approx F'(x)\frac{dx}{dt} = (x^2 - 1)\frac{dx}{dt}.$$

But since $dy/dt = -x/\mu$ from (4), we find $dx/dt = -x/\mu(x^2 - 1)$. Therefore

$$dt \approx -\frac{\mu(x^2 - 1)}{x}\,dx \tag{5}$$

on a slow branch. As you can check (Exercise 7.5.1), the positive branch begins at $x_A = 2$ and ends at $x_B = 1$. Hence

$$T \approx 2\int_2^1 \frac{-\mu}{x}(x^2 - 1)\,dx = 2\mu\left[\frac{x^2}{2} - \ln x\right]_1^2 = \mu\left[3 - 2\ln 2\right], \tag{6}$$

which is $O(\mu)$ as expected. ■

The formula (6) can be refined. With much more work, one can show that $T \approx \mu[3 - 2\ln 2] + 2\alpha\mu^{-1/3} + \ldots$, where $\alpha \approx 2.338$ is the smallest root of $\mathrm{Ai}(-\alpha) = 0$. Here $\mathrm{Ai}(x)$ is a special function called the Airy function. This correction term comes from an estimate of the time required to turn the corner between the jumps and the crawls. See Grimshaw (1990, pp. 161–163) for a readable derivation of this wonderful formula, discovered by Mary Cartwright (1952). See also Stoker (1950) for more about relaxation oscillations.

One last remark: We have seen that a relaxation oscillation has two time scales that operate *sequentially*—a slow buildup is followed by a fast discharge. In the next section we will encounter problems where two time scales operate *concurrently*, and that makes the problems a bit more subtle.

7.6 Weakly Nonlinear Oscillators

This section deals with equations of the form

$$\ddot{x} + x + \varepsilon h(x, \dot{x}) = 0 \qquad (1)$$

where $0 \le \varepsilon << 1$ and $h(x, \dot{x})$ is an arbitrary smooth function. Such equations represent small perturbations of the linear oscillator $\ddot{x} + x = 0$ and are therefore called ***weakly nonlinear oscillators***. Two fundamental examples are the van der Pol equation

$$\ddot{x} + x + \varepsilon(x^2 - 1)\dot{x} = 0, \qquad (2)$$

(now in the limit of small nonlinearity), and the ***Duffing equation***

$$\ddot{x} + x + \varepsilon x^3 = 0. \qquad (3)$$

To illustrate the kinds of phenomena that can arise, Figure 7.6.1 shows a computer-generated solution of the van der Pol equation in the $(x, \dot{x})$ phase plane, for $\varepsilon = 0.1$ and an initial condition close to the origin. The trajectory is a slowly winding spiral; it takes many cycles for the amplitude to grow substantially. Eventually

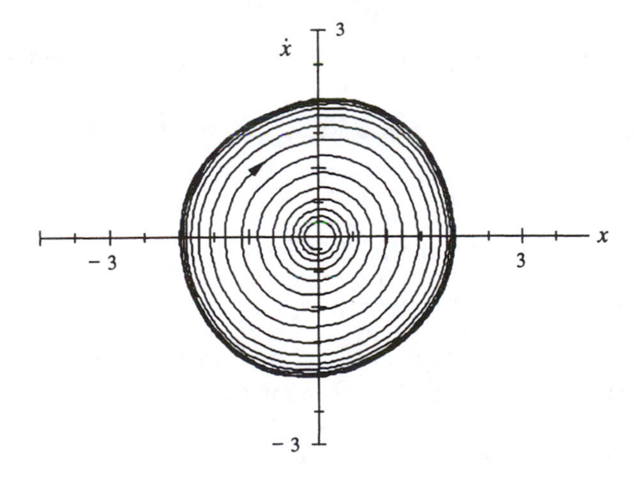

Figure 7.6.1

the trajectory asymptotes to an approximately circular limit cycle whose radius is close to 2.

We'd like to be able to predict the shape, period, and radius of this limit cycle. Our analysis will exploit the fact that the oscillator is "close to" a simple harmonic oscillator, which we understand completely.

Regular Perturbation Theory and Its Failure

As a first approach, we seek solutions of (1) in the form of a power series in ε. Thus if $x(t,\varepsilon)$ is a solution, we expand it as

$$x(t,\varepsilon) = x_0(t) + \varepsilon x_1(t) + \varepsilon^2 x_2(t) + \dots, \tag{4}$$

where the unknown functions $x_k(t)$ are to be determined from the governing equation and the initial conditions. The hope is that all the important information is captured by the first few terms—ideally, the first *two*—and that the higher-order terms represent only tiny corrections. This technique is called ***regular perturbation theory***. It works well on certain classes of problems (for instance, Exercise 7.3.9), but as we'll see, it runs into trouble here.

To expose the source of the difficulties, we begin with a practice problem that can be solved exactly. Consider the weakly damped linear oscillator

$$\ddot{x} + 2\varepsilon\dot{x} + x = 0, \tag{5}$$

with initial conditions

$$x(0) = 0, \qquad \dot{x}(0) = 1. \tag{6}$$

Using the techniques of Chapter 5, we find the exact solution

$$x(t,\varepsilon) = \left(1-\varepsilon^2\right)^{-1/2} e^{-\varepsilon t} \sin\left[\left(1-\varepsilon^2\right)^{1/2} t\right]. \tag{7}$$

Now let's solve the same problem using perturbation theory. Substitution of (4) into (5) yields

$$\frac{d^2}{dt^2}\left(x_0 + \varepsilon x_1 + \dots\right) + 2\varepsilon\frac{d}{dt}\left(x_0 + \varepsilon x_1 + \dots\right) + \left(x_0 + \varepsilon x_1 + \dots\right) = 0. \tag{8}$$

If we group the terms according to powers of ε, we get

$$\left[\ddot{x}_0 + x_0\right] + \varepsilon\left[\ddot{x}_1 + 2\dot{x}_0 + x_1\right] + O(\varepsilon^2) = 0. \tag{9}$$

Since (9) is supposed to hold for *all* sufficiently small ε, the coefficients of each power of ε must vanish separately. Thus we find

$$O(1): \ \ddot{x}_0 + x_0 = 0 \tag{10}$$

$$O(\varepsilon): \ \ddot{x}_1 + 2\dot{x}_0 + x_1 = 0. \tag{11}$$

(We're ignoring the $O(\varepsilon^2)$ and higher equations, in the optimistic spirit mentioned earlier.)

The appropriate initial conditions for these equations come from (6). At $t=0$, (4) implies that $0 = x_0(0) + \varepsilon x_1(0) + \ldots$; this holds for all ε, so

$$x_0(0) = 0, \qquad x_1(0) = 0. \tag{12}$$

By applying a similar argument to $\dot{x}(0)$ we obtain

$$\dot{x}_0(0) = 1, \qquad \dot{x}_1(0) = 0. \tag{13}$$

Now we solve the initial-value problems one by one; they fall like dominoes. The solution of (10), subject to the initial conditions $x_0(0) = 0$, $\dot{x}_0(0) = 1$, is

$$x_0(t) = \sin t. \tag{14}$$

Plugging this solution into (11) gives

$$\ddot{x}_1 + x_1 = -2\cos t. \tag{15}$$

Here's the first sign of trouble: the right-hand side of (15) is a ***resonant*** forcing. The solution of (15) subject to $x_1(0) = 0$, $\dot{x}_1(0) = 0$ is

$$x_1(t) = -t\sin t, \tag{16}$$

which is a ***secular*** term, i.e., a term that *grows* without bound as $t \to \infty$.

In summary, the solution of (5), (6) according to perturbation theory is

$$x(t,\varepsilon) = \sin t - \varepsilon t \sin t + O(\varepsilon^2). \tag{17}$$

How does this compare with the exact solution (7)? In Exercise 7.6.1, you are asked to show that the two formulas agree in the following sense: If (7) is expanded as power series in ε, the first two terms are given by (17). In fact, (17) is the beginning of a *convergent* series expansion for the true solution. For any fixed t, (17) provides a good approximation as long as ε is small enough—specifically, we need $\varepsilon t \ll 1$ so that the correction term (which is actually $O(\varepsilon^2 t^2)$) is negligible.

But normally we are interested in the behavior for *fixed* ε, not fixed t. In that case we can only expect the perturbation approximation to work for times $t \ll O(1/\varepsilon)$. To illustrate this limitation, Figure 7.6.2 plots the exact solution (7) and the perturbation series (17) for $\varepsilon = 0.1$. As expected, the perturbation series works reasonably well if $t \ll \frac{1}{\varepsilon} = 10$, but it breaks down after that.

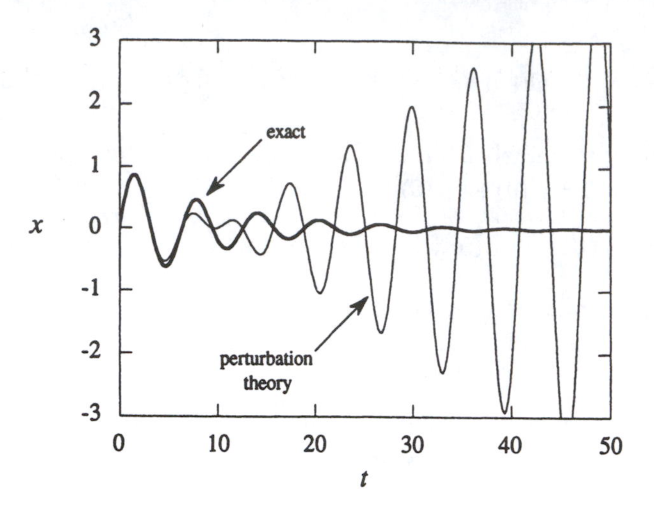

Figure 7.6.2

In many situations we'd like our approximation to capture the true solution's qualitative behavior for all t, or at least for large t. By this criterion, (17) is a failure, as Figure 7.6.2 makes obvious. There are two major problems:

1. The true solution (7) exhibits ***two time scales***: a *fast time* $t \sim O(1)$ for the sinusoidal oscillations and a *slow time* $t \sim 1/\varepsilon$ over which the amplitude decays. Equation (17) completely misrepresents the slow time scale behavior. In particular, because of the secular term $t \sin t$, (17) falsely suggests that the solution grows with time whereas we know from (7) that the amplitude $A = \left(1-\varepsilon^2\right)^{-1/2} e^{-\varepsilon t}$ decays exponentially.

   The discrepancy occurs because $e^{-\varepsilon t} = 1 - \varepsilon t + O(\varepsilon^2 t^2)$, so to this order in ε, it appears (incorrectly) that the amplitude increases with t. To get the correct result, we'd need to calculate an *infinite* number of terms in the series. That's worthless; we want series approximations that work well with just one or two terms.
2. The frequency of the oscillations in (7) is $\omega = \left(1-\varepsilon^2\right)^{1/2} \approx 1 - \frac{1}{2}\varepsilon^2$, which is shifted slightly from the frequency $\omega = 1$ of (17). After a *very* long time $t \sim O(1/\varepsilon^2)$, this frequency error will have a significant cumulative effect. Note that this is a third, *super-slow* time scale!

Two-Timing

The elementary example above reveals a more general truth: There are going to be (at least) two time scales in weakly nonlinear oscillators. We've already met this phenomenon in Figure 7.6.1, where the amplitude of the spiral grew very slowly compared to the cycle time. An analytical method called ***two-timing*** builds in the fact of two time scales from the start, and produces better approximations than regular perturbation theory. In fact, more than two times can be used, but we'll stick to the simplest case.

To apply two-timing to (1), let $\tau = t$ denote the fast $O(1)$ time, and let $T = \varepsilon t$ denote the slow time. We'll treat these two times as if they were *independent* variables. In particular, functions of the slow time T will be regarded as *constants* on the fast time scale τ. It's hard to justify this idea rigorously, but it works! (Here's an analogy: it's like saying that your height is constant on the time scale of a day. Of course, over many months or years your height can change dramatically, especially if you're an infant or a pubescent teenager, but over one day your height stays constant, to a good approximation.)

Now we turn to the mechanics of the method. We expand the solution of (1) as a series

$$x(t,\varepsilon) = x_0(\tau,T) + \varepsilon x_1(\tau,T) + O(\varepsilon^2). \tag{18}$$

The time derivatives in (1) are transformed using the chain rule:

$$\dot{x} = \frac{dx}{dt} = \frac{\partial x}{\partial \tau} + \frac{\partial x}{\partial T}\frac{\partial T}{\partial t} = \frac{\partial x}{\partial \tau} + \varepsilon \frac{\partial x}{\partial T}. \tag{19}$$

A subscript notation for differentiation is more compact; thus we write (19) as

$$\dot{x} = \partial_\tau x + \varepsilon \partial_T x. \tag{20}$$

After substituting (18) into (20) and collecting powers of ε, we find

$$\dot{x} = \partial_\tau x_0 + \varepsilon\left(\partial_T x_0 + \partial_\tau x_1\right) + O(\varepsilon^2). \tag{21}$$

Similarly,

$$\ddot{x} = \partial_{\tau\tau} x_0 + \varepsilon\left(\partial_{\tau\tau} x_1 + 2\partial_{T\tau} x_0\right) + O(\varepsilon^2). \tag{22}$$

To illustrate the method, let's apply it to our earlier test problem.

EXAMPLE 7.6.1:

Use two-timing to approximate the solution to the damped linear oscillator $\ddot{x} + 2\varepsilon\dot{x} + x = 0$, with initial conditions $x(0) = 0$, $\dot{x}(0) = 1$.

Solution: After substituting (21) and (22) for $\dot{x}$ and $\ddot{x}$, we get

$$\partial_{\tau\tau} x_0 + \varepsilon\left(\partial_{\tau\tau} x_1 + 2\partial_{T\tau} x_0\right) + 2\varepsilon\partial_\tau x_0 + x_0 + \varepsilon x_1 + O(\varepsilon^2) = 0. \tag{23}$$

Collecting powers of ε yields a pair of differential equations:

$$O(1): \partial_{\tau\tau} x_0 + x_0 = 0 \tag{24}$$

$$O(\varepsilon): \partial_{\tau\tau} x_1 + 2\partial_{T\tau} x_0 + 2\partial_\tau x_0 + x_1 = 0. \tag{25}$$

Equation (24) is just a simple harmonic oscillator. Its general solution is

$$x_0 = A \sin \tau + B \cos \tau, \tag{26}$$

but now comes the interesting part: *The "constants" A and B are actually functions of the slow time T.* Here we are invoking the above-mentioned ideas that τ and T should be regarded as independent variables, with functions of T behaving like constants on the fast time scale τ.

To determine $A(T)$ and $B(T)$, we need to go to the next order of ε. Substituting (26) into (25) gives

$$\begin{aligned} \partial_{\tau\tau} x_1 + x_1 &= -2\left(\partial_{T\tau} x_0 + \partial_\tau x_0\right) \\ &= -2(A' + A)\cos\tau \ + \ 2(B' + B)\sin\tau \end{aligned} \tag{27}$$

where the prime denotes differentiation with respect to T.

Now we face the same predicament that ruined us after (15). As in that case, the right-hand side of (27) is a resonant forcing that will produce *secular terms* like $\tau \sin \tau$ and $\tau \cos \tau$ in the solution for x_1. These terms would lead to a convergent but useless series expansion for x. Since we want an approximation free from secular terms, *we set the coefficients of the resonant terms to zero*—this manuever is characteristic of all two-timing calculations. Here it yields

$$A' + A = 0 \tag{28}$$

$$B' + B = 0. \tag{29}$$

The solutions of (28) and (29) are

$$\begin{aligned} A(T) &= A(0)e^{-T} \\ B(T) &= B(0)e^{-T}. \end{aligned}$$

The last step is to find the initial values $A(0)$ and $B(0)$. They are determined by (18), (26), and the given initial conditions $x(0) = 0$, $\dot{x}(0) = 1$, as follows. Equation (18) gives $0 = x(0) = x_0(0,0) + \varepsilon x_1(0,0) + O(\varepsilon^2)$. To satisfy this equation for *all* sufficiently small ε, we must have

$$x_0(0,0) = 0 \tag{30}$$

and $x_1(0,0) = 0$. Similarly,

$$1 = \dot{x}(0) = \partial_\tau x_0(0,0) + \varepsilon\left(\partial_T x_0(0,0) + \partial_\tau x_1(0,0)\right) + O(\varepsilon^2)$$

so

$$\partial_\tau x_0(0,0) = 1 \tag{31}$$

and $\partial_T x_0(0,0) + \partial_\tau x_1(0,0) = 0$. Combining (26) and (30) we find $B(0) = 0$; hence $B(T) \equiv 0$. Similarly, (26) and (31) imply $A(0) = 1$, so $A(T) = e^{-T}$. Thus (26) becomes

$$x_0(\tau, T) = e^{-T} \sin \tau . \tag{32}$$

Hence

$$\begin{aligned} x &= e^{-T} \sin \tau + O(\varepsilon) \\ &= e^{-\varepsilon t} \sin t + O(\varepsilon) \end{aligned} \tag{33}$$

is the approximate solution predicted by two-timing. ■

Figure 7.6.3 compares the two-timing solution (33) to the exact solution (7) for $\varepsilon = 0.1$. The two curves are almost indistinguishable, even though ε is not terribly small. This is a characteristic feature of the method—it often works better than it has any right to.

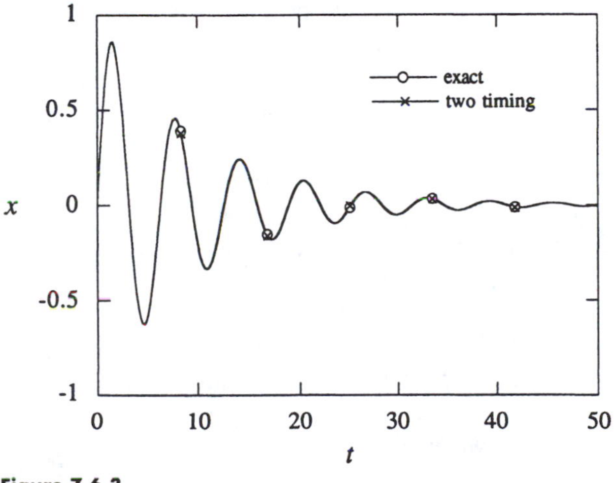

Figure 7.6.3

If we wanted to go further with Example 7.6.1, we could either solve for x_1 and higher-order corrections, or introduce a super-slow time $\Im = \varepsilon^2 t$ to investigate the long-term phase shift caused by the $O(\varepsilon^2)$ error in frequency. But Figure 7.6.3 shows that we already have a good approximation.

OK, enough practice problems! Now that we have calibrated the method, let's unleash it on a genuine nonlinear problem.

EXAMPLE 7.6.2:

Use two-timing to show that the van der Pol oscillator (2) has a stable limit cycle that is nearly circular, with a radius $= 2 + O(\varepsilon)$ and a frequency $\omega = 1 + O(\varepsilon^2)$.

Solution: The equation is $\ddot{x}+x+\varepsilon(x^2-1)\dot{x}=0$. Using (21) and (22) and collecting powers of ε, we find the following equations:

$$O(1): \partial_{\tau\tau}x_0 + x_0 = 0 \tag{34}$$

$$O(\varepsilon): \partial_{\tau\tau}x_1 + x_1 = -2\partial_{\tau T}x_0 - (x_0{}^2-1)\partial_\tau x_0. \tag{35}$$

As always, the $O(1)$ equation is a simple harmonic oscillator. Its general solution can be written as (26), or alternatively, as

$$x_0 = r(T)\cos(\tau+\phi(T)) \tag{36}$$

where $r(T)$ and $\phi(T)$ are the ***slowly-varying amplitude and phase*** of x_0.

To find equations governing r and ϕ, we insert (36) into (35). This yields

$$\begin{aligned}\partial_{\tau\tau}x_1 + x_1 &= -2\left(r'\sin(\tau+\phi)+r\phi'\cos(\tau+\phi)\right)\\ &\quad - r\sin(\tau+\phi)\left[r^2\cos^2(\tau+\phi)-1\right].\end{aligned} \tag{37}$$

As before, we need to avoid resonant terms on the right-hand side. These are terms proportional to $\cos(\tau+\phi)$ and $\sin(\tau+\phi)$. Some terms of this form already appear explicitly in (37). But—and this is the important point—there is also a resonant term lurking in $\sin(\tau+\phi)\cos^2(\tau+\phi)$, because of the trigonometric identity

$$\sin(\tau+\phi)\cos^2(\tau+\phi)=\tfrac{1}{4}\left[\sin(\tau+\phi)+\sin 3(\tau+\phi)\right]. \tag{38}$$

(Exercise 7.6.10 reminds you how to derive such identities, but usually we won't need them—shortcuts are available, as we'll see.) After substituting (38) into (37), we get

$$\begin{aligned}\partial_{\tau\tau}x_1 + x_1 &= \left[-2r'+r-\tfrac{1}{4}r^3\right]\sin(\tau+\phi)\\ &\quad +\left[-2r\phi'\right]\cos(\tau+\phi)-\tfrac{1}{4}r^3\sin 3(\tau+\phi).\end{aligned} \tag{39}$$

To avoid secular terms, we require

$$-2r'+r-\tfrac{1}{4}r^3=0 \tag{40}$$

$$-2r\phi'=0. \tag{41}$$

First consider (40). It may be rewritten as a vector field

$$r'=\tfrac{1}{8}r(4-r^2) \tag{42}$$

on the half-line $r\ge 0$. Following the methods of Chapter 2 or Example 7.1.1, we see that $r^*=0$ is an unstable fixed point and $r^*=2$ is a stable fixed point. Hence $r(T)\to 2$ as $T\to\infty$. Secondly, (41) implies $\phi'=0$, so $\phi(T)=\phi_0$ for some constant ϕ_0. Hence $x_0(\tau,T)\to 2\cos(\tau+\phi_0)$ and therefore

$$x(t) \to 2\cos(t+\phi_0)+O(\varepsilon) \tag{43}$$

as $t \to \infty$. Thus $x(t)$ approaches a stable limit cycle of radius $= 2+O(\varepsilon)$.

To find the frequency implied by (43), let $\theta = t+\phi(T)$ denote the argument of the cosine. Then the angular frequency ω is given by

$$\omega = \frac{d\theta}{dt} = 1+\frac{d\phi}{dT}\frac{dT}{dt} = 1+\varepsilon\phi' = 1, \tag{44}$$

through first order in ε. Hence $\omega = 1+O(\varepsilon^2)$; if we want an explicit formula for this $O(\varepsilon^2)$ correction term, we'd need to introduce a super-slow time $\mathfrak{I} = \varepsilon^2 t$, or we could use the Poincaré–Lindstedt method, as discussed in the exercises. ■

Averaged Equations

The same steps occur again and again in problems about weakly nonlinear oscillators. We can save time by deriving some general formulas.

Consider the equation for a general weakly nonlinear oscillator:

$$\ddot{x}+x+\varepsilon h(x,\dot{x}) = 0. \tag{45}$$

The usual two-timing substitutions give

$$O(1): \partial_{\tau\tau}x_0 + x_0 = 0 \tag{46}$$

$$O(\varepsilon): \partial_{\tau\tau}x_1 + x_1 = -2\partial_{\tau T}x_0 - h \tag{47}$$

where now $h = h(x_0, \partial_\tau x_0)$. As in Example 7.6.2, the solution of the $O(1)$ equation is

$$x_0 = r(T)\cos(\tau+\phi(T)). \tag{48}$$

Our goal is to derive differential equations for r' and ϕ', analogous to (40) and (41). We'll find these equations by insisting, as usual, that there be no terms proportional to $\cos(\tau+\phi)$ and $\sin(\tau+\phi)$ on the right-hand side of (47). Substituting (48) into (47), we see that this right-hand side is

$$2[r'\sin(\tau+\phi)+r\phi'\cos(\tau+\phi)]-h \tag{49}$$

where now $h = h(r\cos(\tau+\phi), -r\sin(\tau+\phi))$.

To extract the terms in h proportional to $\cos(\tau+\phi)$ and $\sin(\tau+\phi)$, we borrow some ideas from Fourier analysis. (If you're unfamiliar with Fourier analysis, don't worry—we'll derive all that we need in Exercise 7.6.12.) Notice that h is a 2π-periodic function of $\tau+\phi$. Let

$$\theta = \tau+\phi.$$

Fourier analysis tells us that $h(\theta)$ can be written as a ***Fourier series***

$$h(\theta)=\sum_{k=0}^{\infty}a_k\cos k\theta+\sum_{k=1}^{\infty}b_k\sin k\theta \tag{50}$$

where the ***Fourier coefficients*** are given by

$$\begin{aligned}a_0&=\tfrac{1}{2\pi}\int_0^{2\pi}h(\theta)\,d\theta\\ a_k&=\tfrac{1}{\pi}\int_0^{2\pi}h(\theta)\cos k\theta\,d\theta,\quad k\ge 1\\ b_k&=\tfrac{1}{\pi}\int_0^{2\pi}h(\theta)\sin k\theta\,d\theta,\quad k\ge 1.\end{aligned} \tag{51}$$

Hence (49) becomes

$$2\left[r'\sin\theta+r\phi'\cos\theta\right]-\sum_{k=0}^{\infty}a_k\cos k\theta-\sum_{k=1}^{\infty}b_k\sin k\theta\ . \tag{52}$$

The only resonant terms in (52) are $\left[2r'-b_1\right]\sin\theta$ and $\left[2r\phi'-a_1\right]\cos\theta$. Therefore, to avoid secular terms we need $r'=b_1/2$ and $r\phi'=a_1/2$. Using the expressions in (51) for a_1 and b_1, we obtain

$$\begin{aligned}r'&=\tfrac{1}{2\pi}\int_0^{2\pi}h(\theta)\sin\theta\,d\theta\equiv\langle h\sin\theta\rangle\\ r\phi'&=\tfrac{1}{2\pi}\int_0^{2\pi}h(\theta)\cos\theta\,d\theta\equiv\langle h\cos\theta\rangle\end{aligned} \tag{53}$$

where the angled brackets $\langle\cdot\rangle$ denote an average over one cycle of θ.

The equations in (53) are called the ***averaged*** or ***slow-time equations***. To use them, we write out $h=h(r\cos(\tau+\phi),\ -r\sin(\tau+\phi))=h(r\cos\theta\ ,\ -r\sin\theta)$ explicitly, and then compute the relevant averages over the fast variable θ, treating the slow variable r as constant. Here are some averages that appear often:

$$\begin{aligned}&\langle\cos\rangle=\langle\sin\rangle=0,\quad \langle\sin\ \cos\rangle=0,\quad \left\langle\cos^3\right\rangle=\left\langle\sin^3\right\rangle=0,\quad \left\langle\cos^{2n+1}\right\rangle=\left\langle\sin^{2n+1}\right\rangle=0,\\ &\left\langle\cos^2\right\rangle=\left\langle\sin^2\right\rangle=\tfrac{1}{2},\quad \left\langle\cos^4\right\rangle=\left\langle\sin^4\right\rangle=\tfrac{3}{8},\quad \left\langle\cos^2\sin^2\right\rangle=\tfrac{1}{8},\\ &\left\langle\cos^{2n}\right\rangle=\left\langle\sin^{2n}\right\rangle=\tfrac{1\cdot3\cdot5\ldots(2n-1)}{2\cdot4\cdot6\ldots(2n)},\quad n\ge1.\end{aligned} \tag{54}$$

Other averages can either be derived from these, or found by direct integration. For instance,

$$\left\langle\cos^2\sin^4\right\rangle=\left\langle(1-\sin^2)\sin^4\right\rangle=\left\langle\sin^4\right\rangle-\left\langle\sin^6\right\rangle=\tfrac{3}{8}-\tfrac{15}{48}=\tfrac{1}{16}$$

and

$$\langle \cos^3 \sin \rangle = \tfrac{1}{2\pi}\int_0^{2\pi} \cos^3\theta \, \sin\theta \, d\theta = -\tfrac{1}{2\pi}\left[\cos^4\theta\right]_0^{2\pi} = 0.$$

EXAMPLE 7.6.3:

Consider the van der Pol equation $\ddot{x} + x + \varepsilon(x^2 - 1)\dot{x} = 0$, subject to the initial conditions $x(0) = 1$, $\dot{x}(0) = 0$. Find the averaged equations, and then solve them to obtain an approximate formula for $x(t, \varepsilon)$. Compare your result to a numerical solution of the full equation, for $\varepsilon = 0.1$.

Solution: The van der Pol equation has $h = (x^2 - 1)\dot{x} = (r^2\cos^2\theta - 1)(-r\sin\theta)$. Hence (53) becomes

$$\begin{aligned} r' &= \langle h \sin\theta \rangle = \left\langle (r^2\cos^2\theta - 1)(-r\sin\theta)\,\sin\theta \right\rangle \\ &= r\left\langle \sin^2\theta \right\rangle - r^3\left\langle \cos^2\theta\,\sin^2\theta \right\rangle \\ &= \tfrac{1}{2}r - \tfrac{1}{8}r^3 \end{aligned}$$

and

$$\begin{aligned} r\phi' &= \langle h\cos\theta \rangle = \left\langle (r^2\cos^2\theta - 1)(-r\sin\theta)\,\cos\theta \right\rangle \\ &= r\left\langle \sin\theta\,\cos\theta \right\rangle - r^3\left\langle \cos^3\theta\,\sin\theta \right\rangle \\ &= 0 - 0 = 0. \end{aligned}$$

These equations match those found in Example 7.6.2, as they should.

The initial conditions $x(0) = 1$ and $\dot{x}(0) = 0$ imply $r(0) \approx \sqrt{x(0)^2 + \dot{x}(0)^2} = 1$ and $\phi(0) \approx \tan^{-1}\left(\dot{x}(0)/x(0)\right) - \tau = 0 - 0 = 0$. Since $\phi' = 0$, we find $\phi(T) \equiv 0$. To find $r(T)$, we solve $r' = \tfrac{1}{2}r - \tfrac{1}{8}r^3$ subject to $r(0) = 1$. The differential equation separates to

$$\int \frac{8\,dr}{r(4 - r^2)} = \int dT.$$

After integrating by partial fractions and using $r(0) = 1$, we find

$$r(T) = 2\left(1 + 3e^{-T}\right)^{-1/2}. \tag{55}$$

Hence

$$\begin{aligned} x(t,\varepsilon) &\sim x_0(\tau, T) + O(\varepsilon) \\ &= \frac{2}{\sqrt{1 + 3e^{-\varepsilon t}}}\cos t + O(\varepsilon). \end{aligned} \tag{56}$$

Equation (56) describes the transient dynamics of the oscillator as it spirals out to its limit cycle. Notice that $r(T) \to 2$ as $T \to \infty$, as in Example 7.6.2.

In Figure 7.6.4 we plot the "exact" solution of the van der Pol equation, obtained by numerical integration for $\varepsilon = 0.1$ and initial conditions $x(0) = 1$, $\dot{x}(0) = 0$. For comparison, the slowly-varying amplitude $r(T)$ predicted by (55) is also shown. The agreement is striking. Alternatively, we could have plotted the whole solution (56) instead of just its envelope; then the two curves would be virtually indistinguishable, like those in Figure 7.6.3. ■

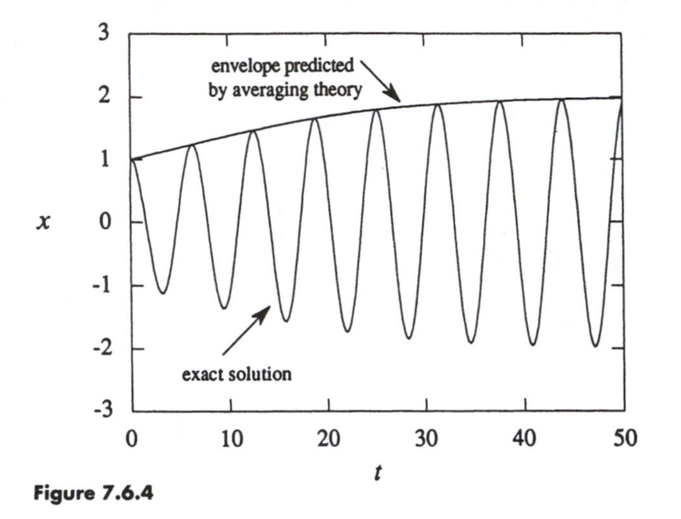

Figure 7.6.4

Now we consider an example in which the frequency of an oscillator depends on its amplitude. This is a common phenomenon, and one that is intrinsically *nonlinear*—it cannot occur for linear oscillators.

EXAMPLE 7.6.4:

Find an approximate relation between the amplitude and frequency of the Duffing oscillator $\ddot{x} + x + \varepsilon x^3 = 0$, where ε can have either sign. Interpret the results physically.

Solution: Here $h = x^3 = r^3 \cos^3 \theta$. Equation (53) becomes

$$r' = \langle h \sin\theta \rangle = r^3 \langle \cos^3 \theta \sin\theta \rangle = 0$$

and

$$r\phi' = \langle h \cos\theta \rangle = r^3 \langle \cos^4 \theta \rangle = \tfrac{3}{8} r^3 .$$

Hence $r(T) \equiv a$, for some constant a, and $\phi' = \frac{3}{8} a^2$. As in Example 7.6.2, the frequency ω is given by

$$\omega = 1 + \varepsilon\phi' = 1 + \tfrac{3}{8}\varepsilon a^2 + O(\varepsilon^2). \qquad (57)$$

Now for the physical interpretation. The Duffing equation describes the undamped motion of a unit mass attached to a nonlinear spring with restoring force $F(x) = -x - \varepsilon x^3$. We can use our intuition about ordinary linear springs if we write $F(x) = -kx$, where the spring stiffness is now dependent on x:

$$k = k(x) = 1 + \varepsilon x^2.$$

Suppose $\varepsilon > 0$. Then the spring gets *stiffer* as the displacement x increases—this is called a ***hardening spring.*** On physical grounds we'd expect it to *increase* the frequency of the oscillations, consistent with (57). For $\varepsilon < 0$ we have a ***softening spring***, exemplified by the pendulum (Exercise 7.6.15).

It also makes sense that $r' = 0$. The Duffing equation is a conservative system and for all ε sufficiently small, it has a *nonlinear center* at the origin (Exercise 6.5.13). Since all orbits close to the origin are periodic, there can be no long-term change in amplitude, consistent with $r' = 0$. ■

Validity of Two-Timing

We conclude with a few comments about the validity of the two-timing method. The rule of thumb is that the one-term approximation x_0 will be within $O(\varepsilon)$ of the true solution x for all times up to and including $t \sim O(1/\varepsilon)$, assuming that both x and x_0 start from the same initial condition. If x is a periodic solution, the situation is even better: x_0 remains within $O(\varepsilon)$ of x for *all* t.

But for precise statements and rigorous results about these matters, and for discussions of the subtleties that can occur, you should consult more advanced treatments, such as Guckenheimer and Holmes (1983) or Grimshaw (1990). Those authors use the *method of averaging*, an alternative approach that yields the same results as two-timing. See Exercise 7.6.25 for an introduction to this powerful technique.

Also, we have been very loose about the sense in which our formulas approximate the true solutions. The relevant notion is that of *asymptotic* approximation. For introductions to asymptotics, see Lin and Segel (1988) or Bender and Orszag (1978).

EXERCISES FOR CHAPTER 7

7.1 Examples

Sketch the phase portrait for each of the following systems. (As usual, r, θ denote polar coordinates.)

7.1.1 $\dot{r} = r^3 - 4r$, $\dot{\theta} = 1$ **7.1.2** $\dot{r} = r(1-r^2)(9-r^2)$, $\dot{\theta} = 1$

7.1.3 $\dot{r} = r(1-r^2)(4-r^2)$, $\dot{\theta} = 2 - r^2$ **7.1.4** $\dot{r} = r\sin r$, $\dot{\theta} = 1$

7.1.5 (From polar to Cartesian coordinates) Show that the system $\dot{r} = r(1-r^2)$, $\dot{\theta} = 1$ is equivalent to

$$\dot{x} = x - y - x(x^2+y^2), \qquad \dot{y} = x + y - y(x^2+y^2),$$

where $x = r\cos\theta$, $y = r\sin\theta$. (Hint: $\dot{x} = \frac{d}{dt}(r\cos\theta) = \dot{r}\cos\theta - r\dot{\theta}\sin\theta$.)

7.1.6 (Circuit for van der Pol oscillator) Figure 1 shows the "tetrode multivibrator" circuit used in the earliest commercial radios and analyzed by van der Pol. In van der Pol's day, the active element was a vacuum tube; today it would be a semiconductor device. It acts like an ordinary resistor when I is high, but like a negative resistor (energy source) when I is low. Its current-voltage characteristic $V = f(I)$ resembles a cubic function, as discussed below.

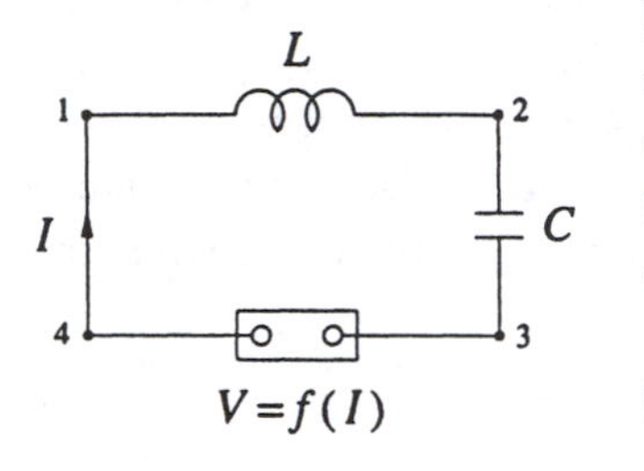

Figure 1

Suppose a source of current is attached to the circuit and then withdrawn. What equations govern the subsequent evolution of the current and the various voltages?

a) Let $V = V_{32} = -V_{23}$ denote the voltage drop from point 3 to point 2 in the circuit. Show that $\dot{V} = -I/C$ and $V = L\dot{I} + f(I)$.

b) Show that the equations in (a) are equivalent to

$$\frac{dw}{d\tau} = -x, \qquad \frac{dx}{d\tau} = w - \mu F(x)$$

where $x = L^{1/2}I$, $w = C^{1/2}V$, $\tau = (LC)^{-1/2}t$, and $F(x) = f(L^{-1/2}x)$.

In Section 7.5, we'll see that this system for (w, x) is equivalent to the van der Pol equation, if $F(x) = \frac{1}{3}x^3 - x$. Thus the circuit produces self-sustained oscillations.

7.1.7 (Waveform) Consider the system $\dot{r} = r(4-r^2)$, $\dot{\theta} = 1$, and let $x(t) = r(t)\cos\theta(t)$. Given the initial condition $x(0) = 0.1$, $y(0) = 0$, sketch the approximate waveform of $x(t)$, *without* obtaining an explicit expression for it.

7.1.8 (A circular limit cycle) Consider $\ddot{x} + a\dot{x}(x^2 + \dot{x}^2 - 1) + x = 0$, where $a > 0$.

a) Find and classify all the fixed points.

b) Show that the system has a circular limit cycle, and find its amplitude and period.

c) Determine the stability of the limit cycle.

d) Give an argument which shows that the limit cycle is unique, i.e., there are no other periodic trajectories.

7.1.9 (Circular pursuit problem) A dog at the center of a circular pond sees a duck swimming along the edge. The dog chases the duck by always swimming straight toward it. In other words, the dog's velocity vector always lies along the line connecting it to the duck. Meanwhile, the duck takes evasive action by swimming around the circumference as fast as it can, always moving counterclockwise.

a) Assuming the pond has unit radius and both animals swim at the same constant speed, derive a pair of differential equations for the path of the dog. (Hint: Use the coordinate system shown in Figure 2 and find equations for $dR/d\theta$ and $d\phi/d\theta$.) Analyze the system. Can you solve it explicitly? Does the dog ever catch the duck?

b) Now suppose the dog swims k times faster than the duck. Derive the differential equations for the dog's path.

c) If $k = \frac{1}{2}$, what does the dog end up doing in the long run?

Note: This problem has a long and intriguing history, dating back to the mid-1800s at least. It is much more difficult than similar ***pursuit problems***—there is no known solution for the path of the dog in part (a), in terms of elementary functions. See Davis (1962, pp. 113–125) for a nice analysis and a guide to the literature.

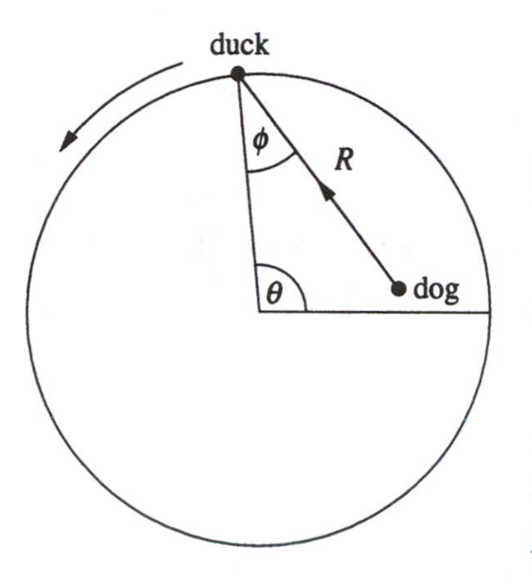

Figure 2

7.2 Ruling Out Closed Orbits

Plot the phase portraits of the following gradient systems $\dot{\mathbf{x}} = -\nabla V$.

7.2.1 $V = x^2 + y^2$ **7.2.2** $V = x^2 - y^2$ **7.2.3** $V = e^x \sin y$

7.2.4 Show that all vector fields on the line are gradient systems. Is the same true of vector fields on the circle?

7.2.5 Let $\dot{x} = f(x,y)$, $\dot{y} = g(x,y)$ be a smooth vector field defined on the phase plane.

a) Show that if this is a gradient system, then $\partial f/\partial y = \partial g/\partial x$.

b) Is the condition in (a) also sufficient?

7.2.6 Given that a system is a gradient system, here's how to find its potential function V. Suppose that $\dot{x} = f(x,y)$, $\dot{y} = g(x,y)$. Then $\dot{\mathbf{x}} = -\nabla V$ implies

$f(x,y) = -\partial V/\partial x$ and $g(x,y) = -\partial V/\partial y$. These two equations may be "partially integrated" to find V. Use this procedure to find V for the following gradient systems.

a) $\dot{x} = y^2 + y\cos x, \quad \dot{y} = 2xy + \sin x$

b) $\dot{x} = 3x^2 - 1 - e^{2y}, \quad \dot{y} = -2xe^{2y}$

7.2.7 Consider the system $\dot{x} = y + 2xy$, $\dot{y} = x + x^2 - y^2$.

a) Show that $\partial f/\partial y = \partial g/\partial x$. (Then Exercise 7.2.5(a) implies this is a gradient system.)

b) Find V.

c) Sketch the phase portrait.

7.2.8 Show that the trajectories of a gradient system always cross the equipotentials at right angles (except at fixed points).

7.2.9 For each of the following systems, decide whether it is a gradient system. If so, find V and sketch the phase portrait. On a separate graph, sketch the equipotentials $V = \text{constant}$. (If the system is not a gradient system, go on to the next question.)

a) $\dot{x} = y + x^2 y, \quad \dot{y} = -x + 2xy$

b) $\dot{x} = 2x, \quad \dot{y} = 8y$

c) $\dot{x} = -2xe^{x^2+y^2}, \quad \dot{y} = -2ye^{x^2+y^2}$

7.2.10 Show that the system $\dot{x} = y - x^3$, $\dot{y} = -x - y^3$ has no closed orbits, by constructing a Liapunov function $V = ax^2 + by^2$ with suitable a, b.

7.2.11 Show that $V = ax^2 + 2bxy + cy^2$ is positive definite if and only if $a > 0$ and $ac - b^2 > 0$. (This is a useful criterion that allows us to test for positive definiteness when the quadratic form V includes a "cross term" $2bxy$.)

7.2.12 Show that $\dot{x} = -x + 2y^3 - 2y^4$, $\dot{y} = -x - y + xy$ has no periodic solutions. (Hint: Choose a, m, and n such that $V = x^m + ay^n$ is a Liapunov function.)

7.2.13 Recall the competition model

$$\dot{N}_1 = r_1 N_1(1 - N_1/K_1) - b_1 N_1 N_2, \qquad \dot{N}_2 = r_2 N_2(1 - N_2/K_2) - b_2 N_1 N_2,$$

of Exercise 6.4.6. Using Dulac's criterion with the weighting function $g = (N_1 N_2)^{-1}$, show that the system has no periodic orbits in the first quadrant $N_1, N_2 > 0$.

7.2.14 Consider $\dot{x} = x^2 - y - 1$, $\dot{y} = y(x - 2)$.

a) Show that there are three fixed points and classify them.

b) By considering the three straight lines through pairs of fixed points, show that there are no closed orbits.
c) Sketch the phase portrait.

7.2.15 Consider the system $\dot{x} = x(2 - x - y)$, $\dot{y} = y(4x - x^2 - 3)$. We know from Example 7.2.4 that this system has no closed orbits.
a) Find the three fixed points and classify them.
b) Sketch the phase portrait.

7.2.16 If R is not simply connected, then the conclusion of Dulac's criterion is no longer valid. Find a counterexample.

7.2.17 Assume the hypotheses of Dulac's criterion, except now suppose that R is topologically equivalent to an annulus, i.e., it has exactly one hole in it. Using Green's theorem, show that there exists *at most* one closed orbit in R. (This result can be useful sometimes as a way of proving that a closed orbit is unique.)

7.3 Poincaré–Bendixson Theorem

7.3.1 Consider $\dot{x} = x - y - x(x^2 + 5y^2)$, $\dot{y} = x + y - y(x^2 + y^2)$.
a) Classify the fixed point at the origin.
b) Rewrite the system in polar coordinates, using $r\dot{r} = x\dot{x} + y\dot{y}$ and $\dot{\theta} = (x\dot{y} - y\dot{x})/r^2$.
c) Determine the circle of maximum radius, r_1, centered on the origin such that all trajectories have a radially *outward* component on it.
d) Determine the circle of minimum radius, r_2, centered on the origin such that all trajectories have a radially *inward* component on it.
e) Prove that the system has a limit cycle somewhere in the trapping region $r_1 \le r \le r_2$.

7.3.2 Using numerical integration, compute the limit cycle of Exercise 7.3.1 and verify that it lies in the trapping region you constructed.

7.3.3 Show that the system $\dot{x} = x - y - x^3$, $\dot{y} = x + y - y^3$ has a periodic solution.

7.3.4 Consider the system

$$\dot{x} = x(1 - 4x^2 - y^2) - \tfrac{1}{2}y(1 + x), \qquad \dot{y} = y(1 - 4x^2 - y^2) + 2x(1 + x).$$

a) Show that the origin is an unstable fixed point.
b) By considering $\dot{V}$, where $V = (1 - 4x^2 - y^2)^2$, show that all trajectories approach the ellipse $4x^2 + y^2 = 1$ as $t \to \infty$.

7.3.5 Show that the system $\dot{x} = -x - y + x(x^2 + 2y^2)$, $\dot{y} = x - y + y(x^2 + 2y^2)$ has at least one periodic solution.

7.3.6 Consider the oscillator equation $\ddot{x} + F(x,\dot{x})\dot{x} + x = 0$, where $F(x,\dot{x}) < 0$ if $r \le a$ and $F(x,\dot{x}) > 0$ if $r \ge b$, where $r^2 = x^2 + \dot{x}^2$.

a) Give a physical interpretation of the assumptions on F.

b) Show that there is at least one closed orbit in the region $a < r < b$.

7.3.7 Consider $\dot{x} = y + ax(1 - 2b - r^2)$, $\dot{y} = -x + ay(1 - r^2)$, where a and b are parameters $(0 < a \le 1,\ 0 \le b < \frac{1}{2})$ and $r^2 = x^2 + y^2$.

a) Rewrite the system in polar coordinates.

b) Prove that there is at least one limit cycle, and that if there are several, they all have the same period $T(a,b)$.

c) Prove that for $b = 0$ there is only one limit cycle.

7.3.8 Recall the system $\dot{r} = r(1 - r^2) + \mu r\cos\theta$, $\dot{\theta} = 1$ of Example 7.3.1. Using the computer, plot the phase portrait for various values of $\mu > 0$. Is there a critical value μ_c at which the closed orbit ceases to exist? If so, estimate it. If not, prove that a closed orbit exists for *all* $\mu > 0$.

7.3.9 (Series approximation for a closed orbit) In Example 7.3.1, we used the Poincaré–Bendixson Theorem to prove that the system $\dot{r} = r(1 - r^2) + \mu r\cos\theta$, $\dot{\theta} = 1$ has a closed orbit in the annulus $\sqrt{1-\mu} < r < \sqrt{1+\mu}$ for all $\mu < 1$.

a) To approximate the shape $r(\theta)$ of the orbit for $\mu \ll 1$, assume a power series solution of the form $r(\theta) = 1 + \mu r_1(\theta) + O(\mu^2)$. Substitute the series into a differential equation for $dr/d\theta$. Neglect all $O(\mu^2)$ terms, and thereby derive a simple differential equation for $r_1(\theta)$. Solve this equation explicitly for $r_1(\theta)$. (The approximation technique used here is called regular perturbation theory; see Section 7.6.)

b) Find the maximum and minimum r on your approximate orbit, and hence show that it lies in the annulus $\sqrt{1-\mu} < r < \sqrt{1+\mu}$, as expected.

c) Use a computer to calculate $r(\theta)$ numerically for various small μ, and plot the results on the same graph as your analytical approximation for $r(\theta)$. How does the maximum error depend on μ?

7.3.10 Consider the two-dimensional system $\dot{\mathbf{x}} = A\mathbf{x} - r^2\mathbf{x}$, where $r = \|\mathbf{x}\|$ and A is a 2×2 constant real matrix with complex eigenvalues $\alpha \pm i\omega$. Prove that there exists at least one limit cycle for $\alpha > 0$ and that there are none for $\alpha < 0$.

7.3.11 (Cycle graphs) Suppose $\dot{\mathbf{x}} = \mathbf{f}(\mathbf{x})$ is a smooth vector field on $\mathbf{R}^2$. An improved version of the Poincaré–Bendixson theorem states that if a trajectory is trapped in a compact region, then it must approach a fixed point, a closed orbit, or something exotic called a ***cycle graph*** (an invariant set containing a finite number of fixed points connected by a finite number of trajectories, all oriented either clockwise or counterclockwise). Cycle graphs are rare in practice; here's a contrived but simple example.

a) Plot the phase portrait for the system

$$\dot{r} = r(1-r^2)\left[r^2 \sin^2\theta + (r^2\cos^2\theta - 1)^2\right]$$

$$\dot{\theta} = r^2 \sin^2\theta + (r^2\cos^2\theta - 1)^2$$

where r, θ are polar coordinates. (Hint: Note the common factor in the two equations; examine where it vanishes.)

b) Sketch x vs. t for a trajectory starting away from the unit circle. What happens as $t \to \infty$?

7.4 Liénard Systems

7.4.1 Show that the equation $\ddot{x} + \mu(x^2 - 1)\dot{x} + \tanh x = 0$, for $\mu > 0$, has exactly one periodic solution, and classify its stability.

7.4.2 Consider the equation $\ddot{x} + \mu(x^4 - 1)\dot{x} + x = 0$.

a) Prove that the system has a unique stable limit cycle if $\mu > 0$.

b) Using a computer, plot the phase portrait for the case $\mu = 1$.

c) If $\mu < 0$, does the system still have a limit cycle? If so, is it stable or unstable?

7.5 Relaxation Oscillations

7.5.1 For the van der Pol oscillator with $\mu >> 1$, show that the positive branch of the cubic nullcline begins at $x_A = 2$ and ends at $x_B = 1$.

7.5.2 In Example 7.5.1, we used a tricky phase plane (often called the ***Liénard plane***) to analyze the van der Pol oscillator for $\mu >> 1$. Try to redo the analysis in the standard phase plane where $\dot{x} = y$, $\dot{y} = -x - \mu(x^2 - 1)$. What is the advantage of the Liénard plane?

7.5.3 Estimate the period of the limit cycle of $\ddot{x} + k(x^2 - 4)\dot{x} + x = 1$ for $k >> 1$.

7.5.4 (Piecewise-linear nullclines) Consider the equation $\ddot{x} + \mu f(x)\dot{x} + x = 0$, where $f(x) = -1$ for $|x| < 1$ and $f(x) = 1$ for $|x| \geq 1$.

a) Show that the system is equivalent to $\dot{x} = \mu(y - F(x))$, $\dot{y} = -x/\mu$, where $F(x)$ is the piecewise-linear function

$$F(x) = \begin{cases} x+2, & x \leq -1 \\ -x, & |x| \leq 1 \\ x-2, & x \geq 1 . \end{cases}$$

b) Graph the nullclines.
c) Show that the system exhibits relaxation oscillations for $\mu >> 1$, and plot the limit cycle in the (x, y) plane.
d) Estimate the period of the limit cycle for $\mu >> 1$.

7.5.5 Consider the equation $\ddot{x} + \mu(|x| - 1)\dot{x} + x = 0$. Find the approximate period of the limit cycle for $\mu >> 1$.

7.5.6 (Biased van der Pol) Suppose the van der Pol oscillator is biased by a constant force: $\ddot{x} + \mu(x^2 - 1)\dot{x} + x = a$, where a can be positive, negative, or zero. (Assume $\mu > 0$ as usual.)
a) Find and classify all the fixed points.
b) Plot the nullclines in the Liénard plane. Show that if they intersect on the *middle* branch of the cubic nullcline, the corresponding fixed point is unstable.
c) For $\mu >> 1$, show that the system has a stable limit cycle if and only if $|a| < a_c$, where a_c is to be determined. (Hint: Use the Liénard plane.)
d) Sketch the phase portrait for a slightly greater than a_c. Show that the system is *excitable* (it has a globally attracting fixed point, but certain disturbances can send the system on a long excursion through phase space before returning to the fixed point; compare Exercise 4.5.3.)

This system is closely related to the Fitzhugh–Nagumo model of neural activity; see Murray (1989) or Edelstein–Keshet (1988) for an introduction.

7.5.7 (Cell cycle) Tyson (1991) proposed an elegant model of the cell division cycle, based on interactions between the proteins cdc2 and cyclin. He showed that the model's mathematical essence is contained in the following set of dimensionless equations:

$$\dot{u} = b(v - u)(\alpha + u^2) - u, \qquad \dot{v} = c - u,$$

where u is proportional to the concentration of the active form of a cdc2-cyclin complex, and v is proportional to the total cyclin concentration (monomers and dimers). The parameters $b >> 1$ and $\alpha << 1$ are fixed and satisfy $8\alpha b < 1$, and c is adjustable.
a) Sketch the nullclines.
b) Show that the system exhibits relaxation oscillations for $c_1 < c < c_2$, where c_1 and c_2 are to be determined approximately. (It is too hard to find c_1 and c_2 exactly, but a good approximation can be achieved if you assume $8\alpha b << 1$.)
c) Show that the system is excitable if c is slightly less than c_1.

7.6 Weakly Nonlinear Oscillators

7.6.1 Show that if (7.6.7) is expanded as a power series in ε, we recover (7.6.17).

7.6.2 (Calibrating regular perturbation theory) Consider the initial value problem $\ddot{x} + x + \varepsilon x = 0$, with $x(0) = 1$, $\dot{x}(0) = 0$.

a) Obtain the exact solution to the problem.

b) Using regular perturbation theory, find x_0, x_1, and x_2 in the series expansion $x(t,\varepsilon) = x_0(t) + \varepsilon x_1(t) + \varepsilon^2 x_2(t) + O(\varepsilon^3)$.

c) Does the perturbation solution contain secular terms? Did you expect to see any? Why?

7.6.3 (More calibration) Consider the initial value problem $\ddot{x} + x = \varepsilon$, with $x(0) = 1$, $\dot{x}(0) = 0$.

a) Solve the problem exactly.

b) Using regular perturbation theory, find x_0, x_1, and x_2 in the series expansion $x(t,\varepsilon) = x_0(t) + \varepsilon x_1(t) + \varepsilon^2 x_2(t) + O(\varepsilon^3)$.

c) Explain why the perturbation solution does or doesn't contain secular terms.

For each of the following systems $\ddot{x} + x + \varepsilon h(x, \dot{x}) = 0$, with $0 < \varepsilon << 1$, calculate the averaged equations (7.6.53) and analyze the long-term behavior of the system. Find the amplitude and frequency of any limit cycles for the original system. If possible, solve the averaged equations explicitly for $x(t,\varepsilon)$, given the initial conditions $x(0) = a$, $\dot{x}(0) = 0$.

7.6.4 $h(x, \dot{x}) = x$

7.6.5 $h(x, \dot{x}) = x\dot{x}^2$

7.6.6 $h(x, \dot{x}) = x\dot{x}$

7.6.7 $h(x, \dot{x}) = (x^4 - 1)\dot{x}$

7.6.8 $h(x, \dot{x}) = (|x| - 1)\dot{x}$

7.6.9 $h(x, \dot{x}) = (x^2 - 1)\dot{x}^3$

7.6.10 Derive the identity $\sin\theta \cos^2\theta = \frac{1}{4}[\sin\theta + \sin 3\theta]$ as follows: Use the complex representations

$$\cos\theta = \frac{e^{i\theta} + e^{-i\theta}}{2}, \qquad \sin\theta = \frac{e^{i\theta} - e^{-i\theta}}{2i},$$

multiply everything out, and then collect terms. This is always the most straightforward method of deriving such identities, and you don't have to remember any others.

7.6.11 (Higher harmonics) Notice the third harmonic $\sin 3(\tau + \phi)$ in Equation (7.6.39). The generation of *higher harmonics* is a characteristic feature of nonlinear systems. To find the effect of such terms, return to Example 7.6.2 and solve for x_1, assuming that the original system had initial conditions $x(0) = 2$, $\dot{x}(0) = 0$.

7.6.12 (Deriving the Fourier coefficients) This exercise leads you through the derivation of the formulas (7.6.51) for the Fourier coefficients. For convenience, let brackets denote the average of a function: $\langle f(\theta)\rangle \equiv \frac{1}{2\pi}\int_0^{2\pi} f(\theta)\,d\theta$ for any 2π-periodic function f. Let k and m be arbitrary integers.

a) Using integration by parts, complex exponentials, trig identities, or otherwise, derive the ***orthogonality relations***

$$\langle \cos k\theta \, \sin m\theta \rangle = 0 \text{, for all } k, m;$$

$$\langle \cos k\theta \, \cos m\theta \rangle = \langle \sin k\theta \, \sin m\theta \rangle = 0 \text{, for all } k \neq m;$$

$$\langle \cos^2 k\theta \rangle = \langle \sin^2 k\theta \rangle = \tfrac{1}{2} \text{, for } k \neq 0.$$

b) To find a_k for $k \neq 0$, multiply both sides of (7.6.50) by $\cos m\theta$ and average both sides term by term over the interval $[0, 2\pi]$. Now using the orthogonality relations from part (a), show that *all the terms on the right-hand side cancel out, except the* $k = m$ *term*! Deduce that $\langle h(\theta)\cos k\theta\rangle = \frac{1}{2}a_k$, which is equivalent to the formula for a_k in (7.6.51).

c) Similarly, derive the formulas for b_k and a_0.

7.6.13 (Exact period of a conservative oscillator) Consider the Duffing oscillator $\ddot{x} + x + \varepsilon x^3 = 0$, where $0 < \varepsilon << 1$, $x(0) = a$, and $\dot{x}(0) = 0$.

a) Using conservation of energy, express the oscillation period $T(\varepsilon)$ as a certain integral.

b) Expand the integrand as a power series in ε, and integrate term by term to obtain an approximate formula $T(\varepsilon) = c_0 + c_1\varepsilon + c_2\varepsilon^2 + O(\varepsilon^3)$. Find c_0, c_1, c_2 and check that c_0, c_1 are consistent with (7.6.57).

7.6.14 (Computer test of two-timing) Consider the equation $\ddot{x} + \varepsilon\dot{x}^3 + x = 0$.

a) Derive the averaged equations.

b) Given the initial conditions $x(0) = a$, $\dot{x}(0) = 0$, solve the averaged equations and thereby find an approximate formula for $x(t, \varepsilon)$.

c) Solve $\ddot{x} + \varepsilon\dot{x}^3 + x = 0$ numerically for $a = 1$, $\varepsilon = 2$, $0 \le t \le 50$, and plot the result on the same graph as your answer to part (b). Notice the impressive agreement, even though ε is not small!

7.6.15 (Pendulum) Consider the pendulum equation $\ddot{x} + \sin x = 0$.

a) Using the method of Example 7.6.4, show that the frequency of small oscillations of amplitude $a << 1$ is given by $\omega \approx 1 - \frac{1}{16}a^2$. (Hint: $\sin x \approx x - \frac{1}{6}x^3$, where $\frac{1}{6}x^3$ is a "small" perturbation.)

b) Is this formula for ω consistent with the exact results obtained in Exercise 6.7.4?

7.6.16 (Amplitude of the van der Pol oscillator via Green's theorem) Here's another way to determine the radius of the nearly circular limit cycle of the van der

Pol oscillator $\ddot{x}+\varepsilon\dot{x}(x^2-1)+x=0$, in the limit $\varepsilon \ll 1$. Assume that the limit cycle is a circle of unknown radius a about the origin, and invoke the normal form of Green's theorem (i.e., the 2-D divergence theorem):

$$\oint_C \mathbf{v}\cdot\mathbf{n}\,d\ell = \iint_A \nabla\cdot\mathbf{v}\,dA$$

where C is the cycle and A is the region enclosed. By substituting $\mathbf{v}=\dot{\mathbf{x}}=(\dot{x},\dot{y})$ and evaluating the integrals, show that $a\approx 2$.

7.6.17 (Playing on a swing) A simple model for a child playing on a swing is

$$\ddot{x}+(1+\varepsilon\gamma+\varepsilon\cos 2t)\sin x=0$$

where ε and γ are parameters, and $0<\varepsilon \ll 1$. The variable x measures the angle between the swing and the downward vertical. The term $1+\varepsilon\gamma+\varepsilon\cos 2t$ models the effects of gravity and the periodic pumping of the child's legs at approximately twice the natural frequency of the swing. The question is: Starting near the fixed point $x=0$, $\dot{x}=0$, can the child get the swing going by pumping her legs this way, or does she need a push?

a) For small x, the equation may be replaced by $\ddot{x}+(1+\varepsilon\gamma+\varepsilon\cos 2t)x=0$. Show that the averaged equations (7.6.53) become

$$r'=\tfrac{1}{4}r\sin 2\phi, \qquad \phi'=\tfrac{1}{2}(\gamma+\tfrac{1}{2}\cos 2\phi),$$

where $x=r\cos\theta=r(T)\cos(t+\phi(T))$, $\dot{x}=-r\sin\theta=-r(T)\sin(t+\phi(T))$, and prime denotes differentiation with respect to slow time $T=\varepsilon t$. Hint: To average terms like $\cos 2t\,\cos\theta\,\sin\theta$ over one cycle of θ, recall that $t=\theta-\phi$ and use trig identities:

$$\begin{aligned}\langle\cos 2t\,\cos\theta\,\sin\theta\rangle &= \tfrac{1}{2}\langle\cos(2\theta-2\phi)\,\sin 2\theta\rangle\\ &= \tfrac{1}{2}\langle(\cos 2\theta\,\cos 2\phi+\sin 2\theta\,\sin 2\phi)\,\sin 2\theta\rangle\\ &= \tfrac{1}{4}\sin 2\phi.\end{aligned}$$

b) Show that the fixed point $r=0$ is unstable to exponentially growing oscillations, i.e., $r(T)=r_0e^{kT}$ with $k>0$, if $|\gamma|<\gamma_c$ where γ_c is to be determined. (Hint: For r near 0, $\phi' \gg r'$ so ϕ equilibrates relatively rapidly.)

c) For $|\gamma|<\gamma_c$, write a formula for the growth rate k in terms of γ.

d) How do the solutions to the averaged equations behave if $|\gamma|>\gamma_c$?

e) Interpret the results physically.

7.6.18 (Mathieu equation and a super-slow time scale) Consider the ***Mathieu equation*** $\ddot{x}+(a+\varepsilon\cos t)x=0$ with $a\approx 1$. Using two-timing with a slow time

$T=\varepsilon^2 t$, show that the solution becomes unbounded as $t\to\infty$ if $1-\frac{1}{12}\varepsilon^2+O(\varepsilon^4)\le a\le 1+\frac{5}{12}\varepsilon^2+O(\varepsilon^4)$.

7.6.19 (Poincaré–Lindstedt method) This exercise guides you through an improved version of perturbation theory known as the ***Poincaré–Lindstedt method***. Consider the Duffing equation $\ddot{x}+x+\varepsilon x^3=0$, where $0<\varepsilon<<1$, $x(0)=a$, and $\dot{x}(0)=0$. We know from phase plane analysis that the true solution $x(t,\varepsilon)$ is periodic; our goal is to find an approximate formula for $x(t,\varepsilon)$ that is valid for all t. The key idea is to regard the frequency ω as *unknown* in advance, and to solve for it by demanding that $x(t,\varepsilon)$ contains no secular terms.

a) Define a new time $\tau=\omega t$ such that the solution has period 2π with respect to τ. Show that the equation transforms to $\omega^2 x''+x+\varepsilon x^3=0$.

b) Let $x(\tau,\varepsilon)=x_0(\tau)+\varepsilon x_1(\tau)+\varepsilon^2 x_2(\tau)+O(\varepsilon^3)$ and $\omega=1+\varepsilon\omega_1+\varepsilon^2\omega_2+O(\varepsilon^3)$. (We know already that $\omega_0=1$ since the solution has frequency $\omega=1$ when $\varepsilon=0$.) Substitute these series into the differential equation and collect powers of ε. Show that

$$O(1):\ x_0''+x_0=0$$
$$O(\varepsilon):\ x_1''+x_1=-2\omega_1 x_0''-x_0^3.$$

c) Show that the initial conditions become $x_0(0)=a$, $\dot{x}_0(0)=0$; $x_k(0)=\dot{x}_k(0)=0$ for all $k>0$.

d) Solve the $O(1)$ equation for x_0.

e) Show that after substitution of x_0 and the use of a trigonometric identity, the $O(\varepsilon)$ equation becomes $x_1''+x_1=(2\omega_1 a-\frac{3}{4}a^3)\cos\tau-\frac{1}{4}a^3\cos 3\tau$. Hence, *to avoid secular terms*, we need $\omega_1=\frac{3}{8}a^2$.

f) Solve for x_1.

Two comments: (1) This exercise shows that the Duffing oscillator has a frequency that depends on amplitude: $\omega=1+\frac{3}{8}\varepsilon a^2+O(\varepsilon^2)$, in agreement with (7.6.57). (2) The Poincaré–Lindstedt method is good for approximating periodic solutions, but that's *all* it can do; if you want to explore transients or nonperiodic solutions, you can't use this method. Use two-timing or averaging theory instead.

7.6.20 Show that if we had used regular perturbation to solve Exercise 7.6.19, we would have obtained $x(t,\varepsilon)=a\cos t+\varepsilon a^3\left[-\frac{3}{8}t\sin t+\frac{1}{32}(\cos 3t-\cos t)\right]+O(\varepsilon^2)$. Why is this solution inferior?

7.6.21 Using the Poincaré–Lindstedt method, show that the frequency of the limit cycle for the van der Pol oscillator $\ddot{x}+\varepsilon(x^2-1)\dot{x}+x=0$ is given by

$\omega = 1 - \frac{1}{16}\varepsilon^2 + O(\varepsilon^3)$.

7.6.22 (Asymmetric spring) Use the Poincaré–Lindstedt method to find the first few terms in the expansion for the solution of $\ddot{x} + x + \varepsilon x^2 = 0$, with $x(0) = a$, $\dot{x}(0) = 0$. Show that the center of oscillation is at $x \approx \frac{1}{2}\varepsilon a^2$, approximately.

7.6.23 Find the approximate relation between amplitude and frequency for the periodic solutions of $\ddot{x} - \varepsilon x \dot{x} + x = 0$.

7.6.24 (Computer algebra) Using Mathematica, Maple, or some other computer algebra package, apply the Poincaré–Lindstedt method to the problem $\ddot{x} + x - \varepsilon x^3 = 0$, with $x(0) = a$, and $\dot{x}(0) = 0$. Find the frequency ω of periodic solutions, up to and including the $O(\varepsilon^3)$ term.

7.6.25 (The method of averaging) Consider the weakly nonlinear oscillator $\ddot{x} + x + \varepsilon h(x, \dot{x}, t) = 0$. Let $x(t) = r(t)\cos(t + \phi(t))$, $\dot{x} = -r(t)\sin(t + \phi(t))$. This change of variables should be regarded as a definition of $r(t)$ and $\phi(t)$.

a) Show that $\dot{r} = \varepsilon h \sin(t + \phi)$, $r\dot{\phi} = \varepsilon h \cos(t + \phi)$. (Hence r and ϕ are slowly varying for $0 < \varepsilon << 1$, and thus $x(t)$ is a sinusoidal oscillation modulated by a slowly drifting amplitude and phase.)

b) Let $\langle r \rangle (t) = \bar{r}(t) = \frac{1}{2\pi}\int_{t-\pi}^{t+\pi} r(\tau)\, d\tau$ denote the running average of r over one cycle of the sinusoidal oscillation. Show that $d\langle r \rangle / dt = \langle dr/dt \rangle$, i.e., it doesn't matter whether we differentiate or time-average first.

c) Show that $d\langle r \rangle / dt = \varepsilon \left\langle h[r\cos(t + \phi),\, -r\sin(t + \phi),\, t] \sin(t + \phi) \right\rangle$.

d) The result of part (c) is exact, but not helpful because the left-hand side involves $\langle r \rangle$ whereas the right-hand side involves r. Now comes the key approximation: replace r and ϕ by their averages over one cycle. Show that $r(t) = \bar{r}(t) + O(\varepsilon)$ and $\phi(t) = \bar{\phi}(t) + O(\varepsilon)$, and therefore

$$d\bar{r}/dt = \varepsilon \left\langle h[\bar{r}\cos(t + \bar{\phi}),\, -\bar{r}\sin(t + \bar{\phi}),\, t] \sin(t + \bar{\phi}) \right\rangle + O(\varepsilon^2)$$

$$\bar{r}\, d\bar{\phi}/dt = \varepsilon \left\langle h[\bar{r}\cos(t + \bar{\phi}),\, -\bar{r}\sin(t + \bar{\phi}),\, t] \cos(t + \bar{\phi}) \right\rangle + O(\varepsilon^2)$$

where the barred quantities are to be treated as constants inside the averages. These equations are just the *averaged equations* (7.6.53), derived by a different approach in the text. It is customary to drop the overbars; one usually doesn't distinguish between slowly varying quantities and their averages.

7.6.26 (Calibrating the method of averaging) Consider the equation $\dot{x} = -\varepsilon x \sin^2 t$, with $0 \le \varepsilon << 1$ and $x = x_0$ at $t = 0$.

a) Find the *exact* solution to the equation.

b) Let $\bar{x}(t) = \frac{1}{2\pi}\int_{t-\pi}^{t+\pi} x(\tau)\,d\tau$. Show that $x(t) = \bar{x}(t) + O(\varepsilon)$. Use the method of averaging to find an approximate differential equation satisfied by $\bar{x}$, and solve it.

c) Compare the results of parts (a) and (b); how large is the error incurred by averaging?

BIFURCATIONS REVISITED

8.0 Introduction

This chapter extends our earlier work on bifurcations (Chapter 3). As we move up from one-dimensional to two-dimensional systems, we still find that fixed points can be created or destroyed or destabilized as parameters are varied—but now the same is true of closed orbits as well. Thus we can begin to *describe the ways in which oscillations can be turned on or off.*

In this broader context, what exactly do we mean by a bifurcation? The usual definition involves the concept of "topological equivalence" (Section 6.3): if the phase portrait changes its topological structure as a parameter is varied, we say that a ***bifurcation*** has occurred. Examples include changes in the number or stability of fixed points, closed orbits, or saddle connections as a parameter is varied.

This chapter is organized as follows: for each bifurcation, we start with a simple prototypical example, and then graduate to more challenging examples, either briefly or in separate sections. Models of genetic switches, chemical oscillators, driven pendula and Josephson junctions are used to illustrate the theory.

8.1 Saddle-Node, Transcritical, and Pitchfork Bifurcations

The bifurcations of fixed points discussed in Chapter 3 have analogs in two dimensions (and indeed, in *all* dimensions). Yet it turns out that nothing really new happens when more dimensions are added—all the action is confined to a one-dimensional subspace along which the bifurcations occur, while in the extra dimensions the flow is either simple attraction or repulsion from that subspace, as we'll see below.

Saddle-Node Bifurcation

The saddle-node bifurcation is the basic mechanism for the creation and destruction of fixed points. Here's the prototypical example in two dimensions:

$$\dot{x} = \mu - x^2$$
$$\dot{y} = -y. \qquad (1)$$

In the x-direction we see the bifurcation behavior discussed in Section 3.1, while in the y-direction the motion is exponentially damped.

Consider the phase portrait as μ varies. For $\mu > 0$, Figure 8.1.1 shows that there are two fixed points, a stable node at $(x^*, y^*) = (\sqrt{\mu}, 0)$ and a saddle at $(-\sqrt{\mu}, 0)$. As μ decreases, the saddle and node approach each other, then collide when $\mu = 0$, and finally disappear when $\mu < 0$.

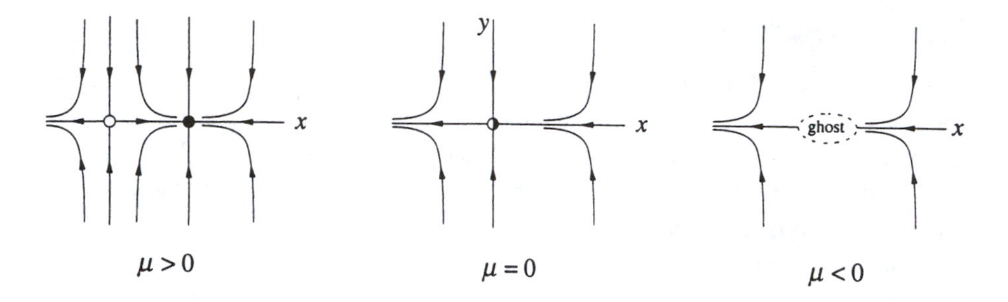

Figure 8.1.1

Even after the fixed points have annihilated each other, they continue to influence the flow—as in Section 4.3, they leave a *ghost*, a bottleneck region that sucks trajectories in and delays them before allowing passage out the other side. For the same reasons as in Section 4.3, the time spent in the bottleneck generically increases as $(\mu - \mu_c)^{-1/2}$, where μ_c is the value at which the saddle-node bifurcation occurs. Some applications of this scaling law in condensed-matter physics are discussed by Strogatz and Westervelt (1989).

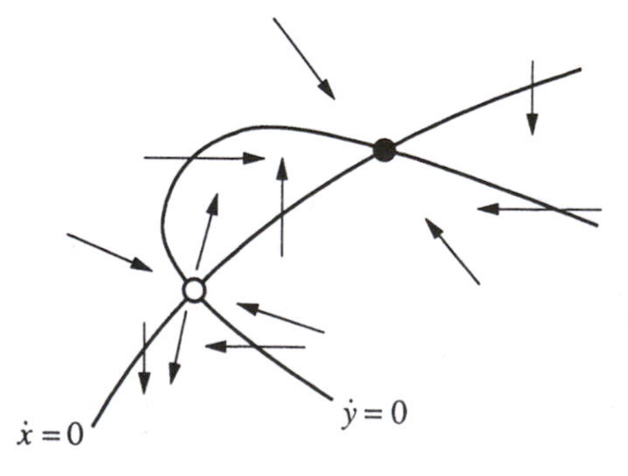

Figure 8.1.2

Figure 8.1.1 is representative of the following more general situation. Consider a two-dimensional system $\dot{x} = f(x, y)$, $\dot{y} = g(x, y)$ that depends on a parameter μ. Suppose that for some value of μ the nullclines intersect as shown in Figure 8.1.2. Notice that each intersection corresponds to a fixed point since $\dot{x}=0$ and $\dot{y}=0$ simultaneously. Thus, to see how the fixed points move as μ changes, we just have to watch the intersections. Now suppose that the nullclines pull away from each other as μ varies, becoming *tangent* at $\mu=\mu_c$. Then the fixed points approach each other and collide when $\mu=\mu_c$; after the nullclines pull apart, there are no intersections and the fixed points disappear with a bang. The point is that *all* saddle-node bifurcations have this character locally.

EXAMPLE 8.1.1:

The following system has been discussed by Griffith (1971) as a model for a genetic control system. The activity of a certain gene is assumed to be directly induced by two copies of the protein for which it codes. In other words, the gene is stimulated by its own product, potentially leading to an autocatalytic feedback process. In dimensionless form, the equations are

$$\dot{x}=-ax+y$$

$$\dot{y}=\frac{x^2}{1+x^2}-by$$

where x and y are proportional to the concentrations of the protein and the messenger RNA from which it is translated, respectively, and a, $b>0$ are parameters that govern the rate of degradation of x and y.

Show that the system has three fixed points when $a<a_c$, where a_c is to be determined. Show that two of these fixed points coalesce in a saddle-node bifurcation when $a=a_c$. Then sketch the phase portrait for $a<a_c$, and give a biological interpretation.

Solution: The nullclines are given by the line $y=ax$ and the sigmoidal curve

$$y=\frac{x^2}{b(1+x^2)}$$

as sketched in Figure 8.1.3. Now suppose we vary a while holding b fixed. This is simple to visualize, since a is the slope of the line. For small a there are three intersections, as in Figure 8.1.3. As a increases, the top two intersections approach each other and collide when the line intersects the curve tangentially. For larger values of a, those fixed points disappear, leaving the origin as the only fixed point.

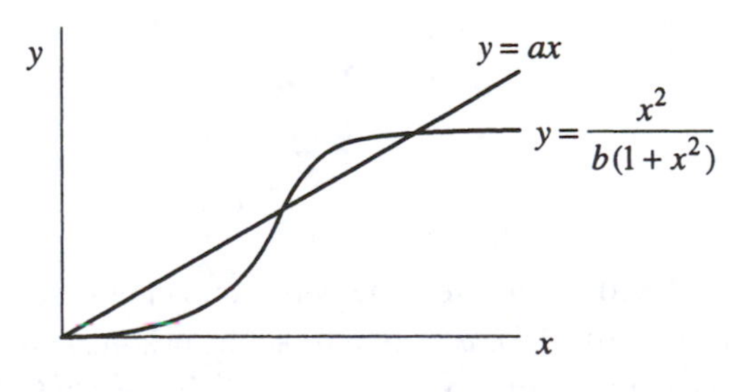

Figure 8.1.3

To find a_c, we compute the

fixed points directly and find where they coalesce. The nullclines intersect when

$$ax = \frac{x^2}{b(1+x^2)}.$$

One solution is $x^* = 0$, in which case $y^* = 0$. The other intersections satisfy the quadratic equation

$$ab(1+x^2) = x \tag{2}$$

which has two solutions

$$x^* = \frac{1 \pm \sqrt{1-4a^2b^2}}{2ab}$$

if $1-4a^2b^2 > 0$, i.e., $2ab < 1$. These solutions coalesce when $2ab = 1$. Hence

$$a_c = 1/2b.$$

For future reference, note that the fixed point $x^* = 1$ at the bifurcation.

The nullclines (Figure 8.1.4) provide a lot of information about the phase portrait for $a < a_c$. The vector field is vertical on the line $y = ax$ and horizontal on the sigmoidal curve. Other arrows can be sketched by noting the signs of $\dot{x}$ and $\dot{y}$. It appears that the middle fixed point is a saddle and the other two are sinks. To confirm this, we turn now to the classification of the fixed points.

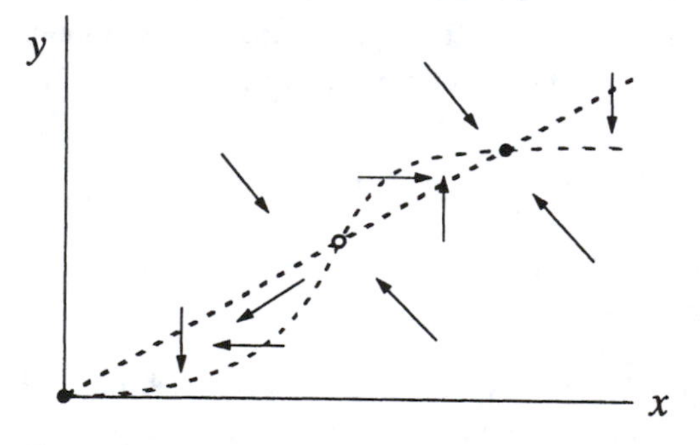

Figure 8.1.4

The Jacobian matrix at (x, y) is

$$A = \begin{pmatrix} -a & 1 \\ \frac{2x}{(1+x^2)^2} & -b \end{pmatrix}.$$

A has trace $\tau = -(a+b) < 0$ so all the fixed points are either sinks or saddles, depending on the value of the determinant Δ. At (0,0), $\Delta = ab > 0$, so the origin is always a stable fixed point. In fact, it is a *stable node*, since $\tau^2 - 4\Delta = (a-b)^2 > 0$

(except in the degenerate case $a = b$, which we disregard). At the other two fixed points, Δ looks messy but it can be simplified using (2). We find

$$\Delta = ab - \frac{2x^*}{\left(1+(x^*)^2\right)^2} = ab\left[1 - \frac{2}{1+(x^*)^2}\right] = ab\left[\frac{(x^*)^2 - 1}{1+(x^*)^2}\right].$$

So $\Delta < 0$ for the "middle" fixed point, which has $0 < x^* < 1$; this is a *saddle point.* The fixed point with $x^* > 1$ is always a *stable node*, since $\Delta < ab$ and therefore $\tau^2 - 4\Delta > (a-b)^2 > 0$.

The phase portrait is plotted in Figure 8.1.5. By looking back at Figure 8.1.4, we can see that the unstable manifold of the saddle is necessarily trapped in the narrow channel between the two nullclines. More importantly, the *stable* manifold separates the plane into two regions, each a basin of attraction for a sink.

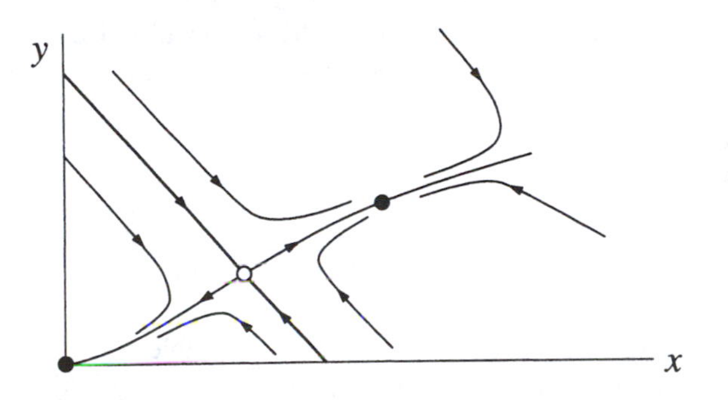

Figure 8.1.5

The biological interpretation is that the system can act like a *biochemical switch*, but only if the mRNA and protein degrade slowly enough—specifically, their decay rates must satisfy $ab < 1/2$. In this case, there are two stable steady states: one at the origin, meaning that the gene is silent and there is no protein around to turn it on; and one where x and y are large, meaning that the gene is active and sustained by the high level of protein. The stable manifold of the saddle acts like a threshold; it determines whether the gene turns on or off, depending on the initial values of x and y. ■

As advertised, the flow in Figure 8.1.5 is qualitatively similar to that in the idealized Figure 8.1.1. All trajectories relax rapidly onto the unstable manifold of the saddle, which plays a completely analogous role to the x-axis in Figure 8.1.1.

Thus, in many respects, the bifurcation is a fundamentally one-dimensional event, with the fixed points sliding toward each other along the unstable manifold like beads on a string. *This is why we spent so much time looking at bifurcations in one-dimensional systems*—they're the building blocks of analogous bifurcations in higher dimensions. (The fundamental role of one-dimensional systems can be justified rigorously by "center manifold theory"—see Wiggins (1990) for an introduction.)

Transcritical and Pitchfork Bifurcations

Using the same idea as above, we can also construct prototypical examples of transcritical and pitchfork bifurcations at a stable fixed point. In the x-direction the dynamics are given by the normal forms discussed in Chapter 3, and in the y-direction the motion is exponentially damped. This yields the following examples:

$$\dot{x} = \mu x - x^2, \quad \dot{y} = -y \quad \text{(transcritical)}$$
$$\dot{x} = \mu x - x^3, \quad \dot{y} = -y \quad \text{(supercritical pitchfork)}$$
$$\dot{x} = \mu x + x^3, \quad \dot{y} = -y \quad \text{(subcritical pitchfork)}$$

The analysis in each case follows the same pattern, so we'll discuss only the supercritical pitchfork, and leave the other two cases as exercises.

EXAMPLE 8.1.2:

Plot the phase portraits for the supercritical pitchfork system $\dot{x} = \mu x - x^3$, $\dot{y} = -y$, for $\mu < 0$, $\mu = 0$, and $\mu > 0$.

Solution: For $\mu < 0$, the only fixed point is a stable node at the origin. For $\mu = 0$, the origin is still stable, but now we have very slow (algebraic) decay along the x-direction instead of exponential decay; this is the phenomenon of "critical slowing down" discussed in Section 3.4 and Exercise 2.4.9. For $\mu > 0$, the origin loses stability and gives birth to two new stable fixed points symmetrically located at $(x^*, y^*) = (\pm\sqrt{\mu}, 0)$. By computing the Jacobian at each point, you can check that the origin is a saddle and the other two fixed points are stable nodes. The phase portraits are shown in Figure 8.1.6. ■

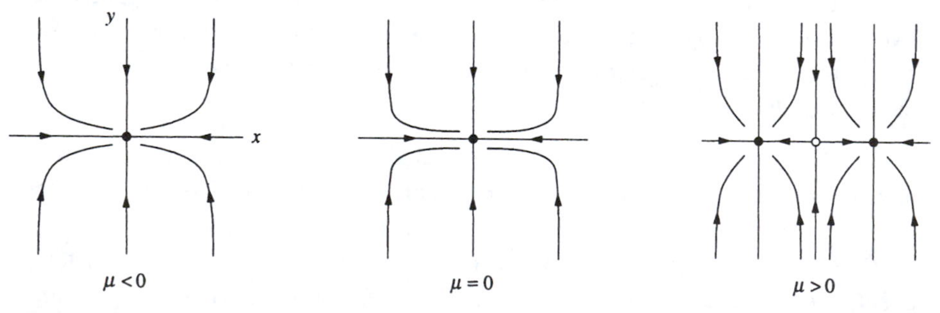

Figure 8.1.6

As mentioned in Chapter 3, pitchfork bifurcations are common in systems that have a symmetry. Here's an example.

EXAMPLE 8.1.3:

Show that a supercritical pitchfork bifurcation occurs at the origin in the system

$$\dot{x} = \mu x + y + \sin x$$
$$\dot{y} = x - y$$

and determine the bifurcation value μ_c. Plot the phase portrait near the origin for μ slightly greater than μ_c.

Solution: The system is invariant under the change of variables $x \to -x$, $y \to -y$, so the phase portrait must be symmetric under reflection through the origin. The origin is a fixed point for all μ, and its Jacobian is

$$A = \begin{pmatrix} \mu+1 & 1 \\ 1 & -1 \end{pmatrix}$$

which has $\tau = \mu$ and $\Delta = -(\mu+2)$. Hence the origin is a stable fixed point if $\mu < -2$ and a saddle if $\mu > -2$. This suggests that a pitchfork bifurcation occurs at $\mu_c = -2$. To confirm this, we seek a symmetric pair of fixed points close to the origin for μ close to μ_c. (Note that at this stage we don't know whether the bifurcation is sub- or supercritical.) The fixed points satisfy $y = x$ and hence $(\mu+1)x + \sin x = 0$. One solution is $x = 0$, but we've found that already. Now suppose x is small and nonzero, and expand the sine as a power series. Then

$$(\mu+1)x + x - \frac{x^3}{3!} + O(x^5) = 0.$$

After dividing through by x and neglecting higher-order terms, we get $\mu + 2 - x^2/6 \approx 0$. Hence there is a pair of fixed points with $x^* \approx \pm\sqrt{6(\mu+2)}$ for μ slightly greater than -2. Thus a *supercritical* pitchfork bifurcation occurs at $\mu_c = -2$. (If the bifurcation had been subcritical, the pair of fixed points would exist when the origin was stable, not after it has become a saddle.) Because the bifurcation is supercritical, we know the new fixed points are stable *without even checking.*

To draw the phase portrait near (0,0) for μ slightly greater than -2, it's helpful to find the eigenvectors of the Jacobian at the origin. This can be done exactly, but a simple approximation is that the Jacobian is close to that *at* the bifurcation. Thus

$$A \approx \begin{pmatrix} -1 & 1 \\ 1 & -1 \end{pmatrix}$$

which has eigenvectors $(1,1)$ and $(1,-1)$, with eigenvalues $\lambda = 0$ and $\lambda = -2$, respectively. For μ slightly greater than -2, the origin becomes a saddle and so the zero eigenvalue becomes slightly positive. This information implies the phase portrait shown in Figure 8.1.7.

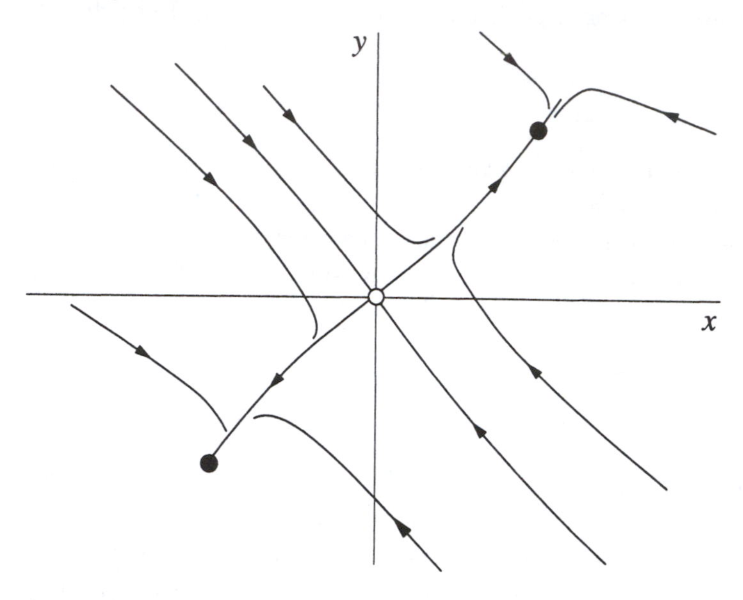

Figure 8.1.7

Note that because of the approximations we've made, this picture is only valid *locally* in both parameter and phase space—if we're not near the origin and if μ is not close to μ_c, all bets are off. ■

In all of the examples above, the bifurcation occurs when $\Delta = 0$, or equivalently, when one of the eigenvalues equals zero. More generally, the saddle-node, transcritical, and pitchfork bifurcations are all examples of ***zero-eigenvalue bifurcations***. (There are other examples, but these are the most common.) Such bifurcations always involve the collision of two or more fixed points.

In the next section we'll consider a fundamentally new kind of bifurcation, one that has no counterpart in one-dimensional systems. It provides a way for a fixed point to lose stability without colliding with any other fixed points.

8.2 Hopf Bifurcations

Suppose a two-dimensional system has a stable fixed point. What are all the possible ways it could lose stability as a parameter μ varies? The eigenvalues of the Jacobian are the key. If the fixed point is stable, the eigenvalues λ_1, λ_2 must both lie in the left half-plane $\text{Re}\,\lambda < 0$. Since the λ's satisfy a quadratic equation with real coefficients, there are two possible pictures: either the eigenvalues are both real and negative (Figure 8.2.1a) or they are complex conjugates (Figure 8.2.1b). To destabilize the fixed point, we need one or both of the eigenvalues to cross into the right half-plane as μ varies.

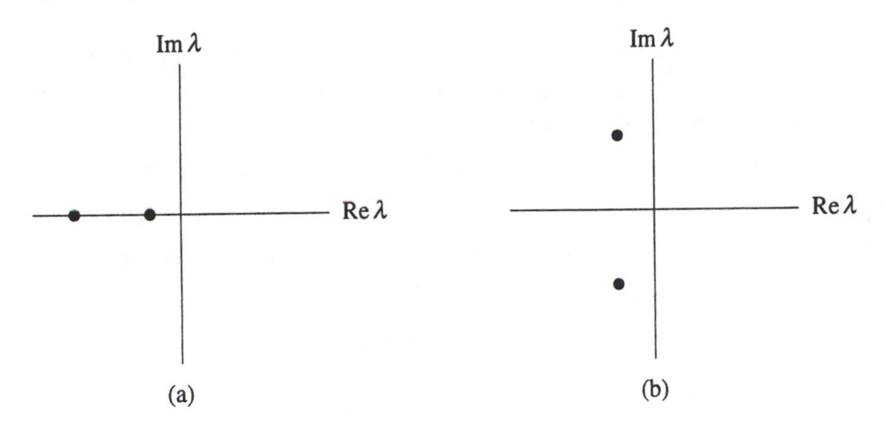

Figure 8.2.1

In Section 8.1 we explored the cases in which a real eigenvalue passes through $\lambda = 0$. These were just our old friends from Chapter 3, namely the saddle-node, transcritical, and pitchfork bifurcations. Now we consider the other possible scenario, in which two complex conjugate eigenvalues simultaneously cross the imaginary axis into the right half-plane.

Supercritical Hopf Bifurcation

Suppose we have a physical system that settles down to equilibrium through exponentially damped oscillations. In other words, small disturbances decay after "ringing" for a while (Figure 8.2.2a). Now suppose that the decay rate depends on a control parameter μ. If the decay becomes slower and slower and finally changes to *growth* at a critical value μ_c, the equilibrium state will lose stability. In many cases the resulting motion is a small-amplitude, sinusoidal, limit cycle oscillation about the former steady state (Figure 8.2.2b). Then we say that the system has undergone a ***supercritical Hopf bifurcation***.

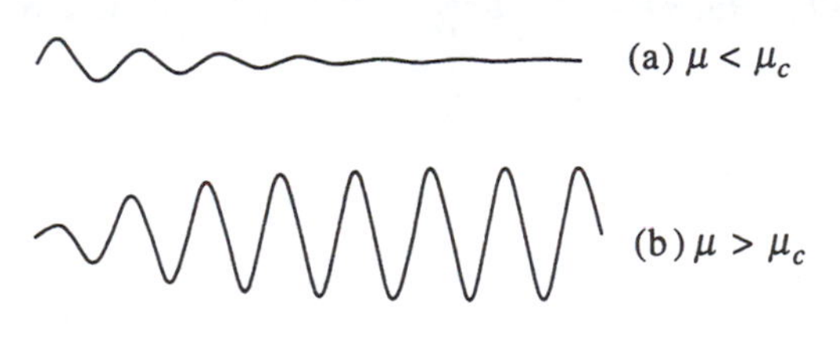

Figure 8.2.2

In terms of the flow in phase space, a supercritical Hopf bifurcation occurs when a stable spiral changes into an unstable spiral surrounded by a small, nearly elliptical limit cycle. Hopf bifurcations can occur in phase spaces of any dimension $n \geq 2$, but as in the rest of this chapter, we'll restrict ourselves to two dimensions.

A simple example of a supercritical Hopf bifurcation is given by the following system:

$$\dot{r} = \mu r - r^3$$
$$\dot{\theta} = \omega + br^2.$$

There are three parameters: μ controls the stability of the fixed point at the origin, ω gives the frequency of infinitesimal oscillations, and b determines the dependence of frequency on amplitude for larger amplitude oscillations.

Figure 8.2.3 plots the phase portraits for μ above and below the bifurcation. When $\mu < 0$ the origin $r = 0$ is a stable spiral whose sense of rotation depends on the sign of ω. For $\mu = 0$ the origin is still a stable spiral, though a very weak one: the decay is only algebraically fast. (This case was shown in Figure 6.3.2. Recall that the linearization wrongly predicts a center at the origin.) Finally, for $\mu > 0$ there is an unstable spiral at the origin and a stable circular limit cycle at $r = \sqrt{\mu}$.

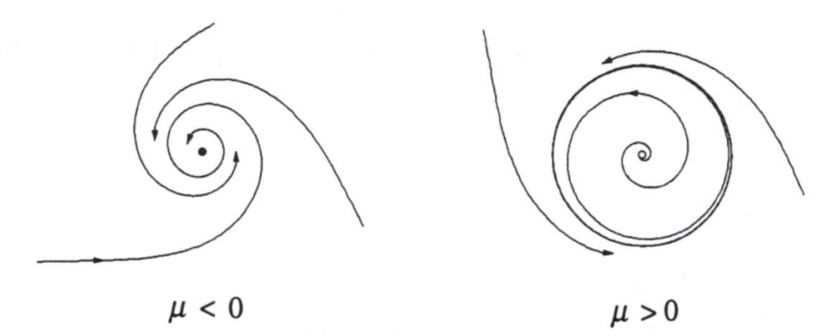

Figure 8.2.3

To see how the eigenvalues behave during the bifurcation, we rewrite the system in Cartesian coordinates; this makes it easier to find the Jacobian. We write $x = r\cos\theta$, $y = r\sin\theta$. Then

$$\begin{aligned}
\dot{x} &= \dot{r}\cos\theta - r\dot{\theta}\sin\theta \\
&= (\mu r - r^3)\cos\theta - r(\omega + br^2)\sin\theta \\
&= \left(\mu - [x^2 + y^2]\right)x \; - \; \left(\omega + b[x^2 + y^2]\right)y \\
&= \mu x - \omega y + \text{cubic terms}
\end{aligned}$$

and similarly

$$\dot{y} = \omega x + \mu y + \text{cubic terms.}$$

So the Jacobian at the origin is

$$A = \begin{pmatrix} \mu & -\omega \\ \omega & \mu \end{pmatrix},$$

which has eigenvalues

$$\lambda = \mu \pm i\omega.$$

As expected, the eigenvalues cross the imaginary axis from left to right as μ increases from negative to positive values.

Rules of Thumb

Our idealized case illustrates two rules that hold *generically* for supercritical Hopf bifurcations:

1. The size of the limit cycle grows continuously from zero, and increases proportional to $\sqrt{\mu - \mu_c}$, for μ close to μ_c.

2. The frequency of the limit cycle is given approximately by $\omega = \text{Im}\,\lambda$, evaluated at $\mu = \mu_c$. This formula is exact at the birth of the limit cycle, and correct within $O(\mu - \mu_c)$ for μ close to μ_c. The period is therefore $T = (2\pi/\text{Im}\,\lambda) + O(\mu - \mu_c)$.

But our idealized example also has some artifactual properties. First, in Hopf bifurcations encountered in practice, the limit cycle is elliptical, not circular, and its shape becomes distorted as μ moves away from the bifurcation point. Our example is only typical topologically, not geometrically. Second, in our idealized case the eigenvalues move on horizontal lines as μ varies, i.e., $\text{Im}\,\lambda$ is strictly independent of μ. Normally, the eigenvalues would follow a curvy path and cross the imaginary axis with nonzero slope (Figure 8.2.4).

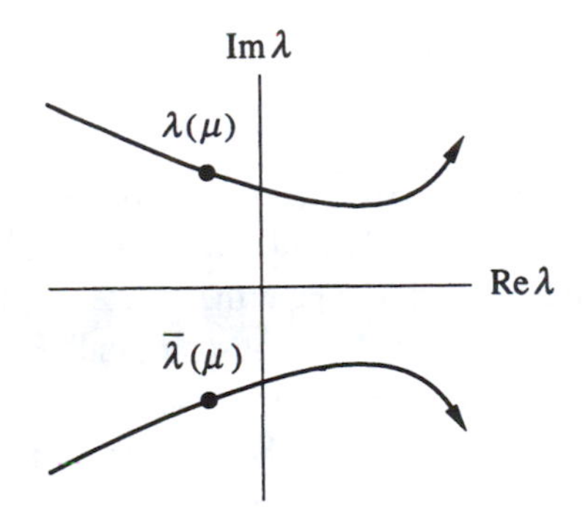

Figure 8.2.4

Subcritical Hopf Bifurcation

Like pitchfork bifurcations, Hopf bifurcations come in both super- and subcritical varieties. The subcritical case is always much more dramatic, and potentially dangerous in engineering applications. After the bifurcation, the trajectories must *jump* to a *distant* attractor, which may be a fixed point, another limit cycle, infinity, or—in three and higher dimensions—a chaotic attractor. We'll see a concrete example of this last, most interesting case when we study the Lorenz equations (Chapter 9).

But for now, consider the two-dimensional example

$$\dot{r} = \mu r + r^3 - r^5$$
$$\dot{\theta} = \omega + br^2.$$

The important difference from the earlier supercritical case is that the cubic term r^3 is now *destabilizing*; it helps to drive trajectories away from the origin.

The phase portraits are shown in Figure 8.2.5. For $\mu < 0$ there are two attractors, a stable limit cycle and a stable fixed point at the origin. Between them lies an unstable cycle, shown as a dashed curve in Figure 8.2.5; it's the player to watch in this scenario. As μ increases, the unstable cycle tightens like a noose around the fixed point. A ***subcritical Hopf bifurcation*** occurs at $\mu = 0$, where the unstable cycle shrinks to zero amplitude and engulfs the origin, rendering it unstable. For $\mu > 0$, the large-amplitude limit cycle is suddenly the only attractor in town. Solutions that used to remain near the origin are now forced to grow into large-amplitude oscillations.

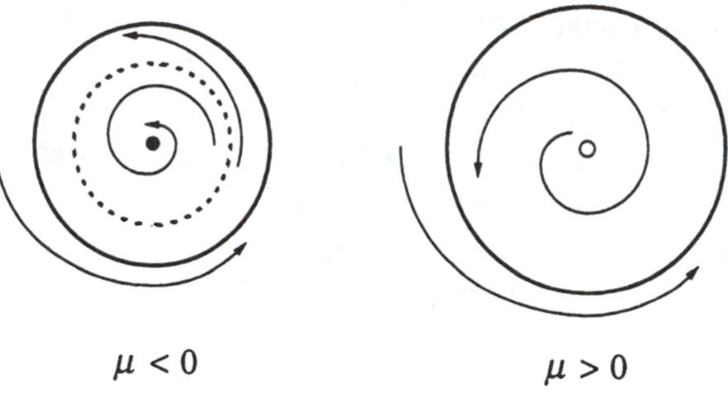

Figure 8.2.5

Note that the system exhibits *hysteresis*: once large-amplitude oscillations have begun, they cannot be turned off by bringing μ back to zero. In fact, the large oscillations will persist until $\mu = -1/4$ where the stable and unstable cycles collide and annihilate. This destruction of the large-amplitude cycle occurs via another type of bifurcation, to be discussed in Section 8.4.

Subcritical Hopf bifurcations occur in the dynamics of nerve cells (Rinzel and Ermentrout 1989), in aeroelastic flutter and other vibrations of airplane wings (Dowell and Ilgamova 1988, Thompson and Stewart 1986), and in instabilities of fluid flows (Drazin and Reid 1981).

Subcritical, Supercritical, or Degenerate Bifurcation?

Given that a Hopf bifurcation occurs, how can we tell if it's sub- or supercritical? The linearization doesn't provide a distinction: in both cases, a pair of eigenvalues moves from the left to the right half-plane.

An analytical criterion exists, but it can be difficult to use (see Exercises 8.2.12–15 for some tractable cases). A quick and dirty approach is to use the computer. If a small, attracting limit cycle appears immediately after the fixed point goes unstable, and if its amplitude shrinks back to zero as the parameter is reversed, the bifurcation is supercritical; otherwise, it's probably subcritical, in which case the nearest attractor might be far from the fixed point, and the system may exhibit hysteresis as the parameter is reversed. Of course, computer experiments are not proofs and you should check the numerics carefully before making any firm conclusions.

Finally, you should also be aware of a ***degenerate Hopf bifurcation***. An example is given by the damped pendulum $\ddot{x}+\mu\dot{x}+\sin x=0$. As we change the damping μ from positive to negative, the fixed point at the origin changes from a stable to an unstable spiral. However at $\mu=0$ we do *not* have a true Hopf bifurcation because there are no limit cycles on either side of the bifurcation. Instead, at $\mu=0$ we have a continuous band of closed orbits surrounding the origin. These are not limit cycles! (Recall that a limit cycle is an *isolated* closed orbit.)

This degenerate case typically arises when a nonconservative system suddenly becomes conservative at the bifurcation point. Then the fixed point becomes a nonlinear center, rather than the weak spiral required by a Hopf bifurcation. See Exercise 8.2.11 for another example.

EXAMPLE 8.2.1:

Consider the system $\dot{x}=\mu x-y+xy^2$, $\dot{y}=x+\mu y+y^3$. Show that a Hopf bifurcation occurs at the origin as μ varies. Is the bifurcation subcritical, supercritical, or degenerate?

Solution: The Jacobian at the origin is $A=\begin{pmatrix}\mu & -1\\ 1 & \mu\end{pmatrix}$, which has $\tau=2\mu$, $\Delta=\mu^2+1>0$, and $\lambda=\mu\pm i$. Hence, as μ increases through zero, the origin changes from a stable spiral to an unstable spiral. This suggests that some kind of Hopf bifurcation takes place at $\mu=0$.

To decide whether the bifurcation is subcritical, supercritical, or degenerate, we use simple reasoning and numerical integration. If we transform the system to polar coordinates, we find that

$$\dot{r}=\mu r+ry^2,$$

as you should check. Hence $\dot{r}\geq\mu r$. This implies that for $\mu>0$, $r(t)$ grows *at least* as fast as $r_0 e^{\mu t}$. In other words, all trajectories are repelled out to infinity! So there are certainly no closed orbits for $\mu > 0$. In particular, the unstable spiral is not surrounded by a stable limit cycle; hence the bifurcation cannot be supercritical.

Could the bifurcation be degenerate? That would require that the origin be a nonlinear center when $\mu = 0$. But $\dot{r}$ is strictly positive away from the x-axis, so closed orbits are still impossible.

By process of elimination, we expect that the bifurcation is *subcritical*. This is confirmed by Figure 8.2.6, which is a computer-generated phase portrait for $\mu = -0.2$.

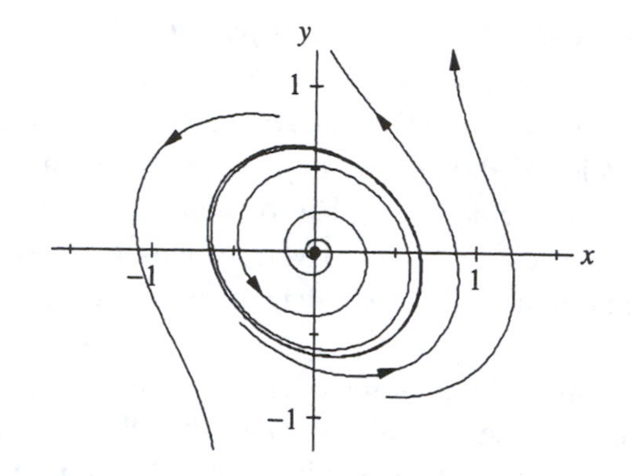

Figure 8.2.6

Note that an *unstable* limit cycle surrounds the stable fixed point, just as we expect in a subcritical bifurcation. Furthermore, the cycle is nearly elliptical and surrounds a gently winding spiral—these are typical features of *either* kind of Hopf bifurcation. ■

8.3 Oscillating Chemical Reactions

For an application of Hopf bifurcations, we now consider a class of experimental systems known as ***chemical oscillators***. These systems are remarkable, both for their spectacular behavior and for the story behind their discovery. After presenting this background information, we analyze a simple model proposed recently for oscillations in the chlorine dioxide–iodine–malonic acid reaction. The definitive reference on chemical oscillations is the book edited by Field and Burger (1985). See also Epstein et al. (1983), Winfree (1987b) and Murray (1989).

Belousov's "Supposedly Discovered Discovery"

In the early 1950s the Russian biochemist Boris Belousov was trying to create a test tube caricature of the Krebs cycle, a metabolic process that occurs in living cells. When he mixed citric acid and bromate ions in a solution of sulfuric acid, and in the presence of a cerium catalyst, he observed to his astonishment that the mixture became yellow, then faded to colorless after about a minute, then returned to yellow a minute later, then became colorless again, and continued to oscillate dozens of times before finally reaching equilibrium after about an hour.

Today it comes as no surprise that chemical reactions can oscillate spontaneously—such reactions have become a standard demonstration in chemistry classes, and you may have seen one yourself. (For recipes, see Winfree (1980).) But in Belousov's day, his discovery was so radical that he couldn't get his work published. It was thought that all solutions of chemical reagents must go *monotonically* to equilibrium, because of the laws of thermodynamics. Belousov's paper was rejected by one journal after another. According to Winfree (1987b, p.161), one editor even added a snide remark about Belousov's "supposedly discovered discovery" to the rejection letter.

Belousov finally managed to publish a brief abstract in the obscure proceedings of a Russian medical meeting (Belousov 1959), although his colleagues weren't aware of it until years later. Nevertheless, word of his amazing reaction circulated among Moscow chemists in the late 1950s, and in 1961 a graduate student named Zhabotinsky was assigned by his adviser to look into it. Zhabotinsky confirmed that Belousov was right all along, and brought this work to light at an international conference in Prague in 1968, one of the few times that Western and Soviet scientists were allowed to meet. At that time there was a great deal of interest in biological and biochemical oscillations (Chance et al. 1973) and the BZ reaction, as it came to be called, was seen as a manageable model of those more complex systems.

The analogy to biology turned out to be surprisingly close: Zaikin and Zhabotinsky (1970) and Winfree (1972) observed beautiful propagating *waves* of oxidation in thin unstirred layers of BZ reagent, and found that these waves annihilate upon collision, just like waves of excitation in neural or cardiac tissue. The waves always take the shape of expanding concentric rings or spirals (Color plate 1). Spiral waves are now recognized to be a ubiquitous feature of chemical, biological, and physical excitable media; in particular, spiral waves and their three-dimensional analogs, "scroll waves" (Front cover illustration) appear to be implicated in certain cardiac arrhythmias, a problem of great medical importance (Winfree 1987b).

Boris Belousov would be pleased to see what he started.

In 1980, he and Zhabotinsky were awarded the Lenin Prize, the Soviet Union's highest medal, for their pioneering work on oscillating reactions. Unfortunately, Belousov had passed away ten years earlier.

For more about the history of the BZ reaction, see Winfree (1984, 1987b). An English translation of Belousov's original paper from 1951 appears in Field and Burger (1985).

Chlorine Dioxide–Iodine–Malonic Acid Reaction

The mechanisms of chemical oscillations can be very complex. The BZ reaction is thought to involve more than twenty elementary reaction steps, but luckily many of them equilibrate rapidly—this allows the kinetics to be reduced to as few as three differential equations. See Tyson (1985) for this reduced system and its analysis.

In a similar spirit, Lengyel et al. (1990) have proposed and analyzed a particularly elegant model of another oscillating reaction, the chlorine dioxide-iodine-malonic acid ($ClO_2 - I_2 - MA$) reaction. Their experiments show that the following three reactions and empirical rate laws capture the behavior of the system:

$$\mathrm{MA} + \mathrm{I_2} \rightarrow \mathrm{IMA} + \mathrm{I^-} + \mathrm{H^+}; \qquad \frac{d[\mathrm{I_2}]}{dt} = -\frac{k_{1a}[\mathrm{MA}][\mathrm{I_2}]}{k_{1b} + [\mathrm{I_2}]} \tag{1}$$

$$\mathrm{ClO_2} + \mathrm{I^-} \rightarrow \mathrm{ClO_2}^- + \tfrac{1}{2}\mathrm{I_2}; \qquad \frac{d[\mathrm{ClO_2}]}{dt} = -k_2[\mathrm{ClO_2}][\mathrm{I^-}] \tag{2}$$

$$\mathrm{ClO_2}^- + 4\mathrm{I^-} + 4\mathrm{H^+} \rightarrow \mathrm{Cl^-} + 2\mathrm{I_2} + 2\mathrm{H_2O};$$

$$\frac{d[\mathrm{ClO_2}^-]}{dt} = -k_{3a}[\mathrm{ClO_2}^-][\mathrm{I^-}][\mathrm{H^+}] - k_{3b}[\mathrm{ClO_2}^-][\mathrm{I_2}]\frac{[\mathrm{I^-}]}{u + [\mathrm{I^-}]^2} \tag{3}$$

Typical values of the concentrations and kinetic parameters are given in Lengyel et al. (1990) and Lengyel and Epstein (1991).

Numerical integrations of (1)–(3) show that the model exhibits oscillations that closely resemble those observed experimentally. However this model is still too complicated to handle analytically. To simplify it, Lengyel et al. (1990) use a result found in their simulations: Three of the reactants (MA, I_2, and ClO_2) vary much more slowly than the intermediates I^- and ClO_2^-, which change by several orders of magnitude during an oscillation period. By approximating the concentrations of the slow reactants as *constants* and making other reasonable simplifications, they reduce the system to a two-variable model. (Of course, since this approximation neglects the slow consumption of the reactants, the model will be unable to account for the eventual approach to equilibrium.) After suitable nondimensionalization, the model becomes

$$\dot{x} = a - x - \frac{4xy}{1+x^2} \tag{4}$$

$$\dot{y} = bx\left(1 - \frac{y}{1+x^2}\right) \tag{5}$$

where x and y are the dimensionless concentrations of I^- and ClO_2^-. The parameters $a, b > 0$ depend on the empirical rate constants and on the concentrations assumed for the slow reactants.

We begin the analysis of (4), (5) by constructing a trapping region and applying the Poincaré–Bendixson theorem. Then we'll show that the chemical oscillations arise from a supercritical Hopf bifurcation.

EXAMPLE 8.3.1:

Prove that the system (4), (5) has a closed orbit in the positive quadrant $x, y > 0$ if a and b satisfy certain constraints, to be determined.

Solution: As in Example 7.3.2, the nullclines help us to construct a trapping region. Equation (4) shows that $\dot{x} = 0$ on the curve

$$y = \frac{(a-x)(1+x^2)}{4x} \tag{6}$$

and (5) shows that $\dot{y} = 0$ on the y-axis and on the parabola $y = 1 + x^2$. These nullclines are sketched in Figure 8.3.1, along with some representative vectors.

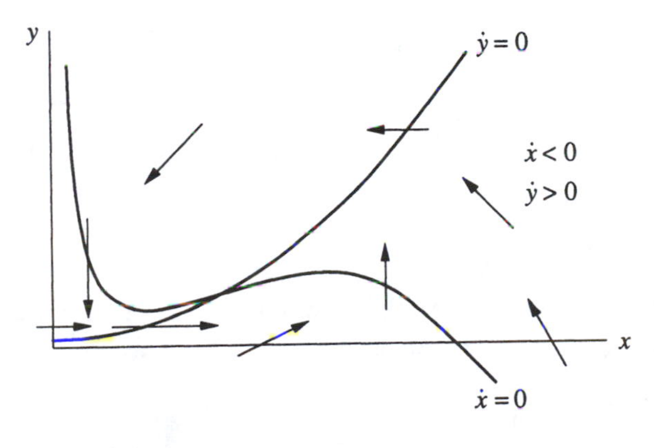

Figure 8.3.1

(We've taken some pedagogical license with Figure 8.3.1; the curvature of the nullcline (6) has been exaggerated to highlight its shape, and to give us more room to draw the vectors.)

Now consider the dashed box shown in Figure 8.3.2. It's a trapping region because all the vectors on the boundary point into the box.

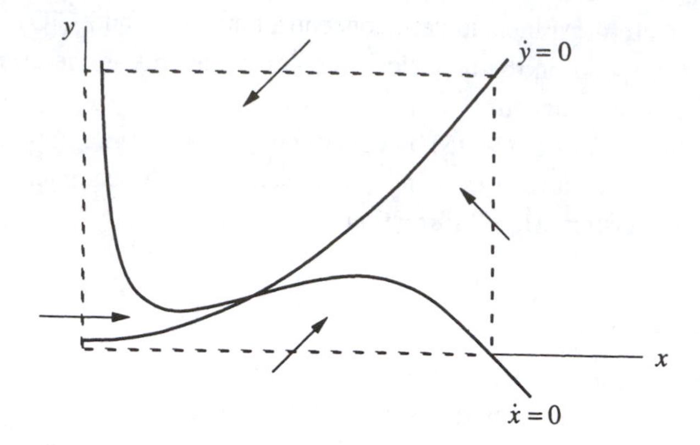

Figure 8.3.2

We can't apply the Poincaré–Bendixson theorem yet, because there's a fixed point

$$x^* = a/5, \qquad y^* = 1 + (x^*)^2 = 1 + (a/5)^2$$

inside the box at the intersection of the nullclines. But now we argue as in Example 7.3.3: if the fixed point turns out to be a *repeller*, we *can* apply the Poincaré–Bendixson theorem to the "punctured" box obtained by removing the fixed point.

All that remains is to see under what conditions (if any) the fixed point is a repeller. The Jacobian at (x^*, y^*) is

$$\frac{1}{1 + (x^*)^2} \begin{pmatrix} 3(x^*)^2 - 5 & -4x^* \\ 2b(x^*)^2 & -bx^* \end{pmatrix}.$$

(We've used the relation $y^* = 1 + (x^*)^2$ to simplify some of the entries in the Jacobian.) The determinant and trace are given by

$$\Delta = \frac{5bx^*}{1 + (x^*)^2} > 0, \qquad \tau = \frac{3(x^*)^2 - 5 - bx^*}{1 + (x^*)^2}.$$

We're in luck—since $\Delta > 0$, the fixed point is never a saddle. Hence (x^*, y^*) is a repeller if $\tau > 0$, i.e., if

$$b < b_c \equiv 3a/5 - 25/a. \tag{7}$$

When (7) holds, the Poincaré–Bendixson theorem implies the existence of a closed orbit somewhere in the punctured box. ■

EXAMPLE 8.3.2:

Using numerical integration, show that a Hopf bifurcation occurs at $b = b_c$ and decide whether the bifurcation is sub- or supercritical.

Solution: The analytical results above show that as b decreases through b_c, the fixed point changes from a stable spiral to an unstable spiral; this is the signature of a Hopf bifurcation. Figure 8.3.3 plots two typical phase portraits. (Here we have chosen $a = 10$; then (7) implies $b_c = 3.5$.) When $b > b_c$, all trajectories spiral into the stable fixed point (Figure 8.3.3a), while for $b < b_c$ they are attracted to a stable limit cycle (Figure 8.3.3b).

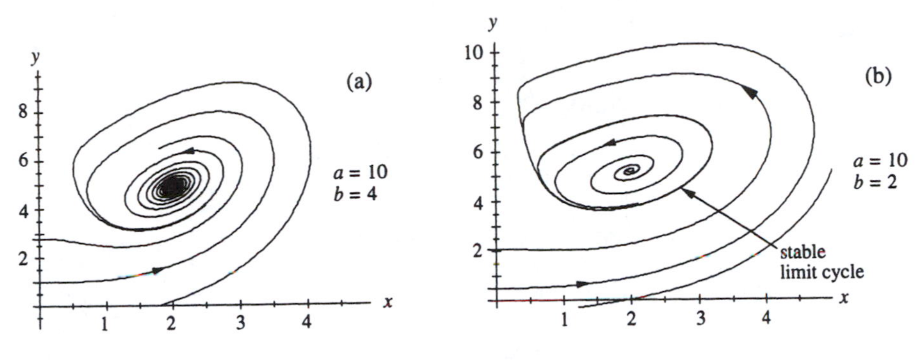

Figure 8.3.3

Hence the bifurcation is *supercritical*—after the fixed point loses stability, it is surrounded by a stable limit cycle. Moreover, by plotting phase portraits as $b \to b_c$ from below, we could confirm that the limit cycle shrinks continuously to a point, as required. ■

Our results are summarized in the stability diagram in Figure 8.3.4. The boundary between the two regions is given by the Hopf bifurcation locus $b = 3a/5 - 25/a$.

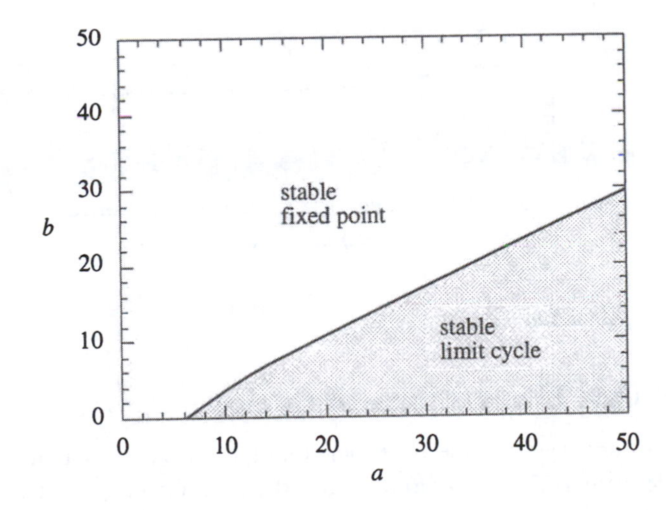

Figure 8.3.4

EXAMPLE 8.3.3:

Approximate the period of the limit cycle for b slightly less than b_c.

Solution: The frequency is approximated by the imaginary part of the eigenvalues at the bifurcation. As usual, the eigenvalues satisfy $\lambda^2 - \tau\lambda + \Delta = 0$. Since $\tau = 0$ and $\Delta > 0$ at $b = b_c$, we find

$$\lambda = \pm i\sqrt{\Delta}.$$

But at b_c,

$$\Delta = \frac{5b_c x^*}{1+(x^*)^2} = \frac{5\left(\frac{3a}{5} - \frac{25}{a}\right)\left(\frac{a}{5}\right)}{1+(a/5)^2} = \frac{15a^2 - 625}{a^2 + 25}.$$

Hence $\omega \approx \Delta^{1/2} = \left[(15a^2 - 625)/(a^2 + 25)\right]^{1/2}$ and therefore

$$\begin{aligned} T &= 2\pi/\omega \\ &= 2\pi\left[(a^2 + 25)/(15a^2 - 625)\right]^{1/2}. \end{aligned}$$

A graph of $T(a)$ is shown in Figure 8.3.5. As $a \to \infty$, $T \to 2\pi/\sqrt{15} \approx 1.63$. ■

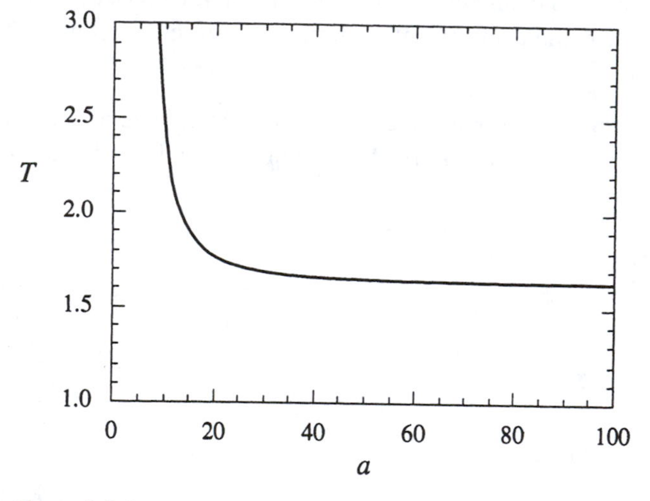

Figure 8.3.5

8.4 Global Bifurcations of Cycles

In two-dimensional systems, there are four common ways in which limit cycles are created or destroyed. The Hopf bifurcation is the most famous, but the other three deserve their day in the sun. They are harder to detect because they involve large regions of the phase plane rather than just the neighborhood of a single fixed point. Hence they are called ***global bifurcations***. In this section we offer some prototypical examples of global bifurcations, and then compare them to one another and to the Hopf bifurcation. A few of their scientific applications are discussed in Sections 8.5 and 8.6 and in the exercises.

Saddle-node Bifurcation of Cycles

A bifurcation in which two limit cycles coalesce and annihilate is called a *fold* or ***saddle-node bifurcation of cycles***, by analogy with the related bifurcation of fixed points. An example occurs in the system

$$\dot{r} = \mu r + r^3 - r^5$$
$$\dot{\theta} = \omega + br^2$$

studied in Section 8.2. There we were interested in the subcritical Hopf bifurcation at $\mu = 0$; now we concentrate on the dynamics for $\mu < 0$.

It is helpful to regard the radial equation $\dot{r} = \mu r + r^3 - r^5$ as a one-dimensional system. As you should check, this system undergoes a saddle-node bifurcation of fixed points at $\mu_c = -1/4$. Now returning to the two-dimensional system, these fixed points correspond to circular *limit cycles*. Figure 8.4.1 plots the "radial phase portraits" and the corresponding behavior in the phase plane.

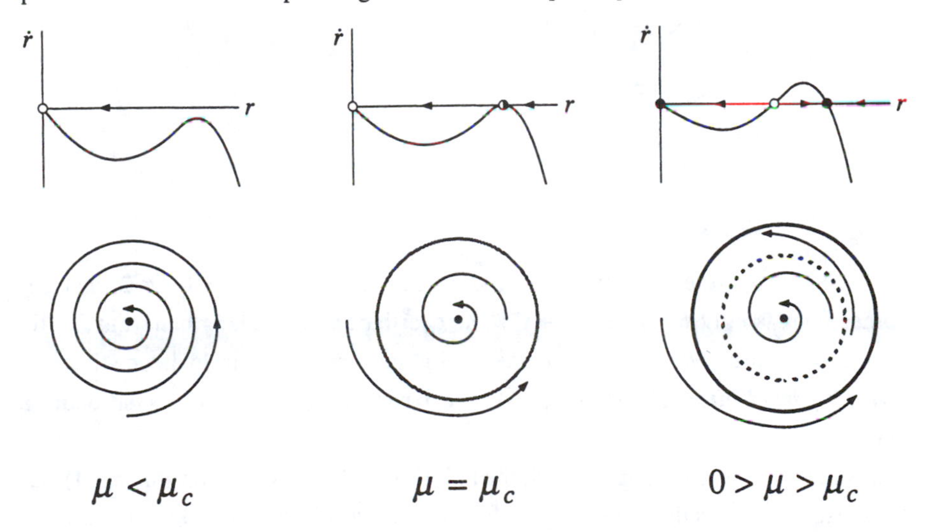

Figure 8.4.1

At μ_c a half-stable cycle is born out of the clear blue sky. As μ increases it splits into a pair of limit cycles, one stable, one unstable. Viewed in the other direction, a stable and unstable cycle collide and disappear as μ decreases through μ_c. Notice that the origin remains stable throughout; it does not participate in this bifurcation.

For future reference, note that at birth the cycle has $O(1)$ amplitude, in contrast to the Hopf bifurcation, where the limit cycle has small amplitude proportional to $(\mu-\mu_c)^{1/2}$.

Infinite-period Bifurcation

Consider the system

$$\dot{r} = r(1-r^2)$$
$$\dot{\theta} = \mu - \sin\theta$$

where $\mu \geq 0$. This system combines two one-dimensional systems that we have studied previously in Chapters 3 and 4. In the radial direction, all trajectories (except $r^* = 0$) approach the unit circle monotonically as $t \to \infty$. In the angular direction, the motion is everywhere counterclockwise if $\mu > 1$, whereas there are two invariant rays defined by $\sin\theta = \mu$ if $\mu < 1$. Hence as μ decreases through $\mu_c = 1$, the phase portraits change as in Figure 8.4.2.

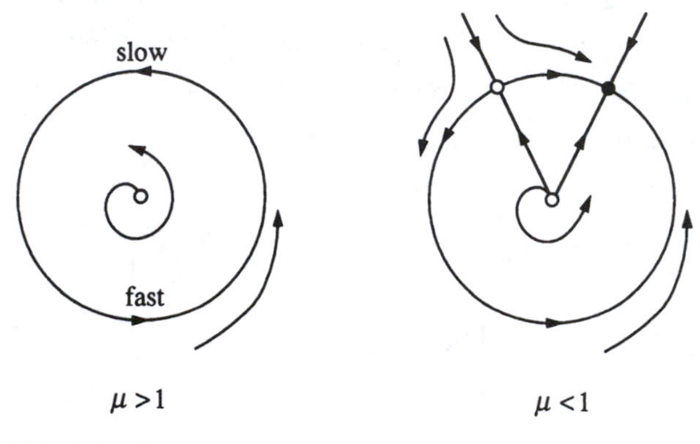

Figure 8.4.2

As μ decreases, the limit cycle $r = 1$ develops a bottleneck at $\theta = \pi/2$ that becomes increasingly severe as $\mu \to 1^+$. The oscillation period lengthens and finally becomes infinite at $\mu_c = 1$, when a fixed point appears on the circle; hence the term ***infinite-period bifurcation***. For $\mu < 1$, the fixed point splits into a saddle and a node.

As the bifurcation is approached, the amplitude of the oscillation stays $O(1)$ but the period increases like $(\mu-\mu_c)^{-1/2}$, for the reasons discussed in Section 4.3.

Homoclinic Bifurcation

In this scenario, part of a limit cycle moves closer and closer to a saddle point. At the bifurcation the cycle touches the saddle point and becomes a homoclinic orbit. This is another kind of infinite-period bifurcation; to avoid confusion, we'll call it a *saddle-loop* or ***homoclinic bifurcation***.

It is hard to find an analytically transparent example, so we resort to the computer. Consider the system

$$\dot{x} = y$$
$$\dot{y} = \mu y + x - x^2 + xy.$$

Figure 8.4.3 plots a series of phase portraits before, during, and after the bifurcation; only the important features are shown.

Numerically, the bifurcation is found to occur at $\mu_c \approx -0.8645$. For $\mu < \mu_c$, say $\mu = -0.92$, a stable limit cycle passes close to a saddle point at the origin (Figure 8.4.3a). As μ increases to μ_c, the limit cycle swells (Figure 8.4.3b) and bangs into the saddle, creating a homoclinic orbit (Figure 8.4.3c). Once $\mu > \mu_c$, the saddle connection breaks and the loop is destroyed (Figure 8.4.3d).

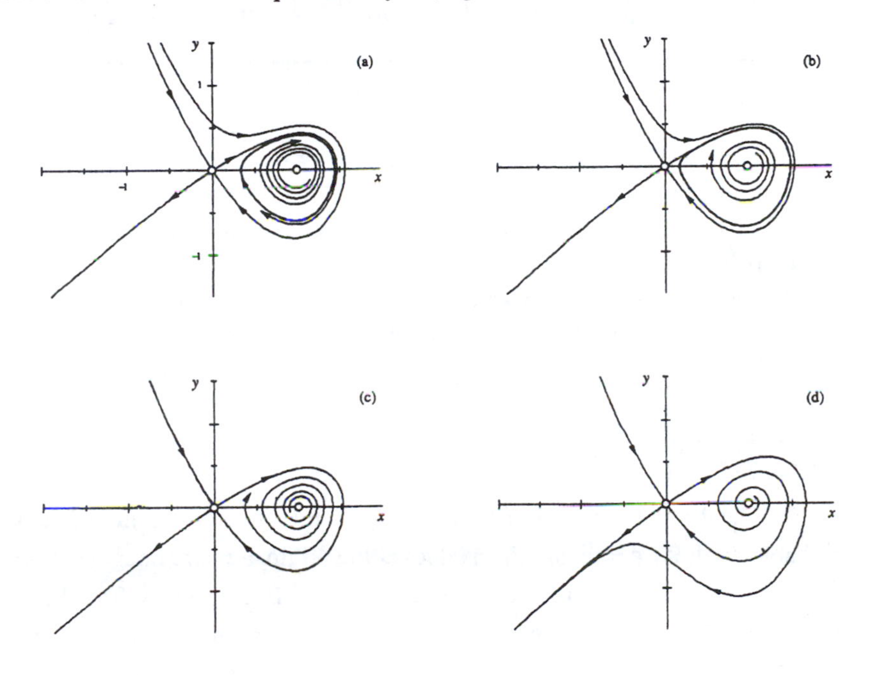

Figure 8.4.3

The key to this bifurcation is the behavior of the unstable manifold of the saddle. Look at the branch of the unstable manifold that leaves the origin to the northeast: after it loops around, it either hits the origin (Figure 8.4.3c) or veers off to one side or the other (Figures 8.4.3a, d).

Scaling Laws

For each of the bifurcations given here, there are characteristic *scaling laws* that govern the amplitude and period of the limit cycle as the bifurcation is approached. Let μ denote some dimensionless measure of the distance from the bifurcation, and assume that $\mu \ll 1$. The generic scaling laws for bifurcations of cycles in two-dimensional systems are given in Table 7.4.1.

	Amplitude of stable limit cycle	Period of cycle
Supercritical Hopf	$O(\mu^{1/2})$	$O(1)$
Saddle-node bifurcation of cycles	$O(1)$	$O(1)$
Infinite-period	$O(1)$	$O(\mu^{-1/2})$
Homoclinic	$O(1)$	$O(\ln \mu)$

Table 7.4.1

All of these laws have been explained previously, except those for the homoclinic bifurcation. The scaling of the period in that case is obtained by estimating the time required for a trajectory to pass by a saddle point (see Exercise 8.4.12 and Gaspard 1990).

Exceptions to these rules can occur, but only if there is some symmetry or other special feature that renders the problem nongeneric, as in the following example.

EXAMPLE 8.4.1:

The van der Pol oscillator $\ddot{x} + \varepsilon \dot{x}(x^2 - 1) + x = 0$ does not seem to fit anywhere in Table 7.4.1. At $\varepsilon = 0$, the eigenvalues at the origin are pure imaginary ($\lambda = \pm i$), suggesting that a Hopf bifurcation occurs at $\varepsilon = 0$. But we know from Section 7.6 that for $0 < \varepsilon \ll 1$, the system has a limit cycle of amplitude $r \approx 2$. Thus the cycle is born "full grown," not with size $O(\varepsilon^{1/2})$ as predicted by the scaling law. What's the explanation?

Solution: The bifurcation at $\varepsilon = 0$ is degenerate. The nonlinear term $\varepsilon \dot{x} x^2$ vanishes at precisely the same parameter value as the eigenvalues cross the imaginary axis. That's a nongeneric coincidence if there ever was one!

We can rescale x to remove this degeneracy. Write the equation as $\ddot{x} + x + \varepsilon x^2 \dot{x} - \varepsilon \dot{x} = 0$. Let $u^2 = \varepsilon x^2$ to remove the ε-dependence of the nonlinear term. Then $u = \varepsilon^{1/2} x$ and the equation becomes

$$\ddot{u}+u+u^2\dot{u}-\varepsilon\dot{u}=0.$$

Now the nonlinear term is not destroyed when the eigenvalues become pure imaginary. From Section 7.6 the limit cycle solution is $x(t,\varepsilon)\approx 2\cos t$ for $0<\varepsilon<<1$. In terms of u this becomes

$$u(t,\varepsilon)\approx\left(2\sqrt{\varepsilon}\right)\cos t.$$

Hence the amplitude grows like $\varepsilon^{1/2}$, just as expected for a Hopf bifurcation. ■

The scaling laws given here were derived by thinking about prototypical examples in *two-dimensional* systems. In higher-dimensional phase spaces, the corresponding bifurcations obey the same scaling laws, but with two caveats: (1) Many *additional* bifurcations of limit cycles become possible; thus our table is no longer exhaustive. (2) The homoclinic bifurcation becomes much more subtle to analyze. It often creates chaotic dynamics in its aftermath (Guckenheimer and Holmes 1983, Wiggins 1990).

All of this begs the question: Why should you care about these scaling laws? Suppose you're an experimental scientist and the system you're studying exhibits a stable limit cycle oscillation. Now suppose you change a control parameter and the oscillation stops. By examining the scaling of the period and amplitude near this bifurcation, you can learn something about the system's dynamics (which are usually not known precisely, if at all). In this way, possible models can be eliminated or supported. For an example in physical chemistry, see Gaspard (1990).

8.5 Hysteresis in the Driven Pendulum and Josephson Junction

This section deals with a physical problem in which both homoclinic and infinite-period bifurcations arise. The problem was introduced back in Sections 4.4 and 4.6. At that time we were studying the dynamics of a damped pendulum driven by a constant torque, or equivalently, its high-tech analog, a superconducting Josephson junction driven by a constant current. Because we weren't ready for two-dimensional systems, we reduced both problems to vector fields on the circle by looking at the heavily *overdamped limit* of negligible mass (for the pendulum) or negligible capacitance (for the Josephson junction).

Now we're ready to tackle the full two-dimensional problem. As we claimed at the end of Section 4.6, for sufficiently weak damping the pendulum and the Josephson junction can exhibit intriguing hysteresis effects, thanks to the coexistence of a stable limit cycle and a stable fixed point. In physical terms, the pendulum can settle into either a rotating solution where it whirls over the top, or a stable rest state where gravity balances the applied torque. The final state depends on the initial conditions. Our goal now is to understand how this bistability comes about.

We will phrase our discussion in terms of the Josephson junction, but will mention the pendulum analog whenever it seems helpful.

Governing Equations

As explained in Section 4.6, the governing equation for the Josephson junction is

$$\frac{\hbar C}{2e}\ddot{\phi}+\frac{\hbar}{2eR}\dot{\phi}+I_c\sin\phi=I_B \tag{1}$$

where $\hbar$ is Planck's constant divided by 2π, e is the charge on the electron, I_B is the constant bias current, C, R, and I_c are the junction's capacitance, resistance, and critical current, and $\phi(t)$ is the phase difference across the junction.

To highlight the role of damping, we nondimensionalize (1) differently from in Section 4.6. Let

$$\tilde{t}=\left(\frac{2eI_c}{\hbar C}\right)^{1/2}t,\qquad I=\frac{I_B}{I_c},\qquad \alpha=\left(\frac{\hbar}{2eI_cR^2C}\right)^{1/2}. \tag{2}$$

Then (1) becomes

$$\phi''+\alpha\phi'+\sin\phi=I \tag{3}$$

where α and I are the dimensionless damping and applied current, and the prime denotes differentiation with respect to $\tilde{t}$. Here $\alpha>0$ on physical grounds, and we may choose $I\geq 0$ without loss of generality (otherwise, redefine $\phi\to-\phi$).

Let $y=\phi'$. Then the system becomes

$$\begin{aligned}\phi'&=y\\ y'&=I-\sin\phi-\alpha y.\end{aligned} \tag{4}$$

As in Section 6.7 the phase space is a *cylinder*, since ϕ is an angular variable and y is a real number (best thought of as an angular velocity).

Fixed Points

The fixed points of (4) satisfy $y^*=0$ and $\sin\phi^*=I$. Hence there are two fixed points on the cylinder if $I<1$, and none if $I>1$. When the fixed points exist, one is a saddle and the other is a sink, since the Jacobian

$$A=\begin{pmatrix}0 & 1\\ -\cos\phi^* & -\alpha\end{pmatrix}$$

has $\tau=-\alpha<0$ and $\Delta=\cos\phi^*=\pm\sqrt{1-I^2}$. When $\Delta>0$, we have a stable node if

$\tau^2 - 4\Delta = \alpha^2 - 4\sqrt{1 - I^2} > 0$, i.e., if the damping is strong enough or if I is close to 1; otherwise the sink is a stable spiral. At $I = 1$ the stable node and the saddle coalesce in a *saddle-node bifurcation of fixed points.*

Existence of a Closed Orbit

What happens when $I > 1$? There are no more fixed points available; something new has to happen. We claim that *all trajectories are attracted to a unique, stable limit cycle.*

The first step is to show that a periodic solution exists. The argument uses a clever idea introduced by Poincaré long ago. Watch carefully—this idea will come up frequently in our later work.

Consider the nullcline $y = \alpha^{-1}(I - \sin\phi)$ where $y' = 0$. The flow is downward above the nullcline and upward below it (Figure 8.5.1).

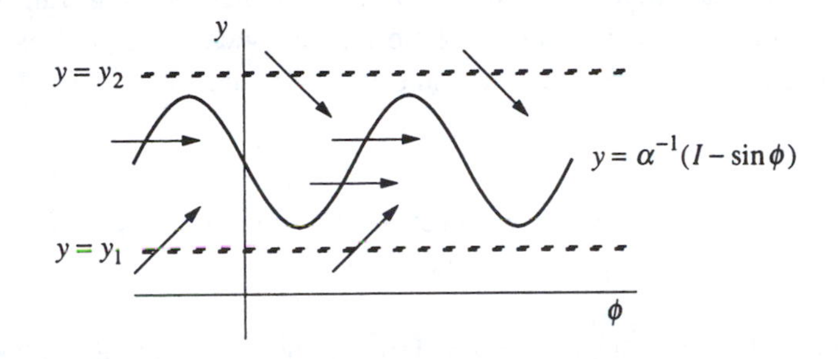

Figure 8.5.1

In particular, all trajectories eventually enter the strip $y_1 \le y \le y_2$ (Figure 8.5.1), and stay in there forever. (Here y_1 and y_2 are any fixed numbers such that $0 < y_1 < (I-1)/\alpha$ and $y_2 > (I+1)/\alpha$.) Inside the strip, the flow is always to the right, because $y > 0$ implies $\phi' > 0$.

Also, since $\phi = 0$ and $\phi = 2\pi$ are equivalent on the cylinder, we may as well confine our attention to the rectangular box $0 \le \phi \le 2\pi$, $y_1 \le y \le y_2$. This box contains all the information about the long-term behavior of the flow (Figure 8.5.2).

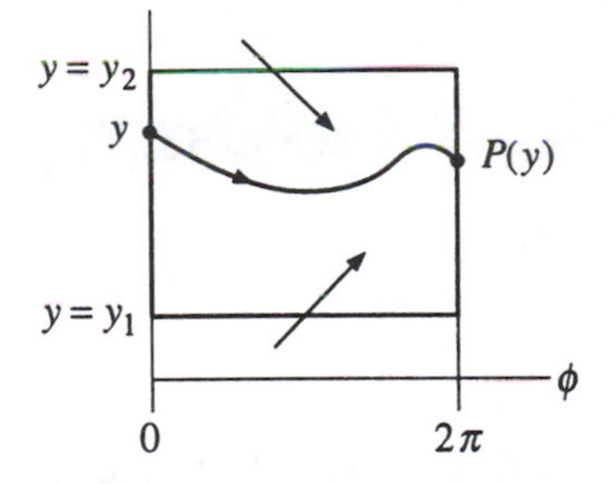

Figure 8.5.2

Now consider a trajectory that starts at a height y on the left side of the box, and follow it until it intersects the right side of the box at some new height $P(y)$, as shown in Figure 8.5.2. The mapping from y to $P(y)$ is called the ***Poincaré map***. It tells us how the height of a trajectory changes after one lap around the cylinder (Figure 8.5.3).

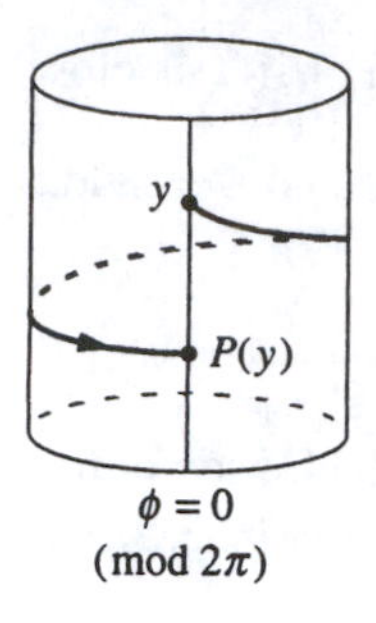

Figure 8.5.3

The Poincaré map is also called the ***first-return map***, because if a trajectory starts at a height y on the line $\phi = 0 \pmod{2\pi}$, then $P(y)$ is its height when it returns to that line for the first time.

Now comes the key point: we can't compute $P(y)$ explicitly, but *if we can show that there's a point* y^* *such that* $P(y^*) = y^*$, *then the corresponding trajectory will be a closed orbit* (because it returns to the same location on the cylinder after one lap).

To show that such a y^* must exist, we need to know what the graph of $P(y)$ looks like, at least roughly. Consider a trajectory that starts at $y = y_1$, $\phi = 0$. We claim that

$$P(y_1) > y_1 .$$

This follows because the flow is strictly upward at first, and the trajectory can never return to the line $y = y_1$, since the flow is everywhere upward on that line (recall Figures 8.5.1 and 8.5.2). By the same kind of argument,

$$P(y_2) < y_2 .$$

Furthermore, $P(y)$ is a *continuous* function. This follows from the theorem that solutions of differential equations depend continuously on initial conditions, if the vector field is smooth enough.

And finally, $P(y)$ is a *monotonic* function. (By drawing pictures, you can convince yourself that if $P(y)$ were not monotonic, two trajectories would cross—and that's forbidden.) Taken together, these results imply that $P(y)$ has the shape shown in Figure 8.5.4.

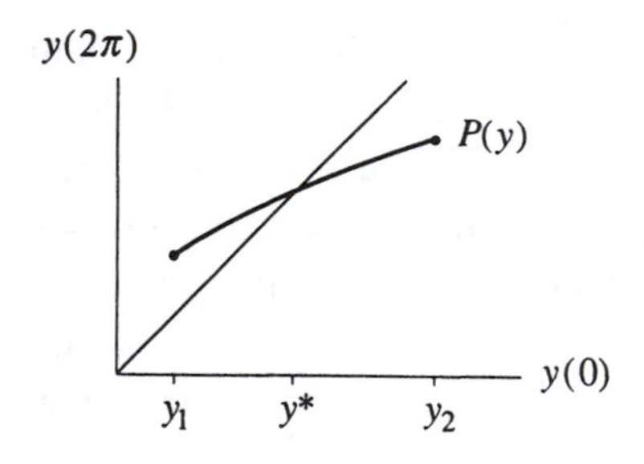

Figure 8.5.4

By the intermediate value theorem (or common sense), the graph of $P(y)$ must cross the 45° diagonal *somewhere*; that intersection is our desired y^*.

Uniqueness of the Limit Cycle

The argument above proves the *existence* of a closed orbit, and almost proves its uniqueness. But we haven't excluded the possibility that $P(y) \equiv y$ on some interval, in which case there would be a band of infinitely many closed orbits.

To nail down the uniqueness part of our claim, we recall from Section 6.7 that there are two topologically different kinds of periodic orbits on a cylinder: ***librations*** and ***rotations*** (Figure 8.5.5).

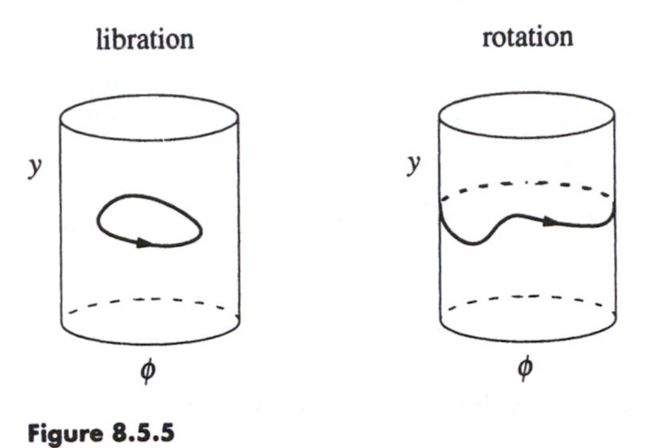

Figure 8.5.5

For $I > 1$, librations are impossible because any libration must encircle a fixed point, by index theory—but there are no fixed points when $I > 1$. Hence we only need to consider rotations.

Suppose there were two different rotations. The phase portrait on the cylinder would have to look like Figure 8.5.6.

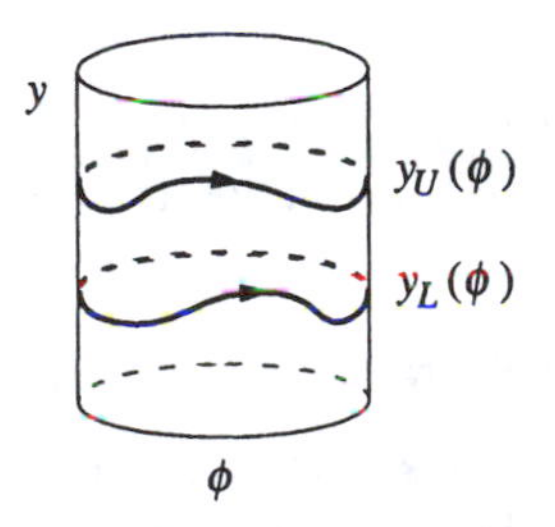

Figure 8.5.6

One of the rotations would have to lie *strictly above* the other because trajectories can't cross. Let $y_U(\phi)$ and $y_L(\phi)$ denote the "upper" and "lower" rotations, where $y_U(\phi) > y_L(\phi)$ for all ϕ.

The existence of two such rotations leads to a contradiction, as shown by the following energy argument. Let

$$E = \tfrac{1}{2} y^2 - \cos\phi. \qquad (5)$$

After one circuit around any rotation $y(\phi)$, the change in energy ΔE must vanish. Hence

$$0 = \Delta E = \int_0^{2\pi} \frac{dE}{d\phi}\, d\phi. \qquad (6)$$

But (5) implies

$$\frac{dE}{d\phi} = y\frac{dy}{d\phi} + \sin\phi \tag{7}$$

and

$$\frac{dy}{d\phi} = \frac{y'}{\phi'} = \frac{I - \sin\phi - \alpha y}{y}, \tag{8}$$

from (4). Substituting (8) into (7) gives $dE/d\phi = I - \alpha y$. Thus (6) implies

$$0 = \int_0^{2\pi} (I - \alpha y)\, d\phi$$

on any rotation $y(\phi)$. Equivalently, any rotation must satisfy

$$\int_0^{2\pi} y(\phi)\, d\phi = \frac{2\pi I}{\alpha}\ . \tag{9}$$

But since $y_U(\phi) > y_L(\phi)$,

$$\int_0^{2\pi} y_U(\phi)\, d\phi > \int_0^{2\pi} y_L(\phi)\, d\phi,$$

and so (9) can't hold for *both* rotations.

This contradiction proves that the rotation for $I > 1$ is unique, as claimed.

Homoclinic Bifurcation

Suppose we slowly decrease I, starting from some value $I > 1$. What happens to the rotating solution? Think about the pendulum: as the driving torque is reduced, the pendulum struggles more and more to make it over the top. At some critical value $I_c < 1$, the torque is insufficient to overcome gravity and damping, and the pendulum can no longer whirl. Then the rotation disappears and all solutions damp out to the rest state.

Our goal now is to visualize the corresponding bifurcation in phase space. In Exercise 8.5.2, you're asked to show (by numerical computation of the phase portrait) that if α is sufficiently small, the stable limit cycle is destroyed in a *homoclinic bifurcation* (Section 8.4). The following schematic drawings summarize the results you should get.

First suppose $I_c < I < 1$. The system is bistable: a sink coexists with a stable limit cycle (Figure 8.5.7).

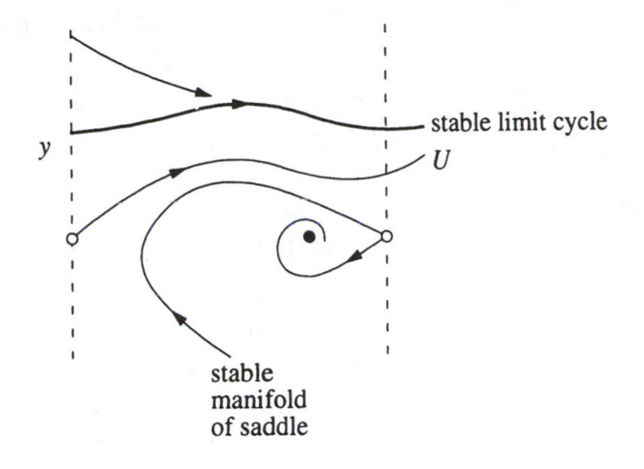

Figure 8.5.7

Keep your eye on the trajectory labeled U in Figure 8.5.7. It is a branch of the unstable manifold of the saddle. As $t \to \infty$, U asymptotically approaches the stable limit cycle.

As I decreases, the stable limit cycle moves down and squeezes U closer to the stable manifold of the saddle. When $I = I_c$, *the limit cycle merges with* U in a homoclinic bifurcation. Now U is a homoclinic orbit—it joins the saddle to itself (Figure 8.5.8).

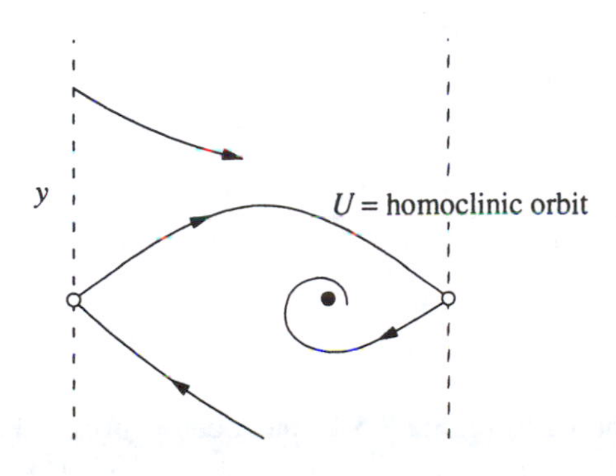

Figure 8.5.8

Finally, when $I < I_c$ the saddle connection breaks and U spirals into the sink (Figure 8.5.9).

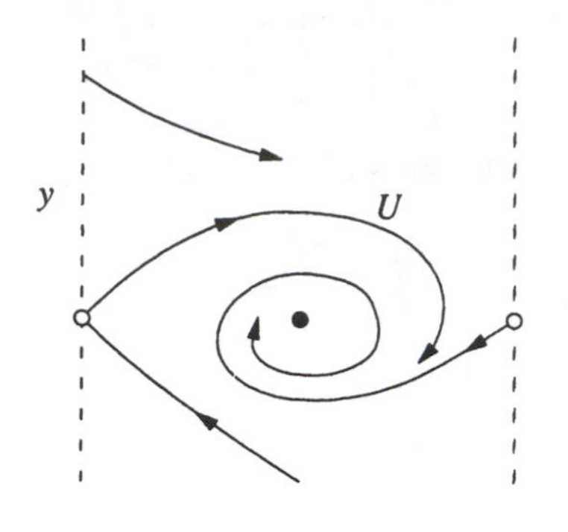

Figure 8.5.9

The scenario described here is valid only if the dimensionless damping α is sufficiently small. We know that something different has to happen for large α. After all, when α is infinite we are in the overdamped limit studied in Section 4.6. Our analysis there showed that the periodic solution is destroyed by an *infinite-period bifurcation* (a saddle and a node are born on the former limit cycle). So it's plausible that an infinite-period bifurcation should also occur if α is large but finite. These intuitive ideas are confirmed by numerical integration (Exercise 8.5.2).

Putting it all together, we arrive at the stability diagram shown in Figure 8.5.10. Three types of bifurcations occur: homoclinic and infinite-period bifurcations of periodic orbits, and a saddle-node bifurcation of fixed points.

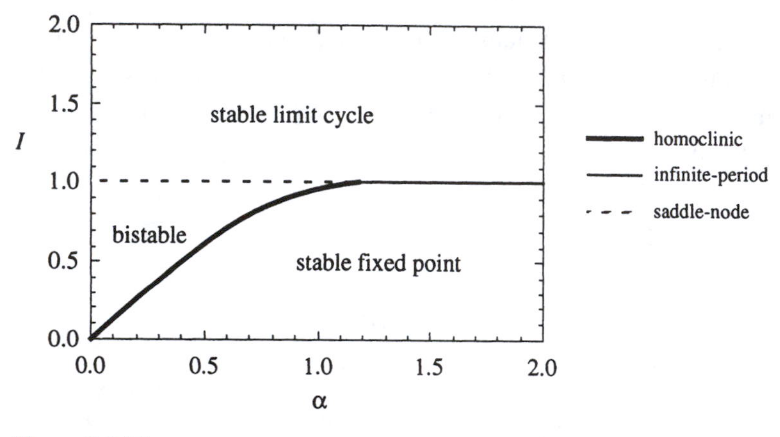

Figure 8.5.10

Our argument leading to Figure 8.5.10 has been heuristic. For rigorous proofs, see Levi et al. (1978). Also, Guckenheimer and Holmes (1983, p. 202) derive an analytical approximation for the homoclinic bifurcation curve for $\alpha \ll 1$, using an advanced technique known as Melnikov's method. They show that the bifurcation curve is tangent to the line $I = 4\alpha/\pi$ as $\alpha \to 0$. Even if α is not so small, this approximation works nicely, thanks to the straightness of the homoclinic bifurcation curve in Figure 8.5.10.

Hysteretic Current-Voltage Curve

Figure 8.5.10 explains why lightly damped Josephson junctions have hysteretic $I-V$ curves. Suppose α is small and I is initially below the homoclinic bifurcation (thick line in Figure 8.5.10). Then the junction will be operating at the stable fixed point, corresponding to the zero-voltage state. As I is increased, nothing changes until I exceeds 1. Then the stable fixed point disappears in a saddle-node bifurcation, and the junction jumps into a nonzero voltage state (the limit cycle).

If I is brought back down, the limit cycle persists below $I = 1$ but its frequency tends to zero continuously as I_c is approached. Specifically, the frequency tends to zero like $\left[\ln(I - I_c)\right]^{-1}$, just as expected from the scaling law discussed in Section 8.4. Now recall from Section 4.6 that the junction's dc-voltage is proportional to its oscillation frequency. Hence, the voltage also returns to zero continuously as $I \to I_c^+$ (Figure 8.5.11).

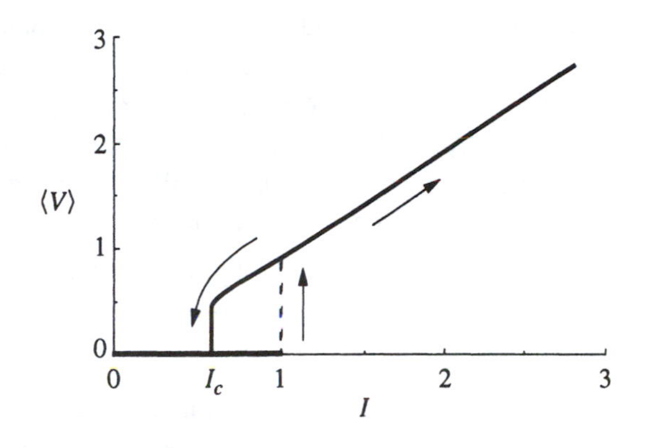

Figure 8.5.11

In practice, the voltage appears to jump discontinuously back to zero, but that is to be expected because $\left[\ln(I - I_c)\right]^{-1}$ has *infinite derivatives of all orders at* I_c! (See Exercise 8.5.1.) The steepness of the curve makes it impossible to resolve the continuous return to zero. For instance, in experiments on pendula, Sullivan and Zimmerman (1971) measured the mechanical analog of the $I - V$ curve—namely, the curve relating the rotation rate to the applied torque. Their data show a jump back to zero rotation rate at the bifurcation.

8.6 Coupled Oscillators and Quasiperiodicity

Besides the plane and the cylinder, another important two-dimensional phase space is the ***torus***. It is the natural phase space for systems of the form

$$\dot{\theta}_1 = f_1(\theta_1, \theta_2)$$
$$\dot{\theta}_2 = f_2(\theta_1, \theta_2)$$

where f_1 and f_2 are periodic in both arguments.

For instance, a simple model of ***coupled oscillators*** is given by

$$\begin{aligned}\dot{\theta}_1 &= \omega_1 + K_1 \sin(\theta_2 - \theta_1)\\ \dot{\theta}_2 &= \omega_2 + K_2 \sin(\theta_1 - \theta_2),\end{aligned} \qquad (1)$$

where θ_1, θ_2 are the *phases* of the oscillators, ω_1, $\omega_2 > 0$ are their *natural frequencies*, and K_1, $K_2 \geq 0$ are *coupling constants*. Equation (1) has been used to model the interaction between human circadian rhythms and the sleep-wake cycle (Strogatz 1986, 1987).

An intuitive way to think about (1) is to imagine two friends jogging on a circular track. Here $\theta_1(t)$, $\theta_2(t)$ represent their positions on the track, and ω_1, ω_2 are proportional to their preferred running speeds. If they were uncoupled, then each would run at his or her preferred speed and the faster one would periodically overtake the slower one (as in Example 4.2.1). But these are *friends*—they want to run around *together*! So they need to compromise, with each adjusting his or her speed as necessary. If their preferred speeds are too different, phase-locking will be impossible and they may want to find new running partners.

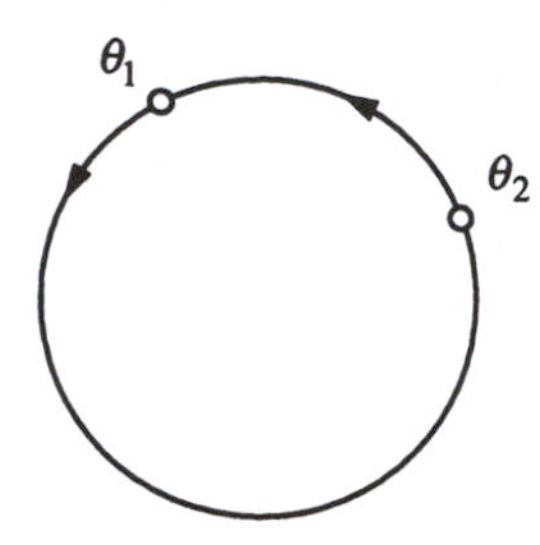

Figure 8.6.1

Here we consider (1) more abstractly, to illustrate some general features of flows on the torus and also to provide an example of a saddle-node bifurcation of cycles (Section 8.4). To visualize the flow, imagine two points running around a circle at instantaneous rates $\dot{\theta}_1$, $\dot{\theta}_2$ (Figure 8.6.1). Alternatively, we could imagine a *single* point tracing out a trajectory on a torus with coordinates θ_1, θ_2 (Figure 8.6.2). The coordinates are analogous to latitude and longitude.

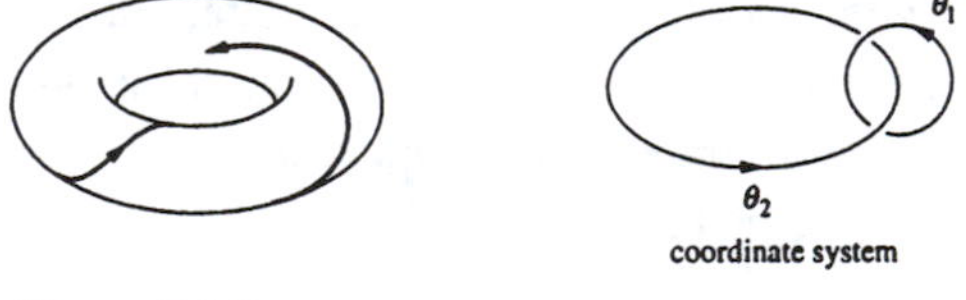

Figure 8.6.2

But since the curved surface of a torus makes it hard to draw phase portraits, we prefer to use an equivalent representation: a *square with periodic boundary conditions*. Then if a trajectory runs off an edge, it magically reappears on the opposite edge, as in some video games (Figure 8.6.3).

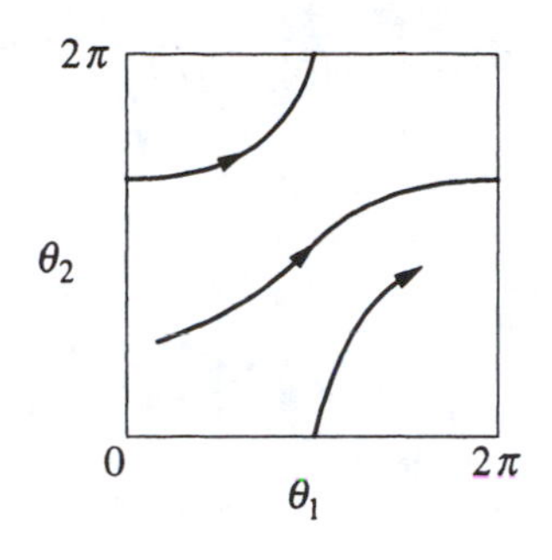

Figure 8.6.3

Uncoupled System

Even the seemingly trivial case of uncoupled oscillators ($K_1, K_2 = 0$) holds some surprises. Then (1) reduces to $\dot{\theta}_1 = \omega_1$, $\dot{\theta}_2 = \omega_2$. The corresponding trajectories on the square are straight lines with constant slope $d\theta_2/d\theta_1 = \omega_2/\omega_1$. There are two qualitatively different cases, depending on whether the slope is a rational or an irrational number.

If the slope is ***rational***, then $\omega_1/\omega_2 = p/q$ for some integers p, q with no common factors. In this case *all trajectories are closed orbits* on the torus, because θ_1 completes p revolutions in the same time that θ_2 completes q revolutions. For example, Figure 8.6.4 shows a trajectory on the square with $p = 3$, $q = 2$.

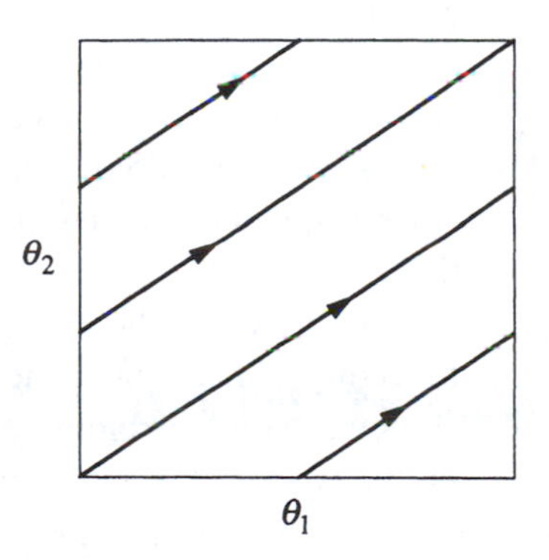

Figure 8.6.4

When plotted on the torus, the same trajectory gives . . . a ***trefoil knot***! Figure 8.6.5 shows a trefoil, alongside a top view of a torus with a trefoil wound around it.

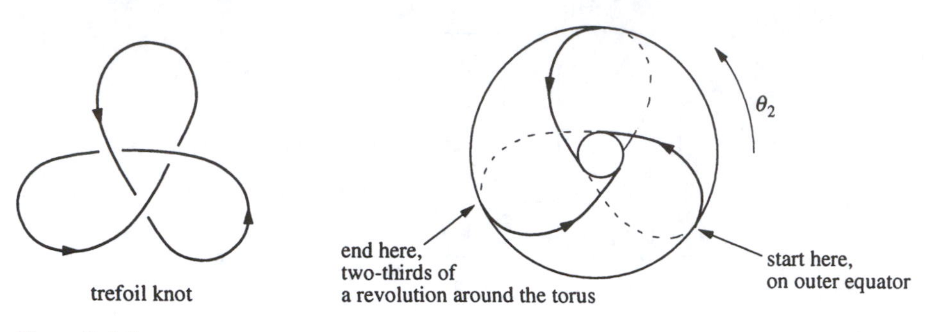

Figure 8.6.5

Do you see why this knot corresponds to $p = 3$, $q = 2$? Follow the knotted trajectory in Figure 8.6.5, and count the number of revolutions made by θ_2 during the time that θ_1 makes one revolution, where θ_1 is latitude and θ_2 is longitude. Starting on the outer equator, the trajectory moves onto the top surface, dives into the hole, travels along the bottom surface, and then reappears on the outer equator, *two-thirds* of the way around the torus. Thus θ_2 makes *two-thirds* of a revolution while θ_1 makes one revolution; hence $p = 3$, $q = 2$.

In fact the trajectories are always knotted if $p, q \geq 2$ have no common factors. The resulting curves are called $p{:}q$ *torus knots*.

The second possibility is that the slope is ***irrational*** (Figure 8.6.6).Then the flow is said to be ***quasiperiodic***. Every trajectory winds around endlessly on the torus, never intersecting itself and yet never quite closing.

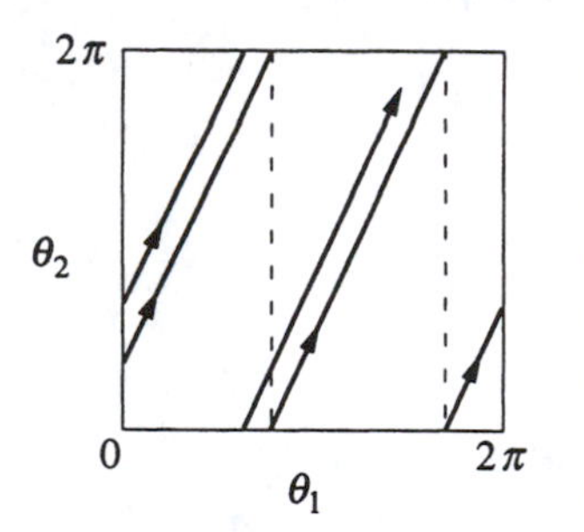

Figure 8.6.6

How can we be sure the trajectories never close? Any closed trajectory necessarily makes an integer number of revolutions in both θ_1 and θ_2; hence the slope would have to be rational, contrary to assumption.

Furthermore, when the slope is irrational, each trajectory is ***dense*** on the torus: in other words, each trajectory comes arbitrarily close to any given point on the torus. This is *not* to say that the trajectory passes *through* each point; it just comes arbitrarily close (Exercise 8.6.3).

Quasiperiodicity is significant because it is a new type of long-term behavior. Unlike the earlier entries (fixed point, closed orbit, homoclinic and heteroclinic orbits and cycles), quasiperiodicity occurs only on the torus.

Coupled System

Now consider (1) in the coupled case where $K_1, K_2 > 0$. The dynamics can be deciphered by looking at the *phase difference* $\phi = \theta_1 - \theta_2$. Then (1) yields

$$\begin{aligned}\dot{\phi} &= \dot{\theta}_1 - \dot{\theta}_2 \\ &= \omega_1 - \omega_2 - (K_1 + K_2)\sin\phi\ , \end{aligned} \tag{2}$$

which is just the nonuniform oscillator studied in Section 4.3. By drawing the standard picture (Figure 8.6.7), we see that there are two fixed points for (2) if $|\omega_1 - \omega_2| < K_1 + K_2$ and none if $|\omega_1 - \omega_2| > K_1 + K_2$. A saddle-node bifurcation occurs when $|\omega_1 - \omega_2| = K_1 + K_2$.

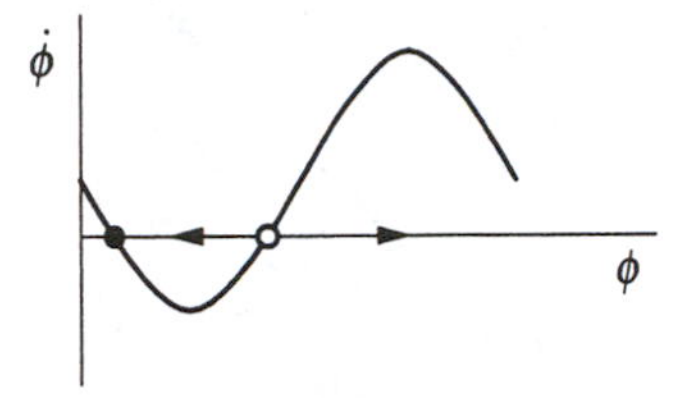

Figure 8.6.7

Suppose for now that there are two fixed points, defined implicitly by

$$\sin\phi^* = \frac{\omega_1 - \omega_2}{K_1 + K_2}\ .$$

As Figure 8.6.7 shows, all trajectories of (2) asymptotically approach the stable fixed point. Therefore, back on the torus, the trajectories of (1) approach a stable ***phase-locked*** solution in which the oscillators are separated by a constant phase difference ϕ^*. The phase-locked solution is *periodic*; in fact, both oscillators run at a constant frequency given by $\omega^* = \dot{\theta}_1 = \dot{\theta}_2 = \omega_2 + K_2 \sin\phi^*$. Substituting for $\sin\phi^*$ yields

$$\omega^* = \frac{K_1\omega_2 + K_2\omega_1}{K_1 + K_2}\ .$$

This is called the ***compromise frequency*** because it lies between the natural frequencies of the two oscillators (Figure 8.6.8).

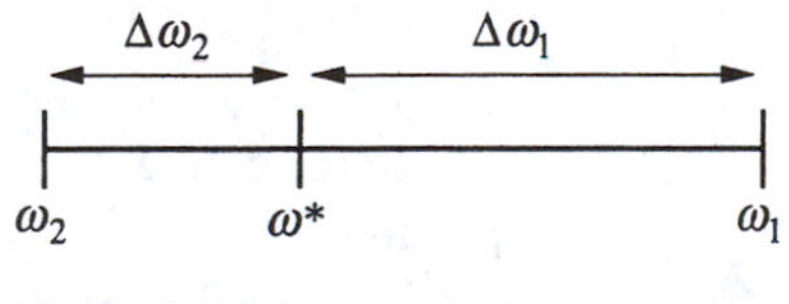

Figure 8.6.8

The compromise is not generally halfway; instead the frequencies are shifted by an amount proportional to the coupling strengths, as shown by the identity

$$\left|\frac{\Delta\omega_1}{\Delta\omega_2}\right| \equiv \left|\frac{\omega_1 - \omega^*}{\omega_2 - \omega^*}\right| = \left|\frac{K_1}{K_2}\right|.$$

Now we're ready to plot the phase portrait on the torus (Figure 8.6.9). The stable and unstable locked solutions appear as diagonal lines of slope 1, since $\dot{\theta}_1 = \dot{\theta}_2 = \omega^*$.

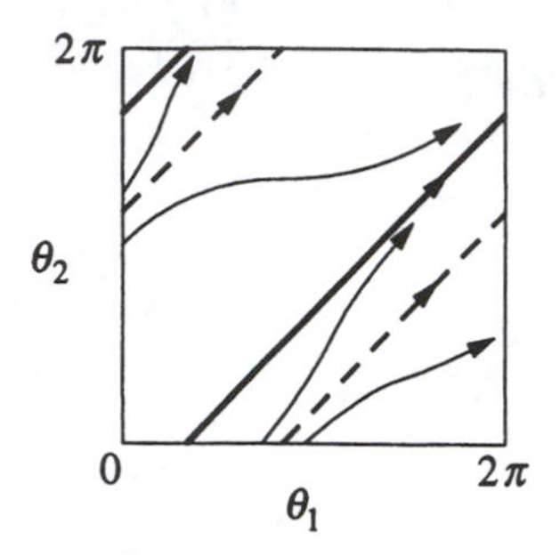

Figure 8.6.9

If we pull the natural frequencies apart, say by detuning one of the oscillators, then the locked solutions approach each other and coalesce when $|\omega_1 - \omega_2| = K_1 + K_2$. Thus the locked solution is destroyed in a *saddle-node bifurcation of cycles* (Section 8.4). After the bifurcation, the flow is like that in the uncoupled case studied earlier: we have either quasiperiodic or rational flow, depending on the parameters. The only difference is that now the trajectories on the square are curvy, not straight.

8.7 Poincaré Maps

In Section 8.5 we used a Poincaré map to prove the existence of a periodic orbit for the driven pendulum and Josephson junction. Now we discuss Poincaré maps more generally.

Poincaré maps are useful for studying swirling flows, such as the flow near a periodic orbit (or as we'll see later, the flow in some chaotic systems). Consider an n-dimensional system $\dot{\mathbf{x}} = \mathbf{f}(\mathbf{x})$. Let S be an $n-1$ dimensional ***surface of section*** (Figure 8.7.1). S is required to be transverse to the flow, i.e., all trajectories starting on S flow through it, not parallel to it.

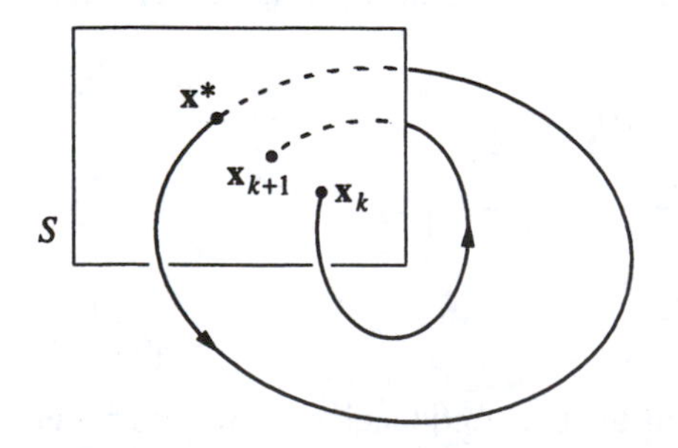

Figure 8.7.1

The ***Poincaré map*** P is a mapping from S to itself, obtained by following trajectories from one intersection with S to the next. If $\mathbf{x}_k \in S$ denotes the kth intersection, then the Poincaré map is defined by

$$\mathbf{x}_{k+1} = P(\mathbf{x}_k).$$

Suppose that $\mathbf{x}^*$ is a ***fixed point*** of P, i.e., $P(\mathbf{x}^*) = \mathbf{x}^*$. Then a trajectory starting at $\mathbf{x}^*$ returns to $\mathbf{x}^*$ after some time T, and is therefore *a closed orbit* for the original system $\dot{\mathbf{x}} = \mathbf{f}(\mathbf{x})$. Moreover, by looking at the behavior of P near this fixed point, we can determine the stability of the closed orbit.

Thus the Poincaré map converts problems about closed orbits (which are difficult) into problems about fixed points of a mapping (which are easier in principle, though not always in practice). The snag is that it's typically impossible to find a formula for P. For the sake of illustration, we begin with two examples for which P can be computed explicitly.

EXAMPLE 8.7.1:

Consider the vector field given in polar coordinates by $\dot{r} = r(1-r^2)$, $\dot{\theta} = 1$. Let S be the positive x-axis, and compute the Poincaré map. Show that the system has a unique periodic orbit and classify its stability.

Solution: Let r_0 be an initial condition on S. Since $\dot{\theta} = 1$, the first return to S occurs after a ***time of flight*** $t = 2\pi$. Then $r_1 = P(r_0)$, where r_1 satisfies

$$\int_{r_0}^{r_1} \frac{dr}{r(1-r^2)} = \int_0^{2\pi} dt = 2\pi.$$

Evaluation of the integral (Exercise 8.7.1) yields $r_1 = \left[1 + e^{-4\pi}(r_0^{-2} - 1)\right]^{-1/2}$. Hence $P(r) = \left[1 + e^{-4\pi}(r^{-2} - 1)\right]^{-1/2}$. The graph of P is plotted in Figure 8.7.2.

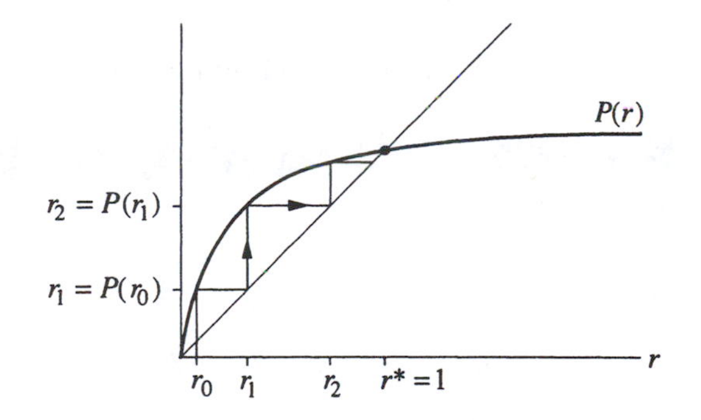

Figure 8.7.2

A fixed point occurs at $r^* = 1$ where the graph intersects the 45° line. The ***cobweb*** construction in Figure 8.7.2 enables us to iterate the map graphically. Given an input r_k, draw a vertical line until it intersects the graph of P; that height is the output r_{k+1}. To iterate, we make r_{k+1} the new input by drawing a horizontal line until it intersects the 45° diagonal line. Then repeat the process. Convince yourself that this construction works; we'll be using it often.

The cobweb shows that the fixed point $r^* = 1$ is stable and unique. No surprise, since we knew from Example 7.1.1 that this system has a stable limit cycle at $r = 1$. ■

EXAMPLE 8.7.2:

A sinusoidally forced *RC*-circuit can be written in dimensionless form as $\dot{x} + x = A \sin \omega t$, where $\omega > 0$. Using a Poincaré map, show that this system has a unique, globally stable limit cycle.

Solution: This is one of the few time-dependent systems we've discussed in this book. Such systems can always be made time-independent by adding a new variable. Here we introduce $\theta = \omega t$ and regard the system as a vector field on a cylinder: $\dot{\theta} = \omega$, $\dot{x} + x = A \sin \theta$. Any vertical line on the cylinder is an appropriate section S; we choose $S = \{(\theta, x) \colon \theta = 0 \bmod 2\pi\}$. Consider an initial condition on S given by $\theta(0) = 0$, $x(0) = x_0$. Then the time of flight between successive intersections is $t = 2\pi/\omega$. In physical terms, we strobe the system once per drive cycle and look at the consecutive values of x.

To compute P, we need to solve the differential equation. Its general solution is a sum of homogeneous and particular solutions: $x(t) = c_1 e^{-t} + c_2 \sin \omega t + c_3 \cos \omega t$. The constants c_2 and c_3 can be found explicitly, but the important point is that they depend on A and ω but *not* on the initial condition x_0; only c_1 depends on x_0. To make the dependence on x_0 explicit, observe that at $t = 0$, $x = x_0 = c_1 + c_3$. Thus

$$x(t) = (x_0 - c_3) e^{-t} + c_2 \sin \omega t + c_3 \cos \omega t.$$

Then P is defined by $x_1 = P(x_0) = x(2\pi/\omega)$. Substitution yields

$$\begin{aligned} P(x_0) &= x(2\pi/\omega) = (x_0 - c_3) e^{-2\pi/\omega} + c_3 \\ &= x_0 e^{-2\pi/\omega} + c_4 \end{aligned}$$

where $c_4 = c_3(1 - e^{-2\pi/\omega})$.

The graph of P is a straight line with a slope $e^{-2\pi/\omega} < 1$ as shown in Figure 8.7.3.

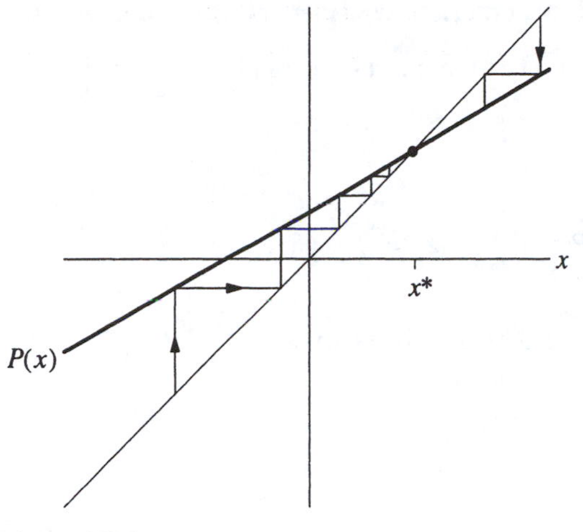

Figure 8.7.3

Since P has slope less than 1, it intersects the diagonal at a unique point. Furthermore, the cobweb shows that the deviation of x_k from the fixed point is reduced by a constant factor with each iteration. Hence the fixed point is unique and globally stable.

In physical terms, the circuit always settles into the same forced oscillation, regardless of the initial conditions. This is a familiar result from elementary physics, looked at in a new way. ■

Linear Stability of Periodic Orbits

Now consider the general case: Given a system $\dot{\mathbf{x}} = \mathbf{f}(\mathbf{x})$ with a closed orbit, how can we tell whether the orbit is stable is not? Equivalently, we ask whether the corresponding fixed point $\mathbf{x}^*$ of the Poincaré map is stable. Let $\mathbf{v}_0$ be an infinitesimal perturbation such that $\mathbf{x}^* + \mathbf{v}_0$ is in S. Then after the first return to S,

$$\begin{aligned} \mathbf{x}^* + \mathbf{v}_1 &= P(\mathbf{x}^* + \mathbf{v}_0) \\ &= P(\mathbf{x}^*) + [DP(\mathbf{x}^*)]\mathbf{v}_0 + O(\|\mathbf{v}_0\|^2) \end{aligned}$$

where $DP(\mathbf{x}^*)$ is an $(n-1) \times (n-1)$ matrix called the ***linearized Poincaré map*** at $\mathbf{x}^*$. Since $\mathbf{x}^* = P(\mathbf{x}^*)$, we get

$$\mathbf{v}_1 = [DP(\mathbf{x}^*)]\mathbf{v}_0$$

assuming that we can neglect the small $O(\|\mathbf{v}_0\|^2)$ terms.

The desired stability criterion is expressed in terms of the eigenvalues λ_j of $DP(\mathbf{x}^*)$: *The closed orbit is linearly stable if and only if* $|\lambda_j| < 1$ *for all* $j = 1, \ldots, n-1$.

To understand this criterion, consider the generic case where there are no repeated eigenvalues. Then there is a basis of eigenvectors $\{\mathbf{e}_j\}$ and so we can write $\mathbf{v}_0 = \sum_{j=1}^{n-1} v_j \mathbf{e}_j$ for some scalars v_j. Hence

$$\mathbf{v}_1 = \left(DP(\mathbf{x}^*)\right)\sum_{j=1}^{n-1} v_j \mathbf{e}_j = \sum_{j=1}^{n-1} v_j \lambda_j \mathbf{e}_j .$$

Iterating the linearized map k times gives

$$\mathbf{v}_k = \sum_{j=1}^{n-1} v_j (\lambda_j)^k \mathbf{e}_j .$$

Hence, if all $|\lambda_j| < 1$, then $\|\mathbf{v}_k\| \to 0$ geometrically fast. This proves that $\mathbf{x}^*$ is linearly stable. Conversely, if $|\lambda_j| > 1$ for some j, then perturbations along $\mathbf{e}_j$ grow, so $\mathbf{x}^*$ is unstable. A borderline case occurs when the largest eigenvalue has magnitude $|\lambda_m| = 1$; this occurs at bifurcations of periodic orbits, and then a nonlinear stability analysis is required.

The λ_j are called the ***characteristic*** or ***Floquet multipliers*** of the periodic orbit. (Strictly speaking, these are the *nontrivial* multipliers; there is always an additional trivial multiplier $\lambda \equiv 1$ corresponding to perturbations *along* the periodic orbit. We have ignored such perturbations since they just amount to time-translation.)

In general, the characteristic multipliers can only be found by numerical integration (see Exercise 8.7.10). The following examples are two of the rare exceptions.

EXAMPLE 8.7.3:

Find the characteristic multiplier for the limit cycle of Example 8.7.1.

Solution: We linearize about the fixed point $r^* = 1$ of the Poincaré map. Let $r = 1 + \eta$, where η is infinitesimal. Then $\dot{r} = \dot{\eta} = (1+\eta)\left(1-(1+\eta)^2\right)$. After neglecting $O(\eta^2)$ terms, we get $\dot{\eta} = -2\eta$. Thus $\eta(t) = \eta_0 e^{-2t}$. After a time of flight $t = 2\pi$, the new perturbation is $\eta_1 = e^{-4\pi}\eta_0$. Hence $e^{-4\pi}$ is the characteristic multiplier. Since $\left|e^{-4\pi}\right| < 1$, the limit cycle is linearly stable. ■

For this simple two-dimensional system, the linearized Poincaré map degenerates to a 1×1 matrix, i.e., a number. Exercise 8.7.1 asks you to show explicitly that

$P'(r^*) = e^{-4\pi}$, as expected from the general theory above.

Our final example comes from a recent analysis of coupled Josephson junctions.

EXAMPLE 8.7.4:

The N-dimensional system

$$\dot{\phi}_i = \Omega + a\sin\phi_i + \tfrac{1}{N}\sum_{j=1}^{N}\sin\phi_j \,, \qquad (1)$$

for $i = 1, \ldots, N$, describes the dynamics of a series array of overdamped Josephson junctions in parallel with a resistive load (Tsang et al. 1991). For technological reasons, there is great interest in the solution where all the junctions oscillate in phase. This *in-phase* solution is given by $\phi_1(t) = \phi_2(t) = \ldots = \phi_N(t) = \phi^*(t)$, where $\phi^*(t)$ denotes the common waveform. Find conditions under which the in-phase solution is periodic, and calculate the characteristic multipliers of this solution.

Solution: For the in-phase solution, all N equations reduce to

$$\frac{d\phi^*}{dt} = \Omega + (a+1)\sin\phi^* \,. \qquad (2)$$

This has a periodic solution (on the circle) if and only if $|\Omega| > |a+1|$. To determine the stability of the in-phase solution, let $\phi_i(t) = \phi^*(t) + \eta_i(t)$, where the $\eta_i(t)$ are infinitesimal perturbations. Then substituting ϕ_i into (1) and dropping quadratic terms in η yields

$$\dot{\eta}_i = [a\cos\phi^*(t)]\eta_i + [\cos\phi^*(t)]\,\tfrac{1}{N}\sum_{j=1}^{N}\eta_j \,. \qquad (3)$$

We don't have $\phi^*(t)$ explicitly, but that doesn't matter, thanks to two tricks. First, the linear system decouples if we change variables to

$$\mu = \tfrac{1}{N}\sum_{j=1}^{N}\eta_j \,,$$

$$\xi_i = \eta_{i+1} - \eta_i, \quad i = 1 \ldots, N-1.$$

Then $\dot{\xi}_i = [a\cos\phi^*(t)]\,\xi_i$. Separation of variables yields

$$\frac{d\xi_i}{\xi_i} = [a\cos\phi^*(t)]\,dt = \frac{[a\cos\phi^*]\,d\phi^*}{\Omega + (a+1)\sin\phi^*} \,,$$

where we've used (2) to eliminate dt. (That was the second trick.)

Now we compute the change in the perturbations after one circuit around the closed orbit ϕ^*:

$$\oint \frac{d\xi_i}{\xi_i} = \int_0^{2\pi} \frac{[a\cos\phi^*]d\phi^*}{\Omega + (a+1)\sin\phi^*}$$

$$\Rightarrow \ln\frac{\xi_i(T)}{\xi_i(0)} = \frac{a}{a+1}\ln\left[\Omega + (a+1)\sin\phi^*\right]_0^{2\pi} = 0.$$

Hence $\xi_i(T) = \xi_i(0)$. Similarly, we can show that $\mu(T) = \mu(0)$. Thus $\eta_i(T) = \eta_i(0)$ for all i; all perturbations are unchanged after one cycle! Therefore all the characteristic multipliers $\lambda_j = 1$. ■

This calculation shows that the in-phase state is (linearly) neutrally stable. That's discouraging technologically—one would like the array to lock into coherent oscillation, thereby greatly increasing the output power over that available from a single junction.

Since the calculation above is based on linearization, you might wonder whether the neglected nonlinear terms could stabilize the in-phase state. In fact they don't: a reversibility argument shows that the in-phase state is not attracting, even if the nonlinear terms are kept (Exercise 8.7.11).

EXERCISES FOR CHAPTER 8

8.1 Saddle-Node, Transcritical, and Pitchfork Bifurcations

8.1.1 For the following prototypical examples, plot the phase portraits as μ varies:

a) $\dot{x} = \mu x - x^2, \quad \dot{y} = -y$ (transcritical bifurcation)

b) $\dot{x} = \mu x + x^3, \quad \dot{y} = -y$ (subcritical pitchfork bifurcation)

For each of the following systems, find the eigenvalues at the stable fixed point as a function of μ, and show that one of the eigenvalues tends to zero as $\mu \to 0$.

8.1.2 $\dot{x} = \mu - x^2, \quad \dot{y} = -y$

8.1.3 $\dot{x} = \mu x - x^2, \quad \dot{y} = -y$

8.1.4 $\dot{x} = \mu x + x^3, \quad \dot{y} = -y$

8.1.5 Prove that at any zero-eigenvalue bifurcation in two dimensions, the nullclines always intersect tangentially. (Hint: Consider the geometrical meaning of the rows in the Jacobian matrix.)

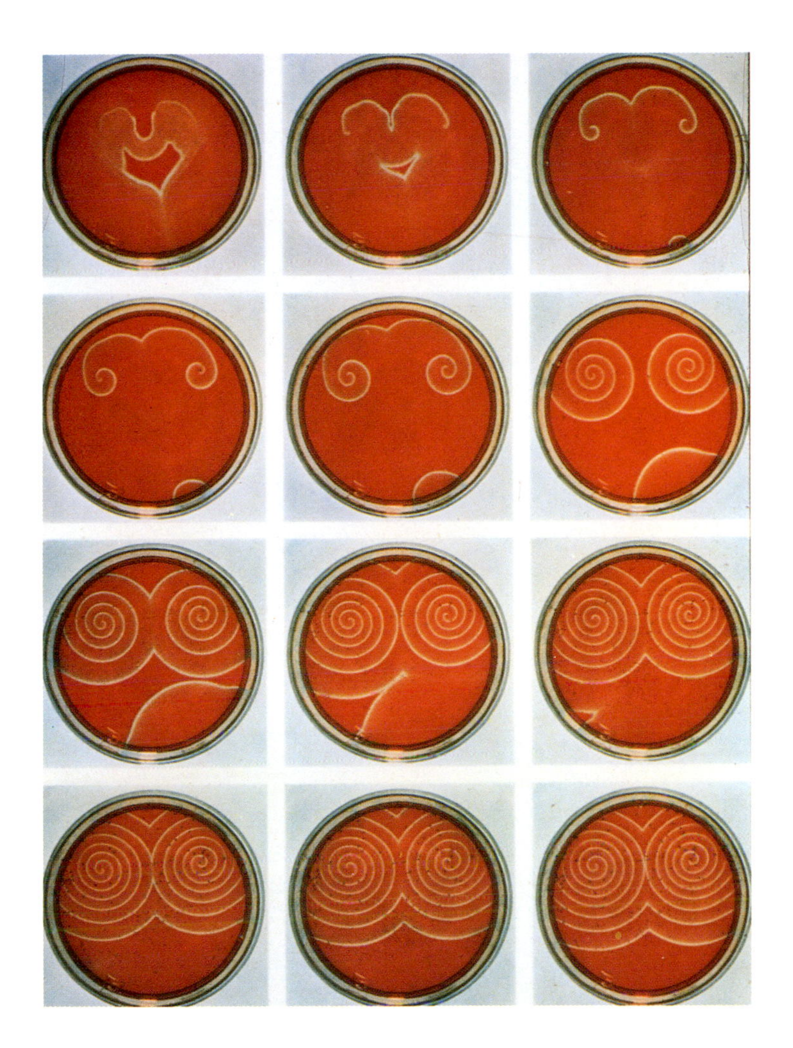

Plate 1: Spiral waves of chemical activity in a shallow dish of the Belousov–Zhabotinsky reaction (Section 8.3). These snapshots read from left to right and top to bottom. The complicated initial condition shown in the upper left was created by touching the liquid with a hot wire, thereby inducing an expanding circular wave of oxidation, and then disrupting this wave by gently rocking the dish. As time evolves, the blue waves propagate by diffusion through the motionless reddish-orange liquid. Whenever two waves collide, they annihilate each other, like grassfires rushing head on. Ultimately the system organizes itself into a pair of counterrotating spirals. Reproduced from Winfree (1974). Photographs by Fritz Goro.

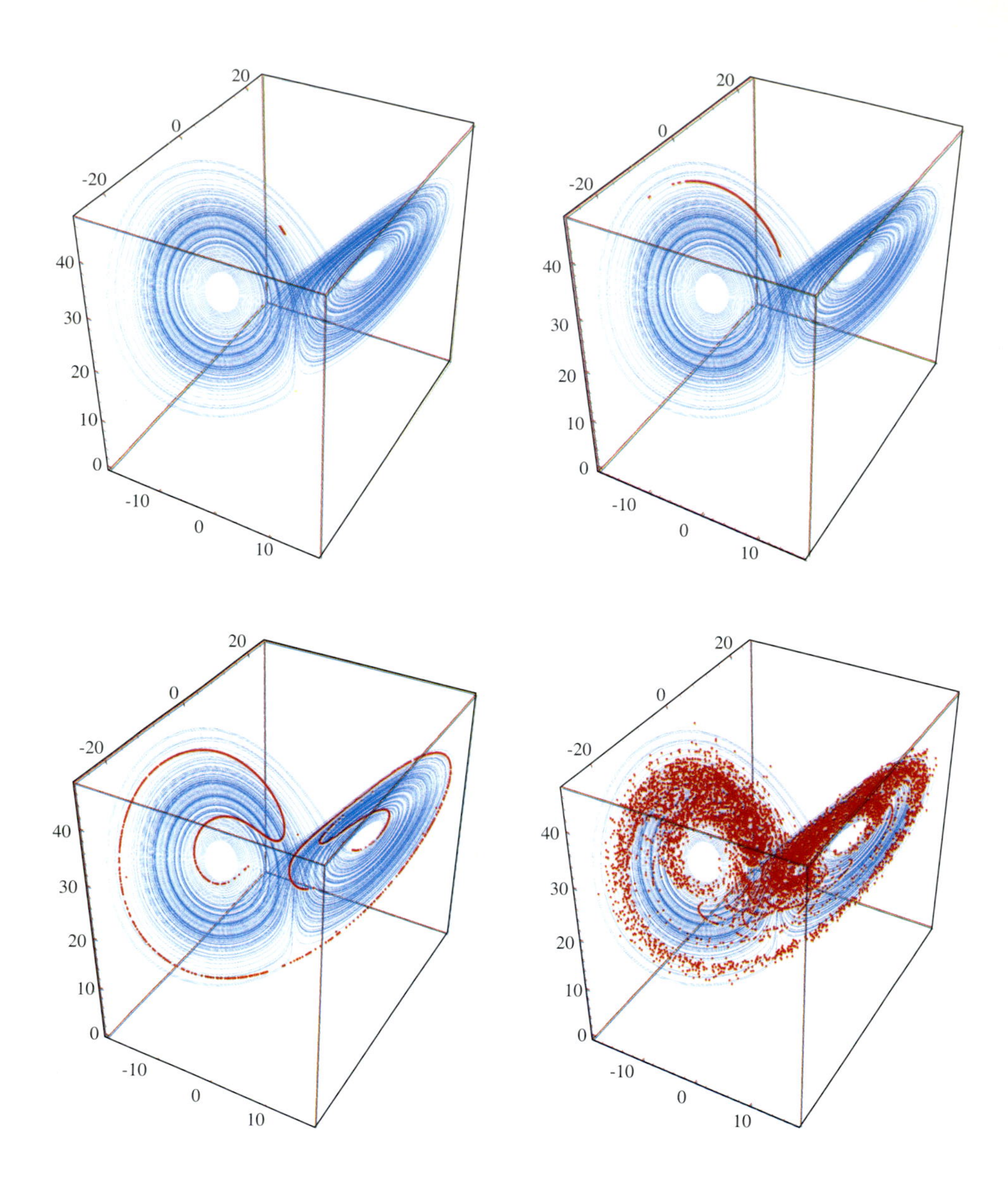

Plate 2: Divergence of nearby trajectories on the Lorenz attractor (Section 9.3). The Lorenz attractor is shown in blue. The red points show the evolution of a small blob of 10,000 nearby initial conditions, at times $t = 3$, 6, 9, and 15. As each point moves according to the Lorenz equations, the blob is stretched into a long thin filament, which then wraps around the attractor. Ultimately the points spread over much of the attractor, showing that the final state could be almost anywhere, even though the initial conditions were almost identical. This sensitive dependence on initial conditions is the signature of a chaotic system.

Plate inspired by a similar illustration in Crutchfield et al. (1986). Numerical integration and computer graphics by Thanos Siapas, using Equation (9.2.1) with parameters $\sigma = 10$, $b = 8/3$, $r = 28$.

Plate 3: Fractal basin boundaries for the periodically forced double-well oscillator

$$x' = y, \qquad y' = x - x^3 - \delta y + F\cos\omega t,$$

with $\delta = 0.25$, $F = 0.25$, $\omega = 1$ (Section 12.5). For these parameter values, the system has two periodic attractors, corresponding to forced oscillations confined to the left or right well.

(a) Color map: The square region $-2.5 \le x, y \le 2.5$ is subdivided into 900×900 cells, and each cell is color-coded according to the x-position of its center point.

(b) Basins of attraction: Each cell is color-coded according to its fate after many drive cycles. Roughly speaking, if the trajectory ends up oscillating in the right well, the original cell is colored red; if it ends up in the left well, it is colored blue. More precisely, given an initial point (x_0, y_0) at the center of a cell, the state $(x(t), y(t))$ is computed at $t = 73 \times 2\pi/\omega$ (that is, after 73 drive cycles), and the original cell is color-coded by the value of $x(t)$. The basins have a complicated shape, and the boundary between them is fractal (Moon and Li 1985). Near the boundary, slight variations in initial conditions can lead to totally different outcomes.

Computations by Thanos Siapas on a Thinking Machines CM-5 parallel computer using a fifth-order Runge–Kutta–Fehlberg method.

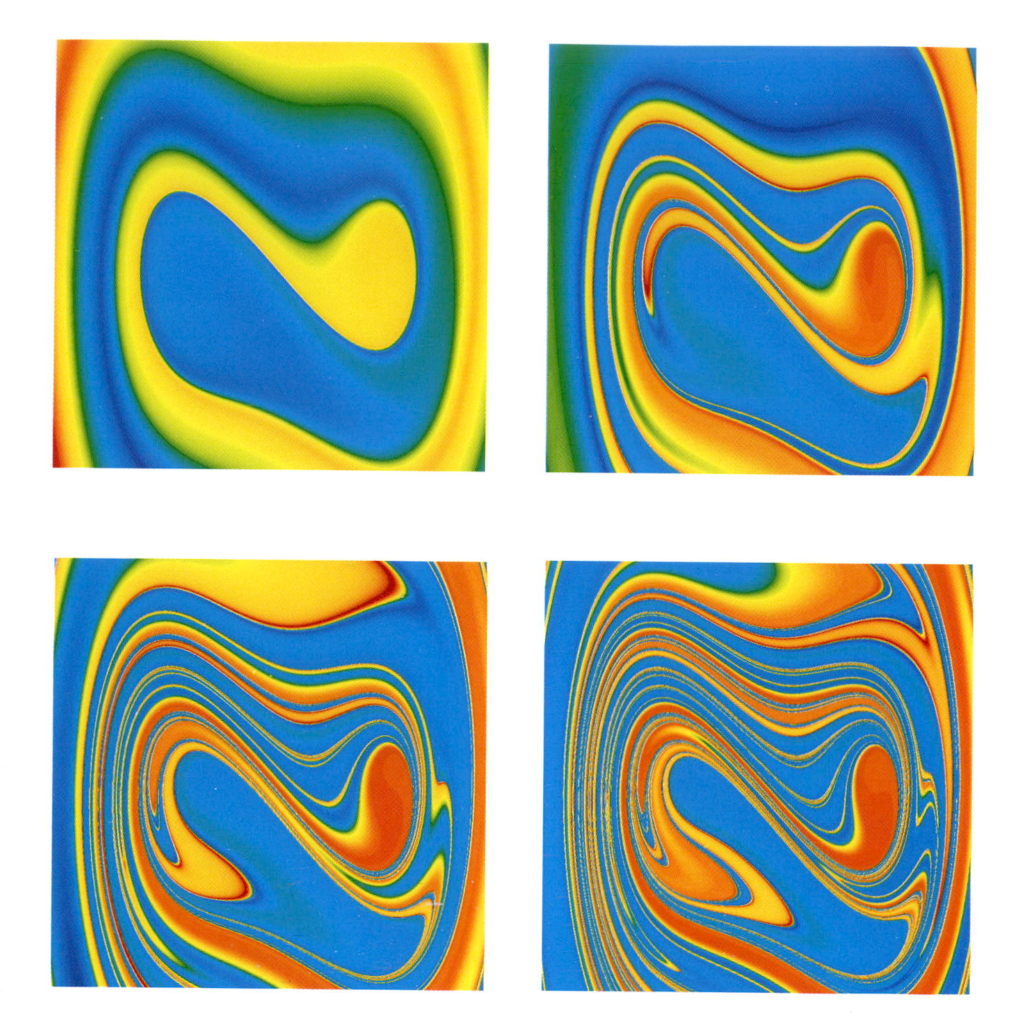

Plate 4: Maps of the short-term behavior of the periodically forced double-well oscillator. Equations, parameters, and color code as in Plate 3. However, instead of showing the system's asymptotic behavior, these plates show the color-coded value of $x(t)$ after only 1, 2, 3, and 4 drive cycles, respectively. The red and blue regions correspond to initial conditions that converge rapidly to one of the two attractors. A rainbow of colors is found near the basin boundary, because those initial conditions lead to trajectories that linger far from either attractor during the time shown.

8.1.6 Consider the system $\dot{x} = y - 2x$, $\dot{y} = \mu + x^2 - y$.

a) Sketch the nullclines.

b) Find and classify the bifurcations that occur as μ varies.

c) Sketch the phase portrait as a function of μ.

8.1.7 Find and classify all bifurcations for the system $\dot{x} = y - ax$, $\dot{y} = -by + x/(1+x)$.

8.1.8 (Bead on rotating hoop, revisited) In Section 3.5, we derived the following dimensionless equation for the motion of a bead on a rotating hoop:

$$\varepsilon \frac{d^2\phi}{d\tau^2} = -\frac{d\phi}{d\tau} - \sin\phi + \gamma \sin\phi \cos\phi .$$

Here $\varepsilon > 0$ is proportional to the mass of the bead, and $\gamma > 0$ is related to the spin rate of the hoop. Previously we restricted our attention to the overdamped limit $\varepsilon \to 0$.

a) Now allow any $\varepsilon > 0$. Find and classify all bifurcations that occur as ε and γ vary.

b) Plot the stability diagram in the positive quadrant of the ε, γ plane.

8.1.9 Plot the stability diagram for the system $\ddot{x} + b\dot{x} - kx + x^3 = 0$, where b and k can be positive, negative, or zero. Label the bifurcation curves in the (b, k) plane.

8.1.10 (Budworms vs. the forest) Ludwig et al. (1978) proposed a model for the effects of spruce budworm on the balsam fir forest. In Section 3.7, we considered the dynamics of the budworm population; now we turn to the dynamics of the forest. The condition of the forest is assumed to be characterized by $S(t)$, the average size of the trees, and $E(t)$, the "energy reserve" (a generalized measure of the forest's health). In the presence of a constant budworm population B, the forest dynamics are given by

$$\dot{S} = r_S S\left(1 - \frac{S}{K_S}\frac{K_E}{E}\right), \qquad \dot{E} = r_E E\left(1 - \frac{E}{K_E}\right) - P\frac{B}{S},$$

where $r_S, r_E, K_S, K_E, P > 0$ are parameters.

a) Interpret the terms in the model biologically.

b) Nondimensionalize the system.

c) Sketch the nullclines. Show that there are two fixed points if B is small, and none if B is large. What type of bifurcation occurs at the critical value of B?

d) Sketch the phase portrait for both large and small values of B.

8.1.11 In a study of isothermal autocatalytic reactions, Gray and Scott (1985) considered a hypothetical reaction whose kinetics are given in dimensionless form by

$$\dot{u} = a(1-u) - uv^2, \qquad \dot{v} = uv^2 - (a+k)v,$$

where $a, k > 0$ are parameters. Show that saddle-node bifurcations occur at $k = -a \pm \frac{1}{2}\sqrt{a}$.

8.1.12 (Interacting bar magnets) Consider the system

$$\dot{\theta}_1 = K\sin(\theta_1 - \theta_2) - \sin\theta_1$$
$$\dot{\theta}_2 = K\sin(\theta_2 - \theta_1) - \sin\theta_2$$

where $K \geq 0$. For a rough physical interpretation, suppose that two bar magnets are confined to a plane, but are free to rotate about a common pin joint, as shown in Figure 1. Let θ_1, θ_2 denote the angular orientations of the north poles of the magnets. Then the term $K\sin(\theta_2 - \theta_1)$ represents a repulsive force that tries to keep the two north poles 180° apart. This repulsion is opposed by the $\sin\theta$ terms, which model external magnets that pull the north poles of both bar magnets to the east. If the inertia of the magnets is negligible compared to viscous damping, then the equations above are a decent approximation to the true dynamics.

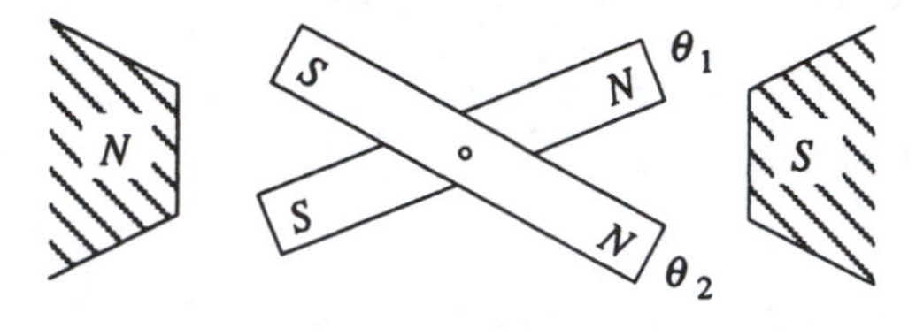

Figure 1

a) Find and classify all the fixed points of the system.
b) Show that a bifurcation occurs at $K = \frac{1}{2}$. What type of bifurcation is it? (Hint: Recall that $\sin(a-b) = \cos b \sin a - \sin b \cos a$.)
c) Show that the system is a "gradient" system, in the sense that $\dot{\theta}_i = -\partial V / \partial \theta_i$ for some potential function $V(\theta_1, \theta_2)$, to be determined.
d) Use part (c) to prove that the system has no periodic orbits.
e) Sketch the phase portrait for $0 < K < \frac{1}{2}$, and then for $K > \frac{1}{2}$.

8.1.13 (Laser model) In Exercise 3.3.1 we introduced the laser model

$$\dot{n} = GnN - kn$$
$$\dot{N} = -GnN - fN + p$$

where $N(t)$ is the number of excited atoms and $n(t)$ is the number of photons in the laser field. The parameter G is the gain coefficient for stimulated emission, k is the decay rate due to loss of photons by mirror transmission, scattering, etc., f is the decay rate for spontaneous emission, and p is the pump strength. All parameters are positive, except p, which can have either sign. For more information, see Milonni and Eberly (1988).

a) Nondimensionalize the system.
b) Find and classify all the fixed points.
c) Sketch all the qualitatively different phase portraits that occur as the dimensionless parameters are varied.
d) Plot the stability diagram for the system. What types of bifurcation occur?

8.2 Hopf Bifurcations

8.2.1 Consider the biased van der Pol oscillator $\ddot{x}+\mu(x^2-1)\dot{x}+x=a$. Find the curves in (μ, a) space at which Hopf bifurcations occur.

The next three exercises deal with the system $\dot{x}=-y+\mu x+xy^2$, $\dot{y}=x+\mu y-x^2$.

8.2.2 By calculating the linearization at the origin, show that the system $\dot{x}=-y+\mu x+xy^2$, $\dot{y}=x+\mu y-x^2$ has pure imaginary eigenvalues when $\mu=0$.

8.2.3 (Computer work) By plotting phase portraits on the computer, show that the system $\dot{x}=-y+\mu x+xy^2$, $\dot{y}=x+\mu y-x^2$ undergoes a Hopf bifurcation at $\mu=0$. Is it subcritical, supercritical, or degenerate?

8.2.4 (A heuristic analysis) The system $\dot{x}=-y+\mu x+xy^2$, $\dot{y}=x+\mu y-x^2$ can be analyzed in a rough, intuitive way as follows.
a) Rewrite the system in polar coordinates.
b) Show that if $r \ll 1$, then $\dot{\theta}\approx 1$ and $\dot{r}\approx \mu r+\frac{1}{8}r^3+\cdots$, where the terms omitted are oscillatory and have essentially zero time-average around one cycle.
c) The formulas in part (b) suggest the presence of an unstable limit cycle of radius $r\approx\sqrt{-8\mu}$ for $\mu<0$. Confirm that prediction numerically. (Since we assumed that $r \ll 1$, the prediction is expected to hold only if $|\mu| \ll 1$.)

The reasoning above is shaky. See Drazin (1992, pp. 188–190) for a proper analysis via the Poincaré–Lindstedt method.

For each of the following systems, a Hopf bifurcation occurs at the origin when $\mu=0$. Using a computer, plot the phase portrait and determine whether the bifurcation is subcritical or supercritical.

8.2.5 $\dot{x}=y+\mu x$, $\dot{y}=-x+\mu y-x^2y$

8.2.6 $\dot{x}=\mu x+y-x^3$, $\dot{y}=-x+\mu y+2y^3$

8.2.7 $\dot{x}=\mu x+y-x^2$, $\dot{y}=-x+\mu y+2x^2$

8.2.8 (Predator-prey model) Odell (1980) considered the system

$$\dot{x}=x[x(1-x)-y], \quad \dot{y}=y(x-a),$$

where $x\geq 0$ is the dimensionless population of the prey, $y \geq 0$ is the dimensionless population of the predator, and $a \geq 0$ is a control parameter.

a) Sketch the nullclines in the first quadrant $x, y \geq 0$.

b) Show that the fixed points are $(0,0)$, $(1,0)$, and $(a, a-a^2)$, and classify them.

c) Sketch the phase portrait for $a > 1$, and show that the predators go extinct.

d) Show that a Hopf bifurcation occurs at $a_c = \frac{1}{2}$. Is it subcritical or supercritical?

e) Estimate the frequency of limit cycle oscillations for a near the bifurcation.

f) Sketch all the topologically different phase portraits for $0 < a < 1$.

The article by Odell (1980) is worth looking up. It is an outstanding pedagogical introduction to the Hopf bifurcation and phase plane analysis in general.

8.2.9 Consider the predator-prey model

$$\dot{x} = x\left(b - x - \frac{y}{1+x}\right), \qquad \dot{y} = y\left(\frac{x}{1+x} - ay\right),$$

where $x, y \geq 0$ are the populations and $a, b > 0$ are parameters.

a) Sketch the nullclines and discuss the bifurcations that occur as b varies.

b) Show that a positive fixed point $x^* > 0$, $y^* > 0$ exists for all $a, b > 0$. (Don't try to find the fixed point explicitly; use a graphical argument instead.)

c) Show that a Hopf bifurcation occurs at the positive fixed point if

$$a = a_c = \frac{4(b-2)}{b^2(b+2)}$$

and $b > 2$. (Hint: A necessary condition for a Hopf bifurcation to occur is $\tau = 0$, where τ is the trace of the Jacobian matrix at the fixed point. Show that $\tau = 0$ if and only if $2x^* = b - 2$. Then use the fixed point conditions to express a_c in terms of x^*. Finally, substitute $x^* = (b-2)/2$ into the expression for a_c and you're done.)

d) Using a computer, check the validity of the expression in (c) and determine whether the bifurcation is subcritical or supercritical. Plot typical phase portraits above and below the Hopf bifurcation.

8.2.10 (Bacterial respiration) Fairén and Velarde (1979) considered a model for respiration in a bacterial culture. The equations are

$$\dot{x} = B - x - \frac{xy}{1+qx^2}, \qquad \dot{y} = A - \frac{xy}{1+qx^2}$$

where x and y are the levels of nutrient and oxygen, respectively, and $A, B, q > 0$ are parameters. Investigate the dynamics of this model. As a start, find all the fixed points and classify them. Then consider the nullclines and try to construct a trapping region. Can you find conditions on A, B, q under which the system has a stable limit cycle? Use numerical integration, the Poincaré–Bendixson theorem, results about Hopf bifurcations, or whatever else seems useful. (This question is deliberately open-ended and could serve as a class project; see how far you can go.)

8.2.11 (Degenerate bifurcation, not Hopf) Consider the damped Duffing oscillator $\ddot{x} + \mu\dot{x} + x - x^3 = 0$.

a) Show that the origin changes from a stable to an unstable spiral as μ decreases though zero.

b) Plot the phase portraits for $\mu > 0$, $\mu = 0$, and $\mu < 0$, and show that the bifurcation at $\mu = 0$ is a degenerate version of the Hopf bifurcation.

8.2.12 (Analytical criterion to decide if a Hopf bifurcation is subcritical or supercritical) Any system at a Hopf bifurcation can be put into the following form by suitable changes of variables:

$$\dot{x} = -\omega y + f(x,y), \qquad \dot{y} = \omega x + g(x,y),$$

where f and g contain only higher-order nonlinear terms that vanish at the origin. As shown by Guckenheimer and Holmes (1983, pp. 152–156), one can decide whether the bifurcation is subcritical or supercritical by calculating the sign of the following quantity:

$$16a = f_{xxx} + f_{xyy} + g_{xxy} + g_{yyy}$$
$$+ \frac{1}{\omega}\left[f_{xy}(f_{xx} + f_{yy}) - g_{xy}(g_{xx} + g_{yy}) - f_{xx}g_{xx} + f_{yy}g_{yy}\right]$$

where the subscripts denote partial derivatives evaluated at $(0,0)$. The criterion is: If $a < 0$, the bifurcation is supercritical; if $a > 0$, the bifurcation is subcritical.

a) Calculate a for the system $\dot{x} = -y + xy^2$, $\dot{y} = x - x^2$.

b) Use part (a) to decide which type of Hopf bifurcation occurs for $\dot{x} = -y + \mu x + xy^2$, $\dot{y} = x + \mu y - x^2$ at $\mu = 0$. (Compare the results of Exercises 8.2.2–8.2.4.)

(You might be wondering what a measures. Roughly speaking, a is the coefficient of the cubic term in the equation $\dot{r} = ar^3$ governing the radial dynamics at the bifurcation. Here r is a slightly transformed version of the usual polar coordinate. For details, see Guckenheimer and Holmes (1983) or Grimshaw (1990).)

For each of the following systems, a Hopf bifurcation occurs at the origin when $\mu = 0$. Use the analytical criterion of Exercise 8.2.12 to decide if the bifurcation is sub- or supercritical. Confirm your conclusions on the computer.

8.2.13 $\dot{x} = y + \mu x$, $\quad \dot{y} = -x + \mu y - x^2 y$

8.2.14 $\dot{x} = \mu x + y - x^3$, $\quad \dot{y} = -x + \mu y + 2y^3$

8.2.15 $\dot{x} = \mu x + y - x^2$, $\quad \dot{y} = -x + \mu y + 2x^2$

8.2.16 In Example 8.2.1, we argued that the system $\dot{x} = \mu x - y + xy^2$,

$\dot{y} = x + \mu y + y^3$ undergoes a subcritical Hopf bifurcation at $\mu = 0$. Use the analytical criterion to confirm that the bifurcation is subcritical.

8.3 Oscillating Chemical Reactions

8.3.1 (Brusselator) The Brusselator is a simple model of a hypothetical chemical oscillator, named after the home of the scientists who proposed it. (This is a common joke played by the chemical oscillator community; there is also the "Oregonator," "Palo Altonator," etc.) In dimensionless form, its kinetics are

$$\dot{x} = 1 - (b+1)x + ax^2 y$$
$$\dot{y} = bx - ax^2 y$$

where $a, b > 0$ are parameters and $x, y \geq 0$ are dimensionless concentrations.

a) Find all the fixed points, and use the Jacobian to classify them.
b) Sketch the nullclines, and thereby construct a trapping region for the flow.
c) Show that a Hopf bifurcation occurs at some parameter value $b = b_c$, where b_c is to be determined.
d) Does the limit cycle exist for $b > b_c$ or $b < b_c$? Explain, using the Poincaré–Bendixson theorem.
e) Find the approximate period of the limit cycle for $b \approx b_c$.

8.3.2 Schnackenberg (1979) considered the following hypothetical model of a chemical oscillator:

$$X \underset{k_{-1}}{\overset{k_1}{\rightleftarrows}} A, \qquad B \xrightarrow{k_2} Y, \qquad 2X + Y \xrightarrow{k_3} 3X.$$

After using the Law of Mass Action and nondimensionalizing, Schnackenberg reduced the system to

$$\dot{x} = a - x + x^2 y$$
$$\dot{y} = b - x^2 y$$

where $a, b > 0$ are parameters and $x, y > 0$ are dimensionless concentrations.

a) Show that all trajectories eventually enter a certain trapping region, to be determined. Make the trapping region as small as possible. (Hint: Examine the ratio $\dot{y}/\dot{x}$ for large x.)
b) Show that the system has a unique fixed point, and classify it.
c) Show that the system undergoes a Hopf bifurcation when $b - a = (a+b)^3$.
d) Is the Hopf bifurcation subcritical or supercritical? Use a computer to decide.
e) Plot the stability diagram in a, b space. (Hint: It is a bit confusing to plot the curve $b - a = (a+b)^3$, since this requires analyzing a cubic. As in Section 3.7, the *parametric form* of the bifurcation curve comes to the rescue. Show that the bifurcation curve can be expressed as

$$a = \tfrac{1}{2}x^*\left(1-(x^*)^2\right), \qquad b = \tfrac{1}{2}x^*\left(1+(x^*)^2\right)$$

where $x^* > 0$ is the x-coordinate of the fixed point. Then plot the bifurcation curve from these parametric equations. This trick is discussed in Murray (1989).)

8.3.3 (Relaxation limit of a chemical oscillator) Analyze the model for the chlorine dioxide–iodine–malonic acid oscillator, (8.3.4), (8.3.5), in the limit $b \ll 1$. Sketch the limit cycle in the phase plane and estimate its period.

8.4 Global Bifurcations of Cycles

8.4.1 Consider the system $\dot{r} = r(1-r^2)$, $\dot{\theta} = \mu - \sin\theta$ for μ slightly greater than 1. Let $x = r\cos\theta$ and $y = r\sin\theta$. Sketch the waveforms of $x(t)$ and $y(t)$. (These are typical of what one might see experimentally for a system on the verge of an infinite-period bifurcation.)

8.4.2 Discuss the bifurcations of the system $\dot{r} = r(\mu - \sin r)$, $\dot{\theta} = 1$ as μ varies.

8.4.3 (Homoclinic bifurcation) Using numerical integration, find the value of μ at which the system $\dot{x} = \mu x + y - x^2$, $\dot{y} = -x + \mu y + 2x^2$ undergoes a homoclinic bifurcation. Sketch the phase portrait just above and below the bifurcation.

8.4.4 (Second-order phase-locked loop) Using a computer, explore the phase portrait of $\ddot{\theta} + (1 - \mu\cos\theta)\dot{\theta} + \sin\theta = 0$ for $\mu \geq 0$. For some values of μ, you should find that the system has a stable limit cycle. Classify the bifurcations that create and destroy the cycle as μ increases from 0.

Exercises 8.4.5–8.4.11 deal with the ***forced Duffing oscillator*** in the limit where the forcing, detuning, damping, and nonlinearity are all weak:

$$\ddot{x} + x + \varepsilon(bx^3 + k\dot{x} - ax - F\cos t) = 0,$$

where $0 < \varepsilon \ll 1$, $b > 0$ is the nonlinearity, $k > 0$ is the damping, a is the detuning, and $F > 0$ is the forcing strength. This system is a small perturbation of a harmonic oscillator, and can therefore be handled with the methods of Section 7.6. We have postponed the problem until now because saddle-node bifurcations of cycles arise in its analysis.

8.4.5 (Averaged equations) Show that the averaged equations (7.6.53) for the system are

$$r' = -\tfrac{1}{2}(kr + F\sin\phi), \qquad \phi' = -\tfrac{1}{8}(4a - 3br^2 + \tfrac{4F}{r}\cos\phi),$$

where $x = r\cos(t+\phi)$, $\dot{x} = -r\sin(t+\phi)$, and prime denotes differentiation with respect to slow time $T = \varepsilon t$, as usual. (If you skipped Section 7.6, accept these equations on faith.)

8.4.6 (Correspondence between averaged and original systems) Show that fixed points for the averaged system correspond to phase-locked periodic solutions for the original forced oscillator. Show further that saddle-node bifurcations of fixed points for the averaged system correspond to saddle-node bifurcations of cycles for the oscillator.

8.4.7 (No periodic solutions for averaged system) Regard (r,ϕ) as polar coordinates in the phase plane. Show that the averaged system has no closed orbits. (Hint: Use Dulac's criterion with $g(r,\phi)\equiv 1$. Let $\mathbf{x}'=(r', r\phi')$. Compute $\nabla\cdot\mathbf{x}'=\frac{1}{r}\frac{\partial}{\partial r}(rr')+\frac{1}{r}\frac{\partial}{\partial \phi}(r\phi')$ and show that it has one sign.)

8.4.8 (No sources for averaged system) The result of the previous exercise shows that we only need to study the fixed points of the averaged system to determine its long-term behavior. Explain why the divergence calculation above also implies that the fixed points cannot be sources; only sinks and saddles are possible.

8.4.9 (Resonance curves and cusp catastrophe) In this exercise you are asked to determine how the equilibrium amplitude of the driven oscillations depends on the other parameters.

a) Show that the fixed points satisfy $r^2\left[k^2+(\frac{3}{4}br^2-a)^2\right]=F^2$.
b) From now on, assume that k and F are fixed. Graph r vs. a for the linear oscillator ($b=0$). This is the familiar resonance curve.
c) Graph r vs. a for the nonlinear oscillator ($b\neq 0$). Show that the curve is single-valued for small nonlinearity, say $b<b_c$, but triple-valued for large nonlinearity ($b>b_c$), and find an explicit formula for b_c. (Thus we obtain the intriguing conclusion that the driven oscillator can have three limit cycles for some values of a and b!)
d) Show that if r is plotted as a surface above the (a,b) plane, the result is a cusp catastrophe surface (recall Section 3.6).

8.4.10 Now for the hard part: analyze the bifurcations of the averaged system.

a) Plot the nullclines $r'=0$ and $\phi'=0$ in the phase plane, and study how their intersections change as the detuning a is increased from negative values to large positive values.
b) Assuming that $b>b_c$, show that as a increases, the number of *stable* fixed points changes from one to two and then back to one again.

8.4.11 (Numerical exploration) Fix the parameters $k=1$, $b=\frac{4}{3}$, $F=2$.

a) Using numerical integration, plot the phase portrait for the averaged system with a increasing from negative to positive values.
b) Show that for $a=2.8$, there are two stable fixed points.
c) Go back to the original forced Duffing equation. Numerically integrate it and plot $x(t)$ as a increases slowly from $a=-1$ to $a=5$, and then decreases

slowly back to $a = -1$. You should see a dramatic hysteresis effect with the limit cycle oscillation suddenly jumping up in amplitude at one value of a, and then back down at another.

8.4.12 (Scaling near a homoclinic bifurcation) To find how the period of a closed orbit scales as a homoclinic bifurcation is approached, we estimate the time it takes for a trajectory to pass by a saddle point (this time is much longer than all others in the problem). Suppose the system is given locally by $\dot{x} \approx \lambda_u x$, $\dot{y} \approx -\lambda_s y$. Let a trajectory pass through the point $(\mu, 1)$, where $\mu << 1$ is the distance from the stable manifold. How long does it take until the trajectory has escaped from the saddle, say out to $x(t) \approx 1$? (See Gaspard (1990) for a detailed discussion.)

8.5 Hysteresis in the Driven Pendulum and Josephson Junction

8.5.1 Show that $\left[\ln(I - I_c)\right]^{-1}$ has infinite derivatives of all orders at I_c. (Hint: Consider $f(I) = (\ln I)^{-1}$ and try to derive a formula for $f^{(n+1)}(I)$ in terms of $f^{(n)}(I)$, where $f^{(n)}(I)$ denotes the nth derivative of $f(I)$.)

8.5.2 Consider the driven pendulum $\phi'' + \alpha\phi' + \sin\phi = I$. By numerical computation of the phase portrait, verify that if α is fixed and sufficiently small, the system's stable limit cycle is destroyed in a homoclinic bifurcation as I decreases. Show that if α is too large, the bifurcation is an infinite-period bifurcation instead.

8.5.3 (Logistic equation with periodically varying carrying capacity) Consider the logistic equation $\dot{N} = rN(1 - N/K(t))$, where the carrying capacity is positive, smooth, and T-periodic in t.

a) Using a Poincaré map argument like that in the text, show that the system has at least one stable limit cycle of period T, contained in the strip $K_{\text{min}} \le N \le K_{\text{max}}$.
b) Is the cycle necessarily unique?

8.6 Coupled Oscillators and Quasiperiodicity

8.6.1 ("Oscillator death" and bifurcations on a torus) In a paper on systems of neural oscillators, Ermentrout and Kopell (1990) illustrated the notion of "oscillator death" with the following model:

$$\dot{\theta}_1 = \omega_1 + \sin\theta_1 \cos\theta_2, \qquad \dot{\theta}_2 = \omega_2 + \sin\theta_2 \cos\theta_1,$$

where $\omega_1, \omega_2 \ge 0$.

a) Sketch all the qualitatively different phase portraits that arise as ω_1, ω_2 vary.
b) Find the curves in ω_1, ω_2 parameter space along which bifurcations occur, and classify the various bifurcations.
c) Plot the stability diagram in ω_1, ω_2 parameter space.

8.6.2 Reconsider the system (8.6.1):

$$\dot{\theta}_1 = \omega_1 + K_1 \sin(\theta_2 - \theta_1), \qquad \dot{\theta}_2 = \omega_2 + K_2 \sin(\theta_1 - \theta_2).$$

a) Show that the system has no fixed points, given that ω_1, $\omega_2 > 0$ and K_1, $K_2 > 0$.
b) Find a conserved quantity for the system. (Hint: Solve for $\sin(\theta_2 - \theta_1)$ in two ways. The existence of a conserved quantity shows that this system is a non-generic flow on the torus; normally there would not be any conserved quantities.)
c) Suppose that $K_1 = K_2$. Show that the system can be nondimensionalized to

$$d\theta_1/d\tau = 1 + a\sin(\theta_2 - \theta_1), \qquad d\theta_2/d\tau = \omega + a\sin(\theta_1 - \theta_2).$$

d) Find the *winding number* $\lim_{\tau\to\infty} \theta_1(\tau)/\theta_2(\tau)$ analytically. (Hint: Evaluate the long-time averages $\langle d(\theta_1 + \theta_2)/d\tau \rangle$ and $\langle d(\theta_1 - \theta_2)/d\tau \rangle$, where the brackets are defined by $\langle f \rangle \equiv \lim_{T\to\infty} \frac{1}{T}\int_0^T f(\tau)\,d\tau$. For another approach, see Guckenheimer and Holmes (1983, p. 299).)

8.6.3 (Irrational flow yields dense orbits) Consider the flow on the torus given by $\dot{\theta}_1 = \omega_1$, $\dot{\theta}_2 = \omega_2$, where ω_1/ω_2 is irrational. Show each trajectory is ***dense***; i.e., given any point p on the torus, any initial condition q, and any $\varepsilon > 0$, there is some $t < \infty$ such that the trajectory starting at q passes within a distance ε of p.

8.6.4 Consider the system

$$\dot{\theta}_1 = E - \sin\theta_1 + K\sin(\theta_2 - \theta_1), \qquad \dot{\theta}_2 = E + \sin\theta_2 + K\sin(\theta_1 - \theta_2)$$

where $E, K \ge 0$.
a) Find and classify all the fixed points.
b) Show that if E is large enough, the system has periodic solutions on the torus. What type of bifurcation creates the periodic solutions?
c) Find the bifurcation curve in (E, K) space at which these periodic solutions are created.

A generalization of this system to $N >> 1$ phases has been proposed as a model of switching in charge-density waves (Strogatz et al. 1988, 1989).

8.6.5 (Plotting Lissajous figures) Using a computer, plot the curve whose parametric equations are $x(t) = \sin t$, $y(t) = \sin \omega t$, for the following rational and irrational values of the parameter ω:

(a) $\omega = 3$ (b) $\omega = \frac{2}{3}$ (c) $\omega = \frac{5}{3}$
(d) $\omega = \sqrt{2}$ (e) $\omega = \pi$ (f) $\omega = \frac{1}{2}(1 + \sqrt{5})$.

The resulting curves are called *Lissajous figures*. In the old days they were displayed on oscilloscopes by using two ac signals of different frequencies as inputs.

8.6.6 (Explaining Lissajous figures) Lissajous figures are one way to visualize the knots and quasiperiodicity discussed in the text. To see this, consider a pair of uncoupled harmonic oscillators described by the four-dimensional system $\ddot{x}+x=0$, $\ddot{y}+\omega^2 y=0$.

a) Show that if $x=A(t)\sin\theta(t)$, $y=B(t)\sin\phi(t)$, then $\dot{A}=\dot{B}=0$ (so A,B are constants) and $\dot{\theta}=1$, $\dot{\phi}=\omega$.
b) Explain why (a) implies that trajectories are typically confined to two-dimensional tori in a four-dimensional phase space.
c) How are the Lissajous figures related to the trajectories of this system?

8.6.7 (Mechanical example of quasiperiodicity) The equations

$$m\ddot{r}=\frac{h^2}{mr^3}-k\,,\qquad \dot{\theta}=\frac{h}{mr^2}$$

govern the motion of a mass m subject to a central force of constant strength $k>0$. Here r,θ are polar coordinates and $h>0$ is a constant (the angular momentum of the particle).

a) Show that the system has a solution $r=r_0$, $\dot{\theta}=\omega_\theta$, corresponding to uniform circular motion at a radius r_0 and frequency ω_θ. Find formulas for r_0 and ω_θ.
b) Find the frequency ω_r of small radial oscillations about the circular orbit.
c) Show that these small radial oscillations correspond to quasiperiodic motion by calculating the winding number ω_r/ω_θ.
d) Show by a geometric argument that the motion is either periodic or quasiperiodic for *any* amplitude of radial oscillation. (To say it in a more interesting way, the motion is never chaotic.)
e) Can you think of a mechanical realization of this system?

8.6.8 Solve the equations of Exercise 8.6.7 on a computer, and plot the particle's path in the plane with polar coordinates r,θ.

8.7 Poincaré Maps

8.7.1 Use partial fractions to evaluate the integral $\int_{r_0}^{r_1}\frac{dr}{r(1-r^2)}$ that arises in Example 8.7.1, and show that $r_1=\left[1+e^{-4\pi}(r_0^{-2}-1)\right]^{-1/2}$. Then confirm that $P'(r^*)=e^{-4\pi}$, as expected from Example 8.7.3.

8.7.2 Consider the vector field on the cylinder given by $\dot{\theta}=1$, $\dot{y}=ay$. Define an appropriate Poincaré map and find a formula for it. Show that the system has a periodic orbit. Classify its stability for all real values of a.

8.7.3 (Overdamped system forced by a square wave) Consider an overdamped linear oscillator (or an RC-circuit) forced by a square wave. The system can be nondimensionalized to $\dot{x}+x=F(t)$, where $F(t)$ is a square wave of period T. To be more specific, suppose

$$F(t)=\begin{cases}+A, & 0<t<T/2\\ -A, & T/2<t<T\end{cases}$$

for $t\in(0,T)$, and then $F(t)$ is periodically repeated for all other t. The goal is to show that all trajectories of the system approach a unique periodic solution. We could try to solve for $x(t)$ but that gets a little messy. Here's an approach based on the Poincaré map—the idea is to "strobe" the system once per cycle.

a) Let $x(0)=x_0$. Show that $x(T)=x_0e^{-T}-A\left(1-e^{-T/2}\right)^2$.

b) Show that the system has a unique periodic solution, and that it satisfies $x_0=-A\tanh(T/4)$.

c) Interpret the limits of $x(T)$ as $T\to 0$ and $T\to\infty$. Explain why they're plausible.

d) Let $x_1=x(T)$, and define the Poincaré map P by $x_1=P(x_0)$. More generally, $x_{n+1}=P(x_n)$. Plot the graph of P.

e) Using a cobweb picture, show that P has a globally stable fixed point. (Hence the original system eventually settles into a periodic response to the forcing.)

8.7.4 A Poincaré map for the system $\dot{x}+x=A\sin\omega t$ was shown Figure 8.7.3, for a particular choice of parameters. Given that $\omega>0$, can you deduce the sign of A? If not, explain why not.

8.7.5 (Another driven overdamped system) By considering an appropriate Poincaré map, prove that the system $\dot{\theta}+\sin\theta=\sin t$ has at least two periodic solutions. Can you say anything about their stability? (Hint: Regard the system as a vector field on a cylinder: $\dot{t}=1$, $\dot{\theta}=\sin t-\sin\theta$. Sketch the nullclines and thereby infer the shape of certain key trajectories that can be used to bound the periodic solutions. For instance, sketch the trajectory that passes through $(t,\theta)=(\frac{\pi}{2},\frac{\pi}{2})$.)

8.7.6 Give a mechanical interpretation of the system $\dot{\theta}+\sin\theta=\sin t$ considered in the previous exercise.

8.7.7 (Computer work) Plot a computer-generated phase portrait of the system $\dot{t}=1$, $\dot{\theta}=\sin t-\sin\theta$. Check that your results agree with your answer to Exercise 8.7.5.

8.7.8 Consider the system $\dot{x}+x=F(t)$, where $F(t)$ is a smooth, T-periodic function. Is it true that the system necessarily has a stable T-periodic solution $x(t)$? If so, prove it; if not, find an F that provides a counterexample.

8.7.9 Consider the vector field given in polar coordinates by $\dot{r} = r - r^2$, $\dot{\theta} = 1$.

a) Compute the Poincaré map from S to itself, where S is the positive x-axis.

b) Show that the system has a unique periodic orbit and classify its stability.

c) Find the characteristic multiplier for the periodic orbit.

8.7.10 Explain how to find Floquet multipliers numerically, starting from perturbations along the coordinate directions.

8.7.11 (Reversibility and the in-phase periodic state of a Josephson array) Use a reversibility argument to prove that the in-phase periodic state of (8.7.1) is not attracting, even if the nonlinear terms are kept.

8.7.12 (Globally coupled oscillators) Consider the following system of N identical oscillators:

$$\dot{\theta}_i = f(\theta_i) + \tfrac{K}{N} \sum_{j=1}^{N} f(\theta_j), \text{ for } i = 1, \ldots, N,$$

where $K > 0$ and $f(\theta)$ is smooth and 2π-periodic. Assume that $f(\theta) > 0$ for all θ so that the in-phase solution is periodic. By calculating the linearized Poincaré map as in Example 8.7.4, show that all the characteristic multipliers equal $+1$.

Thus the neutral stability found in Example 8.7.4 holds for a broader class of oscillator arrays. In particular, the reversibility of the system is not essential. This example is from Tsang et al. (1991).

PART THREE

CHAOS

9

LORENZ EQUATIONS

9.0 Introduction

We begin our study of chaos with the ***Lorenz equations***

$$\begin{aligned}\dot{x} &= \sigma(y-x)\\ \dot{y} &= rx - y - xz\\ \dot{z} &= xy - bz\,.\end{aligned}$$

Here $\sigma, r, b > 0$ are parameters. Ed Lorenz (1963) derived this three-dimensional system from a drastically simplified model of convection rolls in the atmosphere. The same equations also arise in models of lasers and dynamos, and as we'll see in Section 9.1, they *exactly* describe the motion of a certain waterwheel (you might like to build one yourself).

Lorenz discovered that this simple-looking deterministic system could have extremely erratic dynamics: over a wide range of parameters, the solutions oscillate irregularly, never exactly repeating but always remaining in a bounded region of phase space. When he plotted the trajectories in three dimensions, he discovered that they settled onto a complicated set, now called a strange attractor. Unlike stable fixed points and limit cycles, the strange attractor is not a point or a curve or even a surface—it's a fractal, with a fractional dimension between 2 and 3.

In this chapter we'll follow the beautiful chain of reasoning that led Lorenz to his discoveries. Our goal is to get a feel for his strange attractor and the chaotic motion that occurs on it.

Lorenz's paper (Lorenz 1963) is deep, prescient, and surprisingly readable—look it up! It is also reprinted in Cvitanovic (1989a) and Hao (1990). For a captivating history of Lorenz's work and that of other chaotic heroes, see Gleick (1987).

9.1 A Chaotic Waterwheel

A neat mechanical model of the Lorenz equations was invented by Willem Malkus and Lou Howard at MIT in the 1970s. The simplest version is a toy waterwheel with leaky paper cups suspended from its rim (Figure 9.1.1).

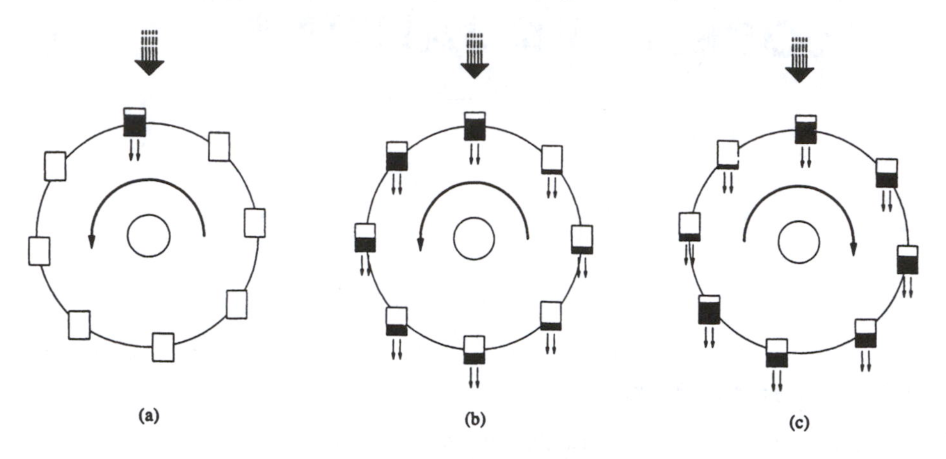

Figure 9.1.1

Water is poured in steadily from the top. If the flow rate is too slow, the top cups never fill up enough to overcome friction, so the wheel remains motionless. For faster inflow, the top cup gets heavy enough to start the wheel turning (Figure 9.1.1a). Eventually the wheel settles into a steady rotation in one direction or the other (Figure 9.1.1b). By symmetry, rotation in either direction is equally possible; the outcome depends on the initial conditions.

By increasing the flow rate still further, we can destabilize the steady rotation. Then the motion becomes chaotic: the wheel rotates one way for a few turns, then some of the cups get too full and the wheel doesn't have enough inertia to carry them over the top, so the wheel slows down and may even reverse its direction (Figure 9.1.1c). Then it spins the other way for a while. The wheel keeps changing direction erratically. Spectators have been known to place bets (small ones, of course) on which way it will be turning after a minute.

Figure 9.1.2 shows Malkus's more sophisticated set-up that we use nowadays at MIT.

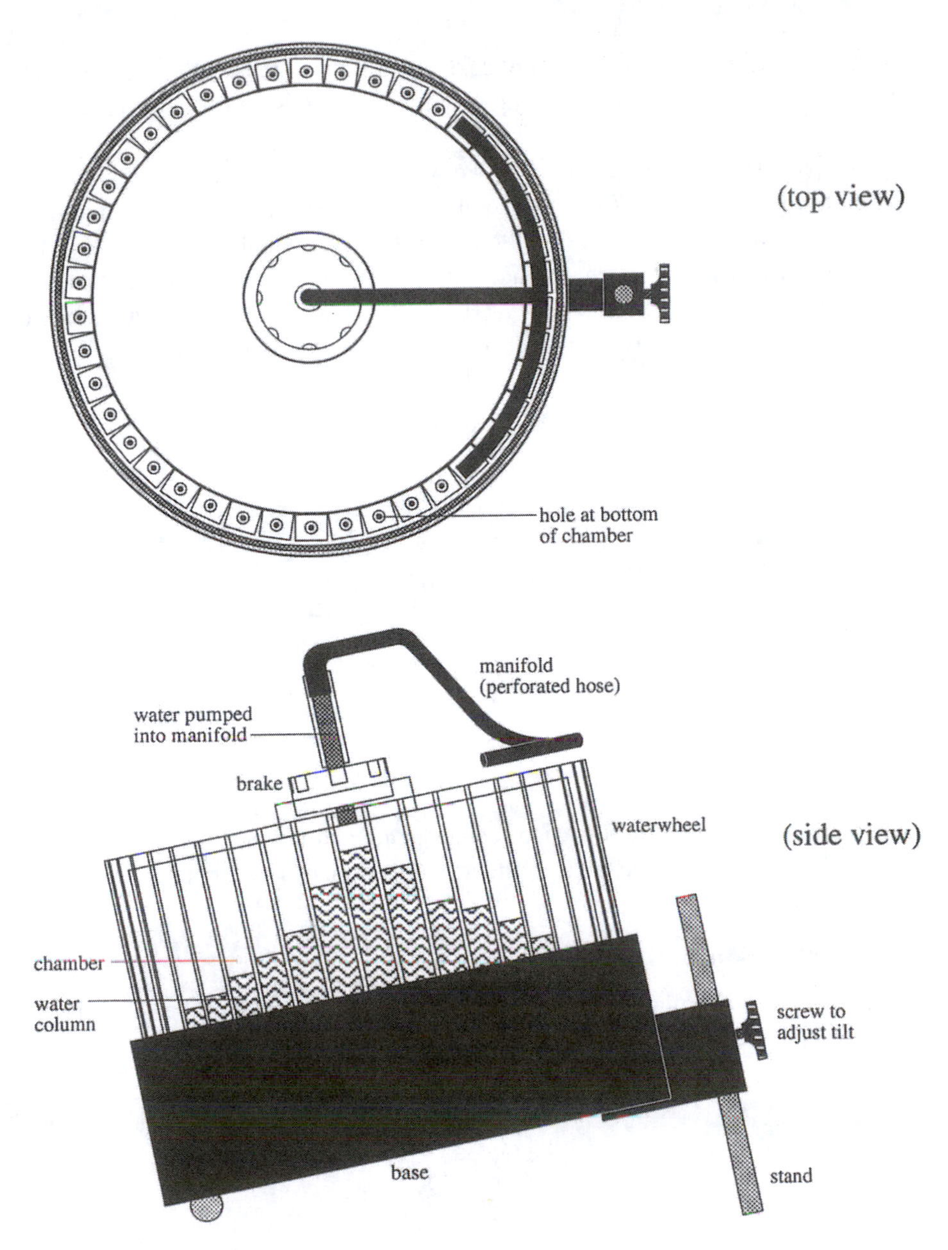

Figure 9.1.2

The wheel sits on a table top. It rotates in a plane that is tilted slightly from the horizontal (unlike an ordinary waterwheel, which rotates in a vertical plane). Water is pumped up into an overhanging manifold and then sprayed out through dozens of small nozzles. The nozzles direct the water into separate chambers around the rim of the wheel. The chambers are transparent, and the water has food coloring in it, so the distribution of water around the rim is easy to see. The water leaks out through a small hole at the bottom of each chamber, and then collects underneath the wheel, where it is pumped back up through the nozzles. This system provides a steady input of water.

The parameters can be changed in two ways. A brake on the wheel can be adjusted to add more or less friction. The tilt of the wheel can be varied by turning a screw that props the wheel up; this alters the effective strength of gravity.

A sensor measures the wheel's angular velocity $\omega(t)$, and sends the data to a strip chart recorder which then plots $\omega(t)$ in real time. Figure 9.1.3 shows a record of $\omega(t)$ when the wheel is rotating chaotically. Notice once again the irregular sequence of reversals.

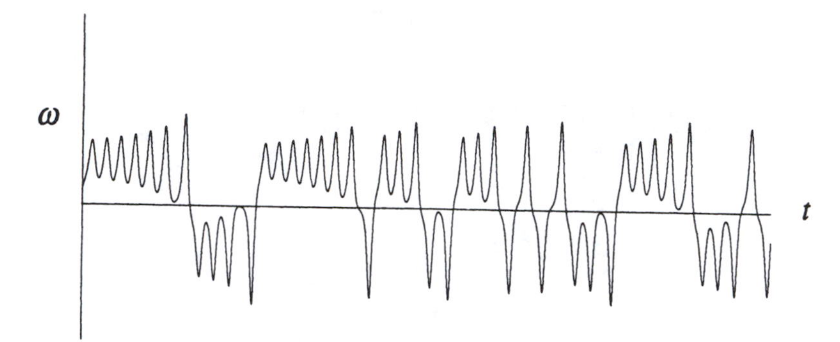

Figure 9.1.3

We want to explain where this chaos comes from, and to understand the bifurcations that cause the wheel to go from static equilibrium to steady rotation to irregular reversals.

Notation

Here are the coordinates, variables and parameters that describe the wheel's motion (Figure 9.1.4):

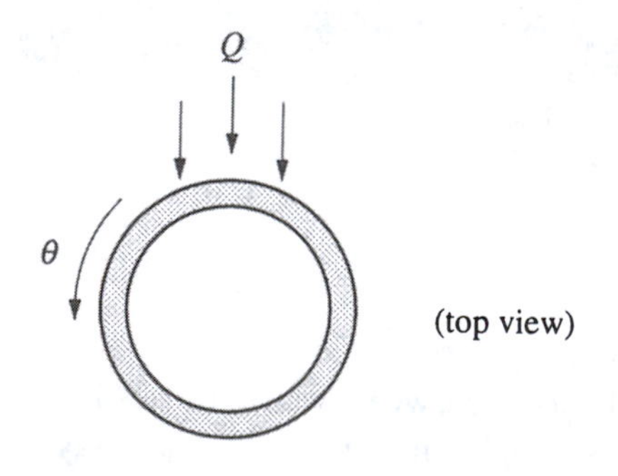

Figure 9.1.4

θ = angle in the lab frame (*not* the frame attached to the wheel)
$\theta = 0 \leftrightarrow$ 12:00 in the lab frame

$\omega(t)$ = angular velocity of the wheel (increases counterclockwise, as does θ)
$m(\theta,t)$ = mass distribution of water around the rim of the wheel, defined

such that the mass between θ_1 and θ_2 is $M(t) = \int_{\theta_1}^{\theta_2} m(\theta,t)\, d\theta$

$Q(\theta)$ = inflow (rate at which water is pumped in by the nozzles above position θ)
r = radius of the wheel
K = leakage rate
ν = rotational damping rate
I = moment of inertia of the wheel

The unknowns are $m(\theta,t)$ and $\omega(t)$. Our first task is to derive equations governing their evolution.

Conservation of Mass

To find the equation for conservation of mass, we use a standard argument. You may have encountered it if you've studied fluids, electrostatics, or chemical engineering. Consider any sector $[\theta_1, \theta_2]$ fixed in space (Figure 9.1.5).

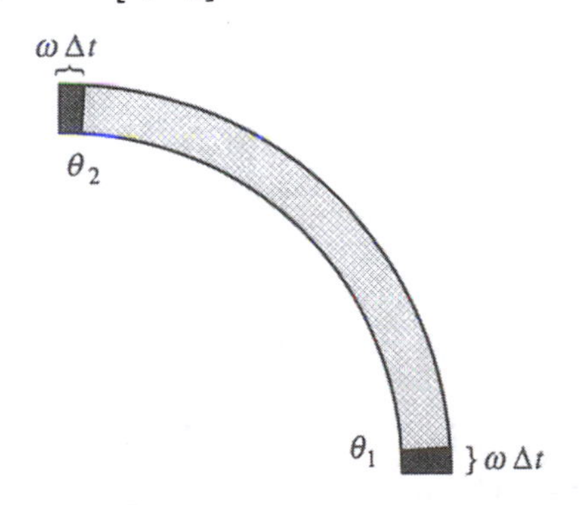

Figure 9.1.5

The mass in that sector is $M(t) = \int_{\theta_1}^{\theta_2} m(\theta,t)\, d\theta$. After an infinitesimal time Δt, what is the change in mass ΔM? There are four contributions:

1. The mass pumped in by the nozzles is $\left[\int_{\theta_1}^{\theta_2} Q d\theta\right]\Delta t$.

2. The mass that leaks out is $\left[-\int_{\theta_1}^{\theta_2} Km\, d\theta\right]\Delta t$. Notice the factor of m in the integral; it implies that leakage occurs at a rate proportional to the mass of water in the chamber—more water implies a larger pressure head and therefore faster leakage. Although this is plausible physically, the fluid mechanics of leakage is complicated, and other rules are conceivable as

well. The real justification for the rule above is that it agrees with direct measurements on the waterwheel itself, to a good approximation. (For experts on fluids: to achieve this linear relation between outflow and pressure head, Malkus attached thin tubes to the holes at the bottom of each chamber. Then the outflow is essentially Poiseuille flow in a pipe.)

3. As the wheel rotates, it carries a new block of water into our observation sector. That block has mass $m(\theta_1)\omega\Delta t$, because it has angular width $\omega\Delta t$ (Figure 9.1.5), and $m(\theta_1)$ is its mass per unit angle.
4. Similarly, the mass carried out of the sector is $-m(\theta_2)\omega\Delta t$.

Hence,

$$\Delta M = \Delta t\left[\int_{\theta_1}^{\theta_2} Q\,d\theta - \int_{\theta_1}^{\theta_2} Km\,d\theta\right] + m(\theta_1)\omega\Delta t - m(\theta_2)\omega\Delta t. \qquad (1)$$

To convert (1) to a differential equation, we put the transport terms inside the integral, using $m(\theta_1)-m(\theta_2) = -\int_{\theta_1}^{\theta_2} \frac{\partial m}{\partial \theta}\,d\theta$. Then we divide by Δt and let $\Delta t \to 0$. The result is

$$\frac{dM}{dt} = \int_{\theta_1}^{\theta_2} \left(Q - Km - \omega\tfrac{\partial m}{\partial \theta}\right) d\theta.$$

But by definition of M,

$$\frac{dM}{dt} = \int_{\theta_1}^{\theta_2} \frac{\partial m}{\partial t}\,d\theta.$$

Hence

$$\int_{\theta_1}^{\theta_2} \frac{\partial m}{\partial t}\,d\theta = \int_{\theta_1}^{\theta_2} \left(Q - Km - \omega\tfrac{\partial m}{\partial \theta}\right) d\theta.$$

Since this holds for *all* θ_1 and θ_2, we must have

$$\frac{\partial m}{\partial t} = Q - Km - \omega\frac{\partial m}{\partial \theta}. \qquad (2)$$

Equation (2) is often called the ***continuity equation***. Notice that it is a *partial* differential equation, unlike all the others considered so far in this book. We'll worry about how to analyze it later; we still need an equation that tells us how $\omega(t)$ evolves.

Torque Balance

The rotation of the wheel is governed by Newton's law $F = ma$, expressed as a balance between the applied torques and the rate of change of angular momentum. Let I denote the moment of inertia of the wheel. Note that in general I depends on

t, because the distribution of water does. But this complication disappears if we wait long enough: as $t \to \infty$, one can show that $I(t) \to$ constant (Exercise 9.1.1). Hence, after the transients decay, the equation of motion is

$$I\dot{\omega} = \text{damping torque} + \text{gravitational torque}.$$

There are two sources of damping: viscous damping due to the heavy oil in the brake, and a more subtle "inertial" damping caused by a spin-up effect—the water enters the wheel at zero angular velocity but is spun up to angular velocity ω before it leaks out. Both of these effects produce torques proportional to ω, so we have

$$\text{damping torque} = -\nu\omega,$$

where $\nu > 0$. The negative sign means that the damping opposes the motion.

The gravitational torque is like that of an inverted pendulum, since water is pumped in at the top of wheel (Figure 9.1.6).

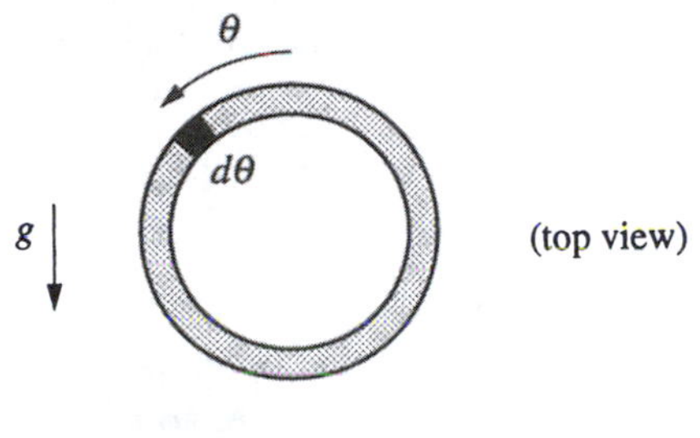

Figure 9.1.6

In an infinitesimal sector $d\theta$, the mass $dM = m d\theta$. This mass element produces a torque

$$d\tau = (dM) gr \sin\theta = mgr \sin\theta \, d\theta.$$

To check that the sign is correct, observe that when $\sin\theta > 0$ the torque tends to *increase* ω, just as in an inverted pendulum. Here g is the effective gravitational constant, given by $g = g_0 \sin\alpha$ where g_0 is the usual gravitational constant and α is the tilt of the wheel from horizontal (Figure 9.1.7).

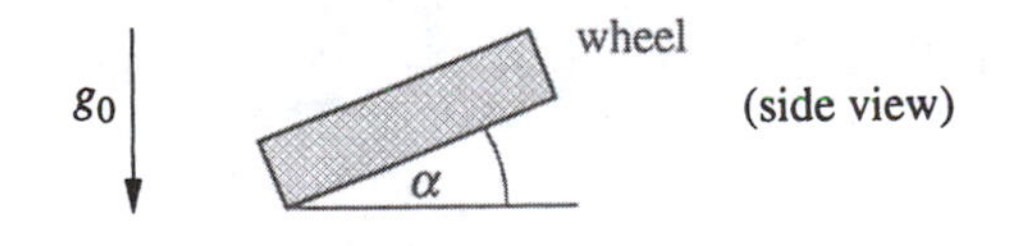

Figure 9.1.7

Integration over all mass elements yields

$$\text{gravitational torque} = gr \int_0^{2\pi} m(\theta, t) \sin\theta \, d\theta.$$

Putting it all together, we obtain the torque balance equation

$$I\dot{\omega} = -\nu\omega + gr\int_0^{2\pi} m(\theta,t)\sin\theta\, d\theta. \tag{3}$$

This is called an *integro-differential equation* because it involves both derivatives and integrals.

Amplitude Equations

Equations (2) and (3) completely specify the evolution of the system. Given the current values of $m(\theta,t)$ and $\omega(t)$, (2) tells us how to update m and (3) tells us how to update ω. So no further equations are needed.

If (2) and (3) truly describe the waterwheel's behavior, there must be some pretty complicated motions hidden in there. How can we extract them? The equations appear much more intimidating than anything we've studied so far.

A miracle occurs if we use Fourier analysis to rewrite the system. Watch!

Since $m(\theta,t)$ is periodic in θ, we can write it as a Fourier series

$$m(\theta,t) = \sum_{n=0}^{\infty} \left[a_n(t)\sin n\theta + b_n(t)\cos n\theta \right]. \tag{4}$$

By substituting this expression into (2) and (3), we'll obtain a set of ***amplitude equations***, ordinary differential equations for the amplitudes a_n, b_n of the different *harmonics* or *modes*. But first we must also write the inflow as a Fourier series:

$$Q(\theta) = \sum_{n=0}^{\infty} q_n \cos n\theta. \tag{5}$$

There are no $\sin n\theta$ terms in the series because water is added *symmetrically* at the top of the wheel; the same inflow occurs at θ and $-\theta$. (In this respect, the waterwheel is unlike an ordinary, real-world waterwheel where asymmetry is used to drive the wheel in the same direction at all times.)

Substituting the series for m and Q into (2), we get

$$\begin{aligned}\frac{\partial}{\partial t}\left[\sum_{n=0}^{\infty} a_n(t)\sin n\theta + b_n(t)\cos n\theta\right] = &-\omega\frac{\partial}{\partial\theta}\left[\sum_{n=0}^{\infty} a_n(t)\sin n\theta + b_n(t)\cos n\theta\right] \\ &+\sum_{n=0}^{\infty} q_n \cos n\theta \\ &-K\left[\sum_{n=0}^{\infty} a_n(t)\sin n\theta + b_n(t)\cos n\theta\right].\end{aligned}$$

Now carry out the differentiations on both sides, and collect terms. By orthogonality of the functions $\sin n\theta$, $\cos n\theta$, we can equate the coefficients of each harmonic separately. For instance, the coefficient of $\sin n\theta$ on the left-hand side is $\dot{a}_n$, and on the right it is $n\omega b_n - Ka_n$. Hence

$$\dot{a}_n = n\omega b_n - Ka_n. \tag{6}$$

Similarly, matching coefficients of $\cos n\theta$ yields

$$\dot{b}_n = -n\omega a_n - Kb_n + q_n. \tag{7}$$

Both (6) and (7) hold for all $n = 0, 1, \ldots$.

Next we rewrite (3) in terms of Fourier series. *Get ready for the miracle.* When we substitute (4) into (3), only one term survives in the integral, by orthogonality:

$$\begin{aligned} I\dot{\omega} &= -\nu\omega + gr\int_0^{2\pi}\left[\sum_{n=0}^{\infty} a_n(t)\sin n\theta + b_n(t)\cos n\theta\right]\sin\theta\, d\theta \\ &= -\nu\omega + gr\int_0^{2\pi} a_1 \sin^2\theta\, d\theta \\ &= -\nu\omega + \pi g r a_1. \end{aligned} \tag{8}$$

Hence, only a_1 enters the differential equation for $\dot{\omega}$. But then (6) and (7) imply that a_1, b_1, *and* ω *form a closed system—these three variables are decoupled from all the other* a_n, b_n, $n \neq 1$! The resulting equations are

$$\begin{aligned} \dot{a}_1 &= \omega b_1 - Ka_1 \\ \dot{b}_1 &= -\omega a_1 - Kb_1 + q_1 \\ \dot{\omega} &= (-\nu\omega + \pi g r a_1)/I. \end{aligned} \tag{9}$$

(If you're curious about the higher modes a_n, b_n, $n \neq 1$, see Exercise 9.1.2.)

We've simplified our problem tremendously: the original pair of integro-partial differential equations (2), (3) has boiled down to the three-dimensional system (9). It turns out that (9) is equivalent to the Lorenz equations! (See Exercise 9.1.3.) Before we turn to that more famous system, let's try to understand a little about (9). No one has ever *fully* understood it—its behavior is fantastically complex—but we can say something.

Fixed Points

We begin by finding the fixed points of (9). For notational convenience, the usual asterisks will be omitted in the intermediate steps.

Setting all the derivatives equal to zero yields

$$a_1 = \omega b_1 / K \tag{10}$$
$$\omega a_1 = q_1 - K b_1 \tag{11}$$
$$a_1 = \nu\omega / \pi g r\,. \tag{12}$$

Now solve for b_1 by eliminating a_1 from (10) and (11):

$$b_1 = \frac{K q_1}{\omega^2 + K^2}\,. \tag{13}$$

Equating (10) and (12) yields $\omega b_1 / K = \nu\omega / \pi g r$. Hence $\omega = 0$ or

$$b_1 = K\nu / \pi g r\,. \tag{14}$$

Thus, there are two kinds of fixed point to consider:

1. If $\omega = 0$, then $a_1 = 0$ and $b_1 = q_1 / K$. This fixed point

$$(a_1{*}, b_1{*}, \omega{*}) = (0,\ q_1/K,\ 0) \tag{15}$$

corresponds to a state of *no rotation*; the wheel is at rest, with inflow balanced by leakage. We're not saying that this state is stable, just that it exists; stability calculations will come later.

2. If $\omega \neq 0$, then (13) and (14) imply $b_1 = K q_1 / (\omega^2 + K^2) = K\nu / \pi g r$. Since $K \neq 0$, we get $q_1 / (\omega^2 + K^2) = \nu / \pi g r$. Hence

$$(\omega{*})^2 = \frac{\pi g r q_1}{\nu} - K^2\,. \tag{16}$$

If the right-hand side of (16) is positive, there are two solutions, $\pm\omega{*}$, corresponding to *steady rotation* in either direction. These solutions exist if and only if

$$\frac{\pi g r q_1}{K^2 \nu} > 1\,. \tag{17}$$

The dimensionless group in (17) is called the ***Rayleigh number***. It measures how hard we're driving the system, relative to the dissipation. More precisely, the ratio in (17) expresses a competition between g and q_1 (gravity and inflow, which tend to spin the wheel), and K and ν (leakage and damping, which tend to stop the wheel). So it makes sense that steady rotation is possible only if the Rayleigh number is large enough.

The Rayleigh number appears in other parts of fluid mechanics, notably convection, in which a layer of fluid is heated from below. There it is proportional to the difference in temperature from bottom to top. For small temperature gradients, heat is conducted vertically but the fluid remains motionless. When the Rayleigh number increases past a critical value, an instability occurs—the hot fluid is less dense and begins to rise, while the cold fluid on top begins to sink. This sets up a pattern of convection rolls, completely analogous to the steady rotation of our waterwheel. With further increases of the Rayleigh number, the rolls become wavy and eventually chaotic.

The analogy to the waterwheel breaks down at still higher Rayleigh numbers, when turbulence develops and the convective motion becomes complex in space as well as time (Drazin and Reid 1981, Bergé et al. 1984, Manneville 1990). In contrast, the waterwheel settles into a pendulum-like pattern of reversals, turning once to the left, then back to the right, and so on indefinitely (see Example 9.5.2).

9.2 Simple Properties of the Lorenz Equations

In this section we'll follow in Lorenz's footsteps. He took the analysis as far as possible using standard techniques, but at a certain stage he found himself confronted with what seemed like a paradox. One by one he had eliminated all the known possibilities for the long-term behavior of his system: he showed that in a certain range of parameters, there could be no stable fixed points and no stable limit cycles, yet he also proved that all trajectories remain confined to a bounded region and are eventually attracted to a set of zero volume. What could that set be? And how do the trajectories move on it? As we'll see in the next section, that set is the strange attractor, and the motion on it is chaotic.

But first we want to see how Lorenz ruled out the more traditional possibilities. As Sherlock Holmes said in *The Sign of Four*, "When you have eliminated the impossible, whatever remains, however improbable, must be the truth."

The Lorenz equations are

$$\begin{aligned} \dot{x} &= \sigma(y-x) \\ \dot{y} &= rx - y - xz \\ \dot{z} &= xy - bz . \end{aligned} \tag{1}$$

Here $\sigma, r, b > 0$ are parameters. σ is the ***Prandtl number***, r is the Rayleigh number, and b has no name. (In the convection problem it is related to the aspect ratio of the rolls.)

Nonlinearity

The system (1) has only two nonlinearities, the quadratic terms xy and xz. This should remind you of the waterwheel equations (9.1.9), which had two nonlinearities, ωa_1 and ωb_1. See Exercise 9.1.3 for the change of variables that transforms the waterwheel equations into the Lorenz equations.

Symmetry

There is an important ***symmetry*** in the Lorenz equations. If we replace $(x,y) \to (-x,-y)$ in (1), the equations stay the same. Hence, if $\left(x(t),\ y(t),\ z(t)\right)$ is a solution, so is $\left(-x(t), -y(t),\ z(t)\right)$. In other words, all solutions are either symmetric themselves, or have a symmetric partner.

Volume Contraction

The Lorenz system is ***dissipative***: volumes in phase space contract under the flow. To see this, we must first ask: how do volumes evolve?

Let's answer the question in general, for any three-dimensional system $\dot{\mathbf{x}} = \mathbf{f}(\mathbf{x})$. Pick an arbitrary closed surface $S(t)$ of volume $V(t)$ in phase space. Think of the points on S as initial conditions for trajectories, and let them evolve for an infinitesimal time dt. Then S evolves into a new surface $S(t+dt)$; what is its volume $V(t+dt)$?

Figure 9.2.1 shows a side view of the volume.

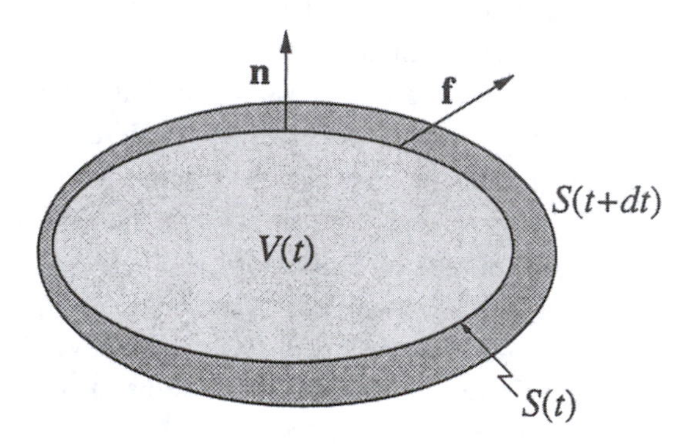

Figure 9.2.1

Let $\mathbf{n}$ denote the outward normal on S. Since $\mathbf{f}$ is the instantaneous velocity of the points, $\mathbf{f} \cdot \mathbf{n}$ is the outward normal component of velocity. Therefore in time dt a patch of area dA sweeps out a volume $(\mathbf{f} \cdot \mathbf{n}\, dt)\, dA$, as shown in Figure 9.2.2.

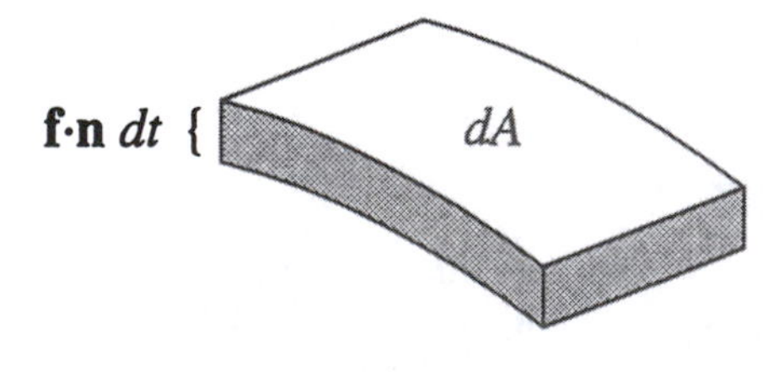

Figure 9.2.2

Hence

$$V(t+dt) = V(t) + (\text{volume swept out by tiny patches of surface, integrated over all patches}),$$

so we obtain

$$V(t+dt) = V(t) + \int_S (\mathbf{f}\cdot\mathbf{n}\,dt)\,dA\,.$$

Hence

$$\dot{V} = \frac{V(t+dt)-V(t)}{dt} = \int_S \mathbf{f}\cdot\mathbf{n}\ dA\,.$$

Finally, we rewrite the integral above by the divergence theorem, and get

$$\dot{V} = \int_V \nabla\cdot\mathbf{f}\ dV\,. \tag{2}$$

For the Lorenz system,

$$\begin{aligned}\nabla\cdot\mathbf{f} &= \frac{\partial}{\partial x}[\sigma(y-x)] + \frac{\partial}{\partial y}[rx-y-xz] + \frac{\partial}{\partial z}[xy-bz]\\ &= -\sigma-1-b<0.\end{aligned}$$

Since the divergence is constant, (2) reduces to $\dot{V} = -(\sigma+1+b)V$, which has solution $V(t) = V(0)e^{-(\sigma+1+b)t}$. Thus *volumes in phase space shrink exponentially fast.*

Hence, if we start with a enormous solid blob of initial conditions, it eventually shrinks to a limiting set of zero volume, like a balloon with the air being sucked out of it. All trajectories starting in the blob end up somewhere in this limiting set; later we'll see it consists of fixed points, limit cycles, or for some parameter values, a strange attractor.

Volume contraction imposes strong constraints on the possible solutions of the Lorenz equations, as illustrated by the next two examples.

EXAMPLE 9.2.1:

Show that there are no quasiperiodic solutions of the Lorenz equations.

Solution: We give a proof by contradiction. If there were a quasiperiodic solution, it would have to lie on the surface of a torus, as discussed in Section 8.6, and this torus would be *invariant* under the flow. Hence the volume inside the torus would be constant in time. But this contradicts the fact that all volumes shrink exponentially fast. ■

EXAMPLE 9.2.2:

Show that it is impossible for the Lorenz system to have either repelling fixed points or repelling closed orbits. (By *repelling,* we mean that *all* trajectories starting near the fixed point or closed orbit are driven away from it.)

Solution: Repellers are incompatible with volume contraction because they are *sources* of volume, in the following sense. Suppose we encase a repeller with a closed surface of initial conditions nearby in phase space. (Specifically, pick a small sphere around a fixed point, or a thin tube around a closed orbit.) A short time later, the surface will have expanded as the corresponding trajectories are driven away. Thus the volume inside the surface would increase. This contradicts the fact that all volumes contract. ■

By process of elimination, we conclude that all fixed points must be sinks or saddles, and closed orbits (if they exist) must be stable or saddle-like. For the case of fixed points, we now verify these general conclusions explicitly.

Fixed Points

Like the waterwheel, the Lorenz system (1) has two types of fixed points. The origin $(x^*, y^*, z^*) = (0,0,0)$ is a fixed point for *all* values of the parameters. It is like the motionless state of the waterwheel. For $r > 1$, there is also a symmetric pair of fixed points $x^* = y^* = \pm\sqrt{b(r-1)}$, $z^* = r-1$. Lorenz called them C^+ and C^-. They represent left- or right-turning convection rolls (analogous to the steady rotations of the waterwheel). As $r \to 1^+$, C^+ and C^- coalesce with the origin in a *pitchfork* bifurcation.

Linear Stability of the Origin

The linearization at the origin is $\dot{x} = \sigma(y-x)$, $\dot{y} = rx - y$, $\dot{z} = -bz$, obtained by omitting the xy and xz nonlinearities in (1). The equation for z is decoupled and shows that $z(t) \to 0$ exponentially fast. The other two directions are governed by the system

$$\begin{pmatrix} \dot{x} \\ \dot{y} \end{pmatrix} = \begin{pmatrix} -\sigma & \sigma \\ r & -1 \end{pmatrix} \begin{pmatrix} x \\ y \end{pmatrix},$$

with trace $\tau = -\sigma - 1 < 0$ and determinant $\Delta = \sigma(1-r)$. If $r > 1$, the origin is a saddle point because $\Delta < 0$. Note that this is *a new type of saddle* for us, since the full system is three-dimensional. Including the decaying z-direction, the saddle has one outgoing and two incoming directions. If $r < 1$, all directions are incoming and the origin is a sink. Specifically, since $\tau^2 - 4\Delta = (\sigma+1)^2 - 4\sigma(1-r) = (\sigma-1)^2 + 4\sigma r > 0$, the origin is a stable node for $r < 1$.

Global Stability of the Origin

Actually, for $r < 1$, we can show that *every* trajectory approaches the origin as $t \to \infty$; the origin is ***globally stable***. Hence there can be no limit cycles or chaos for $r < 1$.

The proof involves the construction of a ***Liapunov function***, a smooth, positive definite function that decreases along trajectories. As discussed in Section 7.2, a Liapunov function is a generalization of an energy function for a classical mechanical system—in the presence of friction or other dissipation, the energy decreases monotonically. There is no systematic way to concoct Liapunov functions, but often it is wise to try expressions involving sums of squares.

Here, consider $V(x,y,z) = \frac{1}{\sigma}x^2 + y^2 + z^2$. The surfaces of constant V are concentric ellipsoids about the origin (Figure 9.2.3).

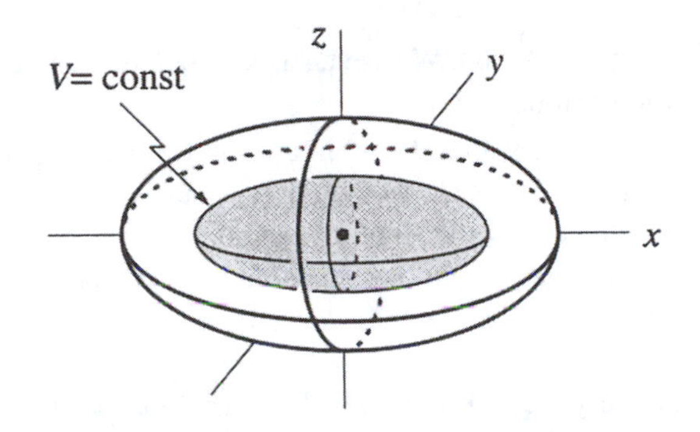

Figure 9.2.3

The idea is to show that if $r < 1$ and $(x,y,z) \neq (0,0,0)$, then $\dot{V} < 0$ along trajectories. This would imply that the trajectory keeps moving to lower V, and hence penetrates smaller and smaller ellipsoids as $t \to \infty$. But V is bounded below by 0, so $V(\mathbf{x}(t)) \to 0$ and hence $\mathbf{x}(t) \to 0$, as desired.

Now calculate:

$$\begin{aligned}\tfrac{1}{2}\dot{V} &= \tfrac{1}{\sigma}x\dot{x} + y\dot{y} + z\dot{z} \\ &= (yx - x^2) + (ryx - y^2 - xzy) + (zxy - bz^2) \\ &= (r+1)xy - x^2 - y^2 - bz^2.\end{aligned}$$

Completing the square in the first two terms gives

$$\tfrac{1}{2}\dot{V} = -\left[x - \tfrac{r+1}{2}y\right]^2 - \left[1 - \left(\tfrac{r+1}{2}\right)^2\right]y^2 - bz^2.$$

We claim that the right-hand side is strictly negative if $r < 1$ and $(x,y,z) \neq (0,0,0)$. It is certainly not positive, since it is a negative sum of squares. But could $\dot{V} = 0$? That would require each of the terms on the right to vanish separately. Hence $y = 0$, $z = 0$, from the second two terms on the right-hand side. (Because of the assumption $r<1$, the coefficient of y^2 is nonzero.) Thus the first term reduces to $-x^2$, which vanishes only if $x=0$.

The upshot is that $\dot{V}=0$ implies $(x,y,z)=(0,0,0)$. Otherwise $\dot{V}<0$. Hence the claim is established, and therefore the origin is globally stable for $r<1$.

Stability of C^+ and C^-

Now suppose $r>1$, so that C^+ and C^- exist. The calculation of their stability is left as Exercise 9.2.1. It turns out that they are linearly stable for

$$1<r<r_H=\frac{\sigma(\sigma+b+3)}{\sigma-b-1}$$

(assuming also that $\sigma-b-1>0$). We use a subscript H because C^+ and C^- lose stability in a Hopf bifurcation at $r=r_H$.

What happens immediately after the bifurcation, for r slightly greater than r_H? You might suppose that C^+ and C^- would each be surrounded by a small stable limit cycle. That would occur if the Hopf bifurcation were supercritical. But actually it's *subcritical*—the limit cycles are *unstable* and exist only for $r<r_H$. This requires a difficult calculation; see Marsden and McCracken (1976) or Drazin (1992, Q8.2 on p. 277).

Here's the intuitive picture. For $r<r_H$ the phase portrait near C^+ is shown schematically in Figure 9.2.4.

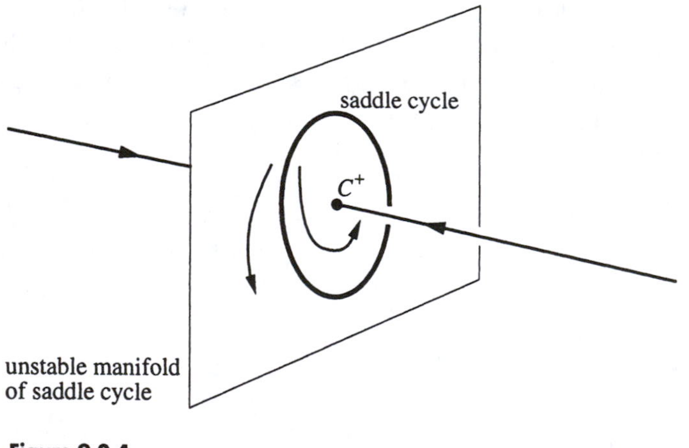

Figure 9.2.4

The fixed point is stable. It is encircled by a ***saddle cycle***, a new type of unstable limit cycle that is possible only in phase spaces of three or more dimensions. The cycle has a two-dimensional unstable manifold (the sheet in Figure 9.2.4), and a two-dimensional stable manifold (not shown). As $r \to r_H$ from below, the cycle shrinks down around the fixed point. At the Hopf bifurcation, the fixed point absorbs the saddle cycle and changes into a saddle point. For $r > r_H$ there are no attractors in the neighborhood.

So for $r > r_H$ trajectories must fly away to a distant attractor. But what can it be? A partial bifurcation diagram for the system, based on the results so far, shows no hint of any stable objects for $r > r_H$ (Figure 9.2.5).

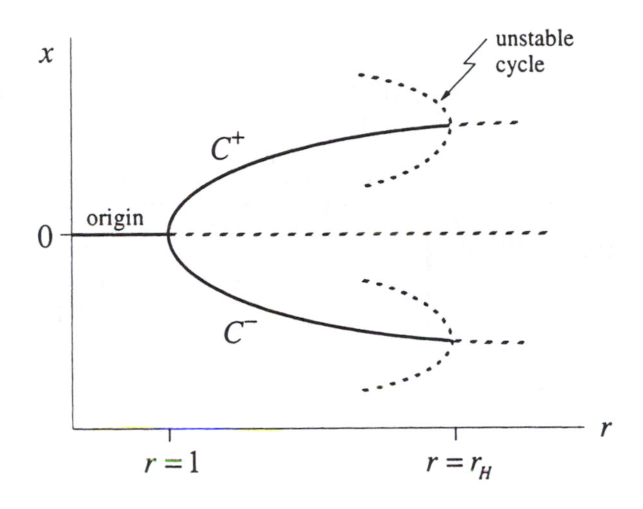

Figure 9.2.5

Could it be that all trajectories are repelled out to infinity? No; we can prove that all trajectories eventually enter and remain in a certain large ellipsoid (Exercise 9.2.2). Could there be some stable limit cycles that we're unaware of? Possibly, but Lorenz gave a persuasive argument that for r slightly greater than r_H, any limit cycles would have to be *unstable* (see Section 9.4).

So the trajectories must have a bizarre kind of long-term behavior. Like balls in a pinball machine, they are repelled from one unstable object after another. At the same time, they are confined to a bounded set of zero volume, yet they manage to move on this set forever without intersecting themselves or others.

In the next section we'll see how the trajectories get out of this conundrum.

9.3 Chaos on a Strange Attractor

Lorenz used numerical integration to see what the trajectories would do in the long run. He studied the particular case $\sigma = 10$, $b = \frac{8}{3}$, $r = 28$. This value of r is just past the Hopf bifurcation value $r_H = \sigma(\sigma + b + 3)/(\sigma - b - 1) \approx 24.74$, so he knew that something strange had to occur. Of course, strange things could occur for another reason—the electromechanical computers of those days were unreliable and difficult to use, so Lorenz had to interpret his numerical results with caution.

He began integrating from the initial condition (0, 1, 0), close to the saddle point at the origin. Figure 9.3.1 plots $y(t)$ for the resulting solution.

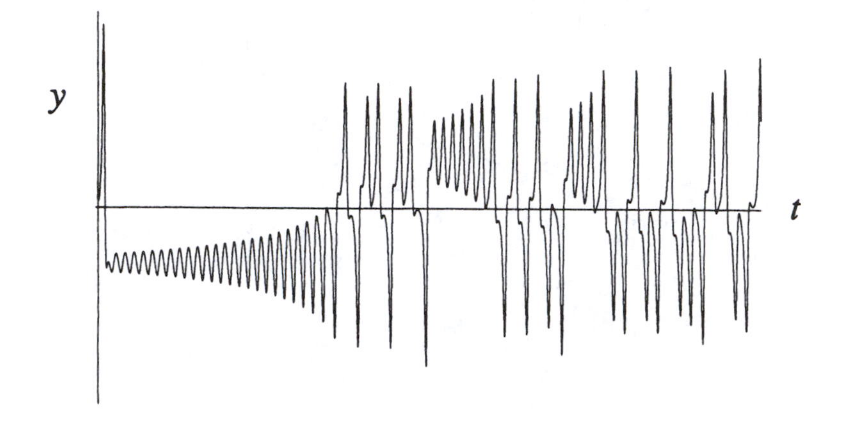

Figure 9.3.1

After an initial transient, the solution settles into an irregular oscillation that persists as $t \to \infty$, but never repeats exactly. The motion is ***aperiodic***.

Lorenz discovered that a wonderful structure emerges if the solution is visualized as a trajectory in phase space. For instance, when $x(t)$ is plotted against $z(t)$, a butterfly pattern appears (Figure 9.3.2).

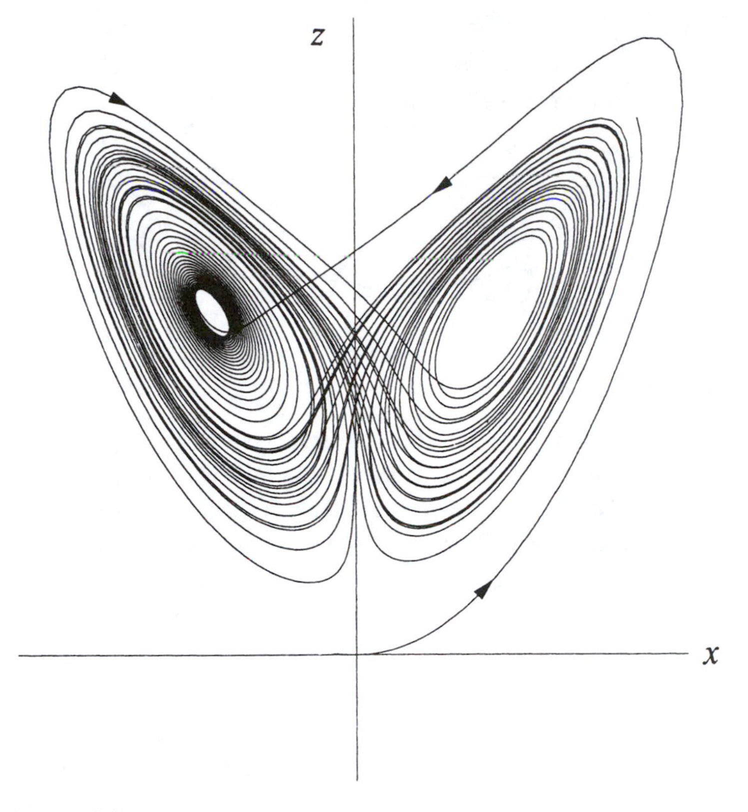

Figure 9.3.2

The trajectory appears to cross itself repeatedly, but that's just an artifact of projecting the three-dimensional trajectory onto a two-dimensional plane. In three dimensions no self-intersections occur.

Let's try to understand Figure 9.3.2 in detail. The trajectory starts near the origin, then swings to the right, and then dives into the center of a spiral on the left. After a very slow spiral outward, the trajectory shoots back over to the right side, spirals around a few times, shoots over to the left, spirals around, and so on indefinitely. The number of circuits made on either side varies unpredictably from one cycle to the next. In fact, the sequence of the number of circuits has many of the characteristics of a *random* sequence. Physically, the switches between left and right correspond to the irregular reversals of the waterwheel that we observed in Section 9.1.

When the trajectory is viewed in all three dimensions, rather than in a two-dimensional projection, it appears to settle onto an exquisitely thin set that looks like a pair of butterfly wings. Figure 9.3.3 shows a schematic of this ***strange attractor*** (a term coined by Ruelle and Takens (1971)). This limiting set is the attracting set of zero volume whose existence was deduced in Section 9.2.

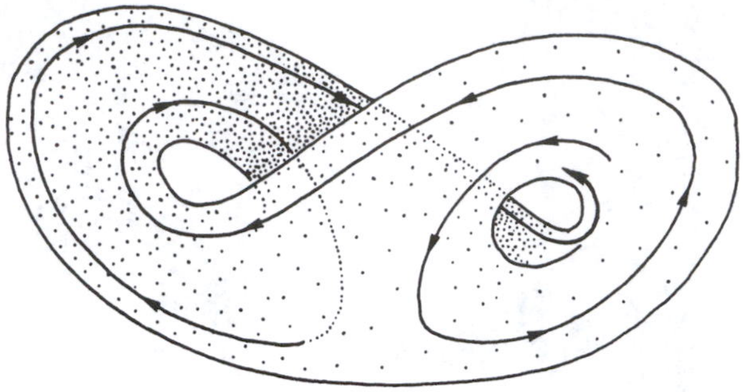

Figure 9.3.3 Abraham and Shaw (1983), p. 88

What is the geometrical structure of the strange attractor? Figure 9.3.3 suggests that it is a pair of surfaces that merge into one in the lower portion of Figure 9.3.3. But how can this be, when the uniqueness theorem (Section 6.2) tells us that trajectories can't cross or merge? Lorenz (1963) gives a lovely explanation—the two surfaces only *appear* to merge. The illusion is caused by the strong volume contraction of the flow, and insufficient numerical resolution. But watch where that idea leads him:

> It would seem, then, that the two surfaces merely appear to merge, and remain distinct surfaces. Following these surfaces along a path parallel to a trajectory, and circling C^+ and C^-, we see that each surface is really a pair of surfaces, so that, where they appear to merge, there are really four surfaces. Continuing this process for another circuit, we see that there are really eight surfaces, etc., and we finally conclude that there is an infinite complex of surfaces, each extremely close to one or the other of two merging surfaces.

Today this "infinite complex of surfaces" would be called a fractal. It is a set of points with zero volume but infinite surface area. In fact, numerical experiments suggest that it has a dimension of about 2.05! (See Example 11.5.1.) The amazing geometric properties of fractals and strange attractors will be discussed in detail in Chapters 11 and 12. But first we want to examine chaos a bit more closely.

Exponential Divergence of Nearby Trajectories

The motion on the attractor exhibits ***sensitive dependence on initial conditions***. This means that two trajectories starting very close together will rapidly diverge from each other, and thereafter have totally different futures. Color Plate 2 vividly illustrates this divergence by plotting the evolution of a small red blob of 10,000 nearby initial conditions. The blob eventually spreads over the whole attractor. Hence nearby trajectories can end up anywhere on the attractor! The practical implication is that long-term prediction becomes impossible in a system like this, where small uncertainties are amplified enormously fast.

Let's make these ideas more precise. Suppose that we let transients decay, so that a trajectory is "on" the attractor. Suppose $\mathbf{x}(t)$ is a point on the attractor at time t, and consider a nearby point, say $\mathbf{x}(t) + \delta(t)$, where δ is a tiny separation vector of initial length $\|\delta_0\| = 10^{-15}$, say (Figure 9.3.4).

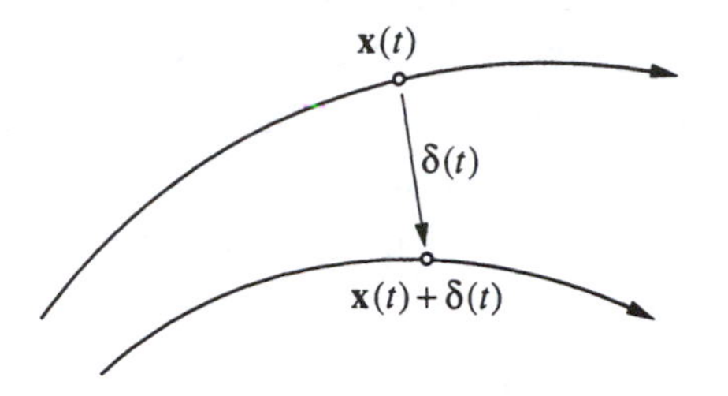

Figure 9.3.4

Now watch how $\delta(t)$ grows. In numerical studies of the Lorenz attractor, one finds that

$$\|\delta(t)\| \sim \|\delta_0\| e^{\lambda t}$$

where $\lambda \approx 0.9$. Hence *neighboring trajectories separate exponentially fast*. Equivalently, if we plot $\ln\|\delta(t)\|$ versus t, we find a curve that is close to a straight line with a positive slope of λ (Figure 9.3.5).

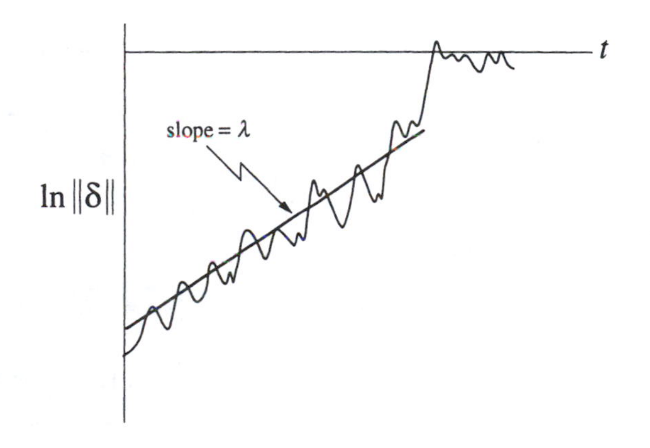

Figure 9.3.5

We need to add some qualifications:

1. The curve is never exactly straight. It has wiggles because the strength of the exponential divergence varies somewhat along the attractor.
2. The exponential divergence must stop when the separation is comparable to the "diameter" of the attractor—the trajectories obviously can't

get any farther apart than that. This explains the leveling off or *saturation* of the curve in Figure 9.3.5.

3. The number λ is often called the ***Liapunov exponent***, although this is a sloppy use of the term, for two reasons:

   First, there are actually n different Liapunov exponents for an n-dimensional system, defined as follows. Consider the evolution of an infinitesimal sphere of perturbed initial conditions. During its evolution, the sphere will become distorted into an infinitesimal ellipsoid. Let $\delta_k(t)$, $k = 1,\ldots,n$, denote the length of the kth principal axis of the ellipsoid. Then $\delta_k(t) \sim \delta_k(0)e^{\lambda_k t}$, where the λ_k are the Liapunov exponents. For large t, the diameter of the ellipsoid is controlled by the most positive λ_k. Thus our λ is actually the *largest* Liapunov exponent.

   Second, λ depends (slightly) on which trajectory we study. We should average over many different points on the same trajectory to get the true value of λ.

When a system has a positive Liapunov exponent, there is a *time horizon* beyond which prediction breaks down, as shown schematically in Figure 9.3.6. (See Lighthill 1986 for a nice discussion.) Suppose we measure the initial conditions of an experimental system very accurately. Of course, no measurement is perfect—there is always some error $\|\delta_0\|$ between our estimate and the true initial state.

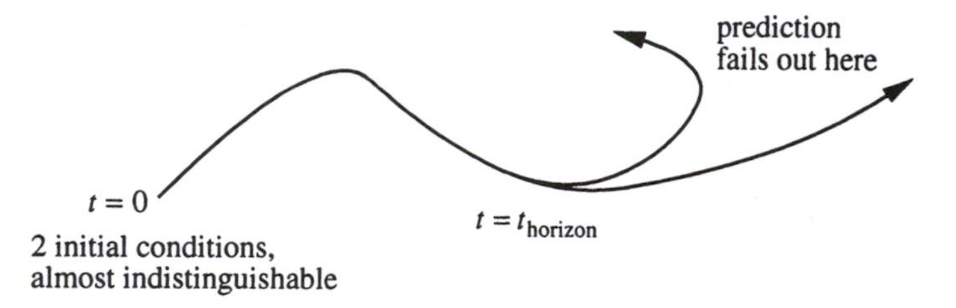

Figure 9.3.6

After a time t, the discrepancy grows to $\|\delta(t)\| \sim \|\delta_0\|e^{\lambda t}$. Let a be a measure of our tolerance, i.e., if a prediction is within a of the true state, we consider it acceptable. Then our prediction becomes intolerable when $\|\delta(t)\| \geq a$, and this occurs after a time

$$t_{\text{horizon}} \sim O\left(\frac{1}{\lambda}\ln\frac{a}{\|\delta_0\|}\right).$$

The logarithmic dependence on $\|\delta_0\|$ is what hurts us. No matter how hard we work to reduce the initial measurement error, we can't predict longer than a few multiples of $1/\lambda$. The next example is intended to give you a quantitative feel for this effect.

EXAMPLE 9.3.1:

Suppose we're trying to predict the future state of a chaotic system to within a tolerance of $a = 10^{-3}$. Given that our estimate of the initial state is uncertain to within $\|\delta_0\| = 10^{-7}$, for about how long can we predict the state of the system, while remaining within the tolerance? Now suppose we buy the finest instrumentation, recruit the best graduate students, etc., and somehow manage to measure the initial state a *million* times better, i.e., we improve our initial error to $\|\delta_0\| = 10^{-13}$. How much longer can we predict?

Solution: The original prediction has

$$t_{\text{horizon}} \approx \frac{1}{\lambda} \ln \frac{10^{-3}}{10^{-7}} = \frac{1}{\lambda} \ln(10^4) = \frac{4 \ln 10}{\lambda}.$$

The improved prediction has

$$t_{\text{horizon}} \approx \frac{1}{\lambda} \ln \frac{10^{-3}}{10^{-13}} = \frac{1}{\lambda} \ln(10^{10}) = \frac{10 \ln 10}{\lambda}.$$

Thus, after a millionfold improvement in our initial uncertainty, we can predict only $10/4 = 2.5$ times longer! ■

Such calculations demonstrate the futility of trying to predict the detailed long-term behavior of a chaotic system. Lorenz suggested that this is what makes long-term weather prediction so difficult.

Defining Chaos

No definition of the term *chaos* is universally accepted yet, but almost everyone would agree on the three ingredients used in the following working definition:

> ***Chaos*** is *aperiodic long-term behavior* in a *deterministic* system that exhibits *sensitive dependence on initial conditions*.

1. "Aperiodic long-term behavior" means that there are trajectories which do not settle down to fixed points, periodic orbits, or quasiperiodic orbits as $t \to \infty$. For practical reasons, we should require that such trajectories are not too rare. For instance, we could insist that there be an open set of initial conditions leading to aperiodic trajectories, or perhaps that such trajectories should occur with nonzero probability, given a random initial condition.

2. "Deterministic" means that the system has no random or noisy inputs or parameters. The irregular behavior arises from the system's nonlinearity, rather than from noisy driving forces.
3. "Sensitive dependence on initial conditions" means that nearby trajectories separate exponentially fast, i.e., the system has a positive Liapunov exponent.

EXAMPLE 9.3.2:

Some people think that chaos is just a fancy word for instability. For instance, the system $\dot{x} = x$ is deterministic and shows exponential separation of nearby trajectories. Should we call this system chaotic?

Solution: No. Trajectories are repelled to infinity, and never return. So infinity acts like an attracting *fixed point*. Chaotic behavior should be aperiodic, and that excludes fixed points as well as periodic behavior. ■

Defining Attractor and Strange Attractor

The term *attractor* is also difficult to define in a rigorous way. We want a definition that is broad enough to include all the natural candidates, but restrictive enough to exclude the imposters. There is still disagreement about what the exact definition should be. See Guckenheimer and Holmes (1983, p. 256), Eckmann and Ruelle (1985), and Milnor (1985) for discussions of the subtleties involved.

Loosely speaking, an attractor is a set to which all neighboring trajectories converge. Stable fixed points and stable limit cycles are examples. More precisely, we define an ***attractor*** to be a closed set A with the following properties:

1. A is an *invariant set*: any trajectory $\mathbf{x}(t)$ that starts in A stays in A for all time.
2. A *attracts an open set of initial conditions*: there is an open set U containing A such that if $\mathbf{x}(0) \in U$, then the distance from $\mathbf{x}(t)$ to A tends to zero as $t \to \infty$. This means that A attracts all trajectories that start sufficiently close to it. The largest such U is called the *basin of attraction* of A.
3. A is *minimal*: there is no proper subset of A that satisfies conditions 1 and 2.

EXAMPLE 9.3.3:

Consider the system $\dot{x} = x - x^3$, $\dot{y} = -y$. Let I denote the interval $-1 \le x \le 1$,

$y = 0$. Is I an invariant set? Does it attract an open set of initial conditions? Is it an attractor?

Solution: The phase portrait is shown in Figure 9.3.7. There are stable fixed points at the endpoints $(\pm 1, 0)$ of I and a saddle point at the origin. Figure 9.3.7 shows that I is an invariant set; any trajectory that starts in I stays in I forever. (In fact the whole x-axis is an invariant set, since if $y(0) = 0$, then $y(t) = 0$ for all t.) So condition 1 is satisfied.

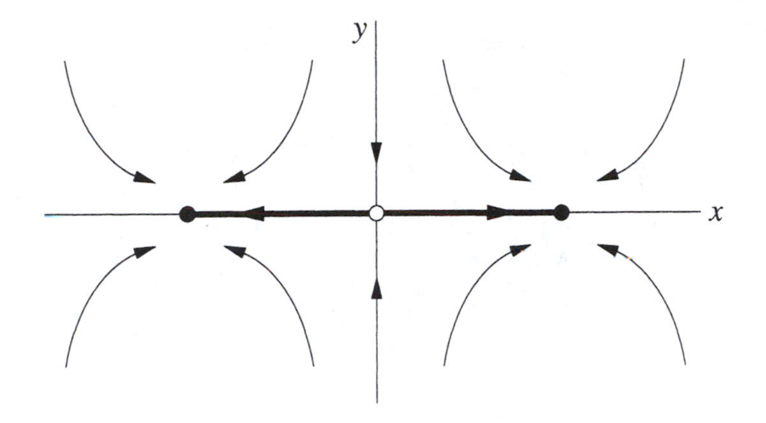

Figure 9.3.7

Moreover, I certainly attracts an open set of initial conditions—it attracts *all* trajectories in the xy plane. So condition 2 is also satisfied.

But I is *not* an attractor because it is not minimal. The stable fixed points $(\pm 1, 0)$ are proper subsets of I that also satisfy properties 1 and 2. These points are the only attractors for the system. ■

There is an important moral to Example 9.3.3. Even if a certain set attracts all trajectories, it may fail to be an attractor because it may not be minimal—it may contain one or more smaller attractors.

The same could be true for the Lorenz equations. Although all trajectories are attracted to a bounded set of zero volume, that set is not necessarily an attractor, since it might not be minimal. To this day, no one has managed to prove that the Lorenz attractor seen in computer experiments is truly an attractor in this technical sense. But everyone believes it is, except for a few purists.

Finally, we define a ***strange attractor*** to be an attractor that exhibits sensitive dependence on initial conditions. Strange attractors were originally called strange because they are often fractal sets. Nowadays this geometric property is regarded as less important than the dynamical property of sensitive dependence on initial conditions. The terms *chaotic attractor* and *fractal attractor* are used when one wishes to emphasize one or the other of those aspects.

9.4 Lorenz Map

Lorenz (1963) found a beautiful way to analyze the dynamics on his strange attractor. He directs our attention to a particular view of the attractor (Figure 9.4.1),

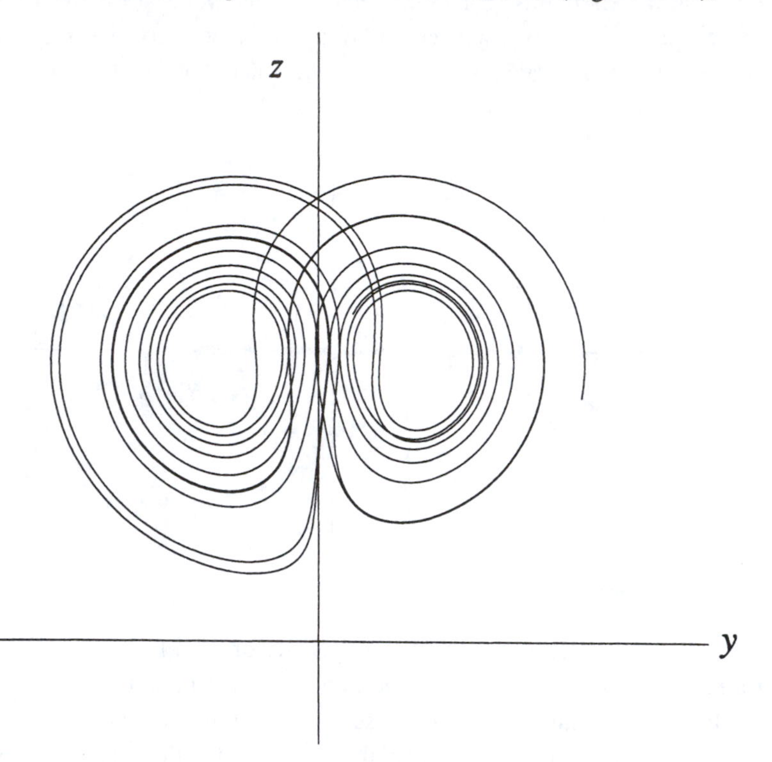

Figure 9.4.1

and then he writes:

> the trajectory apparently leaves one spiral only after exceeding some critical distance from the center. Moreover, the extent to which this distance is exceeded appears to determine the point at which the next spiral is entered; this in turn seems to determine the number of circuits to be executed before changing spirals again. It therefore seems that some single feature of a given circuit should predict the same feature of the following circuit.

The "single feature" that he focuses on is z_n, the nth local maximum of $z(t)$ (Figure 9.4.2).

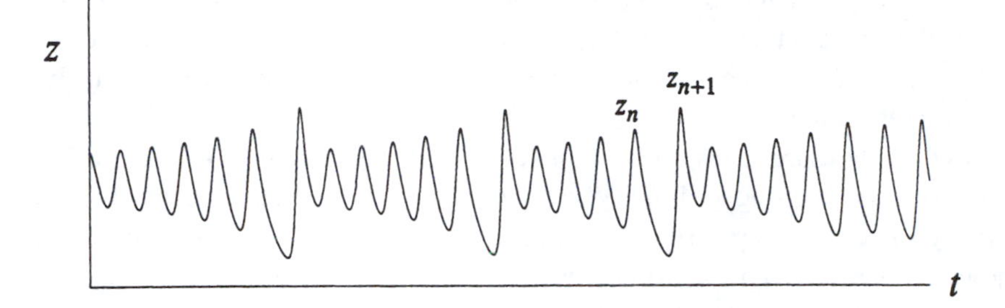

Figure 9.4.2

Lorenz's idea is that z_n should predict z_{n+1}. To check this, he numerically integrated the equations for a long time, then measured the local maxima of $z(t)$, and finally plotted z_{n+1} vs. z_n. As shown in Figure 9.4.3, *the data from the chaotic time series appear to fall neatly on a curve*—there is almost no "thickness" to the graph!

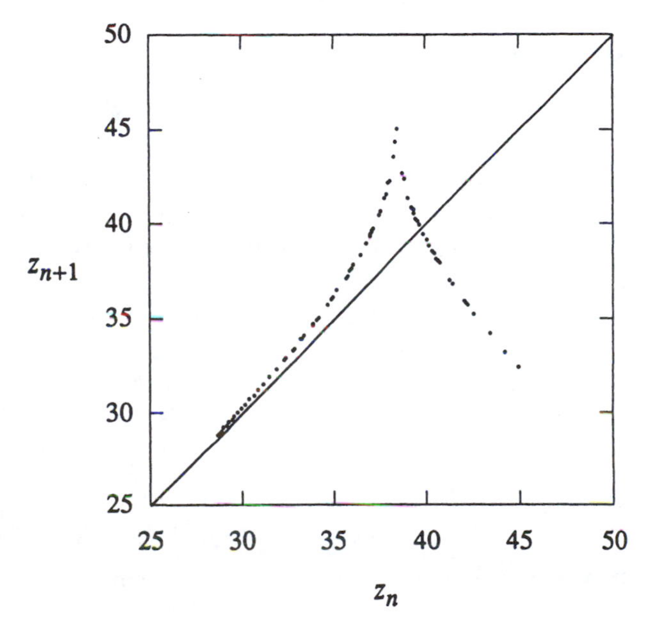

Figure 9.4.3

By this ingenious trick, Lorenz was able to extract order from chaos. The function $z_{n+1} = f(z_n)$ shown in Figure 9.4.3 is now called the ***Lorenz map***. It tells us a lot about the dynamics on the attractor: given z_0, we can predict z_1 by $z_1 = f(z_0)$, and then use that information to predict $z_2 = f(z_1)$, and so on, bootstrapping our way forward in time by iteration. The analysis of this iterated map is going to lead us to a striking conclusion, but first we should make a few clarifications.

First, the graph in Figure 9.4.3 is not actually a curve. It *does* have some thickness. So strictly speaking, $f(z)$ is not a well-defined function, because there can be more than one output z_{n+1} for a given input z_n. On the other hand, the thickness is so small, and there is so much to be gained by treating the graph as a curve, that we will simply make this approximation, keeping in mind that the subsequent analysis is plausible but not rigorous.

Second, the Lorenz map may remind you of a Poincaré map (Section 8.7). In both cases we're trying to simplify the analysis of a differential equation by reducing it to an iterated map of some kind. But there's an important distinction: To construct a Poincaré map for a three-dimensional flow, we compute a trajectory's successive intersections with a two-dimensional surface. The Poincaré map takes a point on that surface, specified by *two* coordinates, and then tells us how those two coordinates change after the first return to the surface. The Lorenz map is different because it characterizes the trajectory by only *one* number, not two. This simpler approach works only if the attractor is very "flat," i.e., close to two-dimensional, as the Lorenz attractor is.

Ruling Out Stable Limit Cycles

How do we know that the Lorenz attractor is not just a stable limit cycle in disguise? Playing devil's advocate, a skeptic might say, "Sure, the trajectories don't ever seem to repeat, but maybe you haven't integrated long enough. Eventually the trajectories *will* settle down into a periodic behavior—it just happens that the period is incredibly long, much longer than you've tried in your computer. Prove me wrong."

So far, no one has been able to refute this argument in a rigorous sense. But by using his map, Lorenz was able to give a plausible counterargument that stable limit cycles do not, in fact, occur for the parameter values he studied.

His argument goes like this: The key observation is that the graph in Figure 9.4.3 satisfies

$$|f'(z)| > 1 \qquad (1)$$

everywhere. This property ultimately implies that if any limit cycles exist, they are necessarily *unstable*.

To see why, we start by analyzing the fixed points of the map f. These are points z^* such that $f(z^*) = z^*$, in which case $z_n = z_{n+1} = z_{n+2} = \ldots$. Figure 9.4.3 shows that there is one fixed point, where the 45° diagonal intersects the graph. It represents a closed orbit that looks like that shown in Figure 9.4.4.

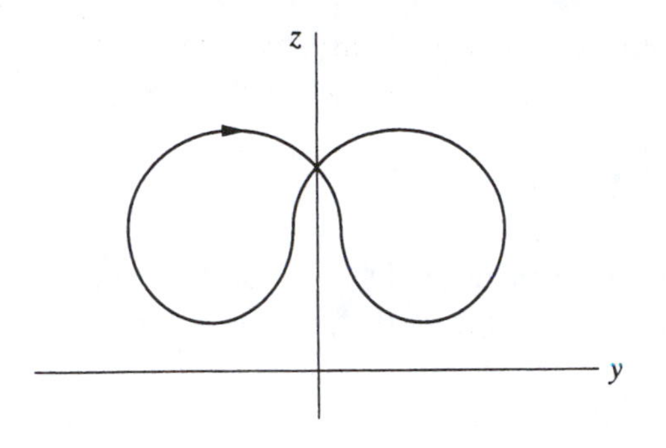

Figure 9.4.4

To show that this closed orbit is unstable, consider a slightly perturbed trajectory that has $z_n = z^* + \eta_n$, where η_n is small. After linearization as usual, we find $\eta_{n+1} \approx f'(z^*)\eta_n$. Since $|f'(z^*)| > 1$, by the key property (1), we get

$$|\eta_{n+1}| > |\eta_n|.$$

Hence the deviation η_n *grows* with each iteration, and so the original closed orbit is unstable.

Now we generalize the argument slightly to show that *all* closed orbits are unstable.

EXAMPLE 9.4.1:

Given the Lorenz map approximation $z_{n+1} = f(z_n)$, with $|f'(z)| > 1$ for all z, show that *all* closed orbits are unstable.

Solution: Think about the sequence $\{z_n\}$ corresponding to an arbitrary closed orbit. It might be a complicated sequence, but since we know that the orbit eventually closes, the sequence must eventually repeat. Hence $z_{n+p} = z_n$, for some integer $p \geq 1$. (Here p is the *period* of the sequence, and z_n is a *period-p point*.)

Now to prove that the corresponding closed orbit is unstable, consider the fate of a small deviation η_n, and look at it after p iterations, when the cycle is complete. We'll show that $|\eta_{n+p}| > |\eta_n|$, which implies that the deviation has grown and the closed orbit is unstable.

To estimate η_{n+p}, go one step at a time. After one iteration, $\eta_{n+1} \approx f'(z_n)\eta_n$, by linearization about z_n. Similarly, after two iterations,

$$\begin{aligned}\eta_{n+2} &\approx f'(z_{n+1})\eta_{n+1} \\ &\approx f'(z_{n+1})\left[f'(z_n)\eta_n\right] \\ &= \left[f'(z_{n+1})f'(z_n)\right]\eta_n .\end{aligned}$$

Hence after p iterations,

$$\eta_{n+p} \approx \left[\prod_{k=0}^{p-1} f'(z_{n+k}) \right] \eta_n \ . \tag{2}$$

In (2), each of the factors in the product has absolute value greater than 1, because $|f'(z)| > 1$ for all z. Hence $|\eta_{n+p}| > |\eta_n|$, which proves that the closed orbit is unstable. ■

9.5 Exploring Parameter Space

So far we have concentrated on the particular parameter values $\sigma = 10$, $b = \frac{8}{3}$, $r = 28$, as in Lorenz (1963). What happens if we change the parameters? It's like a walk through the jungle—one can find exotic limit cycles tied in knots, pairs of limit cycles linked through each other, intermittent chaos, noisy periodicity, as well as strange attractors (Sparrow 1982, Jackson 1990). You should do some exploring on your own, perhaps starting with some of the exercises.

There is a vast three-dimensional parameter space to be explored, and much remains to be discovered. To simplify matters, many investigators have kept $\sigma = 10$ and $b = \frac{8}{3}$ while varying r. In this section we give a glimpse of some of the phenomena observed in numerical experiments. See Sparrow (1982) for the definitive treatment.

The behavior for small values of r is summarized in Figure 9.5.1.

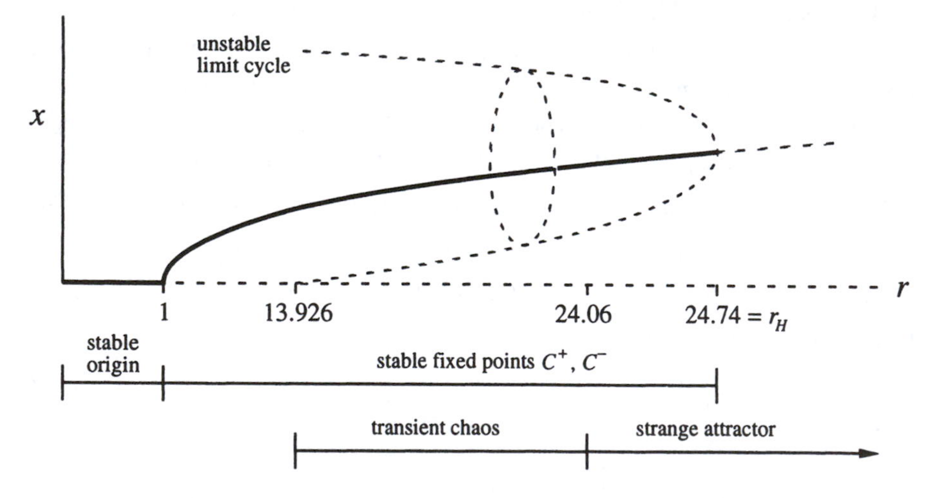

Figure 9.5.1

Much of this picture is familiar. The origin is globally stable for $r < 1$. At $r = 1$ the origin loses stability by a supercritical pitchfork bifurcation, and a symmetric pair of attracting fixed points is born (in our schematic, only one of the pair is shown). At $r_H = 24.74$ the fixed points lose stability by absorbing an unstable limit cycle in a subcritical Hopf bifurcation.

Now for the new results. As we decrease r from r_H, the unstable limit cycles expand and pass precariously close to the saddle point at the origin. At $r \approx 13.926$ the cycles touch the saddle point and become homoclinic orbits; hence we have a *homoclinic bifurcation*. (See Section 8.4 for the much simpler homoclinic bifurcations that occur in two-dimensional systems.) Below $r = 13.926$ there are no limit cycles. Viewed in the other direction, we could say that a pair of unstable limit cycles are created as r increases through $r = 13.926$.

This homoclinic bifurcation has many ramifications for the dynamics, but its analysis is too advanced for us—see Sparrow's (1982) discussion of "homoclinic explosions." The main conclusion is that an amazingly complicated invariant set is born at $r = 13.926$, along with the unstable limit cycles. This set is a thicket of infinitely many saddle-cycles and aperiodic orbits. It is not an attractor and is not observable directly, but it generates sensitive dependence on initial conditions in its neighborhood. Trajectories can get hung up near this set, somewhat like wandering in a maze. Then they rattle around chaotically for a while, but eventually escape and settle down to C^+ or C^-. The time spent wandering near the set gets longer and longer as r increases. Finally, at $r = 24.06$ the time spent wandering becomes infinite and the set becomes a strange attractor (Yorke and Yorke 1979).

EXAMPLE 9.5.1:

Show numerically that the Lorenz equations can exhibit ***transient chaos*** when $r = 21$ (with $\sigma = 10$ and $b = \frac{8}{3}$ as usual).

Solution: After experimenting with a few different initial conditions, it is easy to find solutions like that shown in Figure 9.5.2.

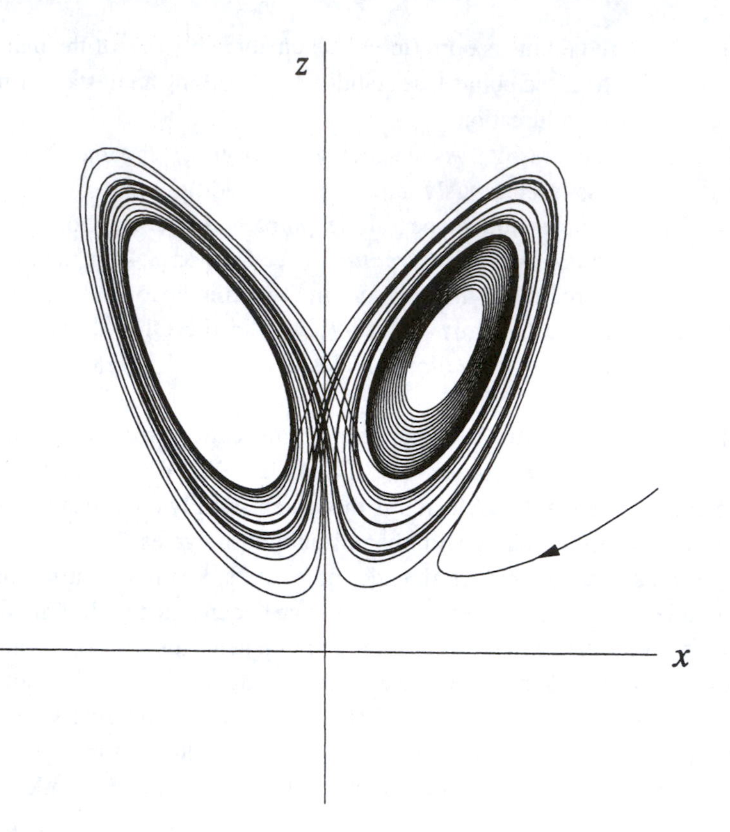

Figure 9.5.2

At first the trajectory seems to be tracing out a strange attractor, but eventually it stays on the right and spirals down toward the stable fixed point C^+. (Recall that both C^+ and C^- are still stable at $r = 21$.) The time series of y vs. t shows the same result: an initially erratic solution ultimately damps down to equilibrium (Figure 9.5.3).

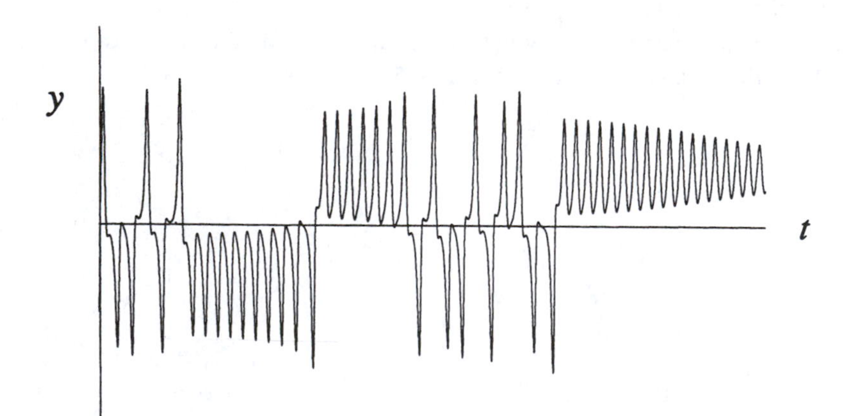

Figure 9.5.3

Other names used for transient chaos are *metastable chaos* (Kaplan and Yorke 1979) or *pre-turbulence* (Yorke and Yorke 1979, Sparrow 1982). ■

By our definition, the dynamics in Example 9.5.1 are not "chaotic," because the long-term behavior is not aperiodic. On the other hand, the dynamics do exhibit sensitive dependence on initial conditions—if we had chosen a slightly different initial condition, the trajectory could easily have ended up at C^- instead of C^+. Thus the system's behavior is unpredictable, at least for certain initial conditions.

Transient chaos shows that a deterministic system can be unpredictable, even if its final states are very simple. In particular, you don't need strange attractors to generate effectively random behavior. Of course, this is familiar from everyday experience—many games of "chance" used in gambling are essentially demonstrations of transient chaos. For instance, think about rolling dice. A crazily-rolling die always stops in one of six stable equilibrium positions. The problem with predicting the outcome is that the final position depends sensitively on the initial orientation and velocity (assuming the initial velocity is large enough).

Before we leave the regime of small r, we note one other interesting implication of Figure 9.5.1: for $24.06 < r < 24.74$, there are *two* types of attractors: fixed points and a strange attractor. This coexistence means that we can have hysteresis between chaos and equilibrium by varying r slowly back and forth past these two endpoints (Exercise 9.5.4). It also means that a large enough perturbation can knock a steadily rotating waterwheel into permanent chaos; this is reminiscent (in spirit, though not detail) of fluid flows that mysteriously become turbulent even though the basic laminar flow is still linearly stable (Drazin and Reid 1981).

The next example shows that the dynamics become simple again when r is sufficiently large.

EXAMPLE 9.5.2:

Describe the long-term dynamics for large values of r, for $\sigma = 10$, $b = \frac{8}{3}$. Interpret the results in terms of the motion of the waterwheel of Section 9.1.

Solution: Numerical simulations indicate that the system has a globally attracting limit cycle for all $r > 313$ (Sparrow 1982). In Figures 9.5.4 and 9.5.5 we plot a typical solution for $r = 350$; note the approach to the limit cycle.

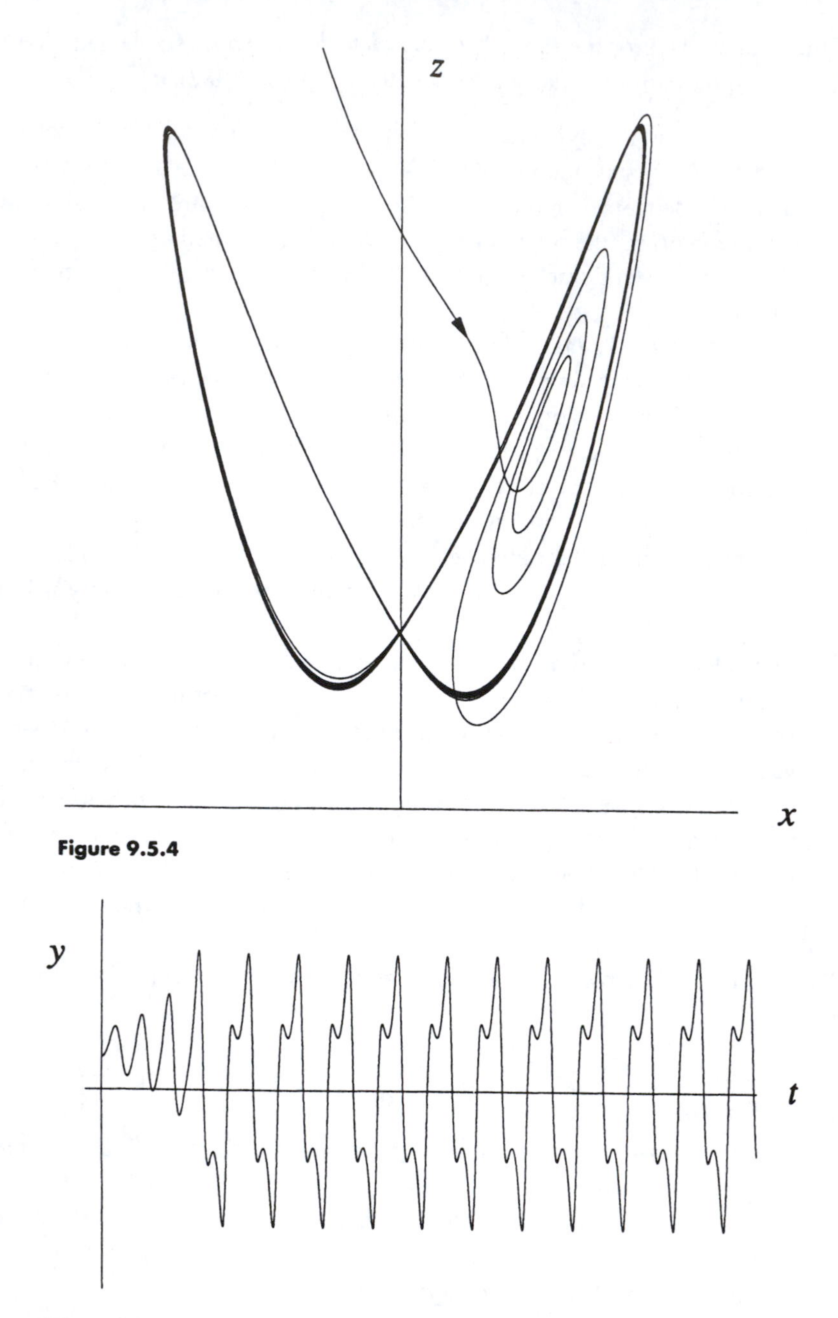

Figure 9.5.4

Figure 9.5.5

This solution predicts that the waterwheel should ultimately rock back and forth like a pendulum, turning once to the right, then back to the left, and so on. This is observed experimentally. ■

In the limit $r \to \infty$ one can obtain many analytical results about the Lorenz equations. For instance, Robbins (1979) used perturbation methods to characterize the limit cycle at large r. For the first steps in her calculation, see Exercise 9.5.5. For more details, see Chapter 7 in Sparrow (1982).

The story is much more complicated for r between 28 and 313. For most values of r one finds chaos, but there are also small windows of periodic behavior interspersed. The three largest windows are $99.524\ldots < r < 100.795\ldots$; $145 < r < 166$; and $r > 214.4$. The alternating pattern of chaotic and periodic regimes resembles that seen in the logistic map (Chapter 10), and so we will defer further discussion until then.

9.6 Using Chaos to Send Secret Messages

One of the most exciting recent developments in nonlinear dynamics is the realization that chaos can be *useful*. Normally one thinks of chaos as a fascinating curiosity at best, and a nuisance at worst, something to be avoided or engineered away. But since about 1990, people have found ways to exploit chaos to do some marvelous and practical things. For an introduction to this new subject, see Vohra et al. (1992).

One application involves "private communications." Suppose you want to send a secret message to a friend or business partner. Naturally you should use a code, so that even if an enemy is eavesdropping, he will have trouble making sense of the message. This is an old problem—people have been making (and breaking) codes for as long as there have been secrets worth keeping.

Kevin Cuomo and Alan Oppenheim (1992, 1993) have implemented a new approach to this problem, building on Pecora and Carroll's (1990) discovery of ***synchronized chaos***. Here's the strategy: When you transmit the message to your friend, you also "mask" it with much louder chaos. An outside listener only hears the chaos, which sounds like meaningless noise. But now suppose that your friend has a magic receiver that perfectly reproduces the chaos—then he can subtract off the chaotic mask and listen to the message!

Cuomo's Demonstration

Kevin Cuomo was a student in my course on nonlinear dynamics, and at the end of the semester he treated our class to a live demonstration of his approach. First he showed us how to make the chaotic mask, using an electronic implementation of the Lorenz equations (Figure 9.6.1). The circuit involves resistors, capacitors, operational amplifiers, and analog multiplier chips.

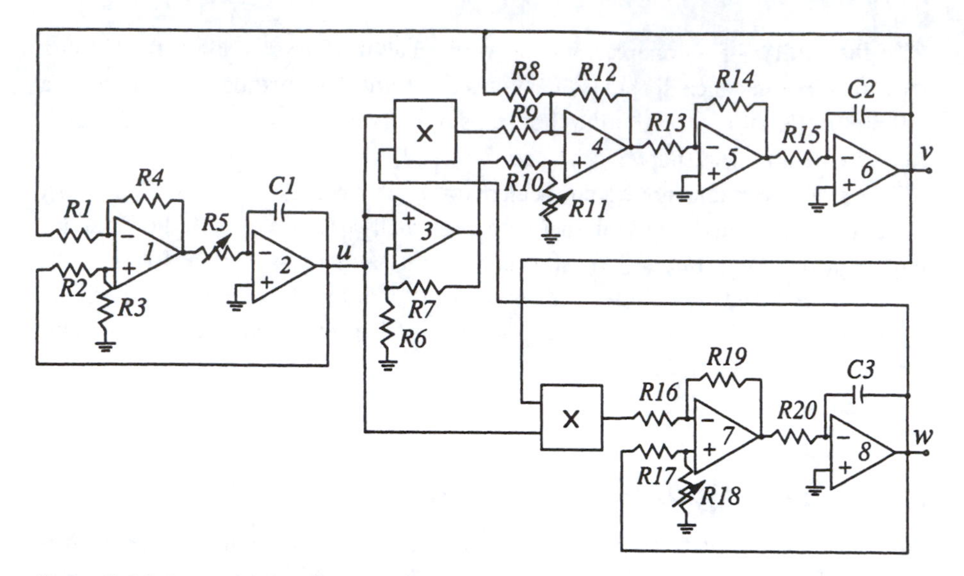

Figure 9.6.1 Cuomo and Oppenheim (1993), p. 66

The voltages u, v, w at three different points in the circuit are proportional to Lorenz's x, y, z. Thus the circuit acts like an analog computer for the Lorenz equations. Oscilloscope traces of $u(t)$ vs. $w(t)$, for example, confirmed that the circuit was following the familiar Lorenz attractor. Then, by hooking up the circuit to a loudspeaker, Cuomo enabled us to *hear* the chaos—it sounds like static on the radio.

The hard part is to make a receiver that can synchronize perfectly to the chaotic transmitter. In Cuomo's set-up, the receiver is an identical Lorenz circuit, driven in a certain clever way by the transmitter. We'll get into the details later, but for now let's content ourselves with the experimental fact that synchronized chaos does occur. Figure 9.6.2 plots the receiver variables $u_r(t)$ and $v_r(t)$ against their transmitter counterparts $u(t)$ and $v(t)$.

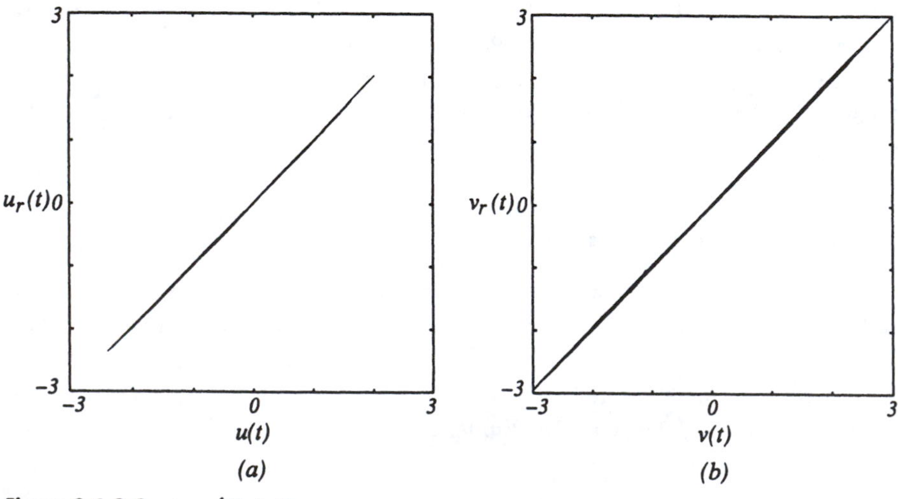

Figure 9.6.2 Courtesy of Kevin Cuomo

The 45° trace on the oscilloscope indicates that the synchronization is nearly perfect, despite the fact that both circuits are running chaotically. The synchronization is also quite stable: the data in Figure 9.6.2 reflect a time span of several minutes, whereas without the drive the circuits would decorrelate in about 1 millisecond.

Cuomo brought the house down when he showed us how to use the circuits to mask a message, which he chose to be a recording of the hit song "Emotions" by Mariah Carey. (One student, apparently with different taste in music, asked "Is that the signal or the noise?") After playing the original version of the song, Cuomo played the masked version. Listening to the hiss, one had absolutely no sense that there was a song buried underneath. Yet when this masked message was sent to the receiver, its output synchronized almost perfectly to the original chaos, and after instant electronic subtraction, we heard Mariah Carey again! The song sounded fuzzy, but easily understandable.

Figures 9.6.3 and 9.6.4 illustrate the system's performance more quantitatively. Figure 9.6.3a is a segment of speech from the sentence "He has the bluest eyes," obtained by sampling the speech waveform at a 48 kHz rate and with 16-bit resolution. This signal was then masked by much louder chaos. The power spectra in Figure 9.6.4 show that the chaos is about 20 decibels louder than the message, with coverage over its whole frequency range. Finally, the unmasked message at the receiver is shown in Figure 9.6.3b. The original speech is recovered with only a tiny amount of distortion (most visible as the increased noise on the flat parts of the record).

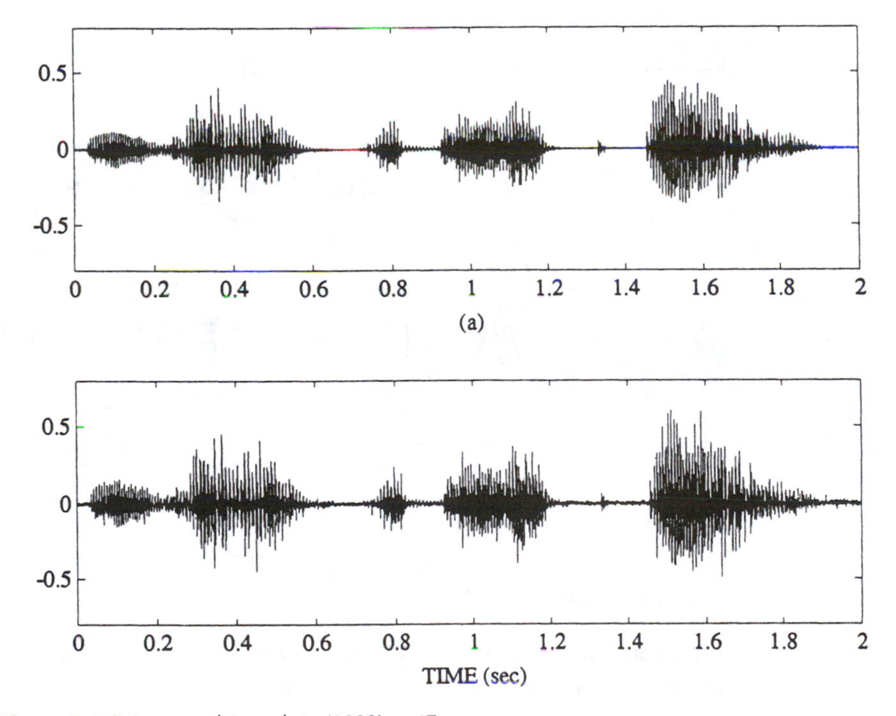

Figure 9.6.3 Cuomo and Oppenheim (1993), p. 67

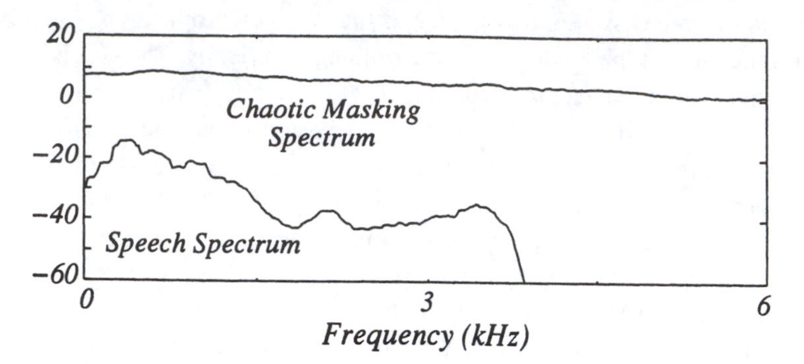

Figure 9.6.4 Cuomo and Oppenheim (1993), p. 68

Proof of Synchronization

The signal-masking method discussed above was made possible by the conceptual breakthrough of Pecora and Carroll (1990). Before their work, many people would have doubted that two chaotic systems could be made to synchronize. After all, chaotic systems are sensitive to slight changes in initial condition, so one might expect any errors between the transmitter and receiver to grow exponentially. But Pecora and Carroll (1990) found a way around these concerns. Cuomo and Oppenheim (1992, 1993) have simplified and clarified the argument; we discuss their approach now.

The receiver circuit is shown in Figure 9.6.5.

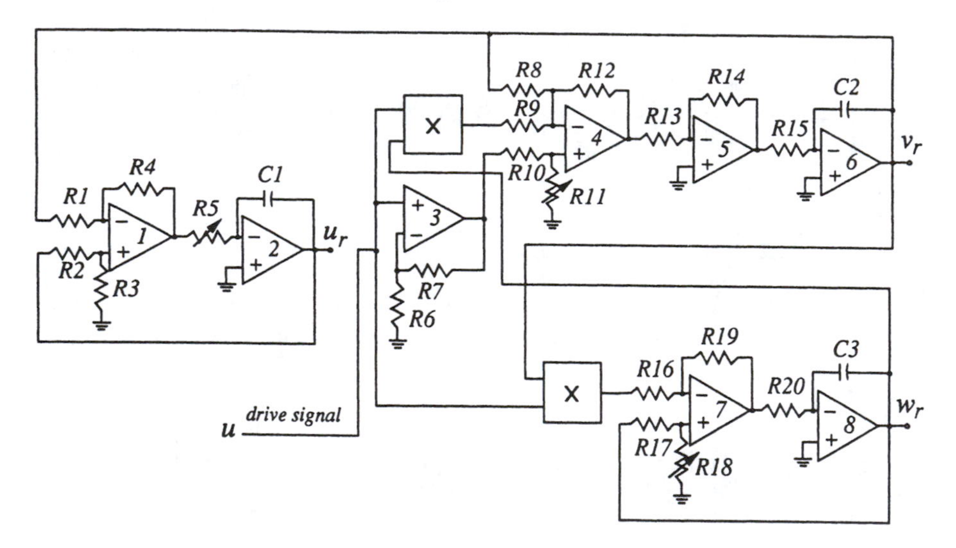

Figure 9.6.5 Courtesy of Kevin Cuomo

It is identical to the transmitter, except that the drive signal $u(t)$ replaces the receiver signal $u_r(t)$ at a crucial place in the circuit (compare Figure 9.6.1). To see what effect this has on the dynamics, we write down the governing equations for both the transmitter and the receiver. Using Kirchhoff's laws and appropriate nondimensionalizations (Cuomo and Oppenheim 1992), we get

$$\begin{aligned}\dot{u} &= \sigma(v-u)\\ \dot{v} &= ru - v - 20uw \\ \dot{w} &= 5uv - bw\end{aligned} \tag{1}$$

as the dynamics of the transmitter. These are just the Lorenz equations, written in terms of scaled variables

$$u = \tfrac{1}{10}x, \qquad v = \tfrac{1}{10}y, \qquad w = \tfrac{1}{20}z.$$

(This scaling is irrelevant mathematically, but it keeps the variables in a more favorable range for electronic implementation, if one unit is supposed to correspond to one volt. Otherwise the wide dynamic range of the solutions exceeds typical power supply limits.)

The receiver variables evolve according to

$$\begin{aligned}\dot{u}_r &= \sigma(v_r - u_r)\\ \dot{v}_r &= ru(t) - v_r - 20u(t)w_r \\ \dot{w}_r &= 5u(t)v_r - bw_r\end{aligned} \tag{2}$$

where we have written $u(t)$ to emphasize that the receiver is driven by the chaotic signal $u(t)$ coming from the transmitter.

The astonishing result is that *the receiver asymptotically approaches perfect synchrony with the transmitter, starting from any initial conditions*! To be precise, let

$$\begin{aligned}\mathbf{d} &= (u, v, w) = \text{state of the transmitter or "driver"}\\ \mathbf{r} &= (u_r, v_r, w_r) = \text{state of the receiver}\\ \mathbf{e} &= \mathbf{d} - \mathbf{r} = \text{error signal}\end{aligned}$$

The claim is that $\mathbf{e}(t) \to \mathbf{0}$ as $t \to \infty$, for all initial conditions.

Why is this astonishing? Because at each instant the receiver has only *partial* information about the state of the transmitter—it is driven solely by $u(t)$, yet somehow it manages to reconstruct the other two transmitter variables $v(t)$ and $w(t)$ as well.

The proof is given in the following example.

EXAMPLE 9.6.1:

By defining an appropriate Liapunov function, show that $\mathbf{e}(t) \to \mathbf{0}$ as $t \to \infty$.

Solution: First we write the equations governing the error dynamics. Subtracting (2) from (1) yields

$$\dot{e}_1 = \sigma(e_2 - e_1)$$
$$\dot{e}_2 = -e_2 - 20u(t)e_3$$
$$\dot{e}_3 = 5u(t)e_2 - be_3$$

This is a linear system for $\mathbf{e}(t)$, but it has a chaotic time-dependent coefficient $u(t)$ in two terms. The idea is to construct a Liapunov function in such a way that *the chaos cancels out*. Here's how: Multiply the second equation by e_2 and the third by $4e_3$ and add. Then

$$\begin{aligned} e_2\dot{e}_2 + 4e_3\dot{e}_3 &= -e_2^2 - 20u(t)e_2e_3 + 20u(t)e_2e_3 - 4be_3^2 \\ &= -e_2^2 - 4be_3^2 \end{aligned} \tag{3}$$

and so the chaotic term disappears!

The left-hand side of (3) is $\frac{1}{2}\frac{d}{dt}\left(e_2^2 + 4e_3^2\right)$. This suggests the form of a Liapunov function. As in Cuomo and Oppenheim (1992), we define the function

$$E(\mathbf{e}, t) = \tfrac{1}{2}\left(\tfrac{1}{\sigma}e_1^2 + e_2^2 + 4e_3^2\right).$$

E is certainly positive definite, since it is a sum of squares (as always, we assume $\sigma > 0$). To show E is a Liapunov function, we must show it decreases along trajectories. We've already computed the time-derivative of the second two terms, so concentrate on the first term, shown in brackets below:

$$\begin{aligned} \dot{E} &= \left[\tfrac{1}{\sigma}e_1\dot{e}_1\right] + e_2\dot{e}_2 + 4e_3\dot{e}_3 \\ &= -\left[e_1^2 - e_1e_2\right] - e_2^2 - 4be_3^2. \end{aligned}$$

Now complete the square for the term in brackets:

$$\begin{aligned} \dot{E} &= -\left[e_1 - \tfrac{1}{2}e_2\right]^2 + \left(\tfrac{1}{2}e_2\right)^2 - e_2^2 - 4be_3^2 \\ &= -\left[e_1 - \tfrac{1}{2}e_2\right]^2 - \tfrac{3}{4}e_2^2 - 4be_3^2. \end{aligned}$$

Hence $\dot{E} \le 0$, with equality only if $\mathbf{e} = \mathbf{0}$. Therefore E is a Liapunov function, and so $\mathbf{e} = \mathbf{0}$ is globally asymptotically stable. ■

A stronger result is possible: one can show that $\mathbf{e}(t)$ decays *exponentially fast* (Cuomo, Oppenheim, and Strogatz 1993; see Exercise 9.6.1). This is important, because rapid synchronization is necessary for the desired application.

We should be clear about what we have and haven't proven. Example 9.6.1 shows only that the receiver will synchronize to the transmitter if the drive signal is $u(t)$. This does *not* prove that the signal-masking approach will work. For that application, the drive is a mixture $u(t) + m(t)$ where $m(t)$ is the message and

$u(t) >> m(t)$ is the mask. We have no proof that the receiver will regenerate $u(t)$ precisely. In fact, it doesn't—that's why Mariah Carey sounded a little fuzzy. So it's still something of a mathematical mystery as to why the approach works as well as it does. But the proof is in the listening!

EXERCISES FOR CHAPTER 9

9.1 A Chaotic Waterwheel

9.1.1 (Waterwheel's moment of inertia approaches a constant) For the waterwheel of Section 9.1, show that $I(t) \to \text{constant}$ as $t \to \infty$, as follows:

a) The total moment of inertia is a sum $I = I_{\text{wheel}} + I_{\text{water}}$, where I_{wheel} depends only on the apparatus itself, and not on the distribution of water around the rim. Express I_{water} in terms of $M = \int_0^{2\pi} m(\theta,t)\,d\theta$.

b) Show that M satisfies $\dot{M} = Q_{\text{total}} - KM$, where $Q_{\text{total}} = \int_0^{2\pi} Q(\theta)\,d\theta$.

c) Show that $I(t) \to \text{constant}$ as $t \to \infty$, and find the value of the constant.

9.1.2 (Behavior of higher modes) In the text, we showed that three of the waterwheel equations decoupled from all the rest. How do the remaining modes behave?

a) If $Q(\theta) = q_1 \cos\theta$, the answer is simple: show that for $n \neq 1$, all modes $a_n, b_n \to 0$ as $t \to \infty$.

b) What do you think happens for a more general $Q(\theta) = \sum_{n=0}^{\infty} q_n \cos n\theta$?

Part (b) is challenging; see how far you can get. For the state of current knowledge, see Kolar and Gumbs (1992).

9.1.3 (Deriving the Lorenz equations from the waterwheel) Find a change of variables that converts the waterwheel equations

$$\dot{a}_1 = \omega b_1 - K a_1$$
$$\dot{b}_1 = -\omega a_1 + q_1 - K b_1$$
$$\dot{\omega} = -\frac{\nu}{I}\omega + \frac{\pi g r}{I} a_1$$

into the Lorenz equations

$$\dot{x} = \sigma(y - x)$$
$$\dot{y} = rx - xz - y$$
$$\dot{z} = xy - bz$$

where $\sigma, b, r > 0$ are parameters. (This can turn into a messy calculation—it helps to be thoughtful and systematic. You should find that x is like ω, y is like a_1, and z is like b_1.) Also, show that when the waterwheel equations are translated into the Lorenz equations, the Lorenz parameter b turns out to be $b = 1$. (So the waterwheel equations are not quite as general as the Lorenz equations.) Express the Prandtl and Rayleigh numbers σ and r in terms of the waterwheel parameters.

9.1.4 (Laser model) As mentioned in Exercise 3.3.2, the Maxwell–Bloch equations for a laser are

$$\dot{E} = \kappa(P - E)$$
$$\dot{P} = \gamma_1(ED - P)$$
$$\dot{D} = \gamma_2(\lambda + 1 - D - \lambda EP).$$

a) Show that the non-lasing state (the fixed point with $E^* = 0$) loses stability above a threshold value of λ, to be determined. Classify the bifurcation at this laser threshold.

b) Find a change of variables that transforms the system into the Lorenz system.

The Lorenz equations also arise in models of geomagnetic dynamos (Robbins 1977) and thermoconvection in a circular tube (Malkus 1972). See Jackson (1990, vol. 2, Sections 7.5 and 7.6) for an introduction to these systems.

9.1.5 (Research project on asymmetric waterwheel) Our derivation of the waterwheel equations assumed that the water is pumped in symmetrically at the top. Investigate the *asymmetric* case. Modify $Q(\theta)$ in (9.1.5) appropriately. Show that a closed set of three equations is still obtained, but that (9.1.9) includes a new term. Redo as much of the analysis in this chapter as possible. You should be able to solve for the fixed points and show that the pitchfork bifurcation is replaced by an imperfect bifurcation (Section 3.6). After that, you're on your own! This problem has not yet been addressed in the literature.

9.2 Simple Properties of the Lorenz Equations

9.2.1 (Parameter where Hopf bifurcation occurs)

a) For the Lorenz equations, show that the characteristic equation for the eigenvalues of the Jacobian matrix at C^+, C^- is

$$\lambda^3 + (\sigma + b + 1)\lambda^2 + (r + \sigma)b\lambda + 2b\sigma(r - 1) = 0.$$

b) By seeking solutions of the form $\lambda = i\omega$, where ω is real, show that there is a pair of pure imaginary eigenvalues when $r = r_H = \sigma\left(\frac{\sigma + b + 3}{\sigma - b - 1}\right)$. Explain why we need to assume $\sigma > b + 1$.

c) Find the third eigenvalue.

9.2.2 (An ellipsoidal trapping region for the Lorenz equations) Show that there is a certain ellipsoidal region E of the form $rx^2 + \sigma y^2 + \sigma(z-2r)^2 \le C$ such that all trajectories of the Lorenz equations eventually enter E and stay in there forever. For a much stiffer challenge, try to obtain the smallest possible value of C with this property.

9.2.3 (A spherical trapping region) Show that all trajectories eventually enter and remain inside a large sphere S of the form $x^2 + y^2 + (z-r-\sigma)^2 = C$, for C sufficiently large. (Hint: Show that $x^2 + y^2 + (z-r-\sigma)^2$ decreases along trajectories for all (x, y, z) outside a certain fixed ellipsoid. Then pick C large enough so that the sphere S encloses this ellipsoid.)

9.2.4 (z-axis is invariant) Show that the z-axis is an invariant line for the Lorenz equations. In other words, a trajectory that starts on the z-axis stays on it forever.

9.2.5 (Stability diagram) Using the analytical results obtained about bifurcations in the Lorenz equations, give a partial sketch of the stability diagram. Specifically, assume $b = 1$ as in the waterwheel, and then plot the pitchfork and Hopf bifurcation curves in the (σ, r) parameter plane. As always, assume $\sigma, r \ge 0$. (For a numerical computation of the stability diagram, including chaotic regions, see Kolar and Gumbs (1992).)

9.2.6 (Rikitake model of geomagnetic reversals) Consider the system

$$\dot{x} = -\nu x + zy$$
$$\dot{y} = -\nu y + (z-a)x$$
$$\dot{z} = 1 - xy$$

where $a, \nu > 0$ are parameters.

a) Show that the system is dissipative.

b) Show that the fixed points may be written in parametric form as $x^* = \pm k$, $y^* = \pm k^{-1}$, $z^* = \nu k^2$, where $\nu(k^2 - k^{-2}) = a$.

c) Classify the fixed points.

These equations were proposed by Rikitake (1958) as a model for the self-generation of the Earth's magnetic field by large current-carrying eddies in the core. Computer experiments show that the model exhibits chaotic solutions for some parameter values. These solutions are loosely analogous to the irregular reversals of the Earth's magnetic field inferred from geological data. See Cox (1982) for the geophysical background.

9.3 Chaos on a Strange Attractor

9.3.1 (Quasiperiodicity ≠ chaos) The trajectories of the quasiperiodic system $\dot{\theta}_1 = \omega_1$, $\dot{\theta}_2 = \omega_2$, ($\omega_1/\omega_2$ irrational) are not periodic.

a) Why isn't this system considered chaotic?
b) Without using a computer, find the largest Liapunov exponent for the system.

(Numerical experiments) For each of the values of r given below, use a computer to explore the dynamics of the Lorenz system, assuming $\sigma = 10$ and $b = 8/3$ as usual. In each case, plot $x(t)$, $y(t)$, and x vs. z. You should investigate the consequences of choosing different initial conditions and lengths of integration. Also, in some cases you may want to ignore the transient behavior, and plot only the sustained long-term behavior.

9.3.2 $r = 10$

9.3.3 $r = 22$ (transient chaos)

9.3.4 $r = 24.5$ (chaos and stable point co-exist)

9.3.5 $r = 100$ (surprise)

9.3.6 $r = 126.52$

9.3.7 $r = 400$

9.3.8 (Practice with the definition of an attractor) Consider the following familiar system in polar coordinates: $\dot{r} = r(1 - r^2)$, $\dot{\theta} = 1$. Let D be the disk $x^2 + y^2 \le 1$.
a) Is D an invariant set?
b) Does D attract an open set of initial conditions?
c) Is D an attractor? If not, why not? If so, find its basin of attraction.
d) Repeat part (c) for the circle $x^2 + y^2 = 1$.

9.3.9 (Exponential divergence) Using numerical integration of two nearby trajectories, estimate the largest Liapunov exponent for the Lorenz system, assuming that the parameters have their standard values $r = 28$, $\sigma = 10$, $b = 8/3$.

9.3.10 (Time horizon) To illustrate the "time horizon" after which prediction becomes impossible, numerically integrate the Lorenz equations for $r = 28$, $\sigma = 10$, $b = 8/3$. Start two trajectories from nearby initial conditions, and plot $x(t)$ for both of them on the same graph.

9.4 Lorenz Map

9.4.1 (Computer work) Using numerical integration, compute the Lorenz map for $r = 28$, $\sigma = 10$, $b = 8/3$.

9.4.2 (Tent map, as model of Lorenz map) Consider the map

$$x_{n+1} = \begin{cases} 2x_n, & 0 \le x_n \le \frac{1}{2} \\ 2 - 2x_n, & \frac{1}{2} \le x_n \le 1 \end{cases}$$

as a simple analytical model of the Lorenz map.
a) Why is it called the "tent map"?
b) Find all the fixed points, and classify their stability.
c) Show that the map has a period-2 orbit. Is it stable or unstable?

d) Can you find any period-3 points? How about period-4? If so, are the corresponding periodic orbits stable or unstable?

9.5 Exploring Parameter Space

(Numerical experiments) For each of the values of r given below, use a computer to explore the dynamics of the Lorenz system, assuming $\sigma = 10$ and $b = 8/3$ as usual. In each case, plot $x(t)$, $y(t)$, and x vs. z.

9.5.1 $r = 166.3$ (intermittent chaos)

9.5.2 $r = 212$ (noisy periodicity)

9.5.3 the interval $145 < r < 166$ (period-doubling)

9.5.4 (Hysteresis between a fixed point and a strange attractor) Consider the Lorenz equations with $\sigma = 10$ and $b = 8/3$. Suppose that we slowly "turn the r knob" up and down. Specifically, let $r = 24.4 + \sin \omega t$, where ω is small compared to typical orbital frequencies on the attractor. Numerically integrate the equations, and plot the solutions in whatever way seems most revealing. You should see a striking hysteresis effect between an equilibrium and a chaotic state.

9.5.5 (Lorenz equations for large r) Consider the Lorenz equations in the limit $r \to \infty$. By taking the limit in a certain way, all the dissipative terms in the equations can be removed (Robbins 1979, Sparrow 1982).

a) Let $\varepsilon = r^{-1/2}$, so that $r \to \infty$ corresponds to $\varepsilon \to 0$. Find a change of variables involving ε such that as $\varepsilon \to 0$, the equations become

$$X' = Y$$
$$Y' = -XZ$$
$$Z' = XY.$$

b) Find two conserved quantities (i.e., constants of the motion) for the new system.
c) Show that the new system is volume-preserving (i.e., the volume of an arbitrary blob of "phase fluid" is conserved by the time-evolution of the system, even though the shape of the blob may change dramatically.)
d) Explain physically why the Lorenz equations might be expected to show some conservative features in the limit $r \to \infty$.
e) Solve the system in part (a) numerically. What is the long-term behavior? Does it agree with the behavior seen in the Lorenz equations for large r?

9.5.6 (Transient chaos) Example 9.5.1 shows that the Lorenz system can exhibit transient chaos for $r = 21$, $\sigma = 10$, $b = \frac{8}{3}$. However, not all trajectories behave this way. Using numerical integration, find three different initial conditions for which there *is* transient chaos, and three others for which there *isn't*. Give a rule of thumb which predicts whether an initial condition will lead to transient chaos or not.

9.6 Using Chaos to Send Secret Messages

9.6.1 (Exponentially fast synchronization) The Liapunov function of Example 9.6.1 shows that the synchronization error $\mathbf{e}(t)$ tends to zero as $t \to \infty$, but it does not provide information about the rate of convergence. Sharpen the argument to show that the synchronization error $\mathbf{e}(t)$ decays exponentially fast.

a) Prove that $V = \frac{1}{2}e_2{}^2 + 2e_3{}^2$ decays exponentially fast, by showing $\dot{V} \le -kV$, for some constant $k > 0$ to be determined.

b) Show that part (a) implies that $e_2(t)$, $e_3(t) \to 0$ exponentially fast.

c) Finally show that $e_1(t) \to 0$ exponentially fast.

9.6.2 (Pecora and Carroll's approach) In the pioneering work of Pecora and Carroll (1990), one of the receiver variables is simply set *equal to* the corresponding transmitter variable. For instance, if $x(t)$ is used as the transmitter drive signal, then the receiver equations are

$$x_r(t) = x(t)$$
$$\dot{y}_r = rx(t) - y_r - x(t)z_r$$
$$\dot{z}_r = x(t)y_r - bz_r$$

where the first equation is *not* a differential equation. Their numerical simulations and a heuristic argument suggested that $y_r(t) \to y(t)$ and $z_r(t) \to z(t)$ as $t \to \infty$, even if there were differences in the initial conditions.

Here is a simple proof of that result, due to He and Vaidya (1992).

a) Show that the error dynamics are

$$e_1 \equiv 0$$
$$\dot{e}_2 = -e_2 - x(t)e_3$$
$$\dot{e}_3 = x(t)e_2 - be_3$$

where $e_1 = x - x_r$, $e_2 = y - y_r$, and $e_3 = z - z_r$.

b) Show that $V = e_2^2 + e_3^2$ is a Liapunov function.

c) What do you conclude?

9.6.3 (Computer experiments on synchronized chaos) Let x, y, z be governed by the Lorenz equations with $r = 60$, $\sigma = 10$, $b = 8/3$. Let x_r, y_r, z_r be governed by the system in Exercise 9.6.2. Choose different initial conditions for y and y_r, and similarly for z and z_r, and then start integrating numerically.

a) Plot $y(t)$ and $y_r(t)$ on the same graph. With any luck, the two time series should eventually merge, even though both are chaotic.

b) Plot the (y, z) projection of both trajectories.

9.6.4 (Some drives don't work) Suppose $z(t)$ were the drive signal in Exercise 9.6.2, instead of $x(t)$. In other words, we replace z_r by $z(t)$ everywhere in the receiver equations, and watch how x_r and y_r evolve.

a) Show numerically that the receiver does *not* synchronize in this case.

b) What if $y(t)$ were the drive?

9.6.5 (Masking) In their signal-masking approach, Cuomo and Oppenheim (1992, 1993) use the following receiver dynamics:

$$\dot{x}_r = \sigma(y_r - x_r)$$
$$\dot{y}_r = rs(t) - y_r - s(t)z_r$$
$$\dot{z}_r = s(t)y_r - bz_r$$

where $s(t) = x(t) + m(t)$, and $m(t)$ is the low-power message added to the much stronger chaotic mask $x(t)$. If the receiver has synchronized with the drive, then $x_r(t) \approx x(t)$ and so $m(t)$ may be recovered as $\hat{m}(t) = s(t) - x_r(t)$. Test this approach numerically, using a sine wave for $m(t)$. How close is the estimate $\hat{m}(t)$ to the actual message $m(t)$? How does the error depend on the frequency of the sine wave?

9.6.6 (Lorenz circuit) Derive the circuit equations for the transmitter circuit shown in Figures 9.6.1.

10

ONE-DIMENSIONAL MAPS

10.0 Introduction

This chapter deals with a new class of dynamical systems in which time is *discrete*, rather than continuous. These systems are known variously as difference equations, recursion relations, iterated maps, or simply ***maps***.

For instance, suppose you repeatedly press the cosine button on your calculator, starting from some number x_0. Then the successive readouts are $x_1 = \cos x_0$, $x_2 = \cos x_1$, and so on. Set your calculator to radian mode and try it. Can you explain the surprising result that emerges after many iterations?

The rule $x_{n+1} = \cos x_n$ is an example of a ***one-dimensional map***, so-called because the points x_n belong to the one-dimensional space of real numbers. The sequence $x_0, x_1, x_2, \ldots$ is called the ***orbit*** starting from x_0.

Maps arise in various ways:

1. *As tools for analyzing differential equations.* We have already encountered maps in this role. For instance, Poincaré maps allowed us to prove the existence of a periodic solution for the driven pendulum and Josephson junction (Section 8.5), and to analyze the stability of periodic solutions in general (Section 8.7). The Lorenz map (Section 9.4) provided strong evidence that the Lorenz attractor is truly strange, and is not just a long-period limit cycle.
2. *As models of natural phenomena.* In some scientific contexts it is natural to regard time as discrete. This is the case in digital electronics, in parts of economics and finance theory, in impulsively driven mechanical systems, and in the study of certain animal populations where successive generations do not overlap.
3. *As simple examples of chaos.* Maps are interesting to study in their own right, as mathematical laboratories for chaos. Indeed, maps are capable

of much wilder behavior than differential equations because the points x_n *hop* along their orbits rather than flow continuously (Figure 10.0.1).

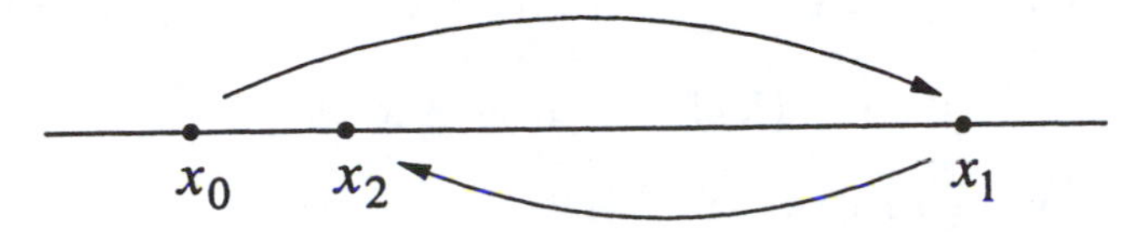

Figure 10.0.1

The study of maps is still in its infancy, but exciting progress has been made in the last twenty years, thanks to the growing availability of calculators, then computers, and now computer graphics. Maps are easy and fast to simulate on digital computers where time is *inherently* discrete. Such computer experiments have revealed a number of unexpected and beautiful patterns, which in turn have stimulated new theoretical developments. Most surprisingly, maps have generated a number of successful predictions about the routes to chaos in semiconductors, convecting fluids, heart cells, lasers, and chemical oscillators.

We discuss some of the properties of maps and the techniques for analyzing them in Sections 10.1–10.5. The emphasis is on period-doubling and chaos in the logistic map. Section 10.6 introduces the amazing idea of universality, and summarizes experimental tests of the theory. Section 10.7 is an attempt to convey the basic ideas of Feigenbaum's renormalization technique.

As usual, our approach will be intuitive. For rigorous treatments of one-dimensional maps, see Devaney (1989) and Collet and Eckmann (1980).

10.1 Fixed Points and Cobwebs

In this section we develop some tools for analyzing one-dimensional maps of the form $x_{n+1} = f(x_n)$, where f is a smooth function from the real line to itself.

A Pedantic Point

When we say "map," do we mean the function f or the difference equation $x_{n+1} = f(x_n)$? Following common usage, we'll call *both* of them maps. If you're disturbed by this, you must be a pure mathematician . . . or should consider becoming one!

Fixed Points and Linear Stability

Suppose x^* satisfies $f(x^*) = x^*$. Then x^* is a ***fixed point***, for if $x_n = x^*$ then $x_{n+1} = f(x_n) = f(x^*) = x^*$; hence the orbit remains at x^* for all future iterations.

To determine the stability of x^*, we consider a nearby orbit $x_n = x^* + \eta_n$ and ask whether the orbit is attracted to or repelled from x^*. That is, does the deviation η_n grow or decay as n increases? Substitution yields

$$x^* + \eta_{n+1} = x_{n+1} = f(x^* + \eta_n) = f(x^*) + f'(x^*)\eta_n + O(\eta_n^2).$$

But since $f(x^*) = x^*$, this equation reduces to

$$\eta_{n+1} = f'(x^*)\eta_n + O(\eta_n^2).$$

Suppose we can safely neglect the $O(\eta_n^2)$ terms. Then we obtain the *linearized map* $\eta_{n+1} = f'(x^*)\eta_n$ with *eigenvalue* or ***multiplier*** $\lambda = f'(x^*)$. The solution of this linear map can be found explicitly by writing a few terms: $\eta_1 = \lambda\eta_0$, $\eta_2 = \lambda\eta_1 = \lambda^2\eta_0$, and so in general $\eta_n = \lambda^n\eta_0$. If $|\lambda| = |f'(x^*)| < 1$, then $\eta_n \to 0$ as $n \to \infty$ and the fixed point x^* is ***linearly stable***. Conversely, if $|f'(x^*)| > 1$ the fixed point is ***unstable***. Although these conclusions about local stability are based on linearization, they can be proven to hold for the original nonlinear map. But the linearization tells us nothing about the ***marginal*** case $|f'(x^*)| = 1$; then the neglected $O(\eta_n^2)$ terms determine the local stability. (All of these results have parallels for differential equations—recall Section 2.4.)

EXAMPLE 10.1.1:

Find the fixed points for the map $x_{n+1} = x_n^2$ and determine their stability.

Solution: The fixed points satisfy $x^* = (x^*)^2$. Hence $x^* = 0$ or $x^* = 1$. The multiplier is $\lambda = f'(x^*) = 2x^*$. The fixed point $x^* = 0$ is stable since $|\lambda| = 0 < 1$, and $x^* = 1$ is unstable since $|\lambda| = 2 > 1$. ■

Try Example 10.1.1 on a hand calculator by pressing the x^2 button over and over. You'll see that for sufficiently small x_0, the convergence to $x^* = 0$ is *extremely* rapid. Fixed points with multiplier $\lambda = 0$ are called ***superstable*** because perturbations decay like $\eta_n \sim \eta_0^{(2^n)}$, which is much faster than the usual $\eta_n \sim \lambda^n\eta_0$ at an ordinary stable point.

Cobwebs

In Section 8.7 we introduced the ***cobweb*** construction for iterating a map (Figure 10.1.1).

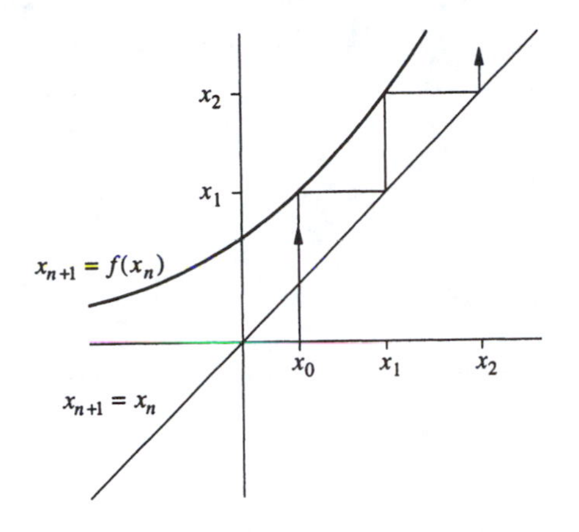

Figure 10.1.1

Given $x_{n+1} = f(x_n)$ and an initial condition x_0, draw a vertical line until it intersects the graph of f; that height is the output x_1. At this stage we could return to the horizontal axis and repeat the procedure to get x_2 from x_1, but it is more convenient simply to trace a horizontal line till it intersects the diagonal line $x_{n+1} = x_n$, and then move vertically to the curve again. Repeat the process n times to generate the first n points in the orbit.

Cobwebs are useful because they allow us to see global behavior at a glance, thereby supplementing the local information available from the linearization. Cobwebs become even more valuable when linear analysis fails, as in the next example.

EXAMPLE 10.1.2:

Consider the map $x_{n+1} = \sin x_n$. Show that the stability of the fixed point $x^* = 0$ is not determined by the linearization. Then use a cobweb to show that $x^* = 0$ is stable—in fact, *globally* stable.

Solution: The multiplier at $x^* = 0$ is $f'(0) = \cos(0) = 1$, which is a marginal case where linear analysis is inconclusive. However, the cobweb of Figure 10.1.2 shows that $x^* = 0$ is locally stable; the orbit slowly rattles down the narrow channel, and heads monotonically for the fixed point. (A similar picture is obtained for $x_0 < 0$.)

To see that the stability is global, we have to show that *all* orbits satisfy $x_n \to 0$. But for any x_0, the first iterate is sent immediately to the interval $-1 \le x_1 \le 1$ since $|\sin x| \le 1$. The cobweb in that interval looks qualitatively like Figure 10.1.2, so convergence is assured. ■

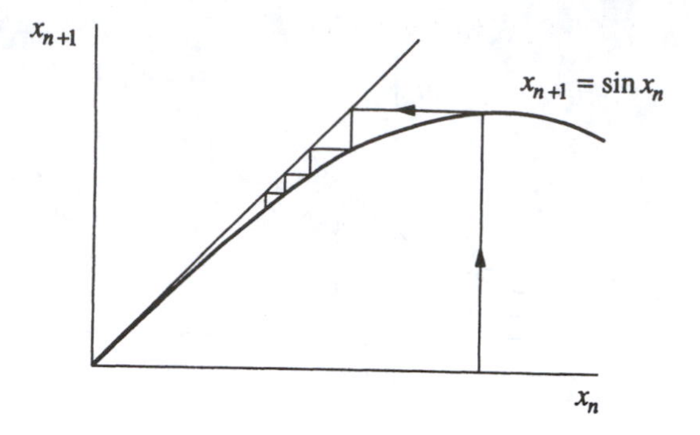

Figure 10.1.2

Finally, let's answer the riddle posed in Section 10.0.

EXAMPLE 10.1.3:

Given $x_{n+1} = \cos x_n$, how does x_n behave as $n \to \infty$?

Solution: If you tried this on your calculator, you found that $x_n \to 0.739\ldots$, no matter where you started. What is this bizarre number? It's the unique solution of the transcendental equation $x = \cos x$, and it corresponds to a fixed point of the map. Figure 10.1.3 shows that a typical orbit spirals into the fixed point $x^* = 0.739\ldots$ as $n \to \infty$. ■

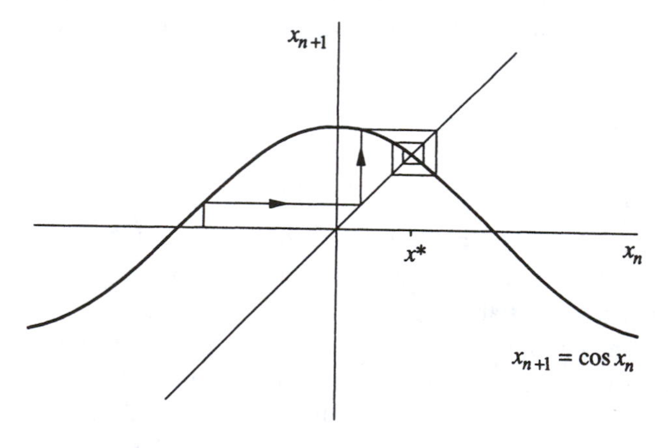

Figure 10.1.3

The spiraling motion implies that x_n converges to x^* through *damped oscillations*. That is characteristic of fixed points with $\lambda < 0$. In contrast, at stable fixed points with $\lambda > 0$ the convergence is monotonic.

10.2 Logistic Map: Numerics

In a fascinating and influential review article, Robert May (1976) emphasized that even simple nonlinear maps could have very complicated dynamics. The article ends memorably with "an evangelical plea for the introduction of these difference equations into elementary mathematics courses, so that students' intuition may be enriched by seeing the wild things that simple nonlinear equations can do."

May illustrated his point with the ***logistic map***

$$x_{n+1} = rx_n(1-x_n), \tag{1}$$

a discrete-time analog of the logistic equation for population growth (Section 2.3). Here $x_n \geq 0$ is a dimensionless measure of the population in the nth generation and $r \geq 0$ is the intrinsic growth rate. As shown in Figure 10.2.1, the graph of (1) is a parabola with a maximum value of $r/4$ at $x = \frac{1}{2}$. We restrict the control parameter r to the range $0 \leq r \leq 4$ so that (1) maps the interval $0 \leq x \leq 1$ into itself. (The behavior is much less interesting for other values of x and r—see Exercise 10.2.1.)

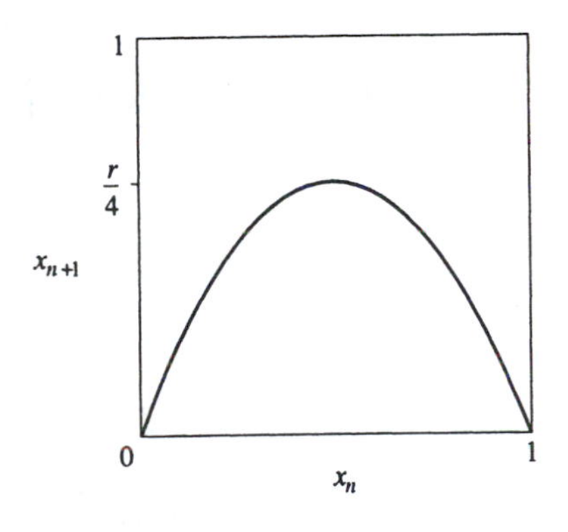

Figure 10.2.1

Period-Doubling

Suppose we fix r, choose some initial population x_0, and then use (1) to generate the subsequent x_n. What happens?

For small growth rate $r < 1$, the population always goes extinct: $x_n \to 0$ as $n \to \infty$. This gloomy result can be proven by cobwebbing (Exercise 10.2.2).

For $1 < r < 3$ the population grows and eventually reaches a nonzero steady state (Figure 10.2.2). The results are plotted here as a ***time series*** of x_n vs. n. To make the sequence clearer, we have connected the discrete points (n, x_n) by line segments, but remember that only the corners of the jagged curves are meaningful.

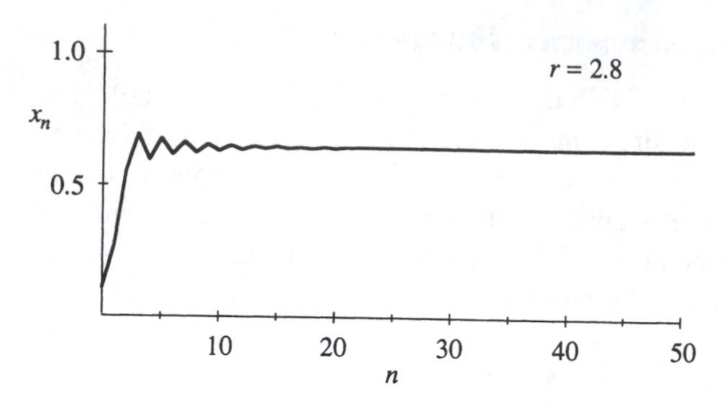

Figure 10.2.2

For larger r, say $r = 3.3$, the population builds up again but now *oscillates* about the former steady state, alternating between a large population in one generation and a smaller population in the next (Figure 10.2.3). This type of oscillation, in which x_n repeats every *two* iterations, is called a ***period-2 cycle***.

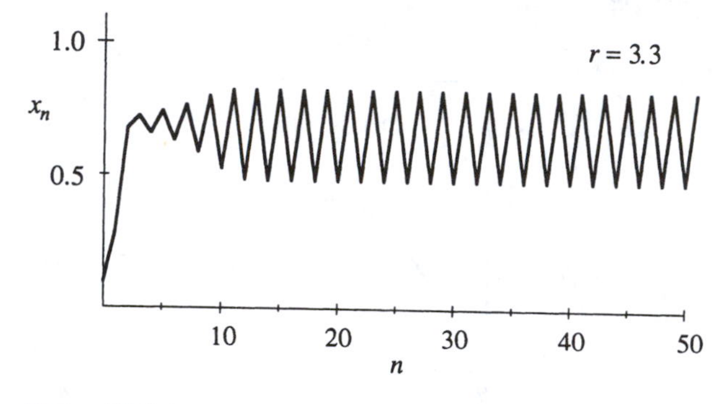

Figure 10.2.3

At still larger r, say $r = 3.5$, the population approaches a cycle that now repeats every *four* generations; the previous cycle has doubled its period to ***period-4*** (Figure 10.2.4).

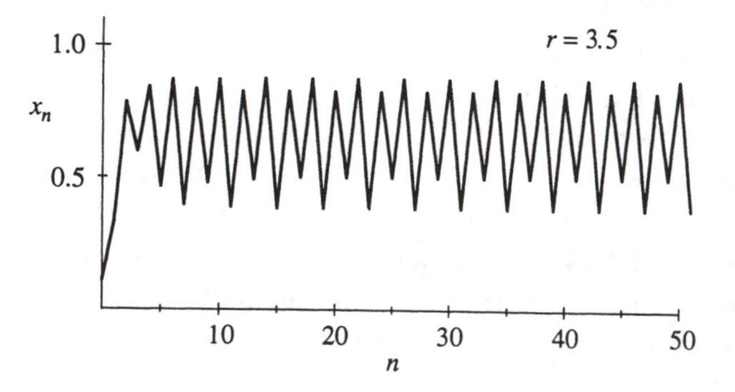

Figure 10.2.4

Further ***period-doublings*** to cycles of period 8, 16, 32, . . . , occur as r increases. Specifically, let r_n denote the value of r where a 2^n-cycle first appears. Then computer experiments reveal that

$r_1 = 3$	(period 2 is born)
$r_2 = 3.449...$	4
$r_3 = 3.54409...$	8
$r_4 = 3.5644...$	16
$r_5 = 3.568759...$	32
$\vdots$	$\vdots$
$r_\infty = 3.569946...$	∞

Note that the successive bifurcations come faster and faster. Ultimately the r_n converge to a limiting value r_∞. The convergence is essentially geometric: in the limit of large n, the distance between successive transitions shrinks by a constant factor

$$\delta = \lim_{n\to\infty} \frac{r_n - r_{n-1}}{r_{n+1} - r_n} = 4.669\ldots.$$

We'll have a lot more to say about this number in Section 10.6.

Chaos and Periodic Windows

According to Gleick (1987, p. 69), May wrote the logistic map on a corridor blackboard as a problem for his graduate students and asked, "*What the Christ happens for* $r > r_\infty$?" The answer turns out to be complicated: For many values of r, the sequence $\{x_n\}$ never settles down to a fixed point or a periodic orbit—instead the long-term behavior is aperiodic, as in Figure 10.2.5. This is a discrete-time version of the chaos we encountered earlier in our study of the Lorenz equations (Chapter 9).

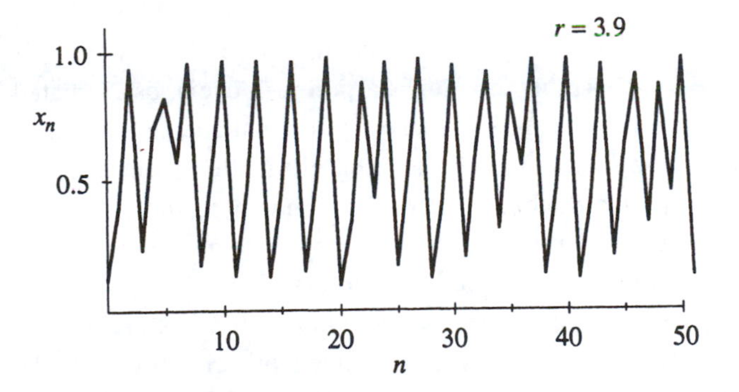

Figure 10.2.5

The corresponding cobweb diagram is impressively complex (Figure 10.2.6).

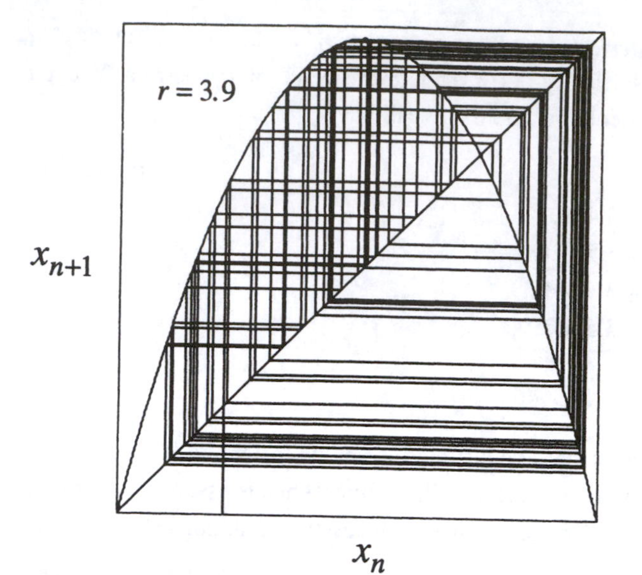

Figure 10.2.6

You might guess that the system would become more and more chaotic as r increases, but in fact the dynamics are more subtle than that. To see the long-term behavior for *all* values of r at once, we plot the ***orbit diagram***, a magnificent picture that has become an icon of nonlinear dynamics (Figure 10.2.7). Figure 10.2.7 plots the system's attractor as a function of r. To generate the orbit diagram for yourself, you'll need to write a computer program with two "loops." First, choose a value of r. Then generate an orbit starting from some random initial condition x_0. Iterate for 300 cycles or so, to allow the system to settle down to its eventual behavior. Once the transients have decayed, plot many points, say $x_{301}, \ldots, x_{600}$ above that r. Then move to an adjacent value of r and repeat, eventually sweeping across the whole picture.

Figure 10.2.7 shows the most interesting part of the diagram, in the region $3.4 \le r \le 4$. At $r = 3.4$, the attractor is a period-2 cycle, as indicated by the two branches. As r increases, both branches split simultaneously, yielding a period-4 cycle. This splitting is the period-doubling bifurcation mentioned earlier. A cascade of further period-doublings occurs as r increases, yielding period-8, period-16, and so on, until at $r = r_\infty \approx 3.57$, the map becomes chaotic and the attractor changes from a finite to an infinite set of points.

For $r > r_\infty$ the orbit diagram reveals an unexpected mixture of order and chaos, with ***periodic windows*** interspersed between chaotic clouds of dots. The large window beginning near $r \approx 3.83$ contains a stable period-3 cycle. A blow-up of part of the period-3 window is shown in the lower panel of Figure 10.2.7. Fantastically, a copy of the orbit diagram reappears in miniature!

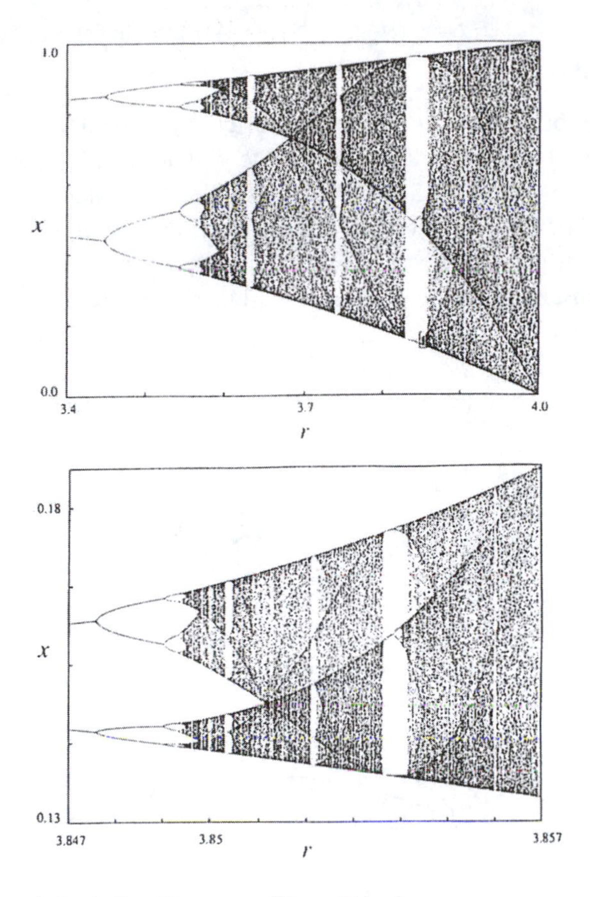

Figure 10.2.7 Campbell (1979), p. 35, courtesy of Roger Eckhardt

10.3 Logistic Map: Analysis

The numerical results of the last section raise many tantalizing questions. Let's try to answer a few of the more straightforward ones.

EXAMPLE 10.3.1:

Consider the logistic map $x_{n+1} = rx_n(1 - x_n)$ for $0 \le x_n \le 1$ and $0 \le r \le 4$. Find all the fixed points and determine their stability.

Solution: The fixed points satisfy $x^* = f(x^*) = rx^*(1 - x^*)$. Hence $x^* = 0$ or $1 = r(1 - x^*)$, i.e., $x^* = 1 - \frac{1}{r}$. The origin is a fixed point for all r, whereas $x^* = 1 - \frac{1}{r}$ is in the range of allowable x only if $r \ge 1$.

Stability depends on the multiplier $f'(x^*) = r - 2rx^*$. Since $f'(0) = r$, the origin is stable for $r < 1$ and unstable for $r > 1$. At the other fixed point,

$f'(x^*) = r - 2r(1-\frac{1}{r}) = 2 - r$. Hence $x^* = 1 - \frac{1}{r}$ is stable for $-1 < (2-r) < 1$, i.e., for $1 < r < 3$. It is unstable for $r > 3$. ■

The results of Example 10.3.1 are clarified by a graphical analysis (Figure 10.3.1). For $r < 1$ the parabola lies below the diagonal, and the origin is the only fixed point. As r increases, the parabola gets taller, becoming tangent to the diagonal at $r = 1$. For $r > 1$ the parabola intersects the diagonal in a second fixed point $x^* = 1 - \frac{1}{r}$, while the origin loses stability. Thus we see that x^* bifurcates from the origin in a ***transcritical bifurcation*** at $r = 1$ (borrowing a term used earlier for differential equations).

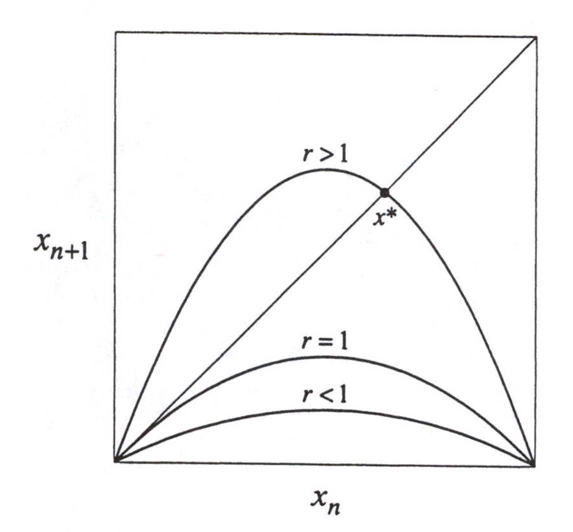

Figure 10.3.1

Figure 10.3.1 also suggests how x^* itself loses stability. As r increases beyond 1, the slope at x^* gets increasingly steep. Example 10.3.1 shows that the critical slope $f'(x^*) = -1$ is attained when $r = 3$. The resulting bifurcation is called a ***flip bifurcation***.

Flip bifurcations are often associated with period-doubling. In the logistic map, the flip bifurcation at $r = 3$ does indeed spawn a 2-cycle, as shown in the next example.

EXAMPLE 10.3.2:

Show that the logistic map has a 2-cycle for all $r > 3$.

Solution: A 2-cycle exists if and only if there are two points p and q such that $f(p) = q$ and $f(q) = p$. Equivalently, such a p must satisfy $f(f(p)) = p$, where $f(x) = rx(1-x)$. Hence p is a fixed point of the ***second-iterate map***

$f^2(x) \equiv f(f(x))$. Since $f(x)$ is a quadratic polynomial, $f^2(x)$ is a *quartic* polynomial. Its graph for $r > 3$ is shown in Figure 10.3.2.

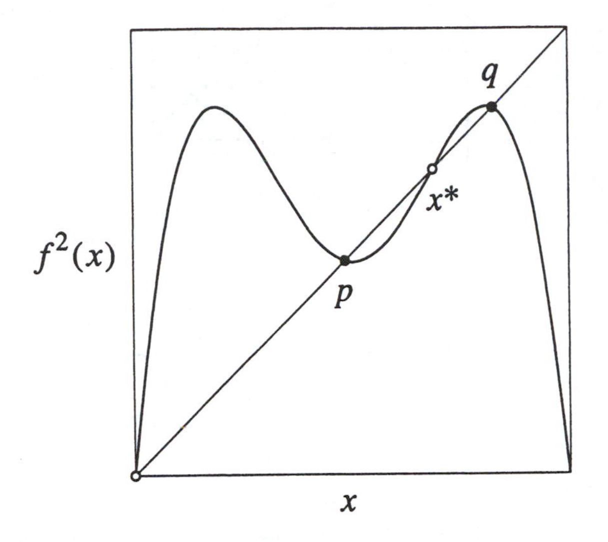

Figure 10.3.2

To find p and q, we need to solve for the points where the graph intersects the diagonal, i.e., we need to solve the fourth-degree equation $f^2(x) = x$. That sounds hard until you realize that the fixed points $x^* = 0$ and $x^* = 1 - \frac{1}{r}$ are trivial solutions of this equation. (They satisfy $f(x^*) = x^*$, so $f^2(x^*) = x^*$ automatically.) After factoring out the fixed points, the problem reduces to solving a quadratic equation.

We outline the algebra involved in the rest of the solution. Expansion of the equation $f^2(x) - x = 0$ gives $r^2 x(1-x)[1 - rx(1-x)] - x = 0$. After factoring out x and $x - (1 - \frac{1}{r})$ by long division, and solving the resulting quadratic equation, we obtain a pair of roots

$$p,\, q = \frac{r + 1 \pm \sqrt{(r-3)(r+1)}}{2r},$$

which are real for $r > 3$. Thus a 2-cycle exists for all $r > 3$, as claimed. At $r = 3$, the roots coincide and equal $x^* = 1 - \frac{1}{r} = \frac{2}{3}$, which shows that the 2-cycle bifurcates *continuously* from x^*. For $r < 3$ the roots are complex, which means that a 2-cycle doesn't exist. ■

A cobweb diagram reveals how flip bifurcations can give rise to period-doubling. Consider any map f, and look at the local picture near a fixed point where $f'(x^*) \approx -1$ (Figure 10.3.3).

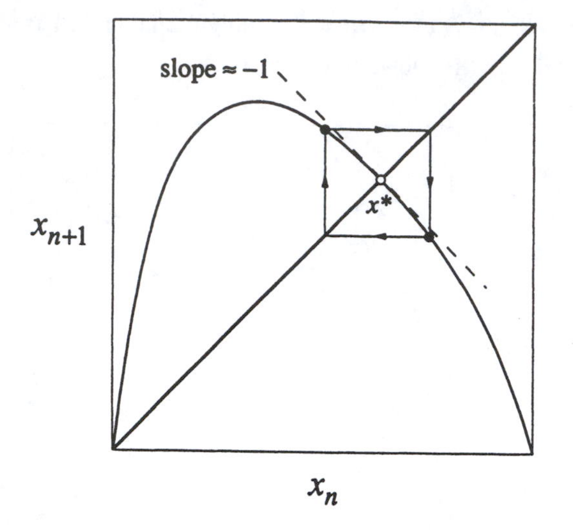

Figure 10.3.3

If the graph of f is concave down near x^*, the cobweb tends to produce a small, stable 2-cycle close to the fixed point. But like pitchfork bifurcations, flip bifurcations can also be subcritical, in which case the 2-cycle exists *below* the bifurcation and is *unstable*—see Exercise 10.3.11.

The next example shows how to determine the stability of a 2-cycle.

EXAMPLE 10.3.3:

Show that the 2-cycle of Example 10.3.2 is stable for $3 < r < 1 + \sqrt{6} = 3.449\ldots$. (This explains the values of r_1 and r_2 found numerically in Section 10.2.)

Solution: Our analysis follows a strategy that is worth remembering: To analyze the stability of a cycle, reduce the problem to a question about the stability of a *fixed point*, as follows. Both p and q are solutions of $f^2(x) = x$, as pointed out in Example 10.3.2; hence p and q are *fixed points of the second-iterate map* $f^2(x)$. The original 2-cycle is stable precisely if p and q are stable fixed points for f^2.

Now we're on familiar ground. To determine whether p is a stable fixed point of f^2, we compute the multiplier

$$\lambda = \frac{d}{dx}\left(f(f(x))\right)_{x=p} = f'(f(p))f'(p) = f'(q)f'(p).$$

(Note that the same λ is obtained at $x = q$, by the symmetry of the final term above. Hence, when the p and q branches bifurcate, they must do so *simultaneously*. We noticed such a simultaneous splitting in our numerical observations of Section 10.2.)

After carrying out the differentiations and substituting for p and q, we obtain

$$\begin{aligned}\lambda &= r(1-2q)\,r(1-2p) \\ &= r^2\left[\,1-2(p+q)+4pq\,\right] \\ &= r^2\left[\,1-2(r+1)/r+4(r+1)/r^2\,\right] \\ &= 4+2r-r^2.\end{aligned}$$

Therefore the 2-cycle is linearly stable for $\left|4+2r-r^2\right|<1$, i.e., for $3<r<1+\sqrt{6}$. ■

Figure 10.3.4 shows a partial ***bifurcation diagram*** for the logistic map, based on our results so far. Bifurcation diagrams are different from orbit diagrams in that *unstable* objects are shown as well; orbit diagrams show only the attractors.

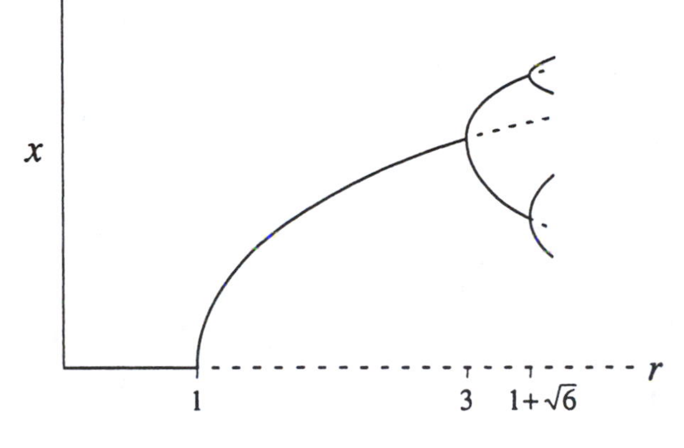

Figure 10.3.4

Our analytical methods are becoming unwieldy. A few more exact results can be obtained (see the exercises), but such results are hard to come by. To elucidate the behavior in the interesting region where $r>r_\infty$, we are going to rely mainly on graphical and numerical arguments.

10.4 Periodic Windows

One of the most intriguing features of the orbit diagram (Figure 10.2.7) is the occurrence of periodic windows for $r>r_\infty$. The period-3 window that occurs near $3.8284\ldots \le r \le 3.8415\ldots$ is the most conspicuous. Suddenly, against a backdrop of chaos, a stable 3-cycle appears out of the blue. Our first goal in this section is to understand how this 3-cycle is created. (The same mechanism accounts for the creation of all the other windows, so it suffices to consider this simplest case.)

First, some notation. Let $f(x)=rx(1-x)$ so that the logistic map is $x_{n+1}=f(x_n)$. Then $x_{n+2}=f(f(x_n))$ or more simply, $x_{n+2}=f^2(x_n)$. Similarly, $x_{n+3}=f^3(x_n)$.

The third-iterate map $f^3(x)$ is the key to understanding the birth of the period-3 cycle. Any point p in a period-3 cycle repeats every three iterates, by definition, so such points satisfy $p = f^3(p)$ and are therefore fixed points of the third-iterate map. Unfortunately, since $f^3(x)$ is an eighth-degree polynomial, we cannot solve for the fixed points explicitly. But a graph provides sufficient insight. Figure 10.4.1 plots $f^3(x)$ for $r = 3.835$.

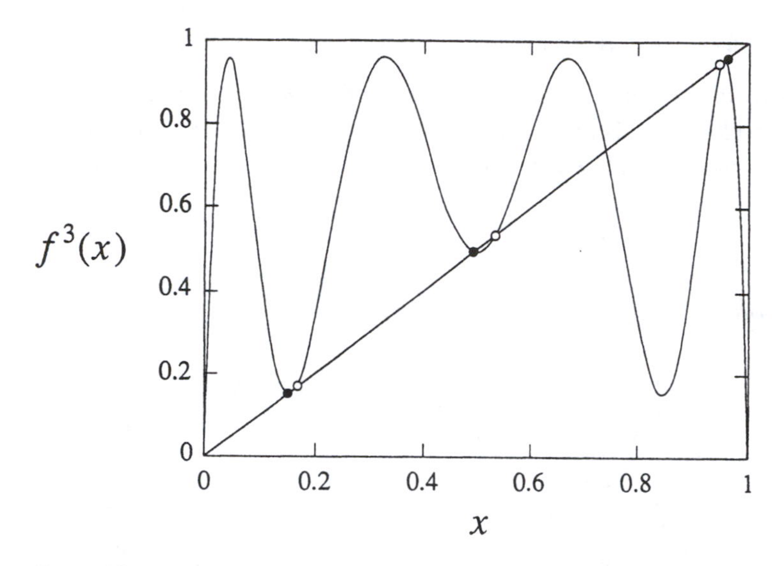

Figure 10.4.1

Intersections between the graph and the diagonal line correspond to solutions of $f^3(x) = x$. There are eight solutions, six of interest to us and marked with dots, and two imposters that are not genuine period-3; they are actually fixed points, or period-1 points for which $f(x^*) = x^*$. The black dots in Figure 10.4.1 correspond to a stable period-3 cycle; note that the slope of $f^3(x)$ is shallow at these points, consistent with the stability of the cycle. In contrast, the slope exceeds 1 at the cycle marked by the open dots; this 3-cycle is therefore unstable.

Now suppose we decrease r toward the chaotic regime. Then the graph in Figure 10.4.1 changes shape—the hills move down and the valleys rise up. The curve therefore pulls away from the diagonal. Figure 10.4.2 shows that when $r = 3.8$, the six marked intersections have vanished. Hence, for some intermediate value between $r = 3.8$ and $r = 3.835$, the graph of $f^3(x)$ must have become *tangent* to the diagonal. At this critical value of r, the stable and unstable period-3 cycles coalesce and annihilate in a ***tangent bifurcation***. This transition defines the beginning of the periodic window.

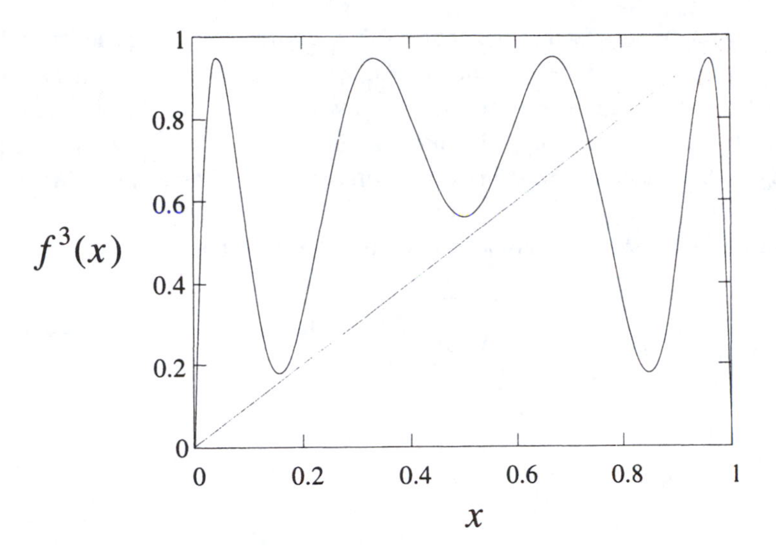

Figure 10.4.2

One can show analytically that the value of r at the tangent bifurcation is $1+\sqrt{8}=3.8284\ldots$ (Myrberg 1958). This beautiful result is often mentioned in textbooks and articles—but always without proof. Given the resemblance of this result to the $1+\sqrt{6}$ encountered in Example 10.3.3, I'd always assumed it should be comparably easy to derive, and once assigned it as a routine homework problem. Oops! It turns out to be a bear. See Exercise 10.4.10 for hints, and Saha and Strogatz (1994) for Partha Saha's solution, the most elementary one my class could find. Maybe you can do better; if so, let me know!

Intermittency

For r just below the period-3 window, the system exhibits an interesting kind of chaos. Figure 10.4.3 shows a typical orbit for $r=3.8282$.

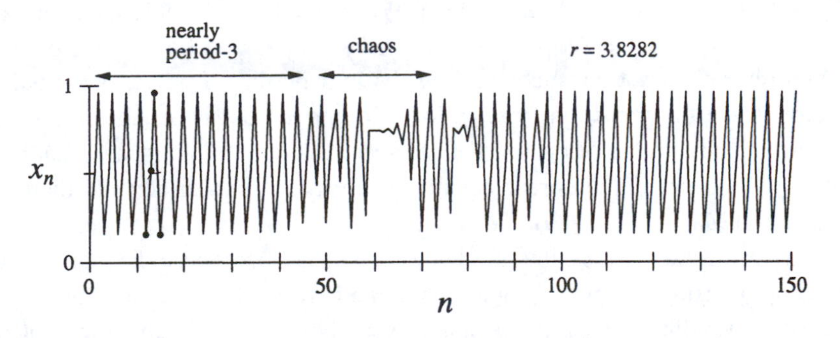

Figure 10.4.3

Part of the orbit looks like a stable 3-cycle, as indicated by the black dots. But this is spooky since the 3-cycle no longer exists! We're seeing the ***ghost*** of the 3-cycle.

We should not be surprised to see ghosts—they *always* occur near saddle-node bifurcations (Sections 4.3 and 8.1) and indeed, a tangent bifurcation is just a saddle-node bifurcation by another name. But the new wrinkle is that the orbit returns to the ghostly 3-cycle repeatedly, with intermittent bouts of chaos between visits. Accordingly, this phenomenon is known as ***intermittency*** (Pomeau and Manneville 1980).

Figure 10.4.4 shows the geometry underlying intermittency.

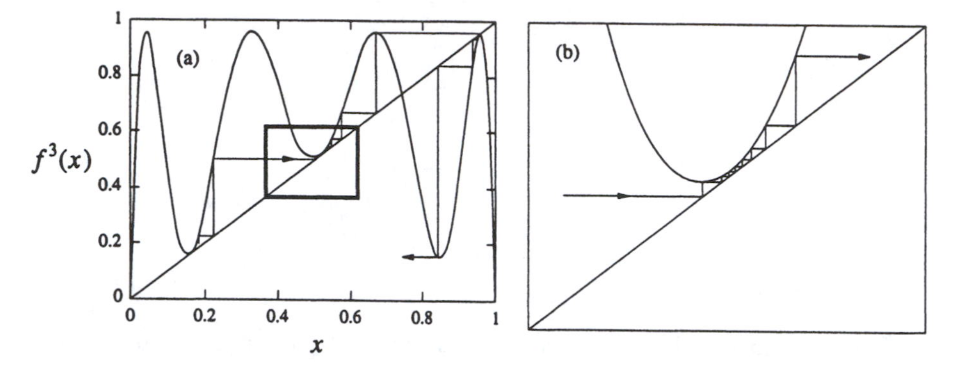

Figure 10.4.4

In Figure 10.4.4a, notice the three narrow channels between the diagonal and the graph of $f^3(x)$. These channels were formed in the aftermath of the tangent bifurcation, as the hills and valleys of $f^3(x)$ pulled away from the diagonal. Now focus on the channel in the small box of Figure 10.4.4a, enlarged in Figure 10.4.4b. The orbit takes many iterations to squeeze through the channel. Hence $f^3(x_n) \approx x_n$ during the passage, and so the orbit looks like a 3-cycle; this explains why we see a ghost.

Eventually, the orbit escapes from the channel. Then it bounces around chaotically until fate sends it back into a channel at some unpredictable later time and place.

Intermittency is not just a curiosity of the logistic map. It arises commonly in systems where the transition from periodic to chaotic behavior takes place by a saddle-node bifurcation of cycles. For instance, Exercise 10.4.8 shows that intermittency can occur in the Lorenz equations. (In fact, it was discovered there; see Pomeau and Manneville 1980).

In experimental systems, intermittency appears as nearly periodic motion interrupted by occasional irregular bursts. The time between bursts is statistically distributed, much like a random variable, even though the system is completely deterministic. As the control parameter is moved farther away from the periodic window, the bursts become more frequent until the system is fully chaotic. This progression is known as the ***intermittency route to chaos***.

Figure 10.4.5 shows an experimental example of the intermittency route to chaos in a laser.

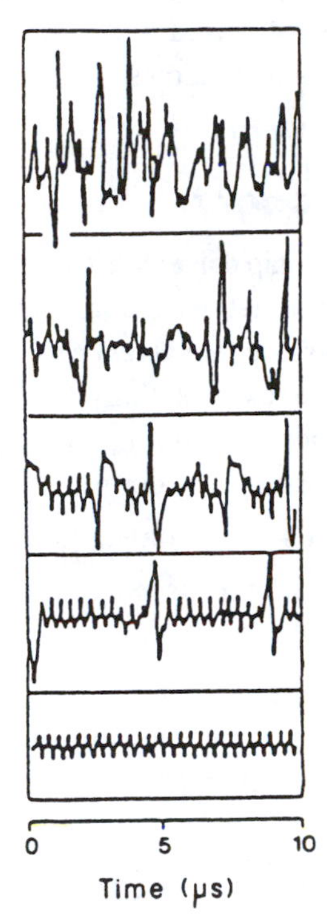

Figure 10.4.5 Harrison and Biswas (1986), p. 396

The intensity of the emitted laser light is plotted as a function of time. In the lowest panel of Figure 10.4.5, the laser is pulsing periodically. A bifurcation to intermittency occurs as the system's control parameter (the tilt of the mirror in the laser cavity) is varied. Moving from bottom to top of Figure 10.4.5, we see that the chaotic bursts occur increasingly often.

For a nice review of intermittency in fluids and chemical reactions, see Bergé et al. (1984). Those authors also review two other types of intermittency (the kind considered here is *Type I intermittency*) and give a much more detailed treatment of intermittency in general.

Period-Doubling in the Window

We commented at the end of Section 10.2 that a copy of the orbit diagram appears in miniature in the period-3 window. The explanation has to do with hills and valleys again. Just after the stable 3-cycle is created in the tangent bifurcation, the slope at the black dots in Figure 10.4.1 is close to +1. As we increase r, the hills rise and the valleys sink. The slope of $f^3(x)$ at the black dots decreases steadily from +1 and eventually reaches −1. When this occurs, a flip bifurcation causes each of the black dots to split in two; the 3-cycle doubles its period and becomes a *6-cycle*. The same mechanism operates here as in the original period-doubling cascade, but now produces orbits of period $3 \cdot 2^n$. A similar period-doubling cascade can be found in *all* of the periodic windows.

10.5 Liapunov Exponent

We have seen that the logistic map can exhibit aperiodic orbits for certain parameter values, but how do we know that this is really chaos? To be called "chaotic," a system should also show *sensitive dependence on initial conditions*, in the sense that neighboring orbits separate exponentially fast, on average. In Section 9.3 we quantified sensitive dependence by defining the Liapunov exponent for a chaotic differential equation. Now we extend the definition to one-dimensional maps.

Here's the intuition. Given some initial condition x_0, consider a nearby point $x_0 + \delta_0$, where the initial separation δ_0 is extremely small. Let δ_n be the separation after n iterates. If $|\delta_n| \approx |\delta_0| e^{n\lambda}$, then λ is called the Liapunov exponent. A positive Liapunov exponent is a signature of chaos.

A more precise and computationally useful formula for λ can be derived. By taking logarithms and noting that $\delta_n = f^n(x_0 + \delta_0) - f^n(x_0)$, we obtain

$$\begin{aligned}
\lambda &\approx \frac{1}{n} \ln \left| \frac{\delta_n}{\delta_0} \right| \\
&= \frac{1}{n} \ln \left| \frac{f^n(x_0 + \delta_0) - f^n(x_0)}{\delta_0} \right| \\
&= \frac{1}{n} \ln \left| (f^n)'(x_0) \right|
\end{aligned}$$

where we've taken the limit $\delta_0 \to 0$ in the last step. The term inside the logarithm can be expanded by the chain rule:

$$(f^n)'(x_0) = \prod_{i=0}^{n-1} f'(x_i) \,.$$

(We've already seen this formula in Example 9.4.1, where it was derived by heuristic reasoning about multipliers, and in Example 10.3.3, for the special case $n = 2$.) Hence

$$\begin{aligned}
\lambda &\approx \frac{1}{n} \ln \left| \prod_{i=0}^{n-1} f'(x_i) \right| \\
&= \frac{1}{n} \sum_{i=0}^{n-1} \ln |f'(x_i)| .
\end{aligned}$$

If this expression has a limit as $n \to \infty$, we define that limit to be the ***Liapunov exponent*** for the orbit starting at x_0:

$$\lambda = \lim_{n\to\infty} \left\{ \frac{1}{n} \sum_{i=0}^{n-1} \ln |f'(x_i)| \right\}.$$

Note that λ depends on x_0. However, it is the same for all x_0 in the basin of attraction of a given attractor. For stable fixed points and cycles, λ is negative; for chaotic attractors, λ is positive.

The next two examples deal with special cases where λ can be found analytically.

EXAMPLE 10.5.1:

Suppose that f has a stable p-cycle containing the point x_0. Show that the Liapunov exponent $\lambda < 0$. If the cycle is superstable, show that $\lambda = -\infty$.

Solution: As usual, we convert questions about p-cycles of f into questions about fixed points of f^p. Since x_0 is an element of a p-cycle, x_0 is a fixed point of f^p. By assumption, the cycle is stable; hence the multiplier $|(f^p)'(x_0)| < 1$. Therefore $\ln |(f^p)'(x_0)| < \ln(1) = 0$, a result that we'll use in a moment.

Next observe that for a p-cycle,

$$\begin{aligned} \lambda &= \lim_{n\to\infty} \left\{ \frac{1}{n} \sum_{i=0}^{n-1} \ln |f'(x_i)| \right\} \\ &= \frac{1}{p} \sum_{i=0}^{p-1} \ln |f'(x_i)| \end{aligned}$$

since the same p terms keep appearing in the infinite sum. Finally, using the chain rule in reverse, we obtain

$$\frac{1}{p} \sum_{i=0}^{p-1} \ln |f'(x_i)| = \frac{1}{p} \ln |(f^p)'(x_0)| < 0,$$

as desired. If the cycle is superstable, then $|(f^p)'(x_0)| = 0$ by definition, and thus $\lambda = \frac{1}{p} \ln(0) = -\infty$. ■

The second example concerns the ***tent map***, defined by

$$f(x) = \begin{cases} rx, & 0 \le x \le \frac{1}{2} \\ r - rx, & \frac{1}{2} \le x \le 1 \end{cases}$$

for $0 \le r \le 2$ and $0 \le x \le 1$ (Figure 10.5.1).

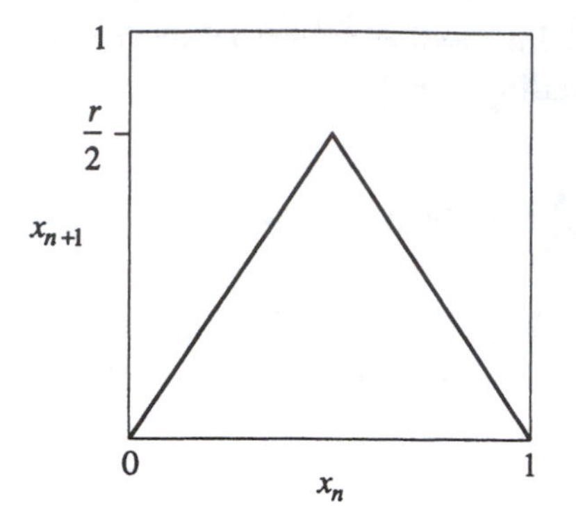

Figure 10.5.1

Because it is piecewise linear, the tent map is far easier to analyze than the logistic map.

EXAMPLE 10.5.2:

Show that $\lambda = \ln r$ for the tent map, independent of the initial condition x_0.

Solution: Since $f'(x) = \pm r$ for all x, we find $\lambda = \lim_{n \to \infty} \left\{ \frac{1}{n} \sum_{i=0}^{n-1} \ln |f'(x_i)| \right\}$ $= \ln r$. ■

Example 10.5.2 suggests that the tent map has chaotic solutions for all $r > 1$, since $\lambda = \ln r > 0$. In fact, the dynamics of the tent map can be understood in detail, even in the chaotic regime; see Devaney (1989).

In general, one needs to use a computer to calculate Liapunov exponents. The next example outlines such a calculation for the logistic map.

EXAMPLE 10.5.3:

Describe a numerical scheme to compute λ for the logistic map $f(x) = rx(1-x)$. Graph the results as a function of the control parameter r, for $3 \le r \le 4$.

Solution: Fix some value of r. Then, starting from a random initial condition, iterate the map long enough to allow transients to decay, say 300 iterates or so. Next compute a large number of additional iterates, say 10,000. You only need to store the current value of x_n, not all the previous iterates. Compute $\ln|f'(x_n)| = \ln|r - 2rx_n|$ and add it to the sum of the previous logarithms. The Liapunov exponent is then obtained by dividing the grand total by 10,000. Repeat this procedure for the next r, and so on. The end result should look like Figure 10.5.2.

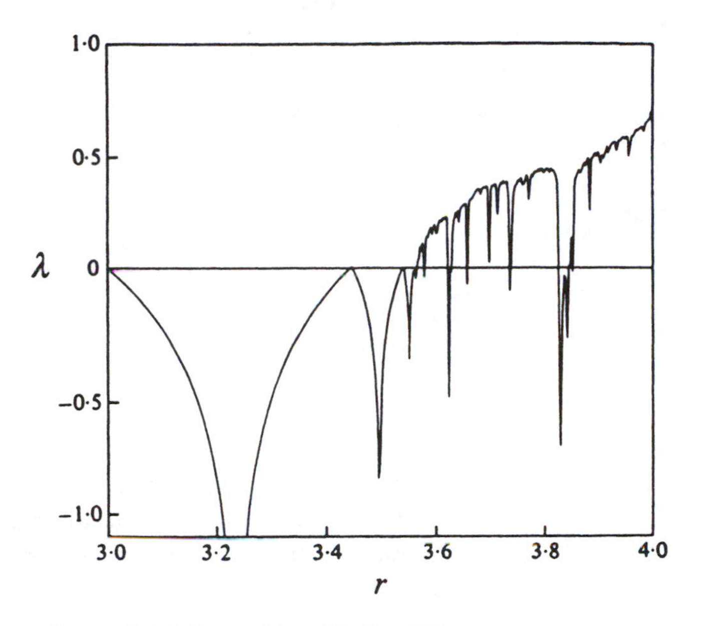

Figure 10.5.2 Olsen and Degn (1985), p. 175

Comparing this graph to the orbit diagram (Figure 10.2.7), we notice that λ remains negative for $r < r_\infty \approx 3.57$, and approaches zero at the period-doubling bifurcations. The negative spikes correspond to the 2^n-cycles. The onset of chaos is visible near $r \approx 3.57$, where λ first becomes positive. For $r > 3.57$ the Liapunov exponent generally increases, except for the dips caused by the windows of periodic behavior. Note the large dip due to the period-3 window near $r = 3.83$. ■

Actually, all the dips in Figure 10.5.2 should drop down to $\lambda = -\infty$, because a superstable cycle is guaranteed to occur somewhere near the middle of each dip, and such cycles have $\lambda = -\infty$, by Example 10.5.1. This part of the spike is too narrow to be resolved in Figure 10.5.2.

10.6 Universality and Experiments

This section deals with some of the most astonishing results in all of nonlinear dynamics. The ideas are best introduced by way of an example.

EXAMPLE 10.6.1:

Plot the graph of the ***sine map*** $x_{n+1} = r \sin \pi x_n$ for $0 \le r \le 1$ and $0 \le x \le 1$, and compare it to the logistic map. Then plot the orbit diagrams for both maps, and list some similarities and differences.

Solution: The graph of the sine map is shown in Figure 10.6.1.

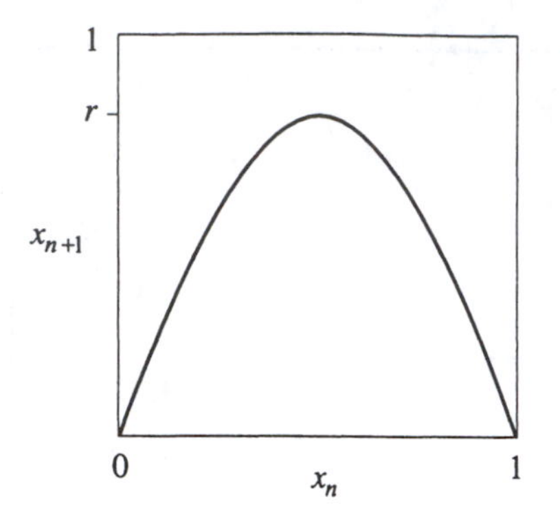

Figure 10.6.1

It has the same shape as the graph of the logistic map. Both curves are smooth, concave down, and have a single maximum. Such maps are called ***unimodal***.

Figure 10.6.2 shows the orbit diagrams for the sine map (top panel) and the logistic map (bottom panel). The resemblance is incredible. Note that both diagrams have the same vertical scale, but that the horizontal axis of the sine map diagram is scaled by a factor of 4. This normalization is appropriate because the maximum of $r \sin \pi x$ is r, whereas that of $rx(1-x)$ is $\frac{1}{4}r$.

Figure 10.6.2 shows that the *qualitative* dynamics of the two maps are identical. They both undergo period-doubling routes to chaos, followed by periodic windows interwoven with chaotic bands. Even more remarkably, the periodic windows occur in the same order, and with the same relative sizes. For instance, the period-3 window is the largest in both cases, and the next largest windows preceding it are period-5 and period-6.

But there are *quantitative* differences. For instance, the period-doubling bifurcations occur later in the logistic map, and the periodic windows are thinner. ■

Qualitative Universality: The U-sequence

Example 10.6.1 illustrates a powerful theorem due to Metropolis et al. (1973). They considered all unimodal maps of the form $x_{n+1} = rf(x_n)$, where $f(x)$ also satisfies $f(0) = f(1) = 0$. (For the precise conditions, see their original paper.) Metropolis et al. proved that as r is varied, the order in which stable periodic solutions appear is *independent* of the unimodal map being iterated. That is, *the periodic attractors always occur in the same sequence,* now called the universal or ***U-sequence***. This amazing result implies that the algebraic form of $f(x)$ is irrelevant; only its overall shape matters.

Up to period 6, the U-sequence is

$$1, 2, 2 \times 2, 6, 5, 3, 2 \times 3, 5, 6, 4, 6, 5, 6.$$

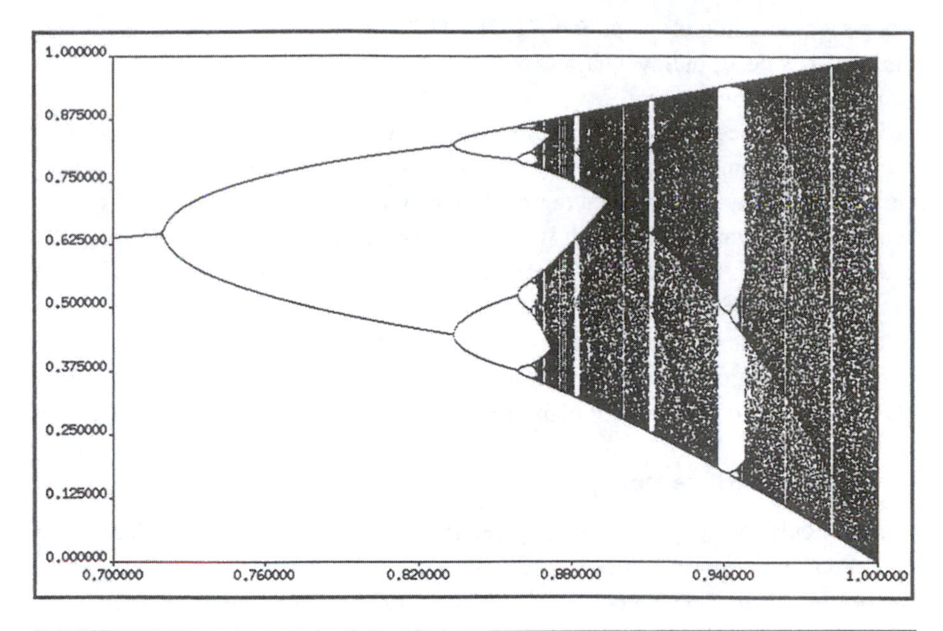

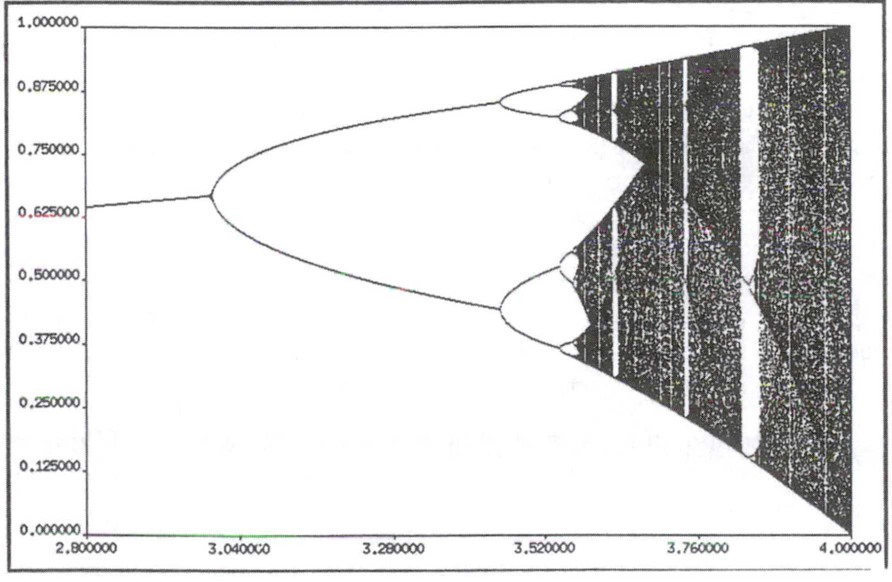

Figure 10.6.2 Courtesy of Andy Christian

The beginning of this sequence is familiar: periods 1, 2, and 2×2 are the first stages in the period-doubling scenario. (The later period-doublings give periods greater than 6, so they are omitted here.) Next, periods 6, 5, 3 correspond to the large windows mentioned in the discussion of Figure 10.6.2. Period 2×3 is the first period-doubling of the period-3 cycle. The later cycles 5, 6, 4, 6, 5, 6 are less familiar; they occur in tiny windows and easy to miss (see Exercise 10.6.5 for their locations in the logistic map).

The U-sequence has been found in experiments on the Belousov–Zhabotinsky chemical reaction. Simoyi et al. (1982) studied the reaction in a continuously stirred flow reactor and found a regime in which periodic and chaotic states alternate as the flow rate is increased. Within the experimental resolution, the periodic states occurred in the exact order predicted by the U-sequence. See Section 12.4 for more details of these experiments.

The U-sequence is qualitative; it dictates the order, but not the precise parameter values, at which periodic attractors occur. We turn now to Mitchell Feigenbaum's celebrated discovery of *quantitative* universality in one-dimensional maps.

Quantitative Universality

You should read the dramatic story behind this work in Gleick (1987), and also see Feigenbaum (1980; reprinted in Cvitanovic 1989a) for his own reminiscences. The original technical papers are Feigenbaum (1978, 1979)—published only after being rejected by other journals. These papers are fairly heavy reading; see Feigenbaum (1980), Schuster (1989) and Cvitanovic (1989b) for more accessible expositions.

Here's a capsule history. Around 1975, Feigenbaum began to study period-doubling in the logistic map. First he developed a complicated (and now forgotten) "generating function theory" to predict r_n, the value of r where a 2^n-cycle first appears. To check his theory numerically, and not being fluent with large computers, he programmed his handheld calculator to compute the first several r_n. As the calculator chugged along, Feigenbaum had time to guess where the next bifurcation would occur. He noticed a simple rule: the r_n converged geometrically, with the distance between successive transitions shrinking by a constant factor of about 4.669.

Feigenbaum (1980) recounts what happened next:

> I spent part of a day trying to fit the convergence rate value, 4.669, to the mathematical constants I knew. The task was fruitless, save for the fact that it made the number memorable.
>
> At this point I was reminded by Paul Stein that period-doubling isn't a unique property of the quadratic map but also occurs, for example, in $x_{n+1} = r \sin \pi x_n$. However my generating function theory rested heavily on the fact that the nonlinearity was simply quadratic and not transcendental. Accordingly, my interest in the problem waned.
>
> Perhaps a month later I decided to compute the r_n's in the transcendental case numerically. This problem was even slower to compute than the quadratic one. Again, it became apparent that the r_n's converged geometrically, and altogether amazingly, the convergence rate was the same 4.669 that I remembered by virtue of my efforts to fit it.

In fact, the same convergence rate appears *no matter what unimodal map is iterated*! In this sense, the number

$$\delta = \lim_{n\to\infty} \frac{r_n - r_{n-1}}{r_{n+1} - r_n} = 4.669\ldots$$

is ***universal***. It is a new mathematical constant, as basic to period-doubling as π is to circles.

Figure 10.6.3 schematically illustrates the meaning of δ. Let $\Delta_n = r_n - r_{n-1}$ denote the distance between consecutive bifurcation values. Then $\Delta_n / \Delta_{n+1} \to \delta$ as $n \to \infty$.

There is also universal scaling in the x-direction. It is harder to state precisely because the pitchforks have varying widths, even at the same value of r. (Look back at the orbit diagrams in Figure 10.6.2 to confirm this.) To take account of this nonuniformity, we define a standard x-scale as follows: Let x_m denote the maximum of f, and let d_n denote the distance from x_m to the *nearest* point in a 2^n-cycle (Figure 10.6.3).

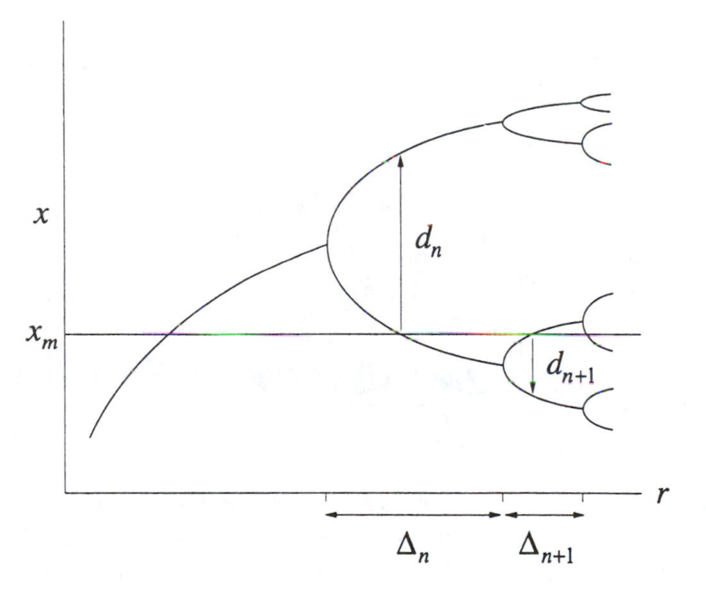

Figure 10.6.3

Then the ratio d_n / d_{n+1} tends to a universal limit as $n \to \infty$:

$$\frac{d_n}{d_{n+1}} \to \alpha = -2.5029\ldots,$$

independent of the precise form of f. Here the negative sign indicates that the nearest point in the 2^n-cycle is alternately above and below x_m, as shown in Figure 10.6.3. Thus the d_n are alternately positive and negative.

Feigenbaum went on to develop a beautiful theory that explained why α and δ are universal (Feigenbaum 1979). He borrowed the idea of renormalization from statistical physics, and thereby found an analogy between α, δ and the universal exponents observed in experiments on second-order phase transitions in magnets, fluids, and other physical systems (Ma 1976). In Section 10.7, we give a brief look at this renormalization theory.

Experimental Tests

Since Feigenbaum's work, sequences of period-doubling bifurcations have been measured in a variety of experimental systems. For instance, in the convection experiment of Libchaber et al. (1982), a box containing liquid mercury is heated from below. The control parameter is the Rayleigh number R, a dimensionless measure of the externally imposed temperature gradient from bottom to top. For R less than a critical value R_c, heat is conducted upward while the fluid remains motionless. But for $R > R_c$, the motionless state becomes unstable and ***convection*** occurs—hot fluid rises on one side, loses its heat at the top, and descends on the other side, setting up a pattern of counterrotating cylindrical ***rolls*** (Figure 10.6.4).

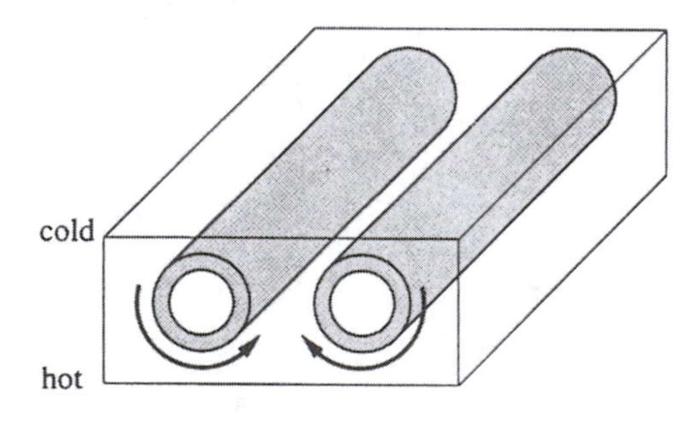

Figure 10.6.4

For R just slightly above R_c, the rolls are straight and the motion is steady. Furthermore, at any fixed location in space, the temperature is constant. With more heating, another instability sets in. A wave propagates back and forth along each roll, causing the temperature to oscillate at each point.

In traditional experiments of this sort, one keeps turning up the heat, causing further instabilities to occur until eventually the roll structure is destroyed and the system becomes turbulent. Libchaber et al. (1982) wanted to be able to increase the heat *without* destabilizing the spatial structure. That's why they chose mercury—

then the roll structure could be stabilized by applying a dc magnetic field to the whole system. Mercury has a high electrical conductivity, so there is a strong tendency for the rolls to align with the field, thereby retaining their spatial organization. There are further niceties in the experimental design, but they need not concern us; see Libchaber et al. (1982) or Bergé et al. (1984).

Now for the experimental results. Figure 10.6.5 shows that this system undergoes a sequence of period-doublings as the Rayleigh number is increased.

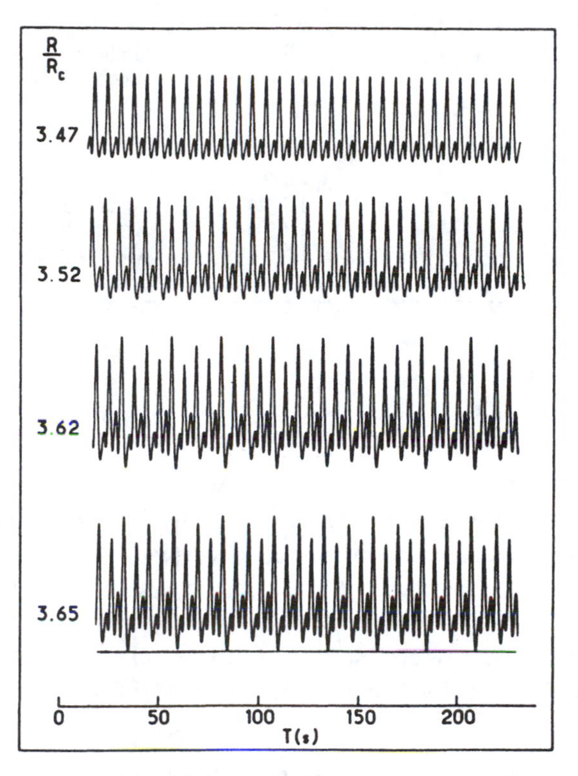

Figure 10.6.5 Libchaber et al. (1982), p. 213

Each time series shows the temperature variations at one point in the fluid. For $R/R_c = 3.47$, the temperature varies periodically. This may be regarded as the basic period-1 state. When R is increased to $R/R_c = 3.52$, the successive temperature maxima are no longer equal; the odd peaks are a little higher than before, and the even peaks are a little lower. This is the period-2 state. Further increases in R generate additional period-doublings, as shown in the lower two time series in Figure 10.6.5.

By carefully measuring the values of R at the period-doubling bifurcations, Libchaber et al. (1982) arrived at a value of $\delta = 4.4 \pm 0.1$, in reasonable agreement with the theoretical result $\delta \approx 4.699$.

Table 10.6.1, adapted from Cvitanovic (1989b), summarizes the results from a few experiments on fluid convection and nonlinear electronic circuits. The experimental estimates of δ are shown along with the errors quoted by the experimentalists; thus 4.3 (8) means 4.3 ± 0.8.

Experiment	Number of period doublings	δ	Authors
Hydrodynamic			
water	4	4.3 (8)	Giglio et al. (1981)
mercury	4	4.4 (1)	Libchaber et al. (1982)
Electronic			
diode	4	4.5 (6)	Linsay (1981)
diode	5	4.3 (1)	Testa et al. (1982)
transistor	4	4.7 (3)	Arecchi and Lisi (1982)
Josephson simul.	3	4.5 (3)	Yeh and Kao (1982)

Table 10.6.1

It is important to understand that these measurements are difficult. Since $\delta \approx 5$, each successive bifurcation requires about a fivefold improvement in the experimenter's ability to measure the external control parameter. Also, experimental noise tends to blur the structure of high-period orbits, so it is hard to tell precisely when a bifurcation has occurred. In practice, one cannot measure more than about five period-doublings. Given these difficulties, the agreement between theory and experiment is impressive.

Period-doubling has also been measured in laser, chemical, and acoustic systems, in addition to those listed here. See Cvitanovic (1989b) for references.

What Do 1-D Maps Have to Do with Science?

The predictive power of Feigenbaum's theory may strike you as mysterious. How can the theory work, given that it includes none of the *physics* of real systems like convecting fluids or electronic circuits? And real systems often have tremendously many degrees of freedom—how can all that complexity be captured by a one-dimensional map? Finally, real systems evolve in continuous time, so how can a theory based on discrete-time maps work so well?

To work toward the answer, let's begin with a system that is simpler than a convecting fluid, yet (seemingly) more complicated than a one-dimensional map. The system is a set of three differential equations concocted by Rössler (1976) to exhibit the simplest possible strange attractor. The ***Rössler system*** is

$$\dot{x} = -y - z$$
$$\dot{y} = x + ay$$
$$\dot{z} = b + z(x - c)$$

where a, b, and c are parameters. This system contains only one nonlinear term, zx, and is even simpler than the Lorenz system (Chapter 9), which has two nonlinearities.

Figure 10.6.6 shows two-dimensional projections of the system's attractor for different values of c (with $a = b = 0.2$ held fixed).

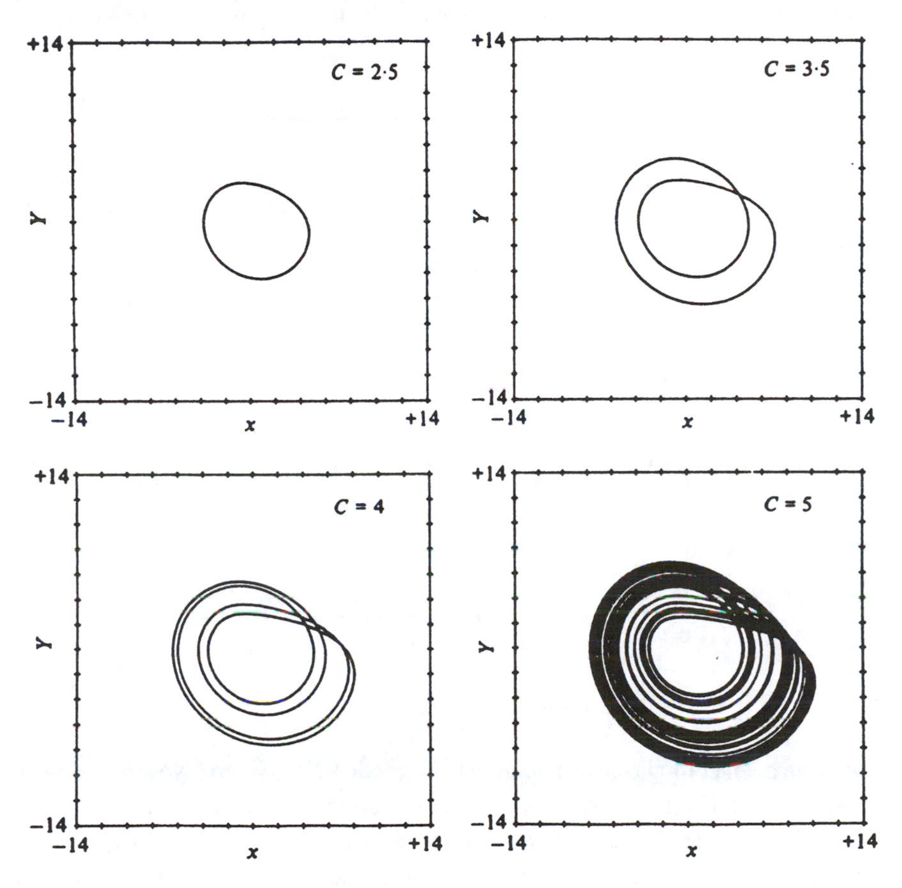

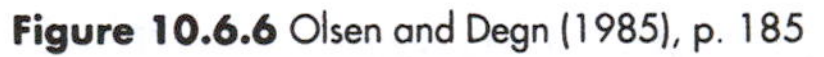
Figure 10.6.6 Olsen and Degn (1985), p. 185

At $c = 2.5$ the attractor is a simple limit cycle. As c is increased to 3.5, the limit cycle goes around twice before closing, and its period is approximately twice that of the original cycle. This is what period-doubling looks like in a continuous-time system! In fact, somewhere between $c = 2.5$ and 3.5, a ***period-doubling bifurcation of cycles*** must have occurred. (As Figure 10.6.6 suggests, such a bifurcation can occur only in three or higher dimensions, since the limit cycle needs room to avoid crossing itself.) Another period-doubling bifurcation creates the four-loop cycle shown at $c = 4$. After an infinite cascade of further period-doublings, one obtains the strange attractor shown at $c = 5$.

To compare these results to those obtained for one-dimensional maps, we use Lorenz's trick for obtaining a map from a flow (Section 9.4). For a given value of c, we record the successive local maxima of $x(t)$ for a trajectory on the strange attractor. Then we plot x_{n+1} vs. x_n, where x_n denotes the nth local maximum. This Lorenz map for $c = 5$ is shown in Figure 10.6.7. The data points fall very nearly on a one-dimensional curve. Note the uncanny resemblance to the logistic map!

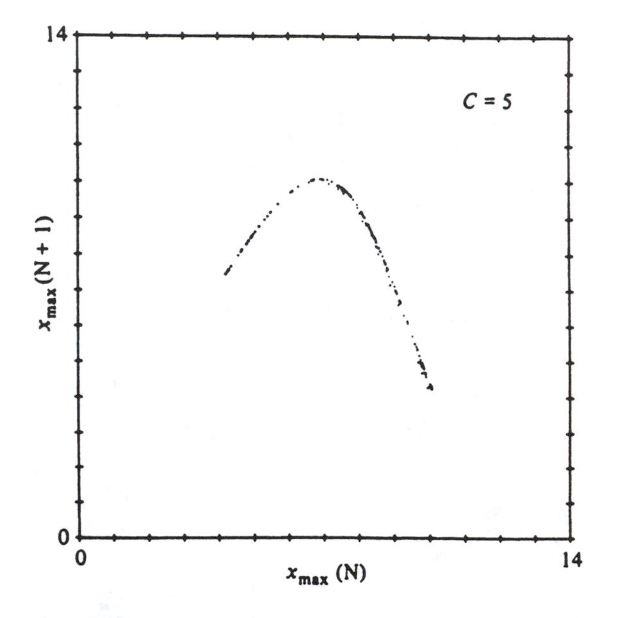

Figure 10.6.7 Olsen and Degn (1985), p. 186

We can even compute an orbit diagram for the Rössler system. Now we allow all values of c, not just those where the system is chaotic. Above each c, we plot *all* the local maxima x_n on the attractor for that value of c. The number of different maxima tells us the "period" of the attractor. For instance, at $c = 3.5$ the attractor is period-2 (Figure 10.6.6), and hence there are two local maxima of $x(t)$. Both of these points are graphed above $c = 3.5$ in Figure 10.6.8. We proceed in this way for all values of c, thereby sweeping out the orbit diagram.

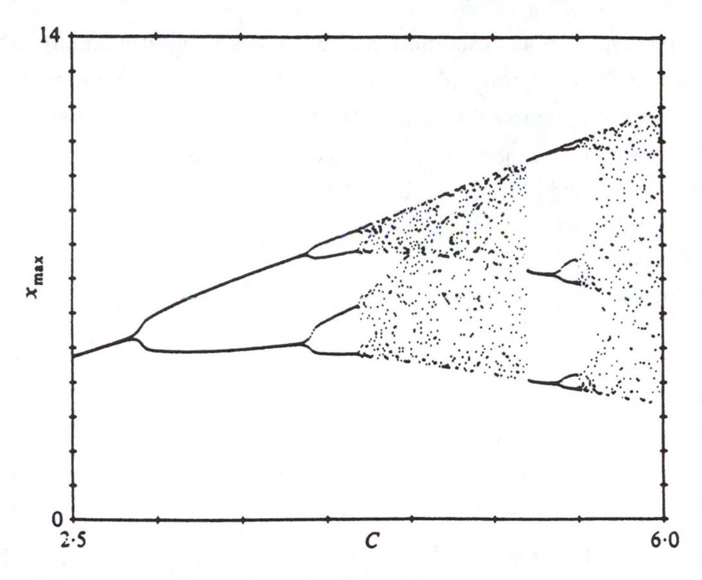

Figure 10.6.8 Olsen and Degn (1985), p. 186

This orbit diagram allows us to keep track of the bifurcations in the Rössler system. We see the period-doubling route to chaos and the large period-3 window—all our old friends are here.

Now we can see why certain physical systems are governed by Feigenbaum's universality theory—if the system's Lorenz map is nearly one-dimensional and unimodal, then the theory applies. This is certainly the case for the Rössler system, and probably for Libchaber's convecting mercury. But not all systems have one-dimensional Lorenz maps. For the Lorenz map to be almost one-dimensional, the strange attractor has to be very flat, i.e., only slightly more than two-dimensional. This requires that the system be highly dissipative; only two or three degrees of freedom are truly active, and the rest follow along slavishly. (Incidentally, that's another reason why Libchaber et al. (1982) applied a magnetic field; it increases the damping in the system, and thereby favors a low-dimensional brand of chaos.)

So while the theory works for some mildly chaotic systems, it does not apply to fully turbulent fluids or fibrillating hearts, where there are many active degrees of freedom corresponding to complicated behavior in space as well as time. We are still a long way from understanding such systems.

10.7 Renormalization

In this section we give an intuitive introduction to Feigenbaum's (1979) renormalization theory for period-doubling. For nice expositions at a higher mathematical level than that presented here, see Feigenbaum (1980), Collet and Eckmann (1980), Schuster (1989), Drazin (1992), and Cvitanovic (1989b).

First we introduce some notation. Let $f(x,r)$ denote a unimodal map that undergoes a period-doubling route to chaos as r increases, and suppose that x_m is the maximum of f. Let r_n denote the value of r at which a 2^n-cycle is born, and let R_n denote the value of r at which the 2^n-cycle is superstable.

Feigenbaum phrased his analysis in terms of the superstable cycles, so let's get some practice with them.

EXAMPLE 10.7.1:

Find R_0 and R_1 for the map $f(x,r) = r - x^2$.

Solution: At R_0 the map has a superstable fixed point, by definition. The fixed point condition is $x^* = R_0 - (x^*)^2$ and the superstability condition is $\lambda = (\partial f/\partial x)_{x=x^*} = 0$. Since $\partial f/\partial x = -2x$, we must have $x^* = 0$, i.e., the fixed point is the maximum of f. Substituting $x^* = 0$ into the fixed point condition yields $R_0 = 0$.

At R_1 the map has a superstable 2-cycle. Let p and q denote the points of the cycle. Superstability requires that the multiplier $\lambda = (-2p)(-2q) = 0$, so the point $x = 0$ must be one of the points in the 2-cycle. Then the period-2 condition $f^2(0, R_1) = 0$ implies $R_1 - (R_1)^2 = 0$. Hence $R_1 = 1$ (since the other root gives a fixed point, not a 2-cycle). ■

Example 10.7.1 illustrates a general rule: A superstable cycle of a unimodal map always contains x_m as one of its points. Consequently, there is a simple graphical way to locate R_n (Figure 10.7.1). We draw a horizontal line at height x_m; then R_n occurs where this line intersects the ***figtree*** portion of the orbit diagram (Feigenbaum = *figtree* in German). Note that R_n lies between r_n and r_{n+1}. Numerical experiments show that the spacing between successive R_n also shrinks by the universal factor $\delta \approx 4.669$.

The renormalization theory is based on the ***self-similarity*** of the figtree—the twigs look like the earlier branches, except they are scaled down in both the x and r directions. This structure reflects the endless repetition of the same dynamical processes; a 2^n-cycle is born, then becomes superstable, and then loses stability in a period-doubling bifurcation.

To express the self-similarity mathematically, we compare f with its second iterate f^2 at corresponding values of r, and then "renormalize" one map into the other. Specifically, look at the graphs of $f(x, R_0)$ and $f^2(x, R_1)$ (Figure 10.7.2, a and b).

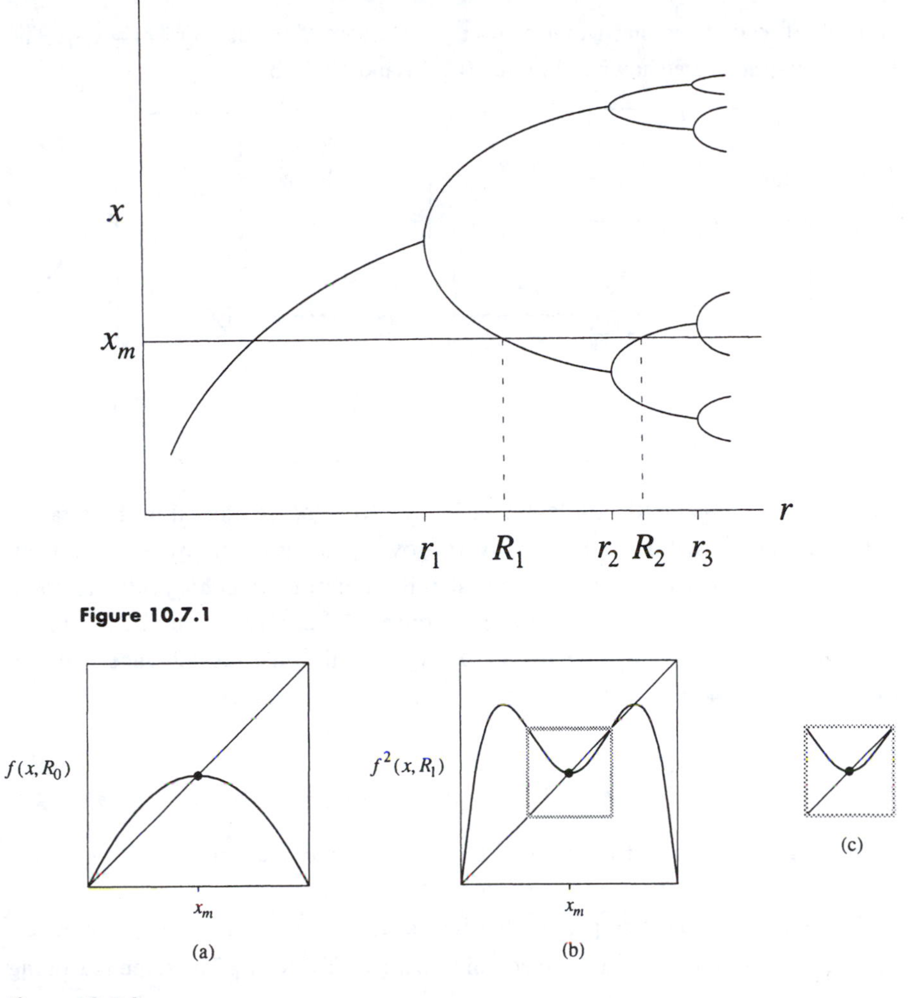

Figure 10.7.1

Figure 10.7.2

This is a fair comparison because the maps have the same stability properties: x_m *is a superstable fixed point for both of them.* Please notice that to obtain Figure 10.7.2b, we took the second iterate of f *and* increased r from R_0 to R_1. This r-shifting is a basic part of the renormalization procedure.

The small box of Figure 10.7.2b is reproduced in Figure 10.7.2c. The key point is that Figure 10.7.2c looks practically identical to Figure 10.7.2a, except for a change of scale and a reversal of both axes. From the point of view of dynamics, the two maps are very similar—cobweb diagrams starting from corresponding points would look almost the same.

Now we need to convert these qualitative observations into formulas. A helpful first step is to translate the origin of x to x_m, by redefining x as $x - x_m$. This redefinition of x dictates that we also subtract x_m from f, since $f(x_n, r) = x_{n+1}$. The translated graphs are shown in Figure 10.7.3a and 10.7.3b.

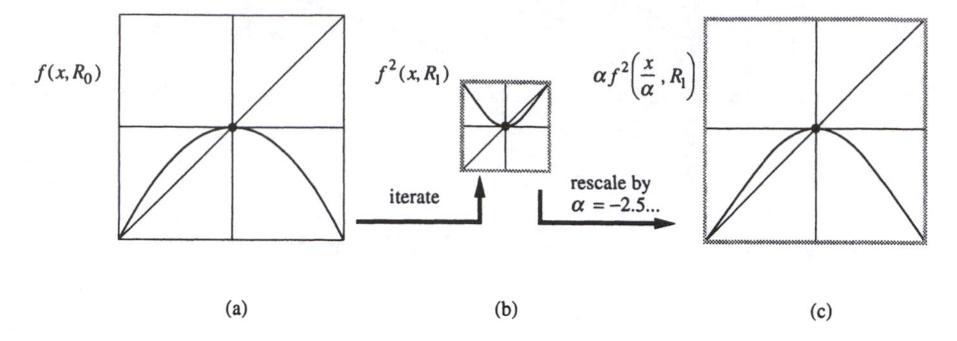

Figure 10.7.3

Next, to make Figure 10.7.3b look like Figure 10.7.3a, we blow it up by a factor $|\alpha| > 1$ in both directions, and also invert it by replacing (x, y) by $(-x, -y)$. Both operations can be accomplished in one step if we define the ***scale factor*** α to be *negative*. As you are asked to show in Exercise 10.7.2, rescaling by α is equivalent to replacing $f^2(x, R_1)$ by $\alpha f^2(x/\alpha, R_1)$. Finally, the resemblance between Figure 10.7.3a and Figure 10.7.3c shows that

$$f(x, R_0) \approx \alpha f^2\left(\frac{x}{\alpha}, R_1\right).$$

In summary, f has been ***renormalized*** by taking its second iterate, rescaling $x \to x/\alpha$, and shifting r to the next superstable value.

There is no reason to stop at f^2. For instance, we can renormalize f^2 to generate f^4; it too has a superstable fixed point if we shift r to R_2. The same reasoning as above yields

$$f^2\left(\frac{x}{\alpha}, R_1\right) \approx \alpha f^4\left(\frac{x}{\alpha^2}, R_2\right).$$

When expressed in terms of the original map $f(x, R_0)$, this equation becomes

$$f(x, R_0) \approx \alpha^2 f^4\left(\frac{x}{\alpha^2}, R_2\right).$$

After renormalizing n times we get

$$f(x, R_0) \approx \alpha^n f^{(2^n)}\left(\frac{x}{\alpha^n}, R_n\right).$$

Feigenbaum found numerically that

$$\lim_{n\to\infty} \alpha^n f^{(2^n)}\left(\frac{x}{\alpha^n}, R_n\right) = g_0(x), \tag{1}$$

where $g_0(x)$ is a ***universal function*** with a superstable fixed point. The limiting function exists only if α is chosen correctly, specifically, $\alpha = -2.5029\ldots$.

Here "universal" means that the limiting function $g_0(x)$ is independent of the original f (almost). This seems incredible at first, but the form of (1) suggests the explanation: $g_0(x)$ depends on f only through its behavior near $x = 0$, since that's all that survives in the argument x/α^n as $n \to \infty$. With each renormalization, we're blowing up a smaller and smaller neighborhood of the maximum of f, so practically all information about the global shape of f is lost.

One caveat: The *order* of the maximum is never forgotten. Hence a more precise statement is that $g_0(x)$ is universal for all f *with a quadratic maximum* (the generic case). A different $g_0(x)$ is found for f's with a fourth-degree maximum, etc.

To obtain other universal functions $g_i(x)$, start with $f(x, R_i)$ instead of $f(x, R_0)$:

$$g_i(x) = \lim_{n\to\infty} \alpha^n f^{(2^n)}\left(\frac{x}{\alpha^n}, R_{n+i}\right).$$

Here $g_i(x)$ is a universal function with superstable 2^i-cycle. The case where we start with $R_i = R_\infty$ (at the onset of chaos) is the most interesting and important, since then

$$f(x, R_\infty) \approx \alpha f^2\left(\frac{x}{\alpha}, R_\infty\right).$$

For once, we don't have to shift r when we renormalize! The limiting function $g_\infty(x)$, usually called $g(x)$, satisfies

$$g(x) = \alpha g^2\left(\frac{x}{\alpha}\right). \tag{2}$$

This is a ***functional equation*** for $g(x)$ and the universal scale factor α. It is self-referential: $g(x)$ is defined in terms of itself.

The functional equation is not complete until we specify boundary conditions on $g(x)$. After the shift of origin, all our unimodal f's have a maximum at $x = 0$, so we require $g'(0) = 0$. Also, we can set $g(0) = 1$ without loss of generality. (This just defines the scale for x; if $g(x)$ is a solution of (2), so is $\mu g(x/\mu)$, with the same α. See Exercise 10.7.3.)

Now we solve for $g(x)$ and α. At $x=0$ the functional equation gives $g(0)=\alpha g(g(0))$. But $g(0)=1$, so $1=\alpha g(1)$. Hence,

$$\alpha = 1/g(1),$$

which shows that α is determined by $g(x)$. No one has ever found a closed form solution for $g(x)$, so we resort to a power series solution

$$g(x)=1+c_2x^2+c_4x^4+\ldots$$

(which assumes that the maximum is quadratic). The coefficients are determined by substituting the power series into (2) and matching like powers of x. Feigenbaum (1979) used a seven-term expansion, and found $c_2 \approx -1.5276$, $c_4 \approx 0.1048$, along with $\alpha \approx -2.5029$. Thus the renormalization theory has succeeded in explaining the value of α observed numerically.

The theory also explains the value of δ. Unfortunately, that part of the story requires more sophisticated apparatus than we are prepared to discuss (operators in function space, Frechet derivatives, etc.). Instead we turn now to a concrete example of renormalization. The calculations are only approximate, but they can be done explicitly, using algebra instead of functional equations.

Renormalization for Pedestrians

The following pedagogical calculation is intended to clarify the renormalization process. As a bonus, it gives closed form approximations for α and δ. Our treatment is modified from May and Oster (1980) and Helleman (1980).

Let $f(x,\mu)$ be any unimodal map that undergoes a period-doubling route to chaos. Suppose that the variables are defined such that the period-2 cycle is born at $x=0$ when $\mu=0$. Then for both x and μ close to 0, the map is approximated by

$$x_{n+1} = -(1+\mu)x_n + ax_n^2 + \ldots,$$

since the eigenvalue is -1 at the bifurcation. (We are going to neglect all higher order terms in x and μ; that's why our results will be only approximate.) Without loss of generality we can set $a=1$ by rescaling $x \to x/a$. So locally our map has the normal form

$$x_{n+1} = -(1+\mu)x_n + x_n^2 + \ldots. \qquad (3)$$

Here's the idea: for $\mu > 0$, there exist period-2 points, say p and q. As μ increases, p and q themselves will eventually period-double. When this happens, the dynamics of f^2 near p will necessarily be approximated by a map *with the same algebraic form as* (3), since all maps have this form near a period-doubling bifurcation. Our strategy is to calculate the map governing the dynamics of f^2 near p, and renormalize it to look like (3). This defines a renormalization iteration, which in turn leads to a prediction of α and δ.

First, we find p and q. By definition of period-2, p is mapped to q and q to p. Hence (3) yields

$$p = -(1+\mu)q + q^2, \qquad q = -(1+\mu)p + p^2.$$

By subtracting one of these equations from the other, and factoring out $p-q$, we find that $p+q=\mu$. Then multiplying the equations together and simplifying yields $pq=-\mu$. Hence

$$p = \frac{\mu+\sqrt{\mu^2+4\mu}}{2}, \qquad q = \frac{\mu-\sqrt{\mu^2+4\mu}}{2}.$$

Now shift the origin to p and look at the local dynamics. Let

$$f(x) = -(1+\mu)x + x^2.$$

Then p is a fixed point of f^2. Expand $p+\eta_{n+1} = f^2(p+\eta_n)$ in powers of the small deviation η_n. After some algebra (Exercise 10.7.10) and neglecting higher order terms as usual, we get

$$\eta_{n+1} = (1-4\mu-\mu^2)\eta_n + C\eta_n^2 + \ldots \tag{4}$$

where

$$C = 4\mu + \mu^2 - 3\sqrt{\mu^2+4\mu}. \tag{5}$$

As promised, the η-map (4) has the same algebraic form as the original map (3)! We can renormalize (4) into (3) by rescaling η and by defining a new μ. (Note: The need for *both* of these steps was anticipated in the abstract version of renormalization discussed earlier. We have to rescale the state variable η *and* shift the bifurcation parameter μ.)

To rescale η, let $\tilde{x}_n = C\eta_n$. Then (4) becomes

$$\tilde{x}_{n+1} = (1-4\mu-\mu^2)\tilde{x}_n + \tilde{x}_n^2 + \ldots. \tag{6}$$

This matches (3) almost perfectly. All that remains is to define a new parameter $\tilde{\mu}$ by $-(1+\tilde{\mu}) = (1-4\mu-\mu^2)$. Then (6) achieves the desired form

$$\tilde{x}_{n+1} = -(1+\tilde{\mu})\tilde{x}_n + \tilde{x}_n^2 + \ldots \quad (7)$$

where the renormalized parameter $\tilde{\mu}$ is given by

$$\tilde{\mu} = \mu^2 + 4\mu - 2. \quad (8)$$

When $\tilde{\mu} = 0$ the renormalized map (7) undergoes a flip bifurcation. Equivalently, the 2-cycle for the original map loses stability and creates a 4-cycle. This brings us to the end of the first period-doubling.

EXAMPLE 10.7.2:

Using (8), calculate the value of μ at which the original map (3) gives birth to a period-4 cycle. Compare your result to the value $r_2 = 1+\sqrt{6}$ found for the logistic map in Example 10.3.3.

Solution: The period-4 solution is born when $\tilde{\mu} = \mu^2 + 4\mu - 2 = 0$. Solving this quadratic equation yields $\mu = -2 + \sqrt{6}$. (The other solution is negative and is not relevant.) Now recall that the origin of μ was defined such that $\mu = 0$ at the birth of period-2, which occurs at $r = 3$ for the logistic map. Hence $r_2 = 3 + (-2+\sqrt{6}) = 1+\sqrt{6}$, which recovers the result obtained in Example 10.3.3. ■

Because (7) has the same form as the original map, we can do the same analysis all over again, now regarding (7) as the fundamental map. In other words, we can renormalize *ad infinitum*! This allows us to bootstrap our way to the onset of chaos, using only the ***renormalization transformation*** (8).

Let μ_k denote the parameter value at which the original map (3) gives birth to a 2^k-cycle. By definition of μ, we have $\mu_1 = 0$; by Example 10.7.2, $\mu_2 = -2+\sqrt{6} \approx 0.449$. In general, the μ_k satisfy

$$\mu_{k-1} = {\mu_k}^2 + 4\mu_k - 2. \quad (9)$$

At first it looks like we have the subscripts backwards, but think about it, using Example 10.7.2 as a guide. To obtain μ_2, we set $\tilde{\mu} = 0$ $(= \mu_1)$ in (8) and then solved for μ. Similarly, to obtain μ_k, we set $\tilde{\mu} = \mu_{k-1}$ in (8) and then solve for μ.

To convert (9) into a forward iteration, solve for μ_k in terms of μ_{k-1}:

$$\mu_k = -2 + \sqrt{6 + \mu_{k-1}}\ . \quad (10)$$

Exercise 10.7.11 asks you to give a cobweb analysis of (10), starting from the initial condition $\mu_1 = 0$. You'll find that $\mu_k \to \mu^*$, where $\mu^* > 0$ is a stable fixed point corresponding to the onset of chaos.

EXAMPLE 10.7.3:

Find μ^*.

Solution: It is slightly easier to work with (9). The fixed point satisfies $\mu^* = (\mu^*)^2 + 4\mu^* - 2$, and is given by

$$\mu^* = \tfrac{1}{2}\left(-3 + \sqrt{17}\right) \approx 0.56. \qquad (11)$$

Incidentally, this gives a remarkably accurate prediction of r_∞ for the logistic map. Recall that $\mu = 0$ corresponds to the birth of period-2, which occurs at $r = 3$ for the logistic map. Thus μ^* corresponds to $r_\infty \approx 3.56$ whereas the actual numerical result is $r_\infty \approx 3.57$! ■

Finally we get to see how δ and α make their entry. For $k \gg 1$, the μ_k should converge geometrically to μ^* at a rate given by the universal constant δ. Hence $\delta \approx (\mu_{k-1} - \mu^*)/(\mu_k - \mu^*)$. As $k \to \infty$, this ratio tends to $0/0$ and therefore may be evaluated by L'Hôpital's rule. The result is

$$\begin{aligned} \delta &\approx \left.\frac{d\mu_{k-1}}{d\mu_k}\right|_{\mu=\mu^*} \\ &= 2\mu^* + 4 \end{aligned}$$

where we have used (9) in calculating the derivative. Finally, we substitute for μ^* using (11) and obtain

$$\delta \approx 1 + \sqrt{17} \approx 5.12.$$

This estimate is about 10 percent larger than the true $\delta \approx 4.67$, which is not bad considering our approximations.

To find the approximate α, note that we used C as a rescaling parameter when we defined $\tilde{x}_n = C\eta_n$. Hence C plays the role of α. Substitution of μ^* into (5) yields

$$C = \frac{1+\sqrt{17}}{2} - 3\left[\frac{1+\sqrt{17}}{2}\right]^{1/2} \approx -2.24,$$

which is also within 10 percent of the actual value $\alpha \approx -2.50$.

EXERCISES FOR CHAPTER 10

Note: Many of these exercises ask you to use a computer. Feel free to write your own programs, or to use commercially available software. The programs in *MacMath* (Hubbard and West 1992) are particularly easy to use.

10.1 Fixed Points and Cobwebs

(Calculator experiments) Use a pocket calculator to explore the following maps. Start with some number and then keep pressing the appropriate function key; what happens? Then try a different number—is the eventual pattern the same? If possible, explain your results mathematically, using a cobweb or some other argument.

10.1.1 $x_{n+1} = \sqrt{x_n}$

10.1.2 $x_{n+1} = x_n^3$

10.1.3 $x_{n+1} = \exp x_n$

10.1.4 $x_{n+1} = \ln x_n$

10.1.5 $x_{n+1} = \cot x_n$

10.1.6 $x_{n+1} = \tan x_n$

10.1.7 $x_{n+1} = \sinh x_n$

10.1.8 $x_{n+1} = \tanh x_n$

10.1.9 Analyze the map $x_{n+1} = 2x_n/(1+x_n)$ for both positive and negative x_n.

10.1.10 Show that the map $x_{n+1} = 1 + \frac{1}{2}\sin x_n$ has a unique fixed point. Is it stable?

10.1.11 (Cubic map) Consider the map $x_{n+1} = 3x_n - x_n^3$.

a) Find all the fixed points and classify their stability.

b) Draw a cobweb starting at $x_0 = 1.9$.

c) Draw a cobweb starting at $x_0 = 2.1$.

d) Try to explain the dramatic difference between the orbits found in parts (b) and (c). For instance, can you prove that the orbit in (b) will remain bounded for all n? Or that $|x_n| \to \infty$ in (c)?

10.1.12 (Newton's method) Suppose you want to find the roots of an equation $g(x) = 0$. Then ***Newton's method*** says you should consider the map $x_{n+1} = f(x_n)$, where

$$f(x_n) = x_n - \frac{g(x_n)}{g'(x_n)} .$$

a) To calibrate the method, write down the "Newton map" $x_{n+1} = f(x_n)$ for the equation $g(x) = x^2 - 4 = 0$.

b) Show that the Newton map has fixed points at $x^* = \pm 2$.

c) Show that these fixed points are *superstable.*

d) Iterate the map numerically, starting from $x_0 = 1$. Notice the extremely rapid convergence to the right answer!

10.1.13 (Newton's method and superstability) Generalize Exercise 10.1.12 as follows. Show that (under appropriate circumstances, to be stated) the roots of an equation $g(x) = 0$ *always* correspond to superstable fixed points of the Newton map $x_{n+1} = f(x_n)$, where $f(x_n) = x_n - g(x_n)/g'(x_n)$. (This explains why Newton's method converges so fast—if it converges at all.)

10.1.14 Prove that $x^* = 0$ is a globally stable fixed point for the map $x_{n+1} = -\sin x_n$. (Hint: Draw the line $x_{n+1} = -x_n$ on your cobweb diagram, in addition to the usual line $x_{n+1} = x_n$.)

10.2 Logistic Map: Numerics

10.2.1 Consider the logistic map for all real x and for any $r > 1$.

a) Show that if $x_n > 1$ for some n, then subsequent iterations diverge toward $-\infty$. (For the application to population biology, this means the population goes extinct.)
b) Given the result of part (a), explain why it is sensible to restrict r and x to the intervals $r \in [0,4]$ and $x \in [0,1]$.

10.2.2 Use a cobweb to show that $x^* = 0$ is globally stable for $0 \le r \le 1$ in the logistic map.

10.2.3 Compute the orbit diagram for the logistic map.

Plot the orbit diagram for each of the following maps. Be sure to use a large enough range for both r and x to include the main features of interest. Also, try different initial conditions, just in case it matters.

10.2.4 $x_{n+1} = x_n e^{-r(1-x_n)}$ (Standard period-doubling route to chaos)

10.2.5 $x_{n+1} = e^{-rx_n}$ (One period-doubling bifurcation and the show is over)

10.2.6 $x_{n+1} = r \cos x_n$ (Period-doubling and chaos galore)

10.2.7 $x_{n+1} = r \tan x_n$ (Nasty mess)

10.2.8 $x_{n+1} = rx_n - x_n^3$ (Attractors sometimes come in symmetric pairs)

10.3 Logistic Map: Analysis

10.3.1 (Superstable fixed point) Find the value of r at which the logistic map has a superstable fixed point.

10.3.2 (Superstable 2-cycle) Let p and q be points in a 2-cycle for the logistic map.

a) Show that if the cycle is *superstable*, then either $p = \frac{1}{2}$ or $q = \frac{1}{2}$. (In other words, the point where the map takes on its maximum must be one of the points in the 2-cycle.)
b) Find the value of r at which the logistic map has a superstable 2-cycle.

10.3.3 Analyze the long-term behavior of the map $x_{n+1} = rx_n/(1+x_n^2)$, where $r > 0$. Find and classify all fixed points as a function of r. Can there be periodic solutions? Chaos?

10.3.4 (Quadratic map) Consider the ***quadratic map*** $x_{n+1} = x_n^2 + c$.

a) Find and classify all the fixed points as a function of c.

b) Find the values of c at which the fixed points bifurcate, and classify those bifurcations.

c) For which values of c is there a stable 2-cycle? When is it superstable?

d) Plot a partial bifurcation diagram for the map. Indicate the fixed points, the 2-cycles, and their stability.

10.3.5 (Conjugacy) Show that the logistic map $x_{n+1} = rx_n(1-x_n)$ can be transformed into the quadratic map $y_{n+1} = y_n^2 + c$ by a linear change of variables, $x_n = ay_n + b$, where a, b are to be determined.

(One says that the logistic and quadratic maps are "conjugate." More generally, a ***conjugacy*** is a change of variables that transforms one map into another. If two maps are conjugate, they are equivalent as far as their dynamics are concerned; you just have to translate from one set of variables to the other. Strictly speaking, the transformation should be a homeomorphism, so that all topological features are preserved.)

10.3.6 (Cubic map) Consider the cubic map $x_{n+1} = f(x_n)$, where $f(x_n) = rx_n - x_n^3$.

a) Find the fixed points. For which values of r do they exist? For which values are they stable?

b) To find the 2-cycles of the map, suppose that $f(p) = q$ and $f(q) = p$. Show that p, q are roots of the equation $x(x^2 - r + 1)(x^2 - r - 1)(x^4 - rx^2 + 1) = 0$ and use this to find all the 2-cycles.

c) Determine the stability of the 2-cycles as a function of r.

d) Plot a partial bifurcation diagram, based on the information obtained.

10.3.7 (A chaotic map that can be analyzed completely) Consider the ***decimal shift map*** on the unit interval given by

$$x_{n+1} = 10x_n \pmod 1 .$$

As usual, "mod 1" means that we look only at the noninteger part of x. For example, $2.63 \pmod 1 = 0.63$.

a) Draw the graph of the map.

b) Find all the fixed points. (Hint: Write x_n in decimal form.)

c) Show that the map has periodic points of all periods, but that all of them are unstable. (For the first part, it suffices to give an explicit example of a period-p point, for each integer $p > 1$.)

d) Show that the map has infinitely many aperiodic orbits.

e) By considering the rate of separation between two nearby orbits, show that the map has sensitive dependence on initial conditions.

10.3.8 (Dense orbit for the decimal shift map) Consider a map of the unit interval into itself. An orbit $\{x_n\}$ is said to be "dense" if it eventually gets arbitrarily close to every point in the interval. Such an orbit has to hop around rather crazily! More precisely, given any $\varepsilon > 0$ and any point $p \in [0,1]$, the orbit $\{x_n\}$ is ***dense*** if there is some finite n such that $|x_n - p| < \varepsilon$.

Explicitly construct a dense orbit for the decimal shift map $x_{n+1} = 10x_n \pmod 1$.

10.3.9 (Binary shift map) Show that the ***binary shift map*** $x_{n+1} = 2x_n \pmod 1$ has sensitive dependence on initial conditions, infinitely many periodic and aperiodic orbits, and a dense orbit. (Hint: Redo Exercises 10.3.7 and 10.3.8, but write x_n as a binary number, not a decimal.)

10.3.10 (Exact solutions for the logistic map with $r = 4$) The previous exercise shows that the orbits of the binary shift map can be wild. Now we are going to see that this same wildness occurs in the logistic map when $r = 4$.

a) Let $\{\theta_n\}$ be an orbit of the binary shift map $\theta_{n+1} = 2\theta_n \pmod 1$, and define a new sequence $\{x_n\}$ by $x_n = \sin^2(\pi\theta_n)$. Show that $x_{n+1} = 4x_n(1 - x_n)$, no matter what θ_0 we started with. Hence any such orbit is an exact solution of the logistic map with $r = 4$!
b) Graph the time series x_n vs. n, for various choices of θ_0.

10.3.11 (Subcritical flip) Let $x_{n+1} = f(x_n)$, where $f(x) = -(1+r)x - x^2 - 2x^3$.
a) Classify the linear stability of the fixed point $x^* = 0$.
b) Show that a flip bifurcation occurs at $x^* = 0$ when $r = 0$.
c) By considering the first few terms in the Taylor series for $f^2(x)$ or otherwise, show that there is an *unstable* 2-cycle for $r < 0$, and that this cycle coalesces with $x^* = 0$ as $r \to 0$ from below.
d) What is the long-term behavior of orbits that start near $x^* = 0$, both for $r < 0$ and $r > 0$?

10.3.12 (Numerics of superstable cycles) Let R_n denote the value of r at which the logistic map has a superstable cycle of period 2^n.
a) Write an implicit but exact formula for R_n in terms of the point $x = \frac{1}{2}$ and the function $f(x,r) = rx(1-x)$.
b) Using a computer and the result of part (a), find $R_2, R_3, \ldots, R_7$ to five significant figures.
c) Evaluate $\dfrac{R_6 - R_5}{R_7 - R_6}$.

10.3.13 (Tantalizing patterns) The orbit diagram of the logistic map (Figure 10.2.7) exhibits some striking features that are rarely discussed in books.

a) There are several smooth, dark tracks of points running through the chaotic part of the diagram. What are these curves? (Hint: Think about $f(x_m, r)$, where $x_m = \frac{1}{2}$ is the point at which f is maximized.)
b) Can you find the exact value of r at the corner of the "big wedge"? (Hint: Several of the dark tracks in part (b) intersect at this corner.)

10.4 Periodic Windows

10.4.1 (Exponential map) Consider the map $x_{n+1} = r \exp x_n$ for $r > 0$.
a) Analyze the map by drawing a cobweb.
b) Show that a tangent bifurcation occurs at $r = 1/e$.
c) Sketch the time series x_n vs. n for r just above and just below $r = 1/e$.

10.4.2 Analyze the map $x_{n+1} = rx_n^2/(1+x_n^2)$. Find and classify all the bifurcations and draw the bifurcation diagram. Can this system exhibit intermittency?

10.4.3 (A superstable 3-cycle) The map $x_{n+1} = 1 - rx_n^2$ has a superstable 3-cycle at a certain value of r. Find a cubic equation for this r.

10.4.4 Approximate the value of r at which the logistic map has a superstable 3-cycle. Please give a numerical approximation that is accurate to at least four places after the decimal point.

10.4.5 (Band merging and crisis) Show numerically that the period-doubling bifurcations of the 3-cycle for the logistic map accumulate near $r = 3.8495\ldots$, to form three small chaotic bands. Show that these chaotic bands merge near $r = 3.857\ldots$ to form a much larger attractor that nearly fills an interval.

This discontinuous jump in the size of an attractor is an example of a ***crisis*** (Grebogi, Ott, and Yorke 1983a).

10.4.6 (A superstable cycle) Consider the logistic map with $r = 3.7389149$. Plot the cobweb diagram, starting from $x_0 = \frac{1}{2}$ (the maximum of the map). You should find a superstable cycle. What is its period?

10.4.7 (Iteration patterns) Superstable cycles for the logistic map can be characterized by a string of R's and L's, as follows. By convention, we start the cycle at $x_0 = \frac{1}{2}$. Then if the nth iterate x_n lies to the right of $x_0 = \frac{1}{2}$, the nth letter in the string is an R; otherwise it's an L. (No letter is used if $x_n = \frac{1}{2}$, since the superstable cycle is then complete.) The string is called the *symbol sequence* or ***iteration pattern*** for the superstable cycle (Metropolis et al. 1973).
a) Show that for the logistic map with $r > 1+\sqrt{5}$, the first two letters are always RL.
b) What is the iteration pattern for the orbit you found in Exercise 10.4.6?

10.4.8 (Intermittency in the Lorenz equations) Solve the Lorenz equations numerically for $\sigma = 10$, $b = \frac{8}{3}$, and r near 166.
a) Show that if $r = 166$, all trajectories are attracted to a stable limit cycle. Plot

both the xz projection of the cycle, and the time series $x(t)$.

b) Show that if $r = 166.2$, the trajectory looks like the old limit cycle for much of the time, but occasionally it is interrupted by chaotic bursts. This is the signature of intermittency.

c) Show that as r increases, the bursts become more frequent and last longer.

10.4.9 (Period-doubling in the Lorenz equations) Solve the Lorenz equations numerically for $\sigma = 10$, $b = \frac{8}{3}$, and $r = 148.5$. You should find a stable limit cycle. Then repeat the experiment for $r = 147.5$ to see a period-doubled version of this cycle. (When plotting your results, discard the initial transient, and use the xy projections of the attractors.)

10.4.10 (The birth of period 3) This is a hard exercise. The goal is to show that the period-3 cycle of the logistic map is born in a tangent bifurcation at $r = 1 + \sqrt{8} = 3.8284\ldots$. Here are a few vague hints. There are four unknowns: the three period-3 points a, b, c and the bifurcation value r. There are also four equations: $f(a) = b$, $f(b) = c$, $f(c) = a$, and the tangent bifurcation condition. Try to eliminate a, b, c (which we don't care about anyway) and get an equation for r alone. It may help to shift coordinates so that the map has its maximum at $x = 0$ rather than $x = \frac{1}{2}$. Also, you may want to change variables again to symmetric polynomials involving sums of products of a, b, c. See Saha and Strogatz (1994) for one solution, probably not the most elegant one!

10.5 Liapunov Exponent

10.5.1 Calculate the Liapunov exponent for the linear map $x_{n+1} = rx_n$.

10.5.2 Calculate the Liapunov exponent for the decimal shift map $x_{n+1} = 10x_n \pmod 1$.

10.5.3 Analyze the dynamics of the tent map for $r \le 1$.

10.5.4 (No windows for the tent map) Prove that, in contrast to the logistic map, the tent map does *not* have periodic windows interspersed with chaos.

10.5.5 Plot the orbit diagram for the tent map.

10.5.6 Using a computer, compute and plot the Liapunov exponent as a function of r for the sine map $x_{n+1} = r \sin \pi x_n$, for $0 \le x_n \le 1$ and $0 \le r \le 1$.

10.5.7 The graph in Figure 10.5.2 suggests that $\lambda = 0$ at each period-doubling bifurcation value r_n. Show analytically that this is correct.

10.6 Universality and Experiments

The first two exercises deal with the sine map $x_{n+1} = r \sin \pi x_n$, where $0 < r \le 1$ and $x \in [0, 1]$. The goal is to learn about some of the practical problems that come up when one tries to estimate δ numerically.

10.6.1 (Naive approach)

a) At each of 200 equally spaced r values, plot x_{700} through x_{1000} vertically above r, starting from some random initial condition x_0. Check your orbit diagram against Figure 10.6.2 to be sure your program is working.

b) Now go to finer resolution near the period-doubling bifurcations, and estimate r_n, for $n = 1, 2, \ldots, 6$. Try to achieve five significant figures of accuracy.

c) Use the numbers from (b) to estimate the Feigenbaum ratio $\dfrac{r_n - r_{n-1}}{r_{n+1} - r_n}$.

(Note: To get accurate estimates in part (b), you need to be clever, or careful, or both. As you probably found, a straightforward approach is hampered by "critical slowing down"—the convergence to a cycle becomes unbearably slow when that cycle is on the verge of period-doubling. This makes it hard to decide precisely where the bifurcation occurs. To achieve the desired accuracy, you may have to use double precision arithmetic, and about 10^4 iterates. But maybe you can find a shortcut by reformulating the problem.)

10.6.2 (Superstable cycles to the rescue) The "critical slowing down" encountered in the previous problem is avoided if we compute R_n instead of r_n. Here R_n denotes the value of r at which the sine map has a superstable cycle of period 2^n.

a) Explain why it should be possible to compute R_n more easily and accurately than r_n.

b) Compute the first six R_n's and use them to estimate δ.

If you're interested in knowing the *best* way to compute δ, see Briggs (1991) for the state of the art.

10.6.3 (Qualitative universality of patterns) The U-sequence dictates the ordering of the windows, but it actually says more: it dictates the *iteration pattern* within each window. (See Exercise 10.4.7 for the definition of iteration patterns.) For instance, consider the large period-6 window for the logistic and sine maps, visible in Figure 10.6.2.

a) For both maps, plot the cobweb for the corresponding superstable 6-cycle, given that it occurs at $r = 3.6275575$ for the logistic map and $r = 0.8811406$ for the sine map. (This cycle acts as a representative for the whole window.)

b) Find the iteration pattern for both cycles, and confirm that they match.

10.6.4 (Period 4) Consider the iteration patterns of all possible period-4 orbits for the logistic map, or any other unimodal map governed by the U-sequence.

a) Show that only two patterns are possible for period-4 orbits: RLL and RLR.

b) Show that the period-4 orbit with pattern RLL always occurs after RLR, i.e., at a larger value of r.

10.6.5 (Unfamiliar later cycles) The final superstable cycles of periods 5, 6, 4, 6, 5, 6 in the logistic map occur at approximately the following values of r:

3.9057065, 3.9375364, 3.9602701, 3.9777664, 3.9902670, 3.9975831 (Metropolis et al. 1973). Notice that they're all near the end of the orbit diagram. They have tiny windows around them and tend to be overlooked.

a) Plot the cobwebs for these cycles.

b) Did you find it hard to obtain the cycles of periods 5 and 6? If so, can you explain why this trouble occurred?

10.6.6 (A trick for locating superstable cycles) Hao and Zheng (1989) give an amusing algorithm for finding a superstable cycle with a specified iteration pattern. The idea works for any unimodal map, but for convenience, consider the map $x_{n+1} = r - x_n^2$, for $0 \le r \le 2$. Define two functions $R(y) = \sqrt{r-y}$, $L(y) = -\sqrt{r-y}$. These are the right and left branches of the inverse map.

a) For instance, suppose we want to find the r corresponding to the superstable 5-cycle with pattern *RLLR*. Then Hao and Zheng show that this amounts to solving the equation $r = RLLR(0)$. Show that when this equation is written out explicitly, it becomes

$$r = \sqrt{r + \sqrt{r + \sqrt{r - \sqrt{r}}}}\,.$$

b) Solve this equation numerically by the iterating the map

$$r_{n+1} = \sqrt{r_n + \sqrt{r_n + \sqrt{r_n - \sqrt{r_n}}}}\,,$$

starting from any reasonable guess, e.g., $r_0 = 2$. Show numerically that r_n converges rapidly to $1.860782522\ldots$.

c) Verify that the answer to (b) yields a cycle with the desired pattern.

10.7 Renormalization

10.7.1 (Hands on the functional equation) The functional equation $g(x) = \alpha g^2(x/\alpha)$ arose in our renormalization analysis of period-doubling. Let's approximate its solution by brute force, assuming that $g(x)$ is even and has a quadratic maximum at $x = 0$.

a) Suppose $g(x) \approx 1 + c_2 x^2$ for small x. Solve for c_2 and α. (Neglect $O(x^4)$ terms.)

b) Now assume $g(x) \approx 1 + c_2 x^2 + c_4 x^4$, and use Mathematica, Maple, Macsyma (or hand calculation) to solve for α, c_2, c_4. Compare your approximate results to the "exact" values $\alpha \approx -2.5029\ldots$, $c_2 \approx -1.527\ldots$, $c_4 \approx 0.1048\ldots$.

10.7.2 Given a map $y_{n+1} = f(y_n)$, rewrite the map in terms of a rescaled variable $x_n = \alpha y_n$. Use this to show that rescaling and inversion converts $f^2(x, R_1)$ into $\alpha f^2(x/\alpha, R_1)$, as claimed in the text.

10.7.3 Show that if g is a solution of the functional equation, so is $\mu g(x/\mu)$, with the same α.

10.7.4 (Wildness of the universal function $g(x)$) Near the origin $g(x)$ is roughly parabolic, but elsewhere it must be rather wild. In fact, the function $g(x)$ has infinitely many wiggles as x ranges over the real line. Verify these statements by demonstrating that $g(x)$ crosses the lines $y=\pm x$ infinitely many times. (Hint: Show that if x^* is a fixed point of $g(x)$, then so is αx^*.)

10.7.5 (Crudest possible estimate of α) Let $f(x,r)=r-x^2$.

a) Write down explicit expressions for $f(x,R_0)$ and $\alpha f^2(x/\alpha, R_1)$.
b) The two functions in (a) are supposed to resemble each other near the origin, if α is chosen correctly. (That's the idea behind Figure 10.7.3.) Show the $O(x^2)$ coefficients of the two functions agree if $\alpha=-2$.

10.7.6 (Improved estimate of α) Redo Exercise 10.7.5 to one higher order: Let $f(x,r)=r-x^2$ again, but now compare $\alpha f^2(x/\alpha, R_1)$ to $\alpha^2 f^4(x/\alpha^2, R_2)$ and match the coefficients of the lowest powers of x. What value of α is obtained in this way?

10.7.7 (Quartic maxima) Develop the renormalization theory for functions with a *fourth-degree* maximum, e.g., $f(x,r)=r-x^4$. What approximate value of α is predicted by the methods of Exercises 10.7.1 and 10.7.5? Estimate the first few terms in the power series for the universal function $g(x)$. By numerical experimentation, estimate the new value of δ for the quartic case.

See Briggs (1991) for precise values of α and δ for this fourth-degree case, as well as for all other integer degrees between 2 and 12.

10.7.8 (Renormalization approach to intermittency: algebraic version) Consider the map $x_{n+1}=f(x_n,r)$, where $f(x_n,r)=-r+x-x^2$. This is the normal form for any map close to a tangent bifurcation.

a) Show that the map undergoes a tangent bifurcation at the origin when $r=0$.
b) Suppose r is small and positive. By drawing a cobweb, show that a typical orbit takes many iterations to pass through the bottleneck at the origin.
c) Let $N(r)$ denote the typical number of iterations of f required for an orbit to get through the bottleneck. Our goal is to see how $N(r)$ scales with r as $r\to 0$. We use a renormalization idea: Near the origin, f^2 looks like a rescaled version of f, and hence it too has a bottleneck there. Show that it takes approximately $\frac{1}{2}N(r)$ iterations for orbits of f^2 to pass through the bottleneck.
d) Expand $f^2(x,r)$ and keep only the terms through $O(x^2)$. Rescale x and r to put this new map into the desired normal form $F(X,R)\approx -R+X-X^2$. Show that this renormalization implies the recursive relation

$$\tfrac{1}{2}N(r) \approx N(4r).$$

e) Show that the equation in (d) has solutions $N(r) = ar^b$ and solve for b.

10.7.9 (Renormalization approach to intermittency: functional version) Show that if the renormalization procedure in Exercise 10.7.8 is done exactly, we are led to the functional equation

$$g(x) = \alpha g^2(x/\alpha)$$

(just as in the case of period-doubling!) but with new boundary conditions appropriate to the tangent bifurcation:

$$g(0) = 0, \qquad g'(0) = 1.$$

Unlike the period-doubling case, this functional equation can be solved *explicitly* (Hirsch et al. 1982).

a) Verify that a solution is $\alpha = 2$, $g(x) = x/(1+ax)$, with a arbitrary.
b) Explain why $\alpha = 2$ is almost obvious, in retrospect. (Hint: Draw cobwebs for both g and g^2 for an orbit passing through the bottleneck. Both cobwebs look like staircases; compare the lengths of their steps.)

10.7.10 Fill in the missing algebraic steps in the concrete renormalization calculation for period-doubling. Let $f(x) = -(1+\mu)x + x^2$. Expand $p + \eta_{n+1} = f^2(p + \eta_n)$ in powers of the small deviation η_n, using the fact that p is a fixed point of f^2. Thereby confirm that (10.7.4) and (10.7.5) are correct.

10.7.11 Give a cobweb analysis of (10.7.10), starting from the initial condition $\mu_1 = 0$. Show that $\mu_k \to \mu^*$, where $\mu^* > 0$ is a stable fixed point corresponding to the onset of chaos.

11

FRACTALS

11.0 Introduction

Back in Chapter 9, we found that the solutions of the Lorenz equations settle down to a complicated set in phase space. This set is the strange attractor. As Lorenz (1963) realized, the geometry of this set must be very peculiar, something like an "infinite complex of surfaces." In this chapter we develop the ideas needed to describe such strange sets more precisely. The tools come from fractal geometry.

Roughly speaking, ***fractals*** are complex geometric shapes with fine structure at arbitrarily small scales. Usually they have some degree of self-similarity. In other words, if we magnify a tiny part of a fractal, we will see features reminiscent of the whole. Sometimes the similarity is exact; more often it is only approximate or statistical.

Fractals are of great interest because of their exquisite combination of beauty, complexity, and endless structure. They are reminiscent of natural objects like mountains, clouds, coastlines, blood vessel networks, and even broccoli, in a way that classical shapes like cones and squares can't match. They have also turned out to be useful in scientific applications ranging from computer graphics and image compression to the structural mechanics of cracks and the fluid mechanics of viscous fingering.

Our goals in this chapter are modest. We want to become familiar with the simplest fractals and to understand the various notions of fractal dimension. These ideas will be used in Chapter 12 to clarify the geometric structure of strange attractors.

Unfortunately, we will not be able to delve into the scientific applications of fractals, nor the lovely mathematical theory behind them. For the clearest introduction to the theory and applications of fractals, see Falconer (1990). The books of Mandelbrot (1982), Peitgen and Richter (1986), Barnsley (1988), Feder (1988), and Schroeder (1991) are also recommended for their many fascinating pictures and examples.

11.1 Countable and Uncountable Sets

This section reviews the parts of set theory that we'll need in later discussions of fractals. You may be familiar with this material already; if not, read on.

Are some infinities larger than others? Surprisingly, the answer is yes. In the late 1800s, Georg Cantor invented a clever way to compare different infinite sets. Two sets X and Y are said to have the same *cardinality* (or number of elements) if there is an invertible mapping that pairs each element $x \in X$ with precisely one $y \in Y$. Such a mapping is called a ***one-to-one correspondence***; it's like a buddy system, where every x has a buddy y, and no one in either set is left out or counted twice.

A familiar infinite set is the set of natural numbers $\mathbf{N} = \{1,2,3,4,...\}$. This set provides a basis for comparison—if another set X can be put into one-to-one correspondence with the natural numbers, then X is said to be ***countable***. Otherwise X is ***uncountable***.

These definitions lead to some surprising conclusions, as the following examples show.

EXAMPLE 11.1.1:

Show that the set of even natural numbers $E = \{2,4,6,...\}$ is countable.

Solution: We need to find a one-to-one correspondence between E and $\mathbf{N}$. Such a correspondence is given by the invertible mapping that pairs each natural number n with the even number $2n$; thus $1 \leftrightarrow 2$, $2 \leftrightarrow 4$, $3 \leftrightarrow 6$, and so on.

Hence there are exactly as many even numbers as natural numbers. You might have thought that there would be only *half* as many, since all the odd numbers are missing! ∎

There is an equivalent characterization of countable sets which is frequently useful. A set X is countable if it can be written as a list $\{x_1, x_2, x_3, ...\}$, with every $x \in X$ appearing somewhere in the list. In other words, given any x, there is some finite n such that $x_n = x$.

A convenient way to exhibit such a list is to give an algorithm that systematically counts the elements of X. This strategy is used in the next two examples.

EXAMPLE 11.1.2:

Show that the integers are countable.

Solution: Here's an algorithm for listing all the integers: We start with 0 and then work in order of increasing absolute value. Thus the list is $\{0,1,-1,2,-2,3,-3,...\}$. Any particular integer appears eventually, so the integers are countable. ∎

EXAMPLE 11.1.3:

Show that the positive rational numbers are countable.

Solution: Here's a *wrong* way: we start listing the numbers $\frac{1}{1}, \frac{1}{2}, \frac{1}{3}, \frac{1}{4} \ldots$ in order. Unfortunately we never finish the $\frac{1}{n}$'s and so numbers like $\frac{2}{3}$ are never counted!

The right way is to make a table where the pq-th entry is p/q. Then the rationals can be counted by the weaving procedure shown in Figure 11.1.1. Any given p/q is reached after a finite number of steps, so the rationals are countable. ■

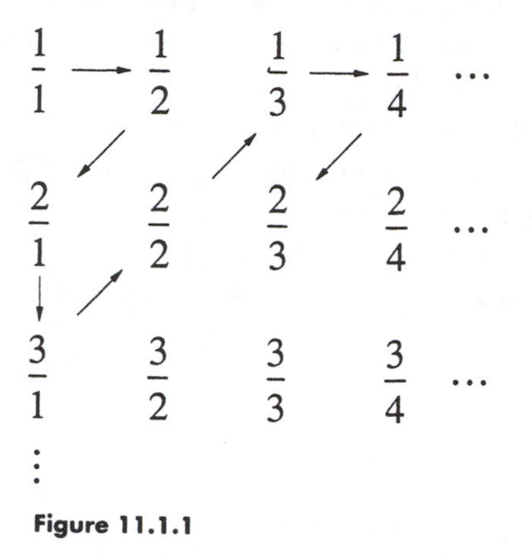

Figure 11.1.1

Now we consider our first example of an uncountable set.

EXAMPLE 11.1.4:

Let X denote the set of all real numbers between 0 and 1. Show that X is uncountable.

Solution: The proof is by contradiction. If X were countable, we could list all the real numbers between 0 and 1 as a set $\{x_1, x_2, x_3, \ldots\}$. Rewrite these numbers in decimal form:

$$x_1 = 0.x_{11}x_{12}x_{13}x_{14}\cdots$$
$$x_2 = 0.x_{21}x_{22}x_{23}x_{24}\cdots$$
$$x_3 = 0.x_{31}x_{32}x_{33}x_{34}\cdots$$
$$\vdots$$

where x_{ij} denotes the jth digit of the real number x_i.

To obtain a contradiction, we'll show that there's a number r between 0 and 1 that is *not* on the list. Hence any list is necessarily incomplete, and so the reals are uncountable.

We construct r as follows: its first digit is *anything other than* x_{11}, the first digit of x_1. Similarly, its second digit is anything other than the second digit of x_2. In general, the nth digit of r is $\bar{x}_{nn}$, defined as any digit other than x_{nn}. Then we claim that the number $r = \bar{x}_{11}\bar{x}_{22}\bar{x}_{33}\cdots$ is not on the list. Why not? It can't be equal to x_1, because it differs from x_1 in the first decimal place. Similarly, r differs from x_2 in the second decimal place, from x_3 in the third decimal place, and so on. Hence r is not on the list, and thus X is uncountable. ∎

This argument (devised by Cantor) is called the ***diagonal argument***, because r is constructed by changing the diagonal entries x_{nn} in the matrix of digits $[x_{ij}]$.

11.2 Cantor Set

Now we turn to another of Cantor's creations, a fractal known as the Cantor set. It is simple and therefore pedagogically useful, but it is also much more than that—as we'll see in Chapter 12, the Cantor set is intimately related to the geometry of strange attractors.

Figure 11.2.1 shows how to construct the Cantor set.

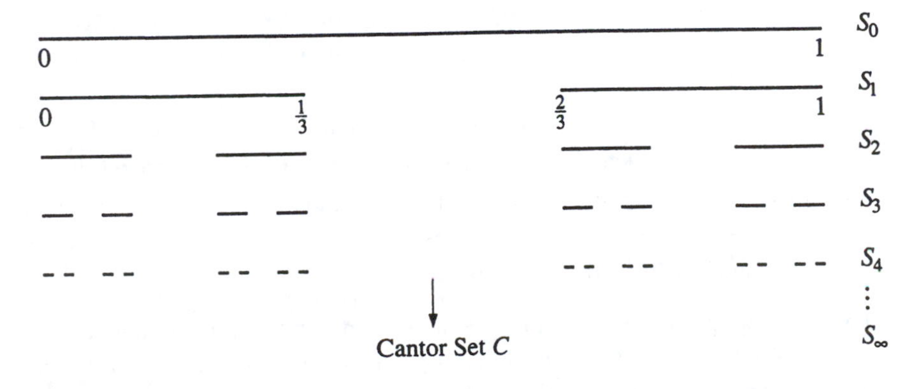

Figure 11.2.1

We start with the closed interval $S_0 = [0,1]$ and remove its open middle third, i.e., we delete the interval $(\frac{1}{3}, \frac{2}{3})$ and leave the endpoints behind. This produces the pair of closed intervals shown as S_1. Then we remove the open middle thirds of *those* two intervals to produce S_2, and so on. The limiting set $C = S_\infty$ is the ***Cantor set***. It is difficult to visualize, but Figure 11.2.1 suggests that it consists of an infinite number of infinitesimal pieces, separated by gaps of various sizes.

Fractal Properties of the Cantor Set

The Cantor set C has several properties that are typical of fractals more generally:

1. *C has structure at arbitrarily small scales.* If we enlarge part of C repeatedly, we continue to see a complex pattern of points separated by gaps of various sizes. This structure is neverending, like worlds within worlds. In contrast, when we look at a smooth curve or surface under repeated magnification, the picture becomes more and more featureless.
2. *C is self-similar.* It contains smaller copies of itself at all scales. For instance, if we take the left part of C (the part contained in the interval $[0, \frac{1}{3}]$) and enlarge it by a factor of three, we get C back again. Similarly, the parts of C in each of the four intervals of S_2 are geometrically similar to C, except scaled down by a factor of nine.

   If you're having trouble seeing the self-similarity, it may help to think about the sets S_n rather than the mind-boggling set S_∞. Focus on the left half of S_2—it looks just like S_1, except three times smaller. Similarly, the left half of S_3 is S_2, reduced by a factor of three. In general, the left half of S_{n+1} looks like *all* of S_n, scaled down by three. Now set $n = \infty$. The conclusion is that the left half of S_∞ looks like S_∞, scaled down by three, just as we claimed earlier.

   Warning: The strict self-similarity of the Cantor set is found only in the simplest fractals. More general fractals are only approximately self-similar.
3. *The dimension of C is not an integer.* As we'll show in Section 11.3, its dimension is actually $\ln 2/\ln 3 \approx 0.63$! The idea of a noninteger dimension is bewildering at first, but it turns out to be a natural generalization of our intuitive ideas about dimension, and provides a very useful tool for quantifying the structure of fractals.

Two other properties of the Cantor set are worth noting, although they are not fractal properties as such: C *has measure zero* and *it consists of uncountably many points.* These properties are clarified in the examples below.

EXAMPLE 11.2.1:

Show that the ***measure*** of the Cantor set is zero, in the sense that it can be covered by intervals whose total length is arbitrarily small.

Solution: Figure 11.2.1 shows that each set S_n completely covers all the sets that come after it in the construction. Hence the Cantor set $C = S_\infty$ is covered by *each* of the sets S_n. So the total length of the Cantor set must be less than the total length of S_n, for any n. Let L_n denote the length of S_n. Then from Figure 11.2.1 we see that $L_0 = 1$, $L_1 = \frac{2}{3}$, $L_2 = (\frac{2}{3})(\frac{2}{3}) = (\frac{2}{3})^2$, and in general, $L_n = (\frac{2}{3})^n$. Since $L_n \to 0$ as $n \to \infty$, the Cantor set has a total length of zero. ■

Example 11.2.1 suggests that the Cantor set is "small" in some sense. On the other hand, it contains tremendously many points—uncountably many, in fact. To see this, we first develop an elegant characterization of the Cantor set.

EXAMPLE 11.2.2:

Show that the Cantor set C consists of all points $c \in [0,1]$ that have no 1's in their base-3 expansion.

Solution: The idea of expanding numbers in different bases may be unfamiliar, unless you were one of those children who was taught "New Math" in elementary school. Now you finally get to see why base-3 is useful!

First let's remember how to write an arbitrary number $x \in [0,1]$ in base-3. We expand in powers of 1/3: thus if $x = \frac{a_1}{3} + \frac{a_2}{3^2} + \frac{a_3}{3^3} + \cdots$, then $x = .a_1 a_2 a_3 \ldots$ in base-3, where the digits a_n are 0, 1, or 2. This expansion has a nice geometric interpretation (Figure 11.2.2).

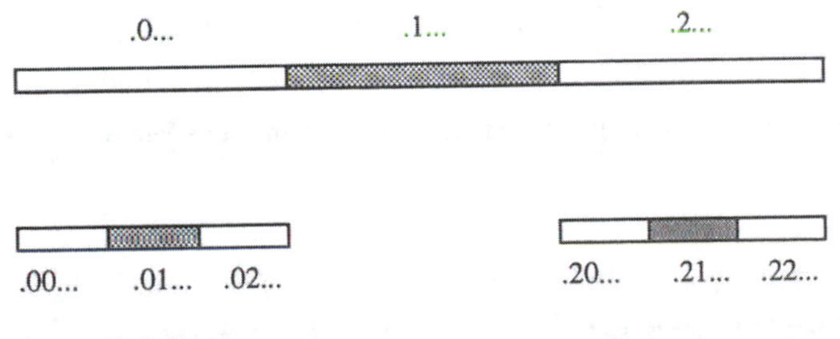

Figure 11.2.2

If we imagine that $[0,1]$ is divided into three equal pieces, then the first digit a_1 tells us whether x is in the left, middle, or right piece. For instance, all numbers with $a_1 = 0$ are in the left piece. (Ordinary base-10 works the same way, except that we divide $[0,1]$ into ten pieces instead of three.) The second digit a_2 provides more refined information: it tells us whether x is in the left, middle, or right third of a given piece. For instance, points of the form $x = .01\ldots$ are in the middle part of the left third of $[0,1]$, as shown in Figure 11.2.2.

Now think about the base-3 expansion of points in the Cantor set C. We deleted the middle third of $[0,1]$ at the first stage of constructing C; this removed all points whose first digit is 1. So those points can't be in C. The points left over (the only ones with a chance of ultimately being in C) must have 0 or 2 as their first digit. Similarly, points whose *second* digit is 1 were deleted at the next stage in the construction. By repeating this argument, we see that C consists of all points whose base-3 expansion contains no 1's, as claimed. ■

There's still a fussy point to be addressed. What about endpoints like $\frac{1}{3} = .1000\ldots$? It's in the Cantor set, yet it has a 1 in its base-3 expansion. Does this contradict what we said above? No, because this point can also be written solely in terms of 0's and 2's, as follows: $\frac{1}{3} = .1000\ldots = .02222\ldots$. By this trick, each point in the Cantor set can be written such that no 1's appear in its base-3 expansion, as claimed.

Now for the payoff.

EXAMPLE 11.2.3:

Show that the Cantor set is uncountable.

Solution: This is just a rewrite of the Cantor diagonal argument of Example 11.1.4, so we'll be brief. Suppose there were a list $\{c_1, c_2, c_3, \ldots\}$ of all points in C. To show that C is uncountable, we produce a point $\bar{c}$ that is in C but not on the list. Let c_{ij} denote the jth digit in the base-3 expansion of c_i. Define $\bar{c} = \bar{c}_{11}\bar{c}_{22}\ldots$, where the overbar means we switch 0's and 2's: thus $\bar{c}_{nn} = 0$ if $c_{nn} = 2$ and $\bar{c}_{nn} = 2$ if $c_{nn} = 0$. Then $\bar{c}$ is in C, since it's written solely with 0's and 2's, but $\bar{c}$ is not on the list, since it differs from c_n in the nth digit. This contradicts the original assumption that the list is complete. Hence C is uncountable. ■

11.3 Dimension of Self-Similar Fractals

What is the "dimension" of a set of points? For familiar geometric objects, the answer is clear—lines and smooth curves are one-dimensional, planes and smooth surfaces are two-dimensional, solids are three-dimensional, and so on. If forced to give a definition, we could say that *the dimension is the minimum number of coordinates needed to describe every point in the set.* For instance, a smooth curve is one-dimensional because every point on it is determined by one number, the arc length from some fixed reference point on the curve.

But when we try to apply this definition to fractals, we quickly run into paradoxes. Consider the ***von Koch curve***, defined recursively in Figure 11.3.1.

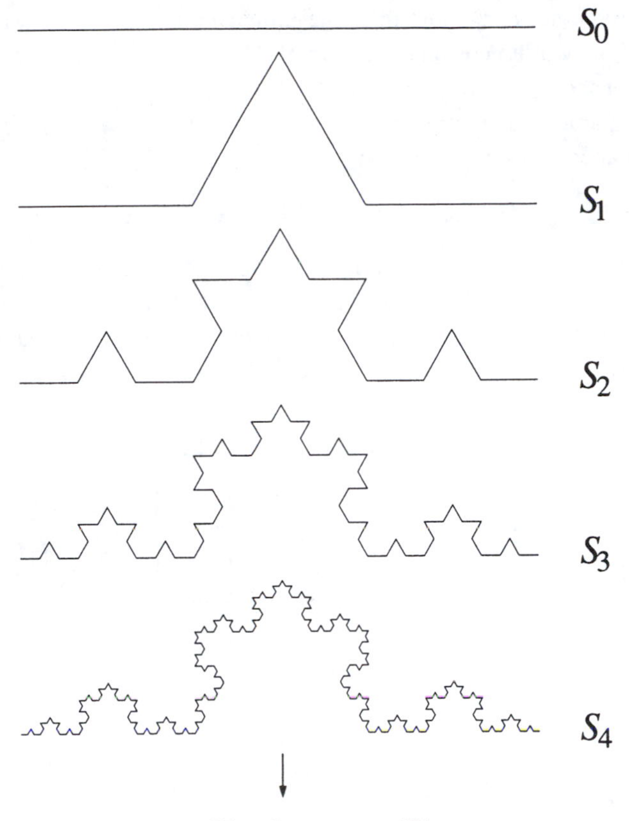

Figure 11.3.1

We start with a line segment S_0. To generate S_1, we delete the middle third of S_0 and replace it with the other two sides of an equilateral triangle. Subsequent stages are generated recursively by the same rule: S_n is obtained by replacing the middle third of each line segment in S_{n-1} by the other two sides of an equilateral triangle. The limiting set $K = S_\infty$ is the von Koch curve.

A Paradox

What is the dimension of the von Koch curve? Since it's a curve, you might be tempted to say it's one-dimensional. But the trouble is that K has *infinite arc length*! To see this, observe that if the length of S_0 is L_0, then the length of S_1 is $L_1 = \frac{4}{3}L_0$, because S_1 contains four segments, each of length $\frac{1}{3}L_0$. The length increases by a factor of $\frac{4}{3}$ at each stage of the construction, so $L_n = (\frac{4}{3})^n L_0 \to \infty$ as $n \to \infty$.

Moreover, the arc length between *any* two points on K is infinite, by similar reasoning. Hence points on K aren't determined by their arc length from a particular point, because every point is infinitely far from every other!

This suggests that K is more than one-dimensional. But would we really want to say that K is two-dimensional? It certainly doesn't seem to have any "area." So the dimension should be *between* 1 and 2, whatever that means.

With this paradox as motivation, we now consider some improved notions of dimension that can cope with fractals.

Similarity Dimension

The simplest fractals are self-similar, i.e., they are made of scaled-down copies of themselves, all the way down to arbitrarily small scales. The dimension of such fractals can be defined by extending an elementary observation about *classical* self-similar sets like line segments, squares, or cubes. For instance, consider the square region shown in Figure 11.3.2.

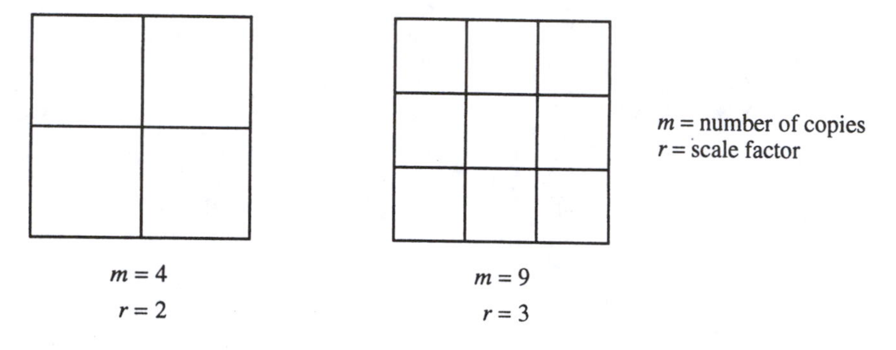

Figure 11.3.2

If we shrink the square by a factor of 2 in each direction, it takes four of the small squares to equal the whole. Or if we scale the original square down by a factor of 3, then nine small squares are required. In general, if we reduce the linear dimensions of the square region by a factor of r, it takes r^2 of the smaller squares to equal the original.

Now suppose we play the same game with a solid cube. The results are different: if we scale the cube down by a factor of 2, it takes eight of the smaller cubes to make up the original. In general, if the cube is scaled down by r, we need r^3 of the smaller cubes to make up the larger one.

The exponents 2 and 3 are no accident; they reflect the two-dimensionality of the square and the three-dimensionality of the cube. This connection between dimensions and exponents suggests the following definition. Suppose that a self-similar set is composed of m copies of itself scaled down by a factor of r. Then the ***similarity dimension*** d is the exponent defined by $m = r^d$, or equivalently,

$$d = \frac{\ln m}{\ln r}.$$

This formula is easy to use, since m and r are usually clear from inspection.

EXAMPLE 11.3.1:

Find the similarity dimension of the Cantor set C.

Solution: As shown in Figure 11.3.3, C is composed of two copies of itself, each scaled down by a factor of 3.

[0, 1]

C

The left half of the Cantor set
is the original Cantor set,
scaled down by a factor of 3

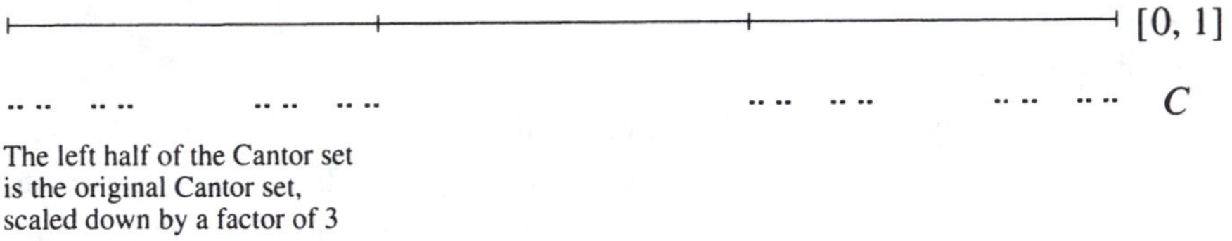

Figure 11.3.3

So $m = 2$ when $r = 3$. Therefore $d = \ln 2/\ln 3 \approx 0.63$. ∎

In the next example we confirm our earlier intuition that the von Koch curve should have a dimension between 1 and 2.

EXAMPLE 11.3.2:

Show that the von Koch curve has a similarity dimension of $\ln 4/\ln 3 \approx 1.26$.

Solution: The curve is made up of four equal pieces, each of which is similar to the original curve but is scaled down by a factor of 3 in both directions. One of these pieces is indicated by the arrows in Figure 11.3.4.

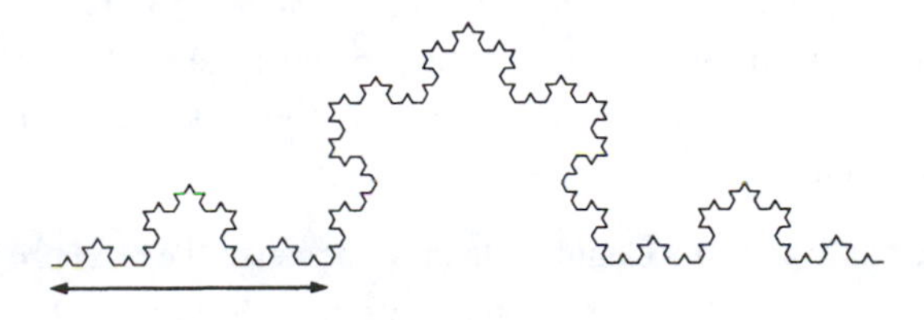

Figure 11.3.4

Hence $m = 4$ when $r = 3$, and therefore $d = \ln 4/\ln 3$. ∎

More General Cantor Sets

Other self-similar fractals can be generated by changing the recursive procedure. For instance, to obtain a new kind of Cantor set, divide an interval into five equal pieces, delete the second and fourth subintervals, and then repeat this process indefinitely (Figure 11.3.5).

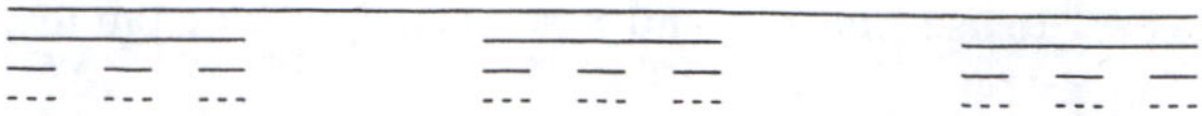

Figure 11.3.5

We call the limiting set the *even-fifths Cantor set*, since the even fifths are removed at each stage. (Similarly, the standard Cantor set of Section 11.2 is often called the ***middle-thirds Cantor set***.)

EXAMPLE 11.3.3:

Find the similarity dimension of the even-fifths Cantor set.

Solution: Let the original interval be denoted S_0, and let S_n denote the nth stage of the construction. If we scale S_n down by a factor of five, we get one third of the set S_{n+1}. Now setting $n = \infty$, we see that the even-fifths Cantor set S_∞ is made of three copies of itself, shrunken by a factor of 5. Hence $m = 3$ when $r = 5$, and so $d = \ln 3 / \ln 5$. ■

There are so many different Cantor-like sets that mathematicians have abstracted their essence in the following definition. A closed set S is called a ***topological Cantor set*** if it satisfies the following properties:

1. S is "totally disconnected." This means that S contains no connected subsets (other than single points). In this sense, all points in S are separated from each other. For the middle-thirds Cantor set and other subsets of the real line, this condition simply says that S contains no intervals.
2. On the other hand, S contains no "isolated points." This means that every point in S has a neighbor arbitrarily close by—given any point $p \in S$ and any small distance $\varepsilon > 0$, there is some other point $q \in S$ within a distance ε of p.

The paradoxical aspects of Cantor sets arise because the first property says that points in S are spread apart, whereas the second property says they're packed together! In Exercise 11.3.6, you're asked to check that the middle-thirds Cantor set satisfies both properties.

Notice that the definition says nothing about self-similarity or dimension. These notions are geometric rather than topological; they depend on concepts of distance, volume, and so on, which are too rigid for some purposes. Topological features are more robust than geometric ones. For instance, if we continuously deform a self-similar Cantor set, we can easily destroy its self-similarity but properties 1 and 2 will persist. When we study strange attractors in Chapter 12, we'll see that the cross sections of strange attractors are often topological Cantor sets, although they are not necessarily self-similar.

11.4 Box Dimension

To deal with fractals that are not self-similar, we need to generalize our notion of dimension still further. Various definitions have been proposed; see Falconer (1990) for a lucid discussion. All the definitions share the idea of "measurement at a scale ε"—roughly speaking, we measure the set in a way that ignores irregularities of size less than ε, and then study how the measurements vary as $\varepsilon \to 0$.

Definition of Box Dimension

One kind of measurement involves covering the set with boxes of size ε (Figure 11.4.1).

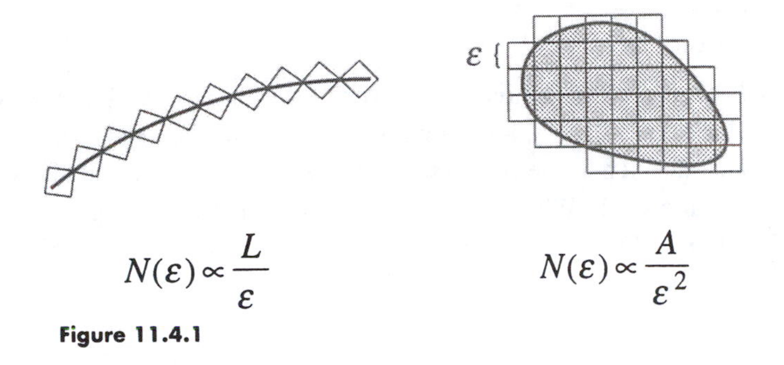

Figure 11.4.1

Let S be a subset of D-dimensional Euclidean space, and let $N(\varepsilon)$ be the minimum number of D-dimensional cubes of side ε needed to cover S. How does $N(\varepsilon)$ depend on ε? To get some intuition, consider the classical sets shown in Figure 11.4.1. For a smooth curve of length L, $N(\varepsilon) \propto L/\varepsilon$; for a planar region of area A bounded by a smooth curve, $N(\varepsilon) \propto A/\varepsilon^2$. The key observation is that the dimension of the set equals the exponent d in the ***power law*** $N(\varepsilon) \propto 1/\varepsilon^d$.

This power law also holds for most fractal sets S, except that d is no longer an integer. By analogy with the classical case, we interpret d as a dimension, usually called the *capacity* or ***box dimension*** of S. An equivalent definition is

$$d = \lim_{\varepsilon \to 0} \frac{\ln N(\varepsilon)}{\ln(1/\varepsilon)}, \text{ if the limit exists.}$$

EXAMPLE 11.4.1:

Find the box dimension of the Cantor set.

Solution: Recall that the Cantor set is covered by each of the sets S_n used in its construction (Figure 11.2.1). Each S_n consists of 2^n intervals of length $(1/3)^n$, so if we pick $\varepsilon = (1/3)^n$, we need all 2^n of these intervals to cover the Cantor set. Hence

$N = 2^n$ when $\varepsilon = (1/3)^n$. Since $\varepsilon \to 0$ as $n \to \infty$, we find

$$d = \lim_{\varepsilon \to 0} \frac{\ln N(\varepsilon)}{\ln(1/\varepsilon)} = \frac{\ln(2^n)}{\ln(3^n)} = \frac{n \ln 2}{n \ln 3} = \frac{\ln 2}{\ln 3}$$

in agreement with the similarity dimension found in Example 11.3.1. ■

This solution illustrates a helpful trick. We used a discrete sequence $\varepsilon = (1/3)^n$ that tends to zero as $n \to \infty$, even though the definition of box dimension says that we should let $\varepsilon \to 0$ continuously. If $\varepsilon \neq (1/3)^n$, the covering will be slightly wasteful—some boxes hang over the edge of the set—but the limiting value of d is the same.

EXAMPLE 11.4.2:

A fractal that is *not* self-similar is constructed as follows. A square region is divided into nine equal squares, and then one of the small squares is selected at random and discarded. Then the process is repeated on each of the eight remaining small squares, and so on. What is the box dimension of the limiting set?

Solution: Figure 11.4.2 shows the first two stages in a typical realization of this random construction.

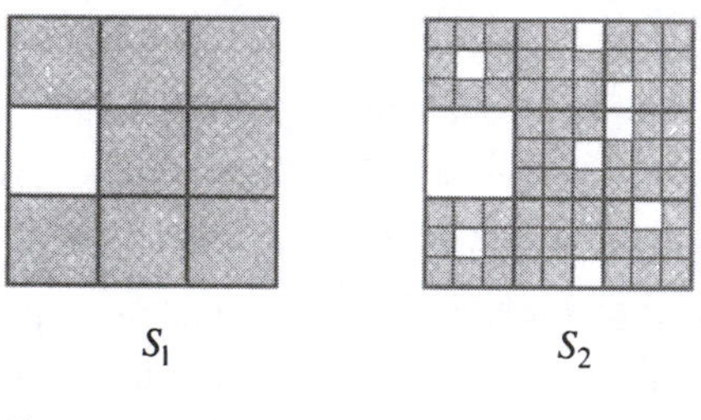

Figure 11.4.2

Pick the unit of length to equal the side of the original square. Then S_1 is covered (with no wastage) by $N = 8$ squares of side $\varepsilon = \frac{1}{3}$. Similarly, S_2 is covered by $N = 8^2$ squares of side $\varepsilon = (\frac{1}{3})^2$. In general, $N = 8^n$ when $\varepsilon = (\frac{1}{3})^n$. Hence

$$d = \lim_{\varepsilon \to 0} \frac{\ln N(\varepsilon)}{\ln(1/\varepsilon)} = \frac{\ln(8^n)}{\ln(3^n)} = \frac{n \ln 8}{n \ln 3} = \frac{\ln 8}{\ln 3}. \blacksquare$$

Critique of Box Dimension

When computing the box dimension, it is not always easy to find a minimal cover. There's an equivalent way to compute the box dimension that avoids this problem. We cover the set with a square mesh of boxes of side ε, count the number of occupied boxes $N(\varepsilon)$, and then compute d as before.

Even with this improvement, the box dimension is rarely used in practice. Its computation requires too much storage space and computer time, compared to other types of fractal dimension (see below). The box dimension also suffers from some mathematical drawbacks. For example, its value is not always what it should be: the set of rational numbers between 0 and 1 can be proven to have a box dimension of 1 (Falconer 1990, p. 44), even though the set has only countably many points.

Falconer (1990) discusses other fractal dimensions, the most important of which is the *Hausdorff dimension*. It is more subtle than the box dimension. The main conceptual difference is that the Hausdorff dimension uses coverings by small sets of *varying* sizes, not just boxes of fixed size ε. It has nicer mathematical properties than the box dimension, but unfortunately it is even harder to compute numerically.

11.5 Pointwise and Correlation Dimensions

Now it's time to return to dynamics. Suppose that we're studying a chaotic system that settles down to a strange attractor in phase space. Given that strange attractors typically have fractal microstructure (as we'll see in Chapter 12), how could we estimate the fractal dimension?

First we generate a set of very many points $\{\mathbf{x}_i, i = 1,\ldots,n\}$ on the attractor by letting the system evolve for a long time (after taking care to discard the initial transient, as usual). To get better statistics, we could repeat this procedure for several different trajectories. In practice, however, almost all trajectories on a strange attractor have the same long-term statistics so it's sufficient to run one trajectory for an extremely long time. Now that we have many points on the attractor, we could try computing the box dimension, but that approach is impractical, as mentioned earlier.

Grassberger and Procaccia (1983) proposed a more efficient approach that has become standard. Fix a point $\mathbf{x}$ on the attractor A. Let $N_{\mathbf{x}}(\varepsilon)$ denote the number of points on A inside a ball of radius ε about $\mathbf{x}$ (Figure 11.5.1).

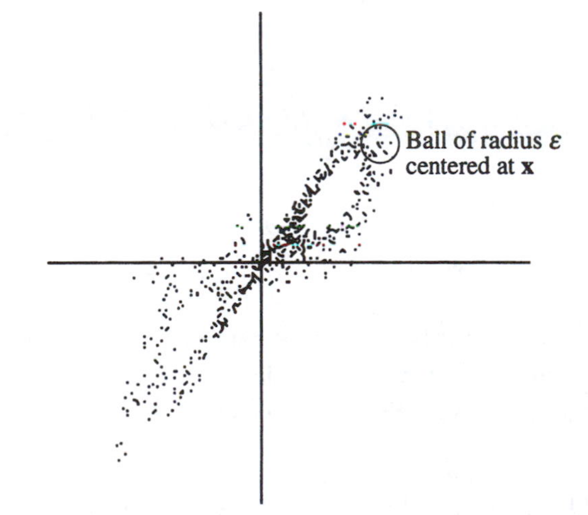

Figure 11.5.1

Most of the points in the ball are unrelated to the immediate portion of the trajectory through $\mathbf{x}$; instead they come from later parts that just happen to pass close to $\mathbf{x}$. Thus $N_{\mathbf{x}}(\varepsilon)$ measures how frequently a typical trajectory visits an ε-neighborhood of $\mathbf{x}$.

Now vary ε. As ε increases, the number of points in the ball typically grows as a power law:

$$N_{\mathbf{x}}(\varepsilon) \propto \varepsilon^d ,$$

where d is called the ***pointwise dimension*** at $\mathbf{x}$. The pointwise dimension can depend significantly on $\mathbf{x}$; it will be smaller in rarefied regions of the attractor. To get an overall dimension of A, one averages $N_{\mathbf{x}}(\varepsilon)$ over many $\mathbf{x}$. The resulting quantity $C(\varepsilon)$ is found empirically to scale as

$$C(\varepsilon) \propto \varepsilon^d ,$$

where d is called the ***correlation dimension***.

The correlation dimension takes account of the density of points on the attractor, and thus differs from the box dimension, which weights all occupied boxes equally, no matter how many points they contain. (Mathematically speaking, the correlation dimension involves an invariant measure supported on a fractal, not just the fractal itself.) In general, $d_{\text{correlation}} \le d_{\text{box}}$, although they are usually very close (Grassberger and Procaccia 1983).

To estimate d, one plots $\log C(\varepsilon)$ vs. $\log \varepsilon$. If the relation $C(\varepsilon) \propto \varepsilon^d$ were valid for all ε, we'd find a straight line of slope d. In practice, the power law holds only over an intermediate range of ε (Figure 11.5.2).

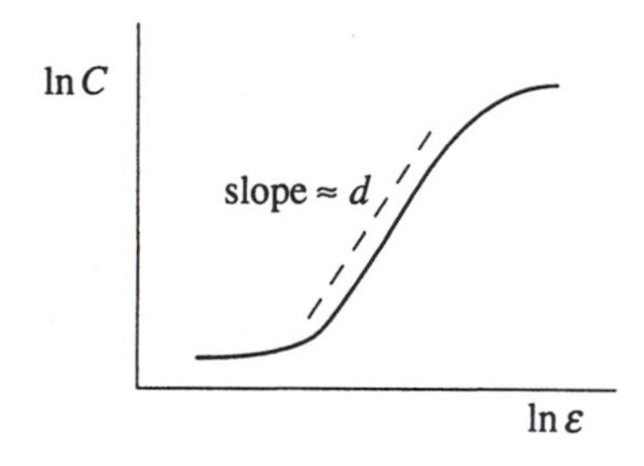

Figure 11.5.2

The curve saturates at large ε because the ε-balls engulf the whole attractor and so $N_{\mathbf{x}}(\varepsilon)$ can grow no further. On the other hand, at extremely small ε, the only point in each ε-ball is $\mathbf{x}$ itself. So the power law is expected to hold only in the *scaling region* where

$$(\text{minimum separation of points on } A) \ll \varepsilon \ll (\text{diameter of } A).$$

EXAMPLE 11.5.1:

Estimate the correlation dimension of the Lorenz attractor, for the standard parameter values $r = 28$, $\sigma = 10$, $b = \frac{8}{3}$.

Solution: Figure 11.5.3 shows the results of Grassberger and Procaccia (1983). (Note that in their notation, the radius of the balls is ℓ and the correlation dimension is ν.) A line of slope $d_{\text{corr}} = 2.05 \pm 0.01$ gives an excellent fit to the data, except for large ε, where the expected saturation occurs.

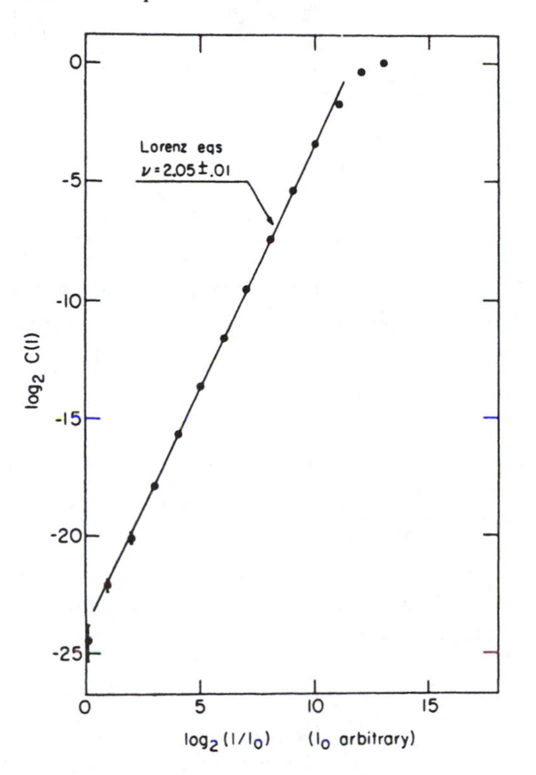

Figure 11.5.3 Grassberger and Procaccia (1983), p. 196

These results were obtained by numerically integrating the system with a Runge–Kutta method. The time step was 0.25, and 15,000 points were computed. Grassberger and Procaccia also report that the convergence was rapid; the correlation dimension could be estimated to within ±5 percent using only a few thousand points. ■

EXAMPLE 11.5.2:

Consider the logistic map $x_{n+1} = rx_n(1 - x_n)$ at the parameter value $r = r_\infty = 3.5699456\ldots$, corresponding to the onset of chaos. Show that the attractor is a Cantor-like set, although it is not strictly self-similar. Then compute its correlation dimension numerically.

Solution: We visualize the attractor by building it up recursively. Roughly speaking, the attractor looks like a 2^n-cycle, for $n >> 1$. Figure 11.5.4 schematically shows some typical 2^n-cycles for small values of n.

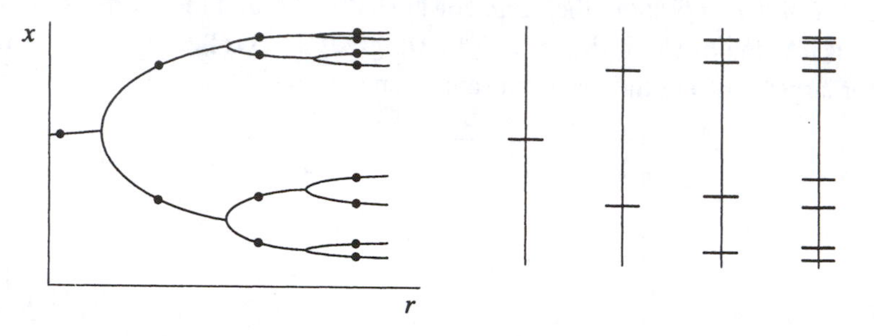

Figure 11.5.4

The dots in the left panel of Figure 11.5.4 represent the superstable 2^n-cycles. The right panel shows the corresponding values of x. As $n \to \infty$, the resulting set approaches a topological Cantor set, with points separated by gaps of various sizes. But the set is not strictly self-similar—the gaps scale by different factors depending on their location. In other words, some of the "wishbones" in the orbit diagram are wider than others at the same r. (We commented on this nonuniformity in Section 10.6, after viewing the computer-generated orbit diagrams of Figure 10.6.2.)

The correlation dimension of the limiting set has been estimated by Grassberger and Procaccia (1983). They generated a single trajectory of 30,000 points, starting from $x_0 = \frac{1}{2}$. Their plot of $\log C(\varepsilon)$ vs. $\log \varepsilon$ is well fit by a straight line of slope $d_{\text{corr}} = 0.500 \pm 0.005$ (Figure 11.5.5).

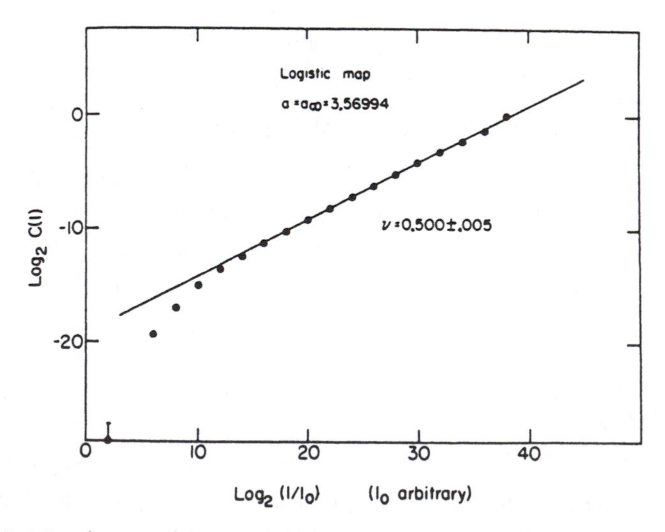

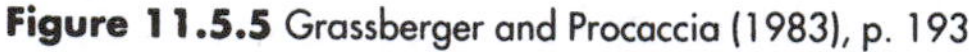

Figure 11.5.5 Grassberger and Procaccia (1983), p. 193

This is smaller than the box dimension $d_{\text{box}} \approx 0.538$ (Grassberger 1981), as expected. ■

For very small ε, the data in Figure 11.5.5 deviate from a straight line. Grassberger and Procaccia (1983) attribute this deviation to residual correlations among the x_n's on their single trajectory. These correlations would be negligible if the map were strongly chaotic, but for a system at the onset of chaos (like this one), the correlations are visible at small scales. To extend the scaling region, one could use a larger number of points or more than one trajectory.

Multifractals

We conclude by mentioning a recent development, although we cannot go into details. In the logistic attractor of Example 11.5.2, the scaling varies from place to place, unlike in the middle-thirds Cantor set, where there is a uniform scaling by $\frac{1}{3}$ everywhere. Thus we cannot completely characterize the logistic attractor by its dimension, or any other single number—we need some kind of distribution function that tells us how the dimension varies across the attractor. Sets of this type are called *multifractals*.

The notion of pointwise dimension allows us to quantify the local variations in scaling. Given a multifractal A, let S_α be the subset of A consisting of all points with pointwise dimension α. If α is a typical scaling factor on A, then it will be represented often, so S_α will be a relatively large set; if α is unusual, then S_α will be a small set. To be more quantitative, we note that each S_α is itself a fractal, so it makes sense to measure its "size" by its fractal dimension. Thus, let $f(\alpha)$ denote the dimension of S_α. Then $f(\alpha)$ is called the *multifractal spectrum* of A or the *spectrum of scaling indices* (Halsey et al. 1986).

Roughly speaking, you can think of the multifractal as an interwoven set of fractals of different dimensions α, where $f(\alpha)$ measures their relative weights. Since very large and very small α are unlikely, the shape of $f(\alpha)$ typically looks like Figure 11.5.6. The maximum value of $f(\alpha)$ turns out to be the box dimension (Halsey et al. 1986).

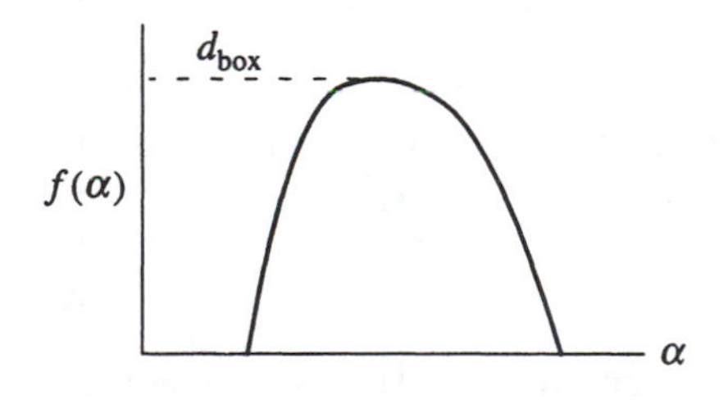

Figure 11.5.6

For systems at the onset of chaos, multifractals lead to a more powerful version of the universality theory mentioned in Section 10.6. The universal quantity is now a *function* $f(\alpha)$, rather than a single number; it therefore offers much more information, and the possibility of more stringent tests. The theory's predictions have been checked for a variety of experimental systems at the onset of chaos, with striking success. See Glazier and Libchaber (1988) for a review. On the other hand, we still lack a rigorous mathematical theory of multifractals; see Falconer (1990) for a discussion of the issues.

EXERCISES FOR CHAPTER 11

11.1 Countable and Uncountable Sets

11.1.1 Why doesn't the diagonal argument used in Example 11.1.4 show that the rationals are also uncountable? (After all, rationals can be represented as decimals.)

11.1.2 Show that the set of odd integers is countable.

11.1.3 Are the irrational numbers countable or uncountable? Prove your answer.

11.1.4 Consider the set of all real numbers whose decimal expansion contains only 2's and 7's. Using Cantor's diagonal argument, show that this set is uncountable.

11.1.5 Consider the set of integer lattice points in three-dimensional space, i.e., points of the form (p,q,r), where p, q, and r are integers. Show that this set is countable.

11.1.6 ($10x$ mod 1) Consider the decimal shift map $x_{n+1} = 10x_n \pmod 1$.

a) Show that the map has countably many periodic orbits, all of which are unstable.
b) Show that the map has uncountably many aperiodic orbits.
c) An "eventually-fixed point" of a map is a point that iterates to a fixed point after a finite number of steps. Thus $x_{n+1} = x_n$ for all $n > N$, where N is some positive integer. Is the number of eventually-fixed points for the decimal shift map countable or uncountable?

11.1.7 Show that the binary shift map $x_{n+1} = 2x_n \pmod 1$ has countably many periodic orbits and uncountably many aperiodic orbits.

11.2 Cantor Set

11.2.1 (Cantor set has measure zero) Here's another way to show that the Cantor set has zero total length. In the first stage of construction of the Cantor set, we removed an interval of length $\frac{1}{3}$ from the unit interval $[0,1]$. At the next stage we removed two intervals, each of length $\frac{1}{9}$. By summing an appropriate infinite series, show that the total length of all the intervals removed is 1, and hence the leftovers (the Cantor set) must have length zero.

11.2.2 Show that the rational numbers have zero measure. (Hint: Make a list of the rationals. Cover the first number with an interval of length ε, cover the second with an interval of length $\frac{1}{2}\varepsilon$. Now take it from there.)

11.2.3 Show that any countable subset of the real line has zero measure. (This generalizes the result of the previous question.)

11.2.4 Consider the set of irrational numbers between 0 and 1.

a) What is the measure of the set?

b) Is it countable or uncountable?

c) Is it totally disconnected?

d) Does it contain any isolated points?

11.2.5 (Base-3 and the Cantor set)

a) Find the base-3 expansion of $1/2$.

b) Find a one-to-one correspondence between the Cantor set C and the interval $[0,1]$. In other words, find an invertible mapping that pairs each point $c \in C$ with precisely one $x \in [0,1]$.

c) Some of my students have thought that the Cantor set is "all endpoints"—they claimed that any point in the set is the endpoint of some sub-interval involved in the construction of the set. Show that this is false by explicitly identifying a point in C that is not an endpoint.

11.2.6 (Devil's staircase) Suppose that we pick a point at random from the Cantor set. What's the probability that this point lies to the left of x, where $0 \le x \le 1$ is some fixed number? The answer is given by a function $P(x)$ called the ***devil's staircase***.

a) It is easiest to visualize $P(x)$ by building it up in stages. First consider the set S_0 in Figure 11.2.1. Let $P_0(x)$ denote the probability that a randomly chosen point in S_0 lies to the left of x. Show that $P_0(x) = x$.

b) Now consider S_1 and define $P_1(x)$ analogously. Draw the graph of $P_1(x)$. (Hint: It should have a plateau in the middle.)

c) Draw the graphs of $P_n(x)$, for $n = 2,3,4$. Be careful about the widths and heights of the plateaus.

d) The limiting function $P_\infty(x)$ is the devil's staircase. Is it continuous? What would a graph of its derivative look like?

Like other fractal concepts, the devil's staircase was long regarded as a mathematical curiosity. But recently it has arisen in physics, in connection with mode-locking of nonlinear oscillators. See Bak (1986) for an entertaining introduction.

11.3 Dimension of Self-Similar Fractals

11.3.1 (Middle-halves Cantor set) Construct a new kind of Cantor set by removing the middle half of each sub-interval, rather than the middle third.

a) Find the similarity dimension of the set.

b) Find the measure of the set.

11.3.2 (Generalized Cantor set) Consider a generalized Cantor set in which we begin by removing an open interval of length $0 < a < 1$ from the middle of $[0,1]$. At subsequent stages, we remove an open middle interval (whose length is the same fraction a) from each of the remaining intervals, and so on. Find the similarity dimension of the limiting set.

11.3.3 (Generalization of even-fifths Cantor set) The "even-sevenths Cantor set" is constructed as follows: divide $[0,1]$ into seven equal pieces; delete pieces 2, 4, and 6; and repeat on sub-intervals.

a) Find the similarity dimension of the set.

b) Generalize the construction to any odd number of pieces, with the even ones deleted. Find the similarity dimension of this generalized Cantor set.

11.3.4 (No odd digits) Find the similarity dimension of the subset of $[0,1]$ consisting of real numbers with only even digits in their decimal expansion.

11.3.5 (No 8's) Find the similarity dimension of the subset of $[0,1]$ consisting of real numbers that can be written without the digit 8 appearing anywhere in their decimal expansion.

11.3.6 Show that the middle-thirds Cantor set contains no intervals. But also show that no point in the set is isolated.

11.3.7 (Snowflake) To construct the famous fractal known as the ***von Koch snowflake curve***, use an equilateral triangle for S_0. Then do the von Koch procedure of Figure 11.3.1 on each of the three sides.

a) Show that S_1 looks like a star of David.

b) Draw S_2 and S_3.

c) The snowflake is the limiting curve $S = S_\infty$. Show that it has infinite arc length.

d) Find the area of the region enclosed by S.

e) Find the similarity dimension of S.

The snowflake curve is continuous but nowhere differentiable—loosely speaking, it is "all corners"!

11.3.8 (Sierpinski carpet) Consider the process shown in Figure 1. The closed unit box is divided into nine equal boxes, and the open central box is deleted. Then this process is repeated for each of the eight remaining sub-boxes, and so on. Figure 1 shows the first two stages.

a) Sketch the next stage S_3.

b) Find the similarity dimension of the limiting fractal, known as the ***Sierpinski carpet***.

c) Show that the Sierpinski carpet has zero area.

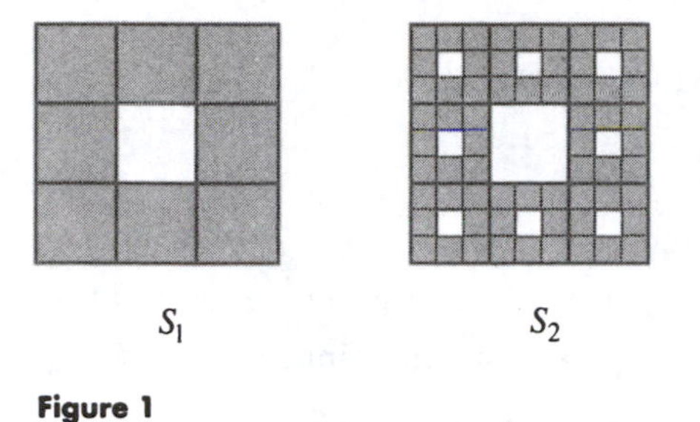

Figure 1

11.3.9 (Sponges) Generalize the previous exercise to three dimensions—start with a solid cube, and divide it into 27 equal sub-cubes. Delete the central cube on each face, along with the central cube. (If you prefer, you could imagine drilling three mutually orthogonal square holes through the centers of the faces.) Infinite iteration of this process yields a fractal called the ***Menger sponge***. Find its similarity dimension. Repeat for the Menger hypersponge in N dimensions, if you dare.

11.3.10 (Fat fractal) A ***fat fractal*** is a fractal with a nonzero measure. Here's a simple example: start with the unit interval $[0,1]$ and delete the open middle $1/2$, $1/4$, $1/8$, etc., of each remaining sub-interval. (Thus a smaller and smaller fraction is removed at each stage, in contrast to the middle-thirds Cantor set, where we always remove $1/3$ of what's left.)

a) Show that the limiting set is a topological Cantor set.

b) Show that the measure of the limiting set is greater than zero. Find its exact value if you can, or else just find a lower bound for it.

Fat fractals answer a fascinating question about the logistic map. Farmer (1985) has shown numerically that the set of parameter values for which chaos occurs is a fat fractal. In particular, if r is chosen at random between r_∞ and $r = 4$, there is about an 89% chance that the map will be chaotic. Farmer's analysis also suggests that the odds of making a mistake (calling an orbit chaotic when it's actually periodic) are about one in a million, if we use double precision arithmetic!

11.4 Box Dimension

Find the box dimension of the following sets.

11.4.1 von Koch snowflake (see Exercise 11.3.7)

11.4.2 Sierpinski carpet (see Exercise 11.3.8)

11.4.3 Menger sponge (see Exercise 11.3.9)

11.4.4 The Cartesian product of the middle-thirds Cantor set with itself.

11.4.5 Menger hypersponge (see Exercise 11.3.9)

11.4.6 (A strange repeller for the tent map) The tent map on the interval $[0,1]$ is defined by $x_{n+1} = f(x_n)$, where

$$f(x) = \begin{cases} rx, & 0 \le x \le \frac{1}{2} \\ r(1-x), & \frac{1}{2} \le x \le 1 \end{cases}$$

and $r > 0$. In this exercise we assume $r > 2$. Then some points get mapped outside the interval $[0,1]$. If $f(x_0) > 1$ then we say that x_0 has "escaped" after one iteration. Similarly, if $f^n(x_0) > 1$ for some finite n, but $f^k(x_0) \in [0,1]$ for all $k < n$, then we say that x_0 has escaped after n iterations.

a) Find the set of initial conditions x_0 that escape after one or two iterations.
b) Describe the set of x_0 that *never* escape.
c) Find the box dimension of the set of x_0 that never escape. (This set is called the invariant set.)
d) Show that the Liapunov exponent is positive at each point in the invariant set.

The invariant set is called a ***strange repeller***, for several reasons: it has a fractal structure; it repels all nearby points that are not in the set; and points in the set hop around chaotically under iteration of the tent map.

11.4.7 (A lopsided fractal) Divide the closed unit interval $[0,1]$ into four quarters. Delete the open second quarter from the left. This produces a set S_1. Repeat this construction indefinitely; i.e., generate S_{n+1} from S_n by deleting the second quarter of each of the intervals in S_n.

a) Sketch the sets $S_1, \ldots, S_4$.
b) Compute the box dimension of the limiting set S_∞.
c) Is S_∞ self-similar?

11.4.8 (A thought question about random fractals) Redo the previous question, except add an element of randomness to the process: to generate S_{n+1} from S_n, flip a coin; if the result is heads, delete the second quarter of every interval in S_n; if tails, delete the third quarter. The limiting set is an example of a ***random fractal***.

a) Can you find the box dimension of this set? Does this question even make sense? In other words, might the answer depend on the particular sequence of heads and tails that happen to come up?
b) Now suppose if tails comes up, we delete the *first* quarter. Could this make a difference? For instance, what if we had a long string of tails?

See Falconer (1990, Chapter 15) for a discussion of random fractals.

11.4.9 (Fractal cheese) A fractal slice of swiss cheese is constructed as follows: The unit square is divided into p^2 squares, and m^2 squares are chosen at random and discarded. (Here $p > m+1$, and p, m are positive integers.) The process is repeated for each remaining square (side = $1/p$). Assuming that this process is repeated indefinitely, find the box dimension of the resulting fractal. (Notice that the resulting fractal may or may not be self-similar, depending on which squares are removed at each stage. Nevertheless, we are still able to calculate the box dimension.)

11.4.10 (Fat fractal) Show that the fat fractal constructed in Exercise 11.3.10 has box dimension equal to 1.

11.5 Pointwise and Correlation Dimensions

11.5.1 (Project) Write a program to compute the correlation dimension of the Lorenz attractor. Reproduce the results in Figure 11.5.3. Then try other values of r. How does the dimension depend on r?

12

STRANGE ATTRACTORS

12.0 Introduction

Our work in the previous three chapters has revealed quite a bit about chaotic systems, but something important is missing: intuition. We know *what* happens but not *why* it happens. For instance, we don't know what causes sensitive dependence on initial conditions, nor how a differential equation can generate a fractal attractor. Our first goal is to understand such things in a simple, geometric way.

These same issues confronted scientists in the mid-1970s. At the time, the only known examples of strange attractors were the Lorenz attractor (1963) and some mathematical constructions of Smale (1967). Thus there was a need for other concrete examples, preferably as transparent as possible. These were supplied by Hénon (1976) and Rössler (1976), using the intuitive concepts of stretching and folding. These topics are discussed in Sections 12.1–12.3. The chapter concludes with experimental examples of strange attractors from chemistry and mechanics. In addition to their inherent interest, these examples illustrate the techniques of attractor reconstruction and Poincaré sections, two standard methods for analyzing experimental data from chaotic systems.

12.1 The Simplest Examples

Strange attractors have two properties that seem hard to reconcile. Trajectories on the attractor remain confined to a bounded region of phase space, yet they separate from their neighbors exponentially fast (at least initially). How can trajectories diverge endlessly and yet stay bounded?

The basic mechanism involves repeated ***stretching and folding***. Consider a small blob of initial conditions in phase space (Figure 12.1.1).

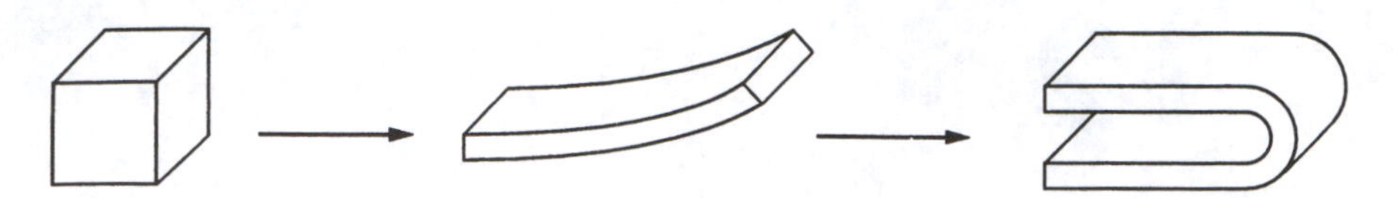

Figure 12.1.1

A strange attractor typically arises when the flow contracts the blob in some directions (reflecting the dissipation in the system) and stretches it in others (leading to sensitive dependence on initial conditions). The stretching cannot go on forever—the distorted blob must be folded back on itself to remain in the bounded region.

To illustrate the effects of stretching and folding, we consider a domestic example.

Making Pastry

Figure 12.1.2 shows a process used to make filo pastry or croissant.

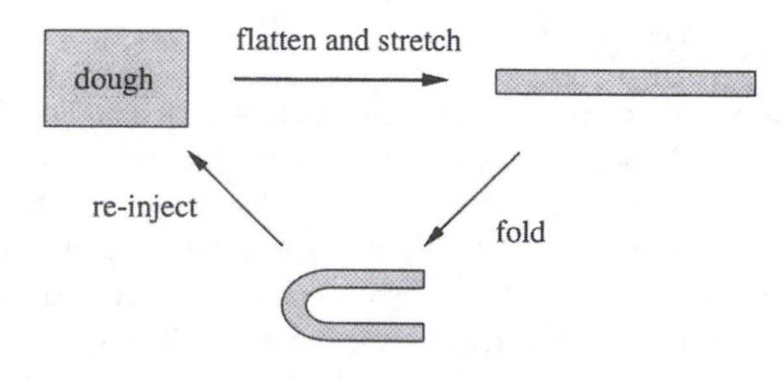

Figure 12.1.2

The dough is rolled out and flattened, then folded over, then rolled out again, and so on. After many repetitions, the end product is a flaky, layered structure—the culinary analog of a fractal attractor.

Furthermore, the process shown in Figure 12.1.2 automatically generates sensitive dependence on initial conditions. Suppose that a small drop of food coloring is put in the dough, representing nearby initial conditions. After many iterations of stretching, folding, and re-injection, the coloring will be spread throughout the dough.

Figure 12.1.3 presents a more detailed view of this ***pastry map***, here modeled as a continuous mapping of a rectangle into itself.

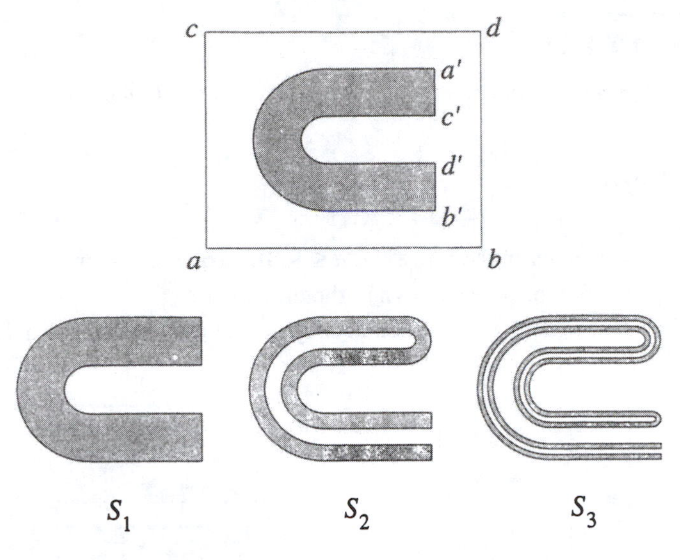

Figure 12.1.3

The rectangle *abcd* is flattened, stretched, and folded into the ***horseshoe*** $a'b'c'd'$, also shown as S_1. In the same way, S_1 is itself flattened, stretched, and folded into S_2, and so on. As we go from one stage to the next, the layers become thinner and there are twice as many of them.

Now try to picture the limiting set S_∞. It consists of infinitely many smooth layers, separated by gaps of various sizes. In fact, a vertical cross section through the middle of S_∞ would resemble a *Cantor set*! Thus S_∞ is (locally) the product of a smooth curve with a Cantor set. The fractal structure of the attractor is a consequence of the stretching and folding that created S_∞ in the first place.

Terminology

The transformation shown in Figure 12.1.3 is normally called a horseshoe map, but we have avoided that name because it encourages confusion with another horseshoe map (the *Smale horseshoe*), which has very different properties. In particular, Smale's horseshoe map does *not* have a strange attractor; its invariant set is more like a strange saddle. The Smale horseshoe is fundamental to rigorous discussions of chaos, but its analysis and significance are best deferred to a more advanced course. See Exercise 12.1.7 for an introduction, and Guckenheimer and Holmes (1983) or Arrowsmith and Place (1990) for detailed treatments.

Because we want to reserve the word *horseshoe* for Smale's mapping, we have used the name *pastry map* for the mapping above. A better name would be "the baker's map" but that name is already taken by the map in the following example.

EXAMPLE 12.1.1:

The ***baker's map*** B of the square $0 \le x \le 1$, $0 \le y \le 1$ to itself is given by

$$(x_{n+1}, y_{n+1}) = \begin{cases} (2x_n, \; ay_n) & \text{for } 0 \le x_n < \frac{1}{2} \\ (2x_n - 1, \; ay_n + \frac{1}{2}) & \text{for } \frac{1}{2} \le x_n \le 1 \end{cases}$$

where a is a parameter in the range $0 < a \le \frac{1}{2}$. Illustrate the geometric action of B by showing its effect on a face drawn in the unit square.

Solution: The reluctant experimental subject is shown in Figure 12.1.4a.

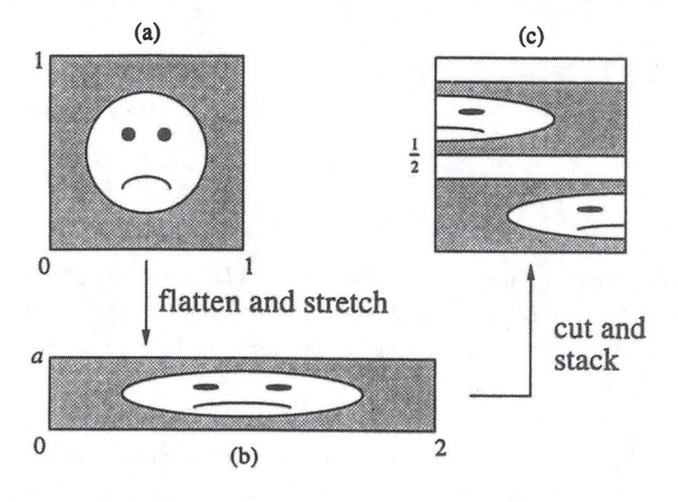

Figure 12.1.4

As we'll see momentarily, the transformation may be regarded as a product of two simpler transformations. First the square is stretched and flattened into a $2 \times a$ rectangle (Figure 12.1.4b). Then the rectangle is cut in half, yielding two $1 \times a$ rectangles, and the right half is stacked on top of the left half such that its base is at the level $y = \frac{1}{2}$ (Figure 12.1.4c).

Why is this procedure equivalent to the formulas for B? First consider the left half of the square, where $0 \le x_n < \frac{1}{2}$. Here $(x_{n+1}, y_{n+1}) = (2x_n, ay_n)$, so the horizontal direction is stretched by 2 and the vertical direction is contracted by a, as claimed. The same is true for the right half of the rectangle, except that the image is shifted left by 1 and up by $\frac{1}{2}$, since $(x_{n+1}, y_{n+1}) = (2x_n, ay_n) + (-1, \frac{1}{2})$. This shift is equivalent to the stacking just claimed. ■

The baker's map exhibits sensitive dependence on initial conditions, thanks to the stretching in the x-direction. It has many chaotic orbits—uncountably many, in fact. These and other dynamical properties of the baker's map are discussed in the exercises.

The next example shows that, like the pastry map, the baker's map has a strange attractor with a Cantor-like cross section.

EXAMPLE 12.1.2:

Show that for $a < \frac{1}{2}$, the baker's map has a fractal attractor A that attracts all orbits. More precisely, show that there is a set A such that for any initial condition (x_0, y_0), the distance from $B^n(x_0, y_0)$ to A converges to zero as $n \to \infty$.

Solution: First we construct the attractor. Let S denote the square $0 \le x \le 1$, $0 \le y \le 1$; this includes all possible initial conditions. The first three images of S under the map B are shown as shaded regions in Figure 12.1.5.

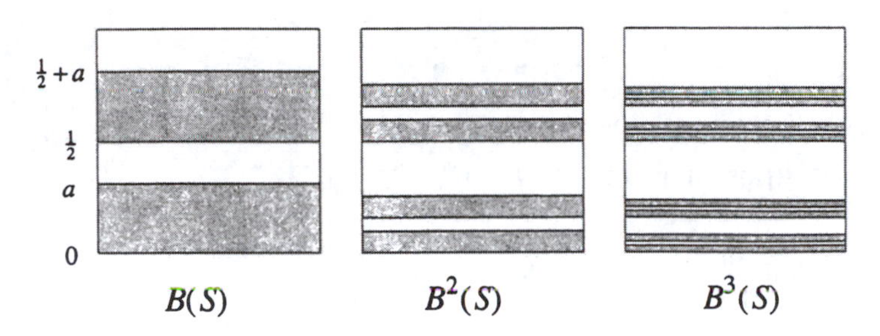

Figure 12.1.5

The first image $B(S)$ consists of two strips of height a, as we know from Example 12.1.1. Then $B(S)$ is flattened, stretched, cut, and stacked to yield $B^2(S)$. Now we have four strips of height a^2. Continuing in this way, we see that $B^n(S)$ consists of 2^n horizontal strips of height a^n. The limiting set $A = B^\infty(S)$ is a fractal. Topologically, it is a Cantor set of line segments.

A technical point: How we can be sure that there actually is a "limiting set"? We invoke a standard theorem from point-set topology. Observe that the successive images of the square are ***nested*** inside each other like Chinese boxes: $B^{n+1}(S) \subset B^n(S)$ for all n. Moreover each $B^n(S)$ is a compact set. The theorem (Munkres 1975) assures us that the countable intersection of a nested family of compact sets is a *non-empty* compact set—this set is our A. Furthermore, $A \subset B^n(S)$ for all n.

The nesting property also helps us to show that A attracts all orbits. The point $B^n(x_0, y_0)$ lies somewhere in one of the strips of $B^n(S)$, and all points in these strips are within a distance a^n of A, because A is contained in $B^n(S)$. Since $a^n \to 0$ as $n \to \infty$, the distance from $B^n(x_0, y_0)$ to A tends to zero as $n \to \infty$, as required. ■

EXAMPLE 12.1.3:

Find the box dimension of the attractor for the baker's map with $a < \frac{1}{2}$.

Solution: The attractor A is approximated by $B^n(S)$, which consists of 2^n strips of height a^n and length 1. Now cover A with square boxes of side $\varepsilon = a^n$ (Figure 12.1.6).

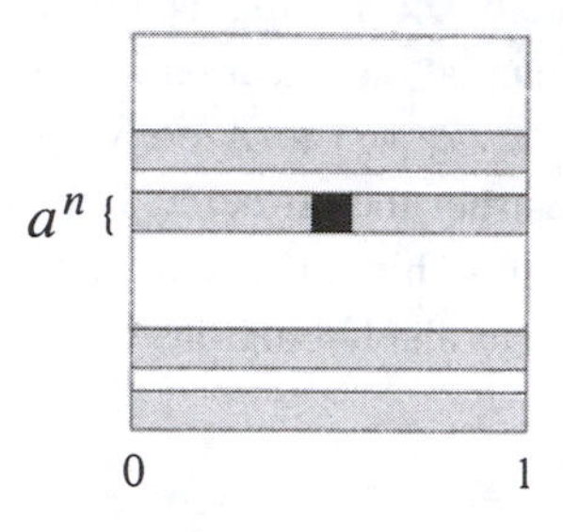

Figure 12.1.6

Since the strips have length 1, it takes about a^{-n} boxes to cover each of them. There are 2^n strips altogether, so $N \approx a^{-n} \times 2^n = (a/2)^{-n}$. Thus

$$d = \lim_{\varepsilon \to 0} \frac{\ln N}{\ln(\frac{1}{\varepsilon})} = \lim_{n \to \infty} \frac{\ln[(a/2)^{-n}]}{\ln(a^{-n})} = 1 + \frac{\ln \frac{1}{2}}{\ln a}.$$

As a check, note that $d \to 2$ as $a \to \frac{1}{2}$; this makes sense because the attractor fills an increasingly large portion of square S as $a \to \frac{1}{2}$. ■

The Importance of Dissipation

For $a < \frac{1}{2}$, the baker's map shrinks areas in phase space. Given any region R in the square,

$$\text{area}(B(R)) < \text{area}(R).$$

This result follows from elementary geometry. The baker's map elongates R by a factor of 2 and flattens it by a factor of a, so $\text{area}(B(R)) = 2a \times \text{area}(R)$. Since $a < \frac{1}{2}$ by assumption, $\text{area}(B(R)) < \text{area}(R)$ as required. (Note that the cutting operation does not change the region's area.)

Area contraction is the analog of the volume contraction that we found for the Lorenz equations in Section 9.2. As in that case, it yields several conclusions. For instance, the attractor A for the baker's map must have zero area. Also, the baker's map cannot have any repelling fixed points, since such points would expand area elements in their neighborhood.

In contrast, when $a = \frac{1}{2}$ the baker's map is ***area-preserving***: $\text{area}(B(R)) = \text{area}(R)$. Now the square S is mapped *onto* itself, with no gaps between the strips. The map has qualitatively different dynamics in this case. Transients never decay—the orbits shuffle around endlessly in the square but never settle down to a lower-dimensional attractor. This is a kind of chaos that we have not seen before!

This distinction between $a < \frac{1}{2}$ and $a = \frac{1}{2}$ exemplifies a broader theme in nonlinear dynamics. In general, if a map or flow contracts volumes in phase space, it is called ***dissipative***. Dissipative systems commonly arise as models of physical situations involving friction, viscosity, or some other process that dissipates energy. In contrast, area-preserving maps are associated with conservative systems, particularly with the Hamiltonian systems of classical mechanics.

The distinction is crucial because *area-preserving maps cannot have attractors* (strange or otherwise). As defined in Section 9.3, an "attractor" should attract all orbits starting in a sufficiently small open set containing it; that requirement is incompatible with area-preservation.

Several of the exercises give a taste of the new phenomena that arise in area-preserving maps. To learn more about the fascinating world of Hamiltonian chaos, see the review articles by Jensen (1987) or Hénon (1983), or the books by Tabor (1989) or Lichtenberg and Lieberman (1992).

12.2 Hénon Map

In this section we discuss another two-dimensional map with a strange attractor. It was devised by the theoretical astronomer Michel Hénon (1976) to illuminate the microstructure of strange attractors.

According to Gleick (1987, p. 149), Hénon became interested in the problem after hearing a lecture by the physicist Yves Pomeau, in which Pomeau described the numerical difficulties he had encountered in trying to resolve the tightly packed sheets of the Lorenz attractor. The difficulties stem from the rapid volume contraction in the Lorenz system: after one circuit around the attractor, a volume in phase space is typically squashed by a factor of about 14,000 (Lorenz 1963).

Hénon had a clever idea. Instead of tackling the Lorenz system directly, he sought a mapping that captured its essential features but which also had an adjustable amount of dissipation. Hénon chose to study mappings rather than differential equations because maps are faster to simulate and their solutions can be followed more accurately and for a longer time.

The ***Hénon map*** is given by

$$x_{n+1} = y_n + 1 - ax_n^2, \quad y_{n+1} = bx_n, \qquad (1)$$

where a and b are adjustable parameters. Hénon (1976) arrived at this map by an elegant line of reasoning. To simulate the stretching and folding that occurs in the Lorenz system, he considered the following chain of transformations (Figure 12.2.1).

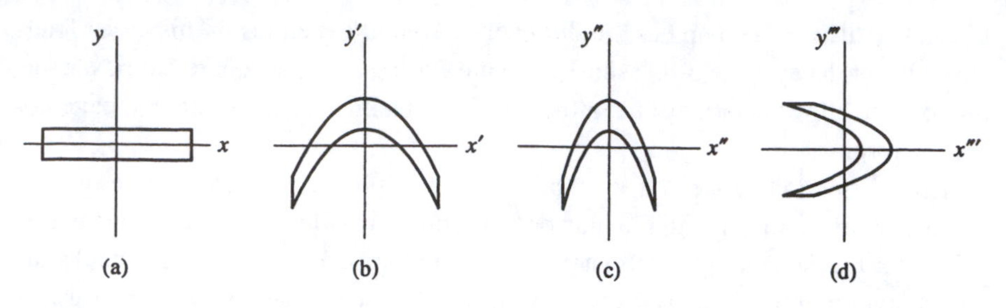

Figure 12.2.1

Start with a rectangular region elongated along the x-axis (Figure 12.2.1a). Stretch and fold the rectangle by applying the transformation

$$T': \; x' = x, \quad y' = 1 + y - ax^2.$$

(The primes denote iteration, not differentiation.) The bottom and top of the rectangle get mapped to parabolas (Figure 12.2.1b). The parameter a controls the folding. Now fold the region even more by contracting Figure 12.2.1b along the x-axis:

$$T'': \; x'' = bx', \quad y'' = y'$$

where $-1 < b < 1$. This produces Figure 12.2.1c. Finally, come back to the orientation along the x-axis by reflecting across the line $y = x$ (Figure 12.2.1d):

$$T''': \; x''' = y'', \quad y''' = x''.$$

Then the composite transformation $T = T''' \, T'' \, T'$ yields the Hénon mapping (1), where we use the notation (x_n, y_n) for (x, y) and (x_{n+1}, y_{n+1}) for (x''', y''').

Elementary Properties of the Hénon Map

As desired, the Hénon map captures several essential properties of the Lorenz system. (These properties will be verified in the examples below and in the exercises.)

1. *The Hénon map is invertible.* This property is the counterpart of the fact that in the Lorenz system, there is a unique trajectory through each point in phase space. In particular, each point has a unique past. In this respect the Hénon map is superior to the logistic map, its one-dimensional analog. The logistic map stretches and folds the unit interval, but it is not invertible since all points (except the maximum) come from *two* pre-images.
2. *The Hénon map is dissipative.* It contracts areas, and does so at the same rate everywhere in phase space. This property is the analog of constant negative divergence in the Lorenz system.

3. *For certain parameter values, the Hénon map has a trapping region.* In other words, there is a region R that gets mapped inside itself (Figure 12.2.2). As in the Lorenz system, the strange attractor is enclosed in the trapping region.

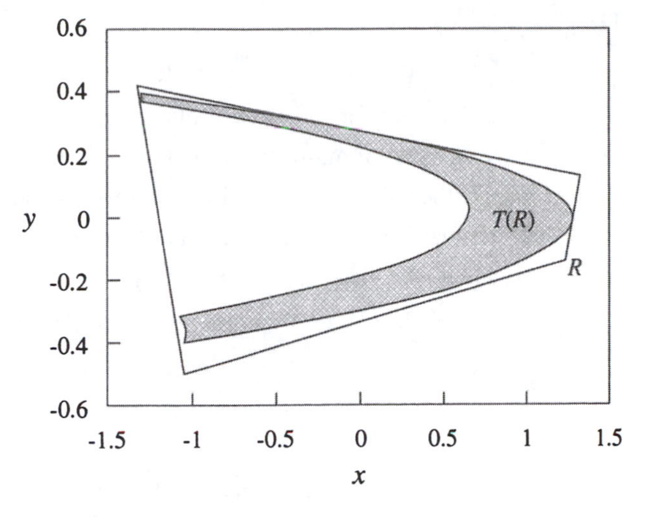

Figure 12.2.2

The next property highlights an important difference between the Hénon map and the Lorenz system.

4. *Some trajectories of the Hénon map escape to infinity.* In contrast, all trajectories of the Lorenz system are bounded; they all eventually enter and stay inside a certain large ellipsoid (Exercise 9.2.2). But it is not surprising that the Hénon map has some unbounded trajectories; far from the origin, the quadratic term in (1) dominates and repels orbits to infinity. Similar behavior occurs in the logistic map—recall that orbits starting outside the unit interval eventually become unbounded.

Now we verify properties 1 and 2. For 3 and 4, see Exercises 12.2.9 and 12.2.10.

EXAMPLE 12.2.1:

Show that the Hénon map T is invertible if $b \neq 0$, and find the inverse T^{-1}.

Solution: We solve (1) for x_n and y_n, given x_{n+1} and y_{n+1}. Algebra yields $x_n = b^{-1} y_{n+1}$, $y_n = x_{n+1} - 1 + ab^{-2}(y_{n+1})^2$. Thus T^{-1} exists for all $b \neq 0$. ■

EXAMPLE 12.2.2:

Show that the Hénon map contracts areas if $-1 < b < 1$.

Solution: To decide whether an arbitrary two-dimensional map $x_{n+1} = f(x_n, y_n)$, $y_{n+1} = g(x_n, y_n)$ is area-contracting, we compute the determinant of its Jacobian matrix

$$\mathbf{J} = \begin{pmatrix} \frac{\partial f}{\partial x} & \frac{\partial f}{\partial y} \\ \frac{\partial g}{\partial x} & \frac{\partial g}{\partial y} \end{pmatrix}.$$

If $|\det \mathbf{J}(x,y)| < 1$ for all (x,y), the map is area-contracting.

This rule follows from a fact of multivariable calculus: if $\mathbf{J}$ is the Jacobian of a two-dimensional map T, then T maps an infinitesimal rectangle at (x,y) with area $dx\,dy$ into an infinitesimal parallelogram with area $|\det \mathbf{J}(x,y)|\,dx\,dy$. Thus if $|\det \mathbf{J}(x,y)| < 1$ everywhere, the map is area-contracting.

For the Hénon map, we have $f(x,y) = 1 - ax^2 + y$ and $g(x,y) = bx$. Therefore

$$\mathbf{J} = \begin{pmatrix} -2ax & 1 \\ b & 0 \end{pmatrix}$$

and $\det \mathbf{J}(x,y) = -b$ for all (x,y). Hence the map is area-contracting for $-1 < b < 1$, as claimed. In particular, the area of any region is reduced by a *constant* factor of $|b|$ with each iteration. ■

Choosing Parameters

The next step is to choose suitable values of the parameters. As Hénon (1976) explains, b should not be too close to zero, or else the area contraction will be excessive and the fine structure of the attractor will be invisible. But if b is too large, the folding won't be strong enough. (Recall that b plays two roles: it controls the dissipation *and* produces extra folding in going from Figure 12.2.1b to Figure 12.2.1c.) A good choice is $b = 0.3$.

To find a good value of a, Hénon had to do some exploring. If a is too small or too large, all trajectories escape to infinity; there is no attractor in these cases. (This is reminiscent of the logistic map, where almost all trajectories escape to infinity unless $0 \le r \le 4$.) For intermediate values of a, the trajectories either escape to infinity or approach an attractor, depending on the initial conditions. As a increases through this range, the attractor changes from a stable fixed point to a stable 2-cycle. The system then undergoes a period-doubling route to chaos, followed by chaos intermingled with periodic windows. Hénon picked $a = 1.4$, well into the chaotic region.

Zooming In on a Strange Attractor

In a striking series of plots, Hénon provided the first direct visualization of the fractal structure of a strange attractor. He set $a = 1.4$, $b = 0.3$ and generated the attractor by computing ten thousand successive iterates of (1), starting from the origin. You really must try this for yourself on a computer. The effect is eerie—the points (x_n, y_n) hop around erratically, but soon the attractor begins to take form, "like a ghost out of the mist" (Gleick 1987, p.150).

The attractor is bent like a boomerang and is made of many parallel curves (Figure 12.2.3a).

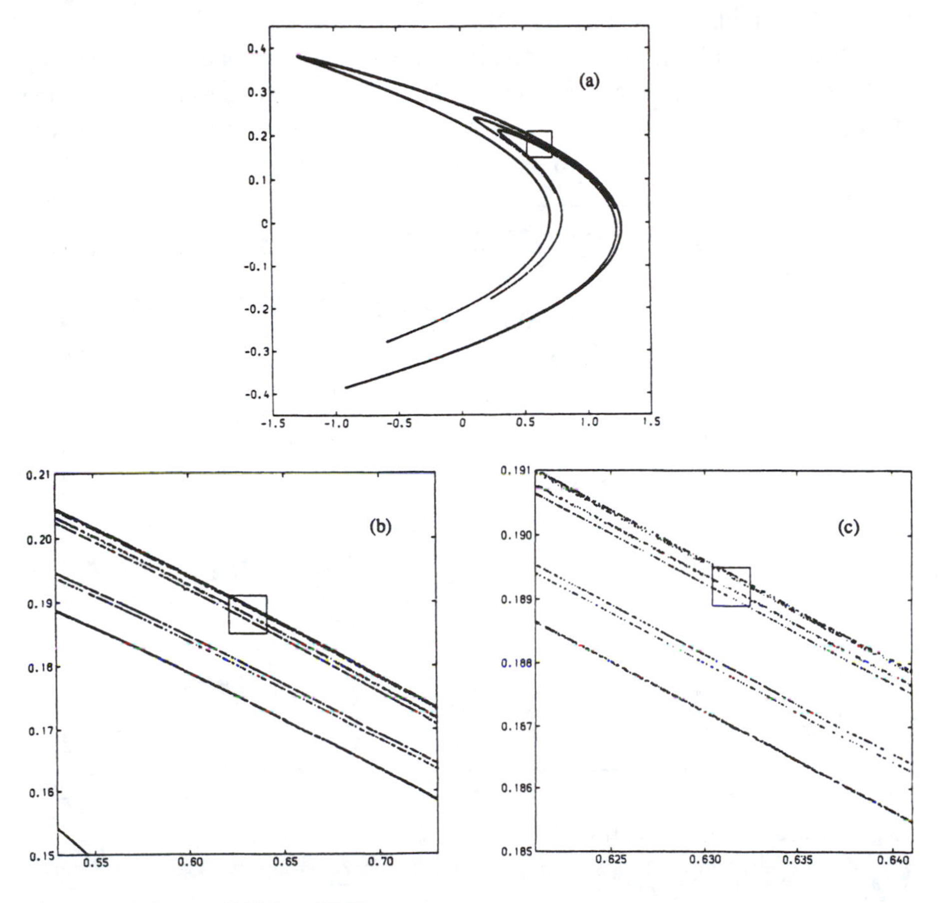

Figure 12.2.3 Hénon (1976), pp 74–76

Figure 12.2.3b is an enlargement of the small square of Figure 12.2.3a. The characteristic fine structure of the attractor begins to emerge. There seem to be six parallel curves: a lone curve near the middle of the frame, then two closely spaced curves above it, and then three more. If we zoom in on those three curves (Figure 12.2.3c), it becomes clear that they are actually six curves, grouped one, two, three, exactly as before! And those curves are themselves made of thinner curves in the same pattern, and so on. The self-similarity continues to arbitrarily small scales.

The Unstable Manifold of the Saddle Point

Figure 12.2.3 suggests that the Hénon attractor is Cantor-like in the transverse direction, but smooth in the longitudinal direction. There's a reason for this. The attractor is closely related to a locally smooth object—the unstable manifold of a saddle point that sits on the edge of the attractor. To be more precise, Benedicks and Carleson (1991) have proven that the attractor is the closure of a branch of the unstable manifold; see also Simó (1979).

Hobson (1993) has recently developed a method for computing this unstable manifold to very high accuracy. As expected, it is indistinguishable from the strange attractor. Hobson also presents some enlargements of less familiar parts of the Hénon attractor, one of which looks like Saturn's rings (Figure 12.2.4).

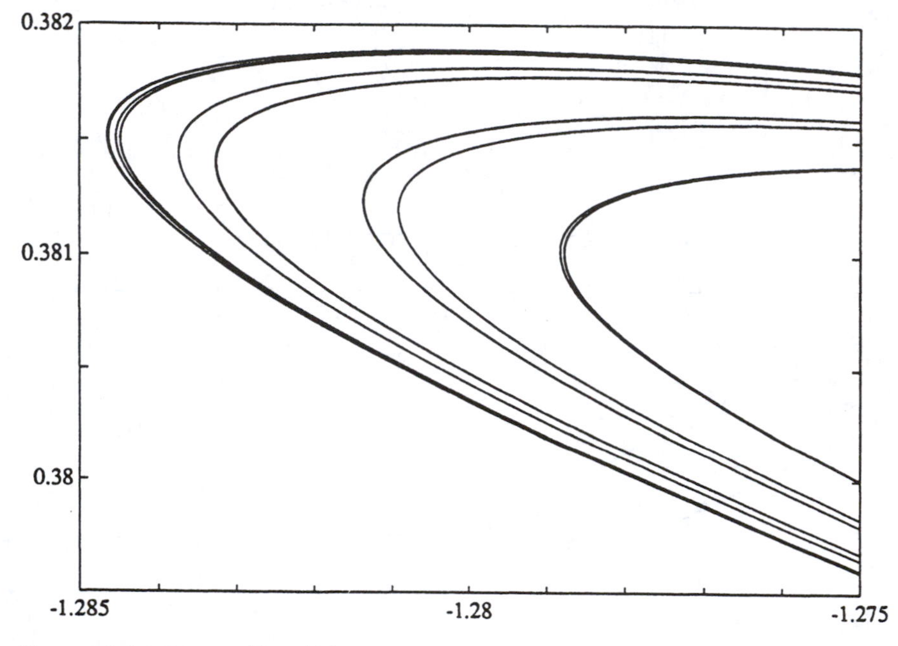

Figure 12.2.4 Courtesy of Dana Hobson

12.3 Rössler System

So far we have used two-dimensional maps to help us understand how stretching and folding can generate strange attractors. Now we return to differential equations.

In the culinary spirit of the pastry map and the baker's map, Otto Rössler (1976) found inspiration in a taffy-pulling machine. By pondering its action, he was led to a system of three differential equations with a simpler strange attractor than Lorenz's. The ***Rössler system*** has only one quadratic nonlinearity xz:

$$\begin{aligned} \dot{x} &= -y - z \\ \dot{y} &= x + ay \\ \dot{z} &= b + z(x - c). \end{aligned} \qquad (1)$$

We first met this system in Section 10.6, where we saw that it undergoes a period-doubling route to chaos as c is increased.

Numerical integration shows that this system has a strange attractor for $a = b = 0.2$, $c = 5.7$ (Figure 12.3.1). A schematic version of the attractor is shown in Figure 12.3.2. Neighboring trajectories separate by spiraling out ("stretching"), then cross without intersecting *by going into the third dimension* ("folding") and then circulate back near their starting places ("re-injection"). We can now see why three dimensions are needed for a flow to be chaotic.

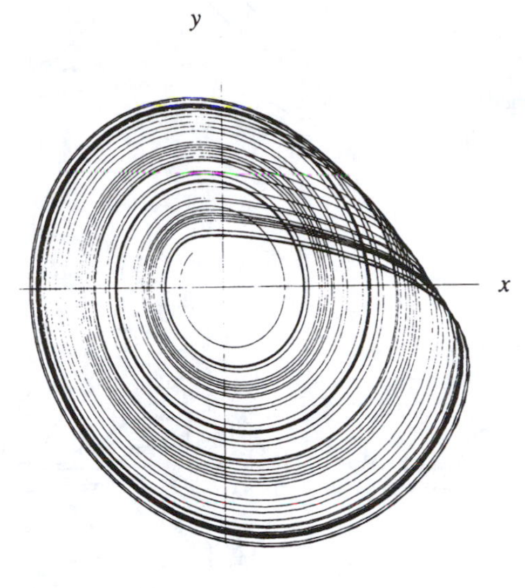

Figure 12.3.1

Let's consider the schematic picture in more detail, following the visual approach of Abraham and Shaw (1983). Our goal is to construct a geometric model of the Rössler attractor, guided by the stretching, folding, and re-injection seen in numerical integrations of the system.

Figure 12.3.3a shows the flow near a typical trajectory. In one direction there's *compression toward* the attractor, and in the other direction there's *divergence along* the attractor. Figure 12.3.3b highlights the sheet on which there's sensitive dependence on initial conditions. These are the expanding directions along which stretching takes place. Next the flow folds the wide part of the sheet in two and then bends it around so that it nearly joins the narrow part (Figure 12.3.4a). Overall, the flow has taken the single sheet and produced *two* sheets after one circuit. Repeating the process, those two sheets produce four (Figure 12.3.4b) and then those produce eight (Figure 12.3.4c), and so on.

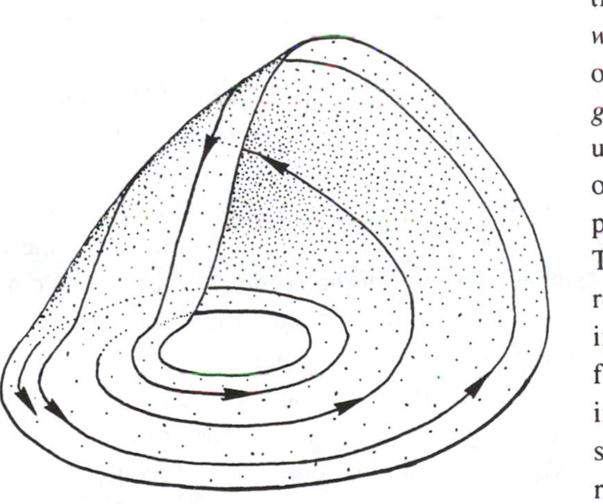

Figure 12.3.2 Abraham and Shaw (1983), p. 121

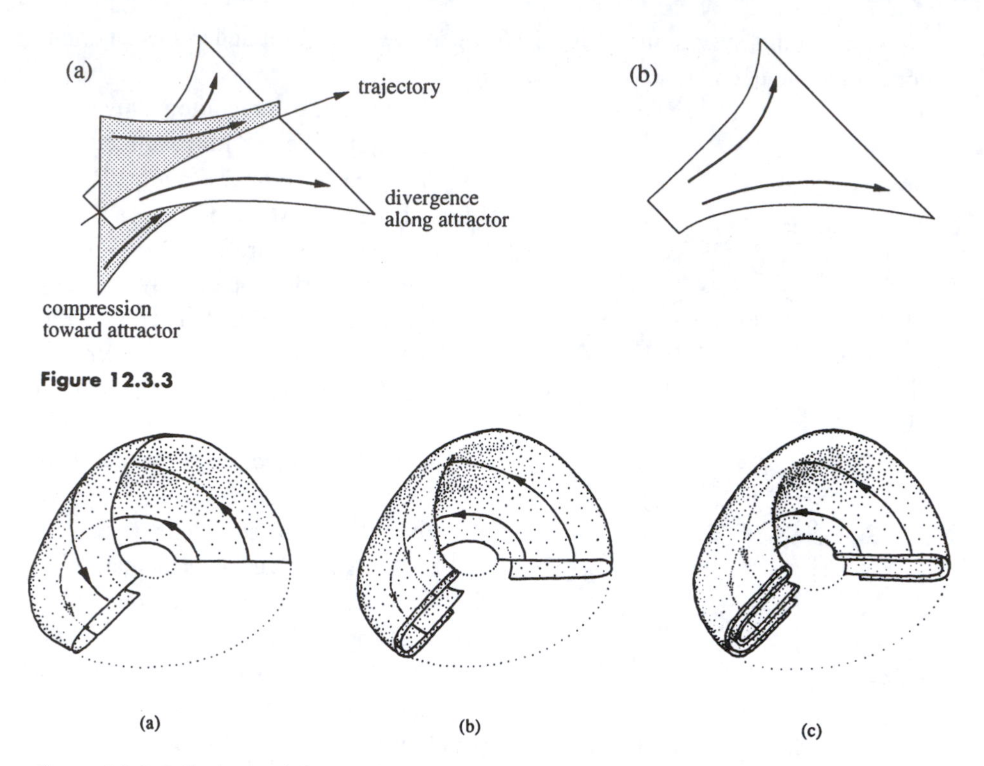

Figure 12.3.3

Figure 12.3.4 Abraham and Shaw (1983), pp 122–123

In effect, the flow is acting like the pastry transformation, and the phase space is acting like the dough! Ultimately the flow generates an infinite complex of tightly packed surfaces: the strange attractor.

Figure 12.3.5 shows a ***Poincaré section*** of the attractor. We slice the attractor with a plane, thereby exposing its cross section. (In the same way, biologists examine complex three-dimensional structures by slicing them and preparing slides.) If we take a further one-dimensional slice or *Lorenz section* through the Poincaré section, we find an infinite set of points separated by gaps of various sizes.

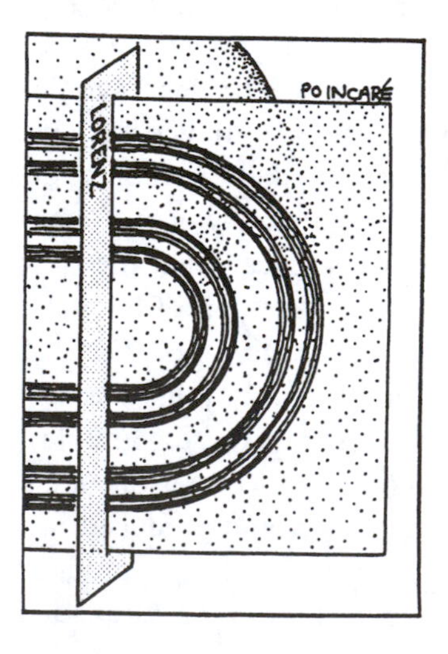

Figure 12.3.5 Abraham and Shaw (1983), p. 123

This pattern of dots and gaps is a topological Cantor set. Since each dot corresponds to one layer of the complex, our model of the Rössler attractor is a *Cantor set of surfaces*. More precisely, the attractor is locally topologically equivalent to the Cartesian product of a ribbon and a Cantor set. This is precisely the structure we would expect, based on our earlier work with the pastry map.

12.4 Chemical Chaos and Attractor Reconstruction

In this section we describe some beautiful experiments on the Belousov–Zhabotinsky chemical reaction. The results show that strange attractors really do occur in nature, not just in mathematics. For more about chemical chaos, see Argoul et al. (1987).

In the BZ reaction, malonic acid is oxidized in an acidic medium by bromate ions, with or without a catalyst (usually cerous or ferrous ions). It has been known since the 1950s that this reaction can exhibit limit-cycle oscillations, as discussed in Section 8.3. By the 1970s, it became natural to inquire whether the BZ reaction could also become *chaotic* under appropriate conditions. Chemical chaos was first reported by Schmitz, Graziani, and Hudson (1977), but their results left room for skepticism—some chemists suspected that the observed complex dynamics might be due instead to uncontrolled fluctuations in experimental control parameters. What was needed was some demonstration that the dynamics obeyed the newly emerging laws of chaos.

The elegant work of Roux, Simoyi, Wolf, and Swinney established the reality of chemical chaos (Simoyi et al. 1982, Roux et al. 1983). They conducted an experiment on the BZ reaction in a "continuous flow stirred tank reactor." In this standard set-up, fresh chemicals are pumped through the reactor at a constant rate to replenish the reactants and to keep the system far from equilibrium. The flow rate acts as a control parameter. The reaction is also stirred continuously to mix the chemicals. This enforces spatial homogeneity, thereby reducing the effective number of degrees of freedom. The behavior of the reaction is monitored by measuring $B(t)$, the concentration of bromide ions.

Figure 12.4.1 shows a time series measured by Roux et al. (1983). At first glance the behavior looks periodic, but it really isn't—the amplitude is erratic. Roux et al. (1983) argued that this aperiodicity corresponds to chaotic motion on a strange attractor, and is not merely random behavior caused by imperfect experimental control.

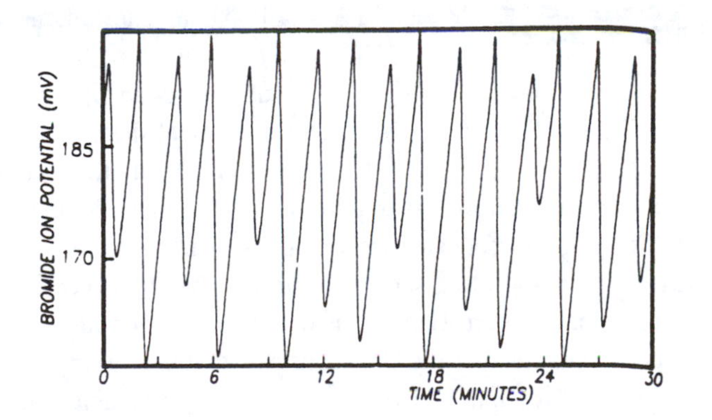

Figure 12.4.1 Roux et al. (1983), p. 258

The first step in their argument is almost magical. Put yourself in their shoes—how could you demonstrate the presence of an underlying strange attractor, given that you only measure a single time series $B(t)$? It seems that there isn't enough information. Ideally, to characterize the motion in phase space, you would like to simultaneously measure the varying concentrations of *all* the other chemical species involved in the reaction. But that's virtually impossible, since there are at least twenty other chemical species, not to mention the ones that are unknown.

Roux et al. (1983) exploited a surprising data-analysis technique, now known as ***attractor reconstruction*** (Packard et al. 1980, Takens 1981). The claim is that for systems governed by an attractor, the dynamics in the full phase space can be reconstructed from measurements of just a *single* time series! Somehow that single variable carries sufficient information about all the others.

The method is based on time delays. For instance, define a two-dimensional vector $\mathbf{x}(t) = \left(B(t), B(t+\tau)\right)$ for some ***delay*** $\tau > 0$. Then the time series $B(t)$ generates a trajectory $\mathbf{x}(t)$ in a two-dimensional phase space. Figure 12.4.2 shows the result of this procedure when applied to the data of Figure 12.4.1, using $\tau = 8.8$ seconds. The experimental data trace out a strange attractor that looks remarkably like the Rössler attractor!

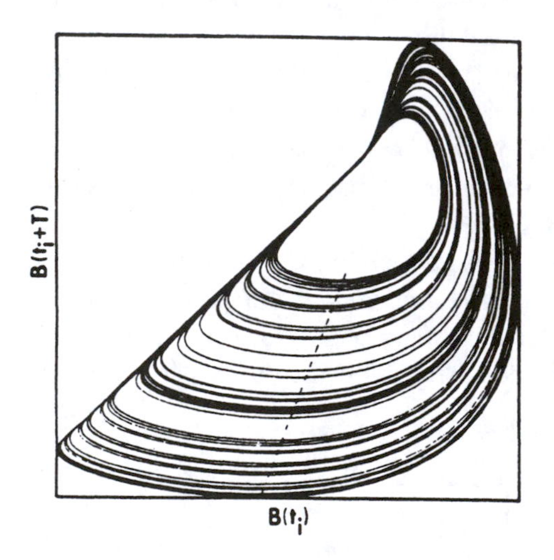

Figure 12.4.2 Roux et al. (1983), p. 262

Roux et al. (1983) also considered the attractor in three dimensions, by defining the three-dimensional vector $\mathbf{x}(t) = \left(B(t), B(t+\tau), B(t+2\tau)\right)$. To obtain a Poincaré section of the attractor, they computed the intersections of the orbits $\mathbf{x}(t)$ with a fixed plane approximately normal to the orbits (shown in projection as a dashed line in Figure 12.4.2). Within the experimental resolution, the data fall on a one-dimensional curve. Hence the chaotic trajectories are confined to an approximately two-dimensional sheet.

Roux et al. then constructed an approximate one-dimensional map that governs the dynamics on the attractor. Let $X_1, X_2, \ldots, X_n, X_{n+1}, \ldots$ denote successive values of $B(t+\tau)$ at points where the orbit $\mathbf{x}(t)$ crosses the dashed line shown in Figure 12.4.2. A plot of X_{n+1} vs. X_n yields the result shown in Figure 12.4.3. The data fall on a smooth one-dimensional map, within experimental resolution. This confirms that the observed aperiodic behavior is governed by *deterministic* laws: Given X_n, the map determines X_{n+1}.

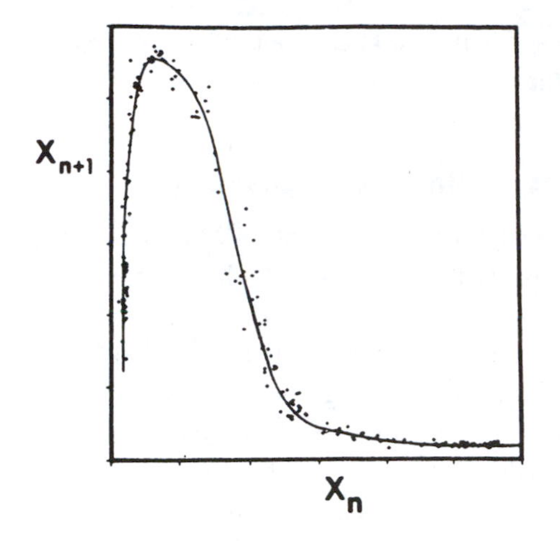

Figure 12.4.3 Roux et al. (1983), p. 262

Furthermore, the map is unimodal, like the logistic map. This suggests that the chaotic state shown in Figure 12.4.1 may be reached by a period-doubling scenario. Indeed such period-doublings were found experimentally (Coffman et al. 1987), as shown in Figure 12.4.4.

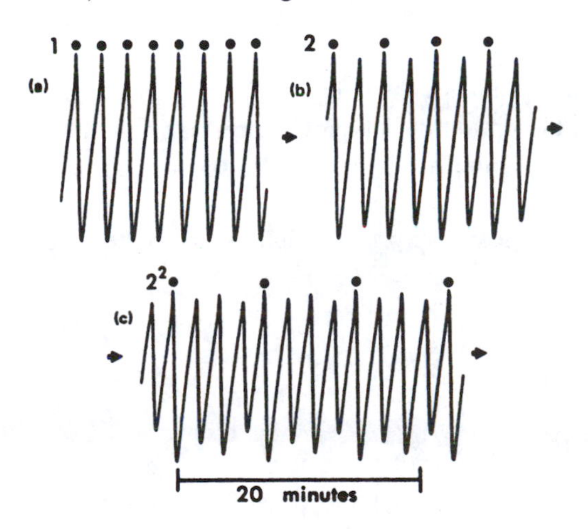

Figure 12.4.4 Coffman et al. (1987), p. 123

The final nail in the coffin was the demonstration that the chemical system obeys the *U-sequence* expected for unimodal maps (Section 10.6). In the regime past the onset of chaos, Roux et al. (1983) observed many distinct periodic windows. As the flow rate was varied, the periodic states occurred in precisely the order predicted by universality theory.

Taken together, these results demonstrate that deterministic chaos can occur in a nonequilibrium chemical system. The most remarkable thing is that the results can be understood (to a large extent) in terms of one-dimensional maps, even though the chemical kinetics are at least twenty-dimensional. Such is the power of universality theory.

But let's not get carried away. The universality theory works only because the attractor is nearly a two-dimensional surface. This low dimensionality results from the continuous stirring of the reaction, along with strong dissipation in the kinetics themselves. Higher-dimensional phenomena like chemical turbulence remain beyond the limits of the theory.

Comments on Attractor Reconstruction

The key to the analysis of Roux et al. (1983) is the attractor reconstruction. There are at least two issues to worry about when implementing the method.

First, how does one choose the ***embedding dimension***, i.e., the number of delays? Should the time series be converted to a vector with two components, or three, or more? Roughly speaking, one needs enough delays so that the underlying attractor can disentangle itself in phase space. The usual approach is to increase the embedding dimension and then compute the correlation dimensions of the resulting attractors. The computed values will keep increasing until the embedding dimension is large enough; then there's enough room for the attractor and the estimated correlation dimension will level off at the "true" value.

Unfortunately, the method breaks down once the embedding dimension is too large; the sparsity of data in phase space causes statistical sampling problems. This limits our ability to estimate the dimension of high-dimensional attractors. For further discussion, see Grassberger and Procaccia (1983), Eckmann and Ruelle (1985), and Moon (1992).

A second issue concerns the optimal value of the delay τ. For real data (which are always contaminated by noise), the optimum is typically around one-tenth to one-half the mean orbital period around the attractor. See Fraser and Swinney (1986) for details.

The following simple example suggests why some delays are better than others.

EXAMPLE 12.4.1:

Suppose that an experimental system has a limit-cycle attractor. Given that one of its variables has a time series $x(t) = \sin t$, plot the time-delayed trajectory $\mathbf{x}(t) = (x(t), x(t+\tau))$ for different values of τ. Which value of τ would be best if the data were noisy?

Solution: Figure 12.4.5 shows $\mathbf{x}(t)$ for three values of τ. For $0 < \tau < \frac{\pi}{2}$, the trajectory is an ellipse with its long axis on the diagonal (Figure 12.4.5a). When $\tau = \frac{\pi}{2}$, $\mathbf{x}(t)$ traces out a circle (Figure 12.4.5b). This makes sense since $x(t) = \sin t$ and $y(t) = \sin(t + \frac{\pi}{2}) = \cos t$; these are the parametric equations of a circle. For larger τ we find ellipses again, but now with their long axes along the line $y = -x$ (Figure 12.4.5c).

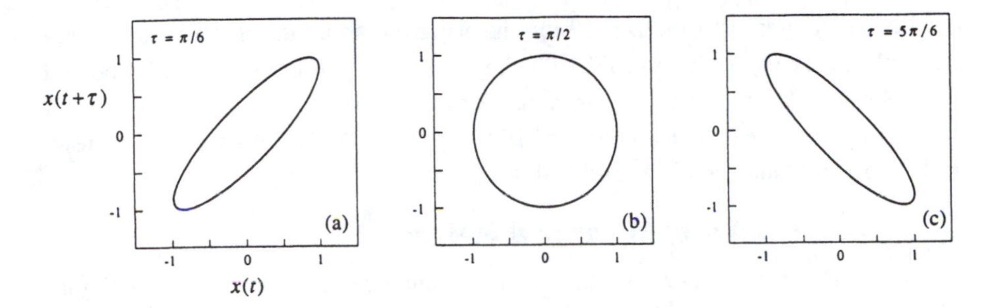

Figure 12.4.5

Note that in each case the method gives a closed curve, which is a topologically faithful reconstruction of the system's underlying attractor (a limit cycle).

For this system the optimum delay is $\tau = \frac{\pi}{2}$, i.e., one-quarter of the natural orbital period, since the reconstructed attractor is then as "open" as possible. Narrower cigar-shaped attractors would be more easily blurred by noise. ■

In the exercises, you're asked to do similar calibrations of the method using quasiperiodic data as well as time series from the Lorenz and Rössler attractors.

Many people find it mysterious that information about the attractor can be extracted from a single time series. Even Ed Lorenz is impressed by the method. When my dynamics class asked him to name the development in nonlinear dynamics that surprised him the most, he cited attractor reconstruction.

In principle, attractor reconstruction can distinguish low-dimensional chaos from noise: as we increase the embedding dimension, the computed correlation dimension levels off for chaos, but keeps increasing for noise (see Eckmann and Ruelle (1985) for examples). Armed with this technique, many optimists have asked questions like, Is there any evidence for deterministic chaos in stock market prices, brain waves, heart rhythms, or sunspots? If so, there may be simple laws waiting to be discovered (and in the case of the stock market, fortunes to be made). Beware: Much of this research is dubious. For a sensible discussion, along with a state-of-the-art method for distinguishing chaos from noise, see Kaplan and Glass (1993).

12.5 Forced Double-Well Oscillator

So far, all of our examples of strange attractors have come from autonomous systems, in which the governing equations have no explicit time-dependence. As soon as we consider forced oscillators and other *nonautonomous* systems, strange attractors start turning up everywhere. That is why we have ignored driven systems until now—we simply didn't have the tools to deal with them.

This section provides a glimpse of some of the phenomena that arise in a particular forced oscillator, the driven double-well oscillator studied by Francis Moon and his colleagues at Cornell. For more information about this system, see Moon and Holmes (1979), Holmes (1979), Guckenheimer and Holmes (1983), Moon and Li (1985), and Moon (1992). For introductions to the vast subject of forced nonlinear oscillations, see Jordan and Smith (1987), Moon (1992), Thompson and Stewart (1986), and Guckenheimer and Holmes (1983).

Magneto-Elastic Mechanical System

Moon and Holmes (1979) studied the mechanical system shown in Figure 12.5.1.

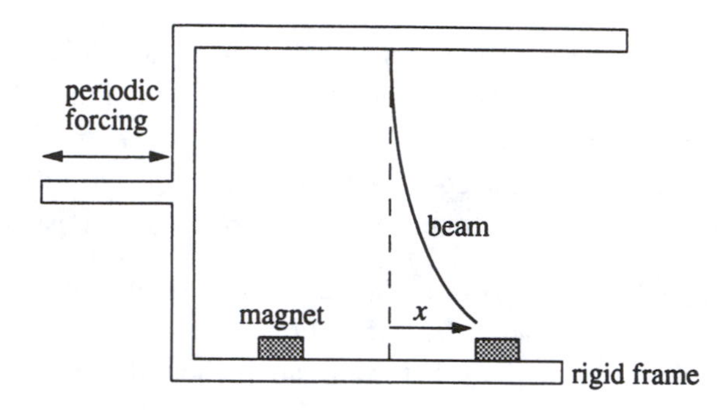

Figure 12.5.1

A slender steel beam is clamped in a rigid framework. Two permanent magnets at the base pull the beam in opposite directions. The magnets are so strong that the beam buckles to one side or the other; either configuration is locally stable. These buckled states are separated by an energy barrier, corresponding to the unstable equilibrium in which the beam is straight and poised halfway between the magnets.

To drive the system out of its stable equilibrium, the whole apparatus is shaken from side to side with an electromagnetic vibration generator. The goal is to understand the forced vibrations of the beam as measured by $x(t)$, the displacement of the tip from the midline of the magnets.

For weak forcing, the beam is observed to vibrate slightly while staying near one or the other magnet, but as the forcing is slowly increased, there is a sudden point at which the beam begins whipping back and forth erratically. The irregular motion is sustained and can be observed for hours—tens of thousands of drive cycles.

Double-Well Analog

The magneto-elastic system is representative of a wide class of driven bistable systems. An easier system to visualize is a damped particle in a double-well potential (Figure 12.5.2). Here the two wells correspond to the two buckled states of the beam, separated by the hump at $x = 0$.

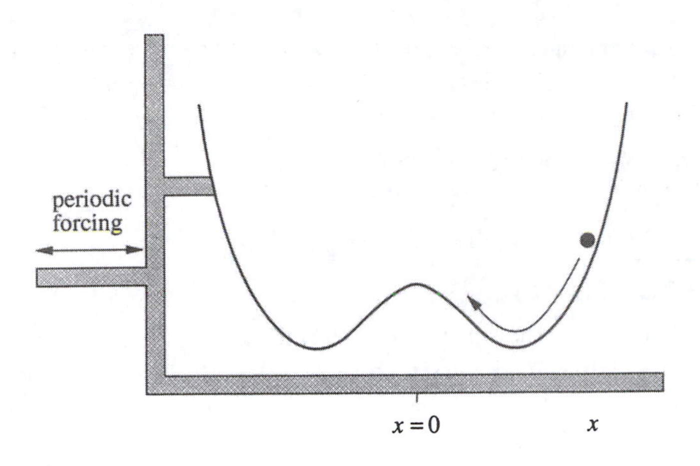

Figure 12.5.2

Suppose the well is shaken periodically from side to side. On physical grounds, what might we expect? If the shaking is weak, the particle should stay near the bottom of a well, jiggling slightly. For stronger shaking, the particle's excursions become larger. We can imagine that there are (at least) *two* types of stable oscillation: a small-amplitude, low-energy oscillation about the bottom of a well; and a large-amplitude, high-energy oscillation in which the particle goes back and forth over the hump, sampling one well and then the other. The choice between these oscillations probably depends on the initial conditions. Finally, when the shaking is extremely strong, the particle is always flung back and forth across the hump, for any initial conditions.

We can also anticipate an intermediate case that seems complicated. If the particle has barely enough energy to climb to the top of the hump, and if the forcing and damping are balanced in a way that keeps the system in this precarious state, then the particle may sometimes fall one way, sometimes the other, depending on the precise timing of the forcing. This case seems potentially chaotic.

Model and Simulations

Moon and Holmes (1979) modeled their system with the dimensionless equation

$$\ddot{x} + \delta \dot{x} - x + x^3 = F \cos \omega t \tag{1}$$

where $\delta > 0$ is the damping constant, F is the forcing strength, and ω is the forcing frequency. Equation (1) can also be viewed as Newton's law for a particle in a double-well potential of the form $V(x) = \frac{1}{4}x^4 - \frac{1}{2}x^2$. In both cases, the force $F \cos \omega t$ is an inertial force that arises from the oscillation of the coordinate system; recall that x is defined as the displacement relative to the *moving* frame, not the lab frame.

The mathematical analysis of (1) requires some advanced techniques from global bifurcation theory; see Holmes (1979) or Section 2.2 of Guckenheimer and Holmes (1983). Our more modest goal is to gain some insight into (1) through numerical simulations.

In all the simulations below, we fix

$$\delta = 0.25\,,\ \omega = 1\,,$$

while varying the forcing strength F.

EXAMPLE 12.5.1:

By plotting $x(t)$, show that (1) has several stable limit cycles for $F = 0.18$.

Solution: Using *MacMath* (Hubbard and West 1992), we obtain the time series shown in Figure 12.5.3.

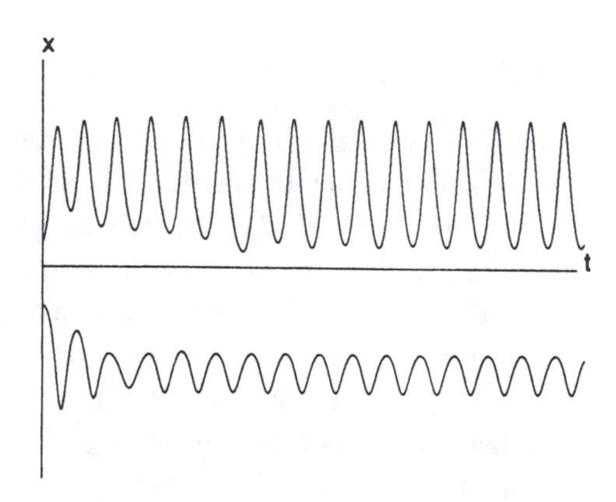

Figure 12.5.3

The solutions converge straightforwardly to periodic solutions. There are two other limit cycles in addition to the two shown here. There might be others, but they are harder to detect. Physically, all these solutions correspond to oscillations confined to a single well. ■

The next example shows that at much larger forcing, the dynamics become complicated.

EXAMPLE 12.5.2:

Compute $x(t)$ and the velocity $y(t) = \dot{x}(t)$, for $F = 0.40$ and initial conditions $(x_0, y_0) = (0,0)$. Then plot $x(t)$ vs. $y(t)$.

Solution: The aperiodic appearance of $x(t)$ and $y(t)$ (Figure 12.5.4) suggests that the system is chaotic, at least for these initial conditions. Note that x changes sign repeatedly; the particle crosses the hump repeatedly, as expected for strong forcing.

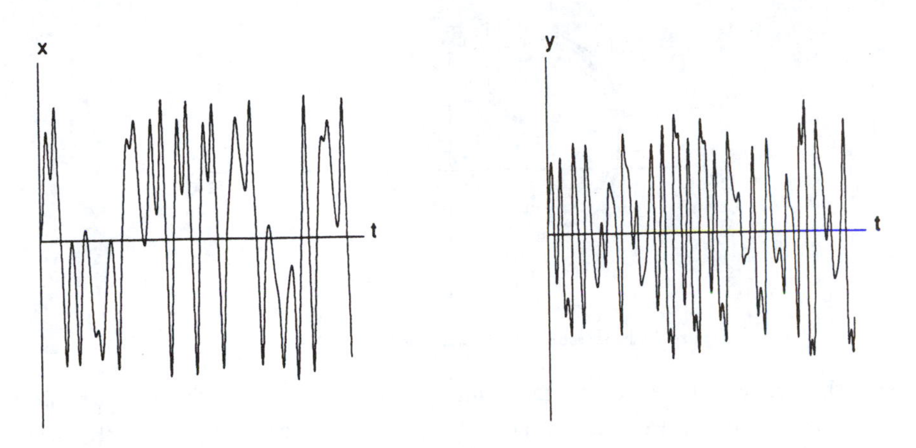

Figure 12.5.4

The plot of $x(t)$ vs. $y(t)$ is messy and hard to interpret (Figure 12.5.5). ∎

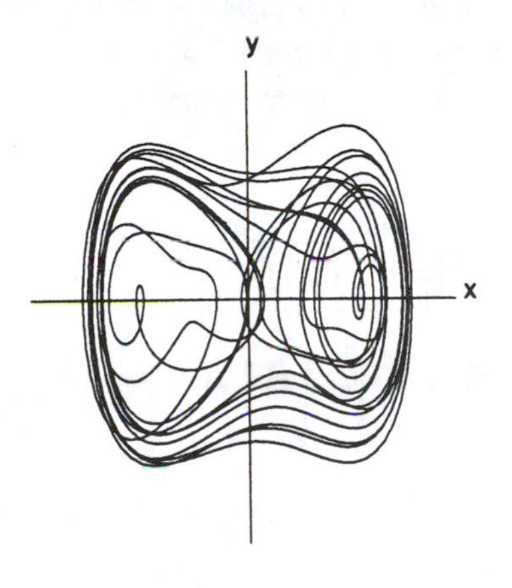

Figure 12.5.5

Note that Figure 12.5.5 is not a true phase portrait, because the system is nonautonomous. As we mentioned in Section 1.2, the state of the system is given by (x, y, t), not (x, y) alone, since all three variables are needed to compute the system's subsequent evolution. Figure 12.5.5 should be regarded as a two-dimensional projection of a three-dimensional trajectory. The tangled appearance of the projection is typical for nonautonomous systems.

Much more insight can be gained from a ***Poincaré section***, obtained by plotting $(x(t), y(t))$ whenever t is an integer multiple of 2π. In physical terms, we "strobe" the system at the same phase in each drive cycle. Figure 12.5.6 shows the Poincaré section for the system of Example 12.5.1.

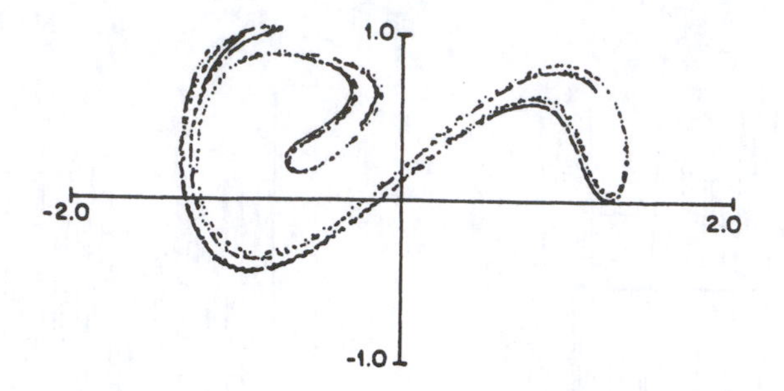

Figure 12.5.6 Guckenheimer and Holmes (1983), p. 90

Now the tangle resolves itself—the points fall on a fractal set, which we interpret as a cross section of a strange attractor for (1). The successive points $(x(t), y(t))$ are found to hop erratically over the attractor, and the system exhibits sensitive dependence on initial conditions, just as we'd expect.

These results suggest that the model is capable of reproducing the sustained chaos observed in the beam experiments. Figure 12.5.7 shows that there is good qualitative agreement between the experimental data (Figure 12.5.7a) and numerical simulations (Figure 12.5.7b).

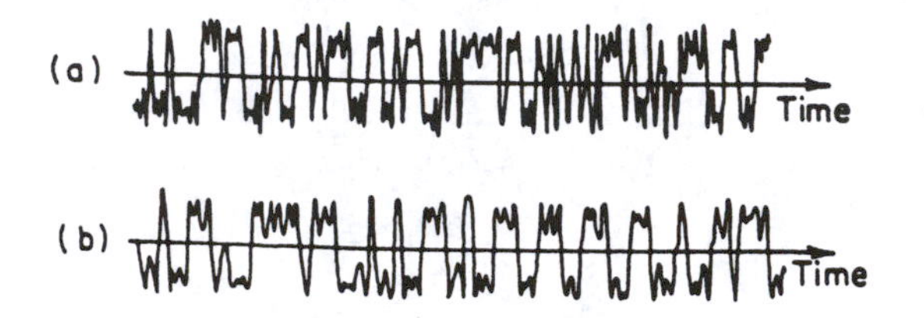

Figure 12.5.7 Guckenheimer and Holmes (1983), p. 84

Transient Chaos

Even when (1) has no strange attractors, it can still exhibit complicated dynamics (Moon and Li 1985). For instance, consider a regime in which two or more stable limit cycles coexist. Then, as shown in the next example, there can be ***transient chaos*** before the system settles down. Furthermore the choice of final state depends sensitively on initial conditions (Grebogi et al. 1983b).

EXAMPLE 12.5.3:

For $F = 0.25$, find two nearby trajectories that both exhibit transient chaos before finally converging to *different* periodic attractors.

Solution: To find suitable initial conditions, we could use trial and error, or we could guess that transient chaos might occur near the ghost of the strange attractor of Figure 12.5.6. For instance, the point $(x_0, y_0) = (0.2, 0.1)$ leads to the time series shown in Figure 12.5.8a.

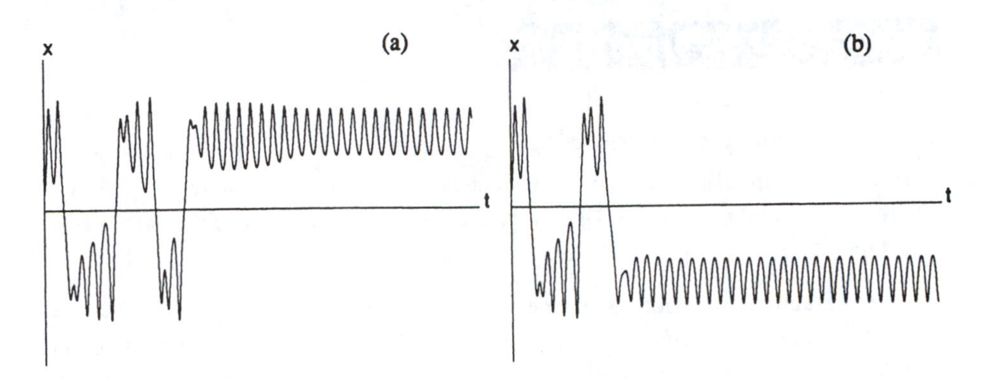

Figure 12.5.8

After a chaotic transient, the solution approaches a periodic state with $x > 0$. Physically, this solution describes a particle that goes back and forth over the hump a few times before settling into small oscillations at the bottom of the well on the right. But if we change x_0 slightly to $x_0 = 0.195$, the particle eventually oscillates in the *left* well (Figure 12.5.8b). ■

Fractal Basin Boundaries

Example 12.5.3 shows that it can be hard to predict the final state of the system, even when that state is simple. This sensitivity to initial conditions is conveyed more vividly by the following graphical method. Each initial condition in a 900×900 grid is color-coded according to its fate. If the trajectory starting at (x_0, y_0) ends up in the left well, we place a blue dot at (x_0, y_0); if the trajectory ends up in the right well, we place a red dot.

Color plate 3 shows the computer-generated result for (1). The blue and red regions are essentially cross sections of the basins of attraction for the two attractors, to the accuracy of the grid. Color plate 3 shows large patches in which all the points are colored red, and others in which all the points are colored blue. In between, however, the slightest change in initial conditions leads to alternations in the final state reached. In fact, if we magnify these regions, we see further intermingling of red and blue, down to arbitrarily small scales. Thus *the boundary between the basins is a fractal.* Near the basin boundary, long-term prediction becomes essentially impossible, because the final state of the system is exquisitely sensitive to tiny changes in initial condition (Color plate 4).

EXERCISES FOR CHAPTER 12

12.1 The Simplest Examples

12.1.1 (Uncoupled linear map) Consider the linear map $x_{n+1} = ax_n$, $y_{n+1} = by_n$, where a, b are real parameters. Draw all the possible patterns of orbits near the origin, depending on the signs and sizes of a and b.

12.1.2 (Stability criterion) Consider the linear map $x_{n+1} = ax_n + by_n$, $y_{n+1} = cx_n + dy_n$, where a, b, c, d are real parameters. Find conditions on the parameters which ensure that the origin is globally asymptotically stable, i.e., $(x_n, y_n) \to (0,0)$ as $n \to \infty$, for all initial conditions.

12.1.3 Sketch the face of Figure 12.1.4 after one more iteration of the baker's map.

12.1.4 (Vertical gaps) Let B be the baker's map with $a < \frac{1}{2}$. Figure 12.1.5 shows that the set $B^2(S)$ consists of horizontal strips separated by vertical gaps of different sizes.

a) Find the size of the largest and smallest gaps in the set $B^2(S)$.
b) Redo part (a) for $B^3(S)$.
c) Finally, answer the question in general for $B^n(S)$.

12.1.5 (Area-preserving baker's map) Consider the dynamics of the baker's map in the area-preserving case $a = \frac{1}{2}$.

a) Given that $(x, y) = (.a_1a_2a_3 \ldots, .b_1b_2b_3 \ldots)$ is the *binary* representation of an arbitrary point in the square, write down the binary representation of $B(x, y)$. (Hint: The answer should look nice.)
b) Using part (a) or otherwise, show that B has a period-2 orbit, and sketch its location in the unit square.
c) Show that B has countably many periodic orbits.
d) Show that B has uncountably many aperiodic orbits.
e) Are there any dense orbits? If so, write one down explicitly. If not, explain why not.

12.1.6 Study the baker's map on a computer for the case $a = \frac{1}{3}$. Starting from a random initial condition, plot the first ten iterates and label them.

12.1.7 (Smale horseshoe) Figure 1 illustrates the mapping known as the ***Smale horseshoe*** (Smale 1967).

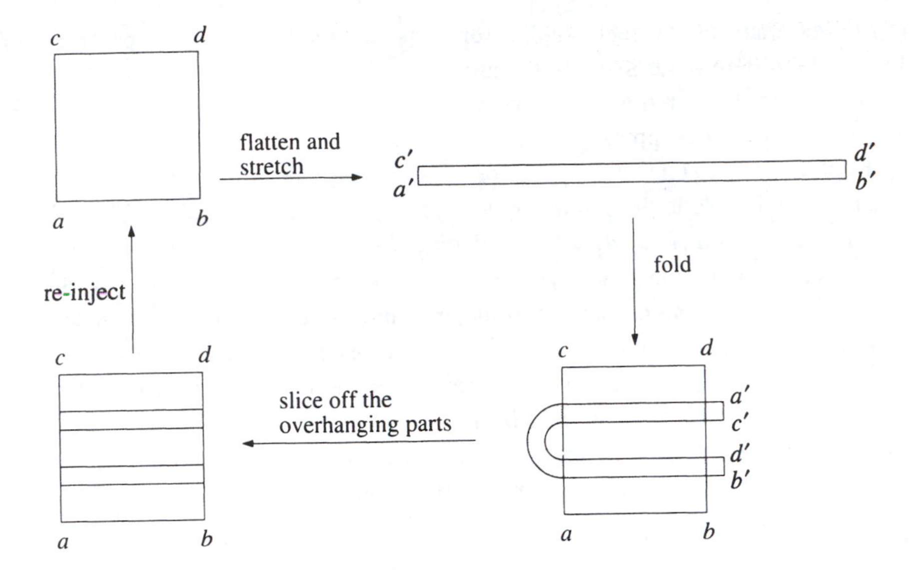

Figure 1

Notice the crucial difference between this map and that shown in Figure 12.1.3: here the horseshoe *hangs over the edge* of the original square. The overhanging parts are lopped off before the next iteration proceeds.

a) The square at the lower left of Figure 1 contains two shaded horizontal strips. Find the points in the original square that map to these strips. (These are the points that survive one iteration, in the sense that they still remain in the square.)

b) Show that after the next round of the mapping, the square on the lower left contains *four* horizontal strips. Find where *they* came from in the original square. (These are the points that survive two iterations.)

c) Describe the set of points in the original square that survive forever.

The horseshoe arises naturally in the analysis of *transient chaos* in differential equations. Roughly speaking, the Poincaré map of such systems can often be approximated by the horseshoe. During the time the orbit remains in a certain region corresponding to the square above, the stretching and folding of the map causes chaos. However, almost all orbits get mapped out of this region eventually (into the "overhang"), and then they escape to some distant part of phase space; this is why the chaos is only *transient*. See Guckenheimer and Holmes (1983) or Arrowsmith and Place (1990) for introductions to the mathematics of horseshoes.

12.1.8 (Hénon's area-preserving quadratic map) The map

$$x_{n+1} = x_n \cos\alpha - (y_n - x_n^2)\sin\alpha$$
$$y_{n+1} = x_n \sin\alpha + (y_n - x_n^2)\cos\alpha$$

illustrates many of the remarkable properties of area-preserving maps (Hénon 1969, 1983). Here $0 \le \alpha \le \pi$ is a parameter.

a) Verify that the map is area-preserving.
b) Find the inverse mapping.
c) Explore the map on the computer for various α. For instance, try $\cos\alpha = 0.24$, and use initial conditions in the square $-1 \le x, y \le 1$. You should be able to find a lovely chain of five ***islands*** surrounding the five points of a period-5 cycle. Then zoom in on the neighborhood of the point $x = 0.57$, $y = 0.16$. You'll see smaller islands, and maybe even smaller islands around them! The complexity extends all the way down to finer and finer scales. If you modify the parameter to $\cos\alpha = 0.22$, you'll still see a prominent chain of five islands, but it's now surrounded by a noticeable ***chaotic sea***.

This mixture of regularity and chaos is typical for area-preserving maps (and for Hamiltonian systems, their continuous-time counterpart).

12.1.9 (The standard map) The map

$$x_{n+1} = x_n + y_{n+1}, \qquad y_{n+1} = y_n + k \sin x_n$$

is called the ***standard map*** because it arises in many different physical contexts, ranging from the dynamics of periodically kicked oscillators to the motion of charged particles perturbed by a broad spectrum of oscillating fields (Jensen 1987, Lichtenberg and Lieberman 1992). The variables x, y, and the governing equations are all to be evaluated modulo 2π. The nonlinearity parameter $k \ge 0$ is a measure of how hard the system is being driven.

a) Show that the map is area-preserving for all k.
b) Plot various orbits for $k = 0$. (This corresponds to the *integrable* limit of the system.)
c) Using a computer, plot the phase portrait for $k = 0.5$. Most orbits should still look regular.
d) Show that for $k = 1$, the phase portrait contains both islands and chaos.
e) Show that at $k = 2$, the chaotic sea has engulfed almost all the islands.

12.2 Hénon Map

12.2.1 Show that the product mapping $T''' \, T'' \, T'$ is equivalent to the formulas in (12.2.1), as claimed in the text.

12.2.2 Show that the transformations T' and T''' are area-preserving, but T'' is not.

12.2.3 Redraw Figure 12.2.1, using an ellipse instead of a rectangle as the test shape.

a) Sketch the successive images of the ellipse under the maps T', T'', T'''.
b) Represent the ellipse parametrically and draw accurate plots on a computer.

The next three exercises deal with the fixed points of the Hénon map.

12.2.4 Find all the fixed points of the Hénon map and show that they exist only if $a > a_0$, where a_0 is to be determined.

12.2.5 Calculate the Jacobian matrix of the Hénon map and find its eigenvalues.

12.2.6 A fixed point of a map is linearly stable if and only if all eigenvalues of the Jacobian satisfy $|\lambda| < 1$. Determine the stability of the fixed points of the Hénon map, as a function of a and b. Show that one fixed point is always unstable, while the other is stable for a slightly larger than a_0. Show that this fixed point loses stability in a flip bifurcation ($\lambda = -1$) at $a_1 = \frac{3}{4}(1-b)^2$.

12.2.7 (2-cycle) Consider the Hénon map with $-1 < b < 1$. Show that the map has a 2-cycle for $a > a_1 = \frac{3}{4}(1-b)^2$. For which values of a is the 2-cycle stable?

12.2.8 (Numerical experiments) Explore numerically what happens in the Hénon map for other values of a, still keeping $b = 0.3$.

a) Show that period-doubling can occur, leading to the onset of chaos at $a \approx 1.06$.

b) Describe the attractor for $a = 1.3$.

12.2.9 (Invariant set for the Hénon map) Consider the Hénon map T with the standard parameter values $a = 1.4$, $b = 0.3$. Let Q denote the quadrilateral with vertices $(-1.33, 0.42)$, $(1.32, 0.133)$, $(1.245, -0.14)$, $(-1.06, -0.5)$.

a) Plot Q and its image $T(Q)$. (Hint: Represent the edges of Q using the parametric equations for a line segment. These segments are mapped to arcs of parabolas.)

b) Prove $T(Q)$ is contained in Q.

12.2.10 Some orbits of the Hénon map escape to infinity. Find one that you can prove diverges.

12.2.11 Show that for a certain choice of parameters, the Hénon map reduces to an effectively one-dimensional map.

12.2.12 Suppose we change the sign of b. Is there any difference in the dynamics?

12.2.13 (Computer project) Explore the area-preserving Hénon map ($b = 1$).

The following exercises deal with the ***Lozi map***

$$x_{n+1} = 1 + y_n - a|x_n|, \qquad y_{n+1} = bx_n,$$

where a, b are real parameters, with $-1 < b < 1$ (Lozi 1978). Note its similarity to the Hénon map. The Lozi map is notable for being one of the first systems *proven* to have a strange attractor (Misiurewicz 1980). This has only recently been achieved for the Hénon map (Benedicks and Carleson 1991) and is still an unsolved problem for the Lorenz equations.

12.2.14 In the style of Figure 12.2.1, plot the image of a rectangle under the Lozi map.

12.2.15 Show that the Lozi map contracts areas if $-1 < b < 1$.

12.2.16 Find and classify the fixed points of the Lozi map.

12.2.17 Find and classify the 2-cycles of the Lozi map.

12.2.18 Show numerically that the Lozi map has a strange attractor when $a = 1.7$, $b = 0.5$.

12.3 Rössler System

12.3.1 (Numerical experiments) Explore the Rössler system numerically. Fix $b = 2$, $c = 4$, and increase a in small steps from 0 to 0.4.

a) Find the approximate value of a at the Hopf bifurcation and at the first period-doubling bifurcation.
b) For each a, plot the attractor, using whatever projection looks best. Also plot the time series $z(t)$.

12.3.2 (Analysis) Find the fixed points of the Rössler system, and state when they exist. Try to classify them. Plot a partial bifurcation diagram of x^* vs. c, for fixed a, b. Can you find a trapping region for the system?

12.3.3 The Rössler system has only one nonlinear term, yet it is much harder to analyze than the Lorenz system, which has two. What makes the Rössler system less tractable?

12.4 Chemical Chaos and Attractor Reconstruction

12.4.1 Prove that the time-delayed trajectory in Figure 12.4.5 traces an ellipse for $0 < \tau < \frac{\pi}{2}$.

12.4.2 (Quasiperiodic data) Plot the time-delayed trajectory $(x(t), x(t+\tau))$ for the signal $x(t) = 3\sin t + \sin(\sqrt{2}\,t)$, for various values of τ. Does the reconstructed attractor appear to be a torus as expected? Which τ seems optimal? Try repeating the process with three-dimensional embeddings; i.e., use $(x(t), x(t+\tau), x(t+2\tau))$.

12.4.3 Numerically integrate the Rössler system for $a = 0.4$, $b = 2$, $c = 4$, and obtain a long time series for $x(t)$. Then use the attractor-reconstruction method for various values of the delay and plot $(x(t), x(t+\tau))$. Find a value of τ for which the reconstructed attractor looks similar to the actual Rössler attractor. How does that τ compare to typical orbital periods of the system?

12.4.4 Redo the previous exercise for the Lorenz equations with the standard parameters $r = 28$, $b = 8/3$, $\sigma = 10$.

12.5 Forced Double-Well Oscillator

12.5.1 (Basins for the unforced oscillator) Sketch the basins for the weakly damped double-well oscillator (12.5.1) in the unforced case when $F = 0$. How does their shape depend on the size of the damping? What happens to the basins as the damping tends to zero? What implications does this have for the predictability of the unforced system?

12.5.2 (Coexisting chaos and limit cycle) Consider the double-well oscillator (12.5.1) with parameters $\delta = 0.15$, $F = 0.3$, and $\omega = 1$. Show numerically that the system has at least two coexisting attractors: a large limit cycle and a smaller strange attractor. Plot both in a Poincaré section.

12.5.3 (Ueda attractor) Consider the system $\ddot{x} + k\dot{x} + x^3 = B\cos t$, with $k = 0.1$, $B = 12$. Show numerically that the system has a strange attractor, and plot its Poincaré section.

12.5.4 (Chaos in the damped driven pendulum) Consider the forced pendulum $\ddot{\theta} + b\dot{\theta} + \sin\theta = F\cos t$, with $b = 0.22$, $F = 2.7$ (Grebogi et al. 1987).

a) Starting from any reasonable initial condition, use numerical integration to compute $\dot{\theta}(t)$. Show that the time series has an erratic appearance, and interpret it in terms of the pendulum's motion.
b) Plot the Poincaré section by strobing the system whenever $t = 2\pi k$, where k is an integer.
c) Zoom in on part of the strange attractor found in part (b). Enlarge a region that reveals the Cantor-like cross section of the attractor.

12.5.5 (Fractal basin boundaries in the damped driven pendulum) Consider the pendulum of the previous exercise, but now let $b = 0.2$, $F = 2$ (Grebogi et al. 1987).

a) Show that there are two stable fixed points in the Poincaré section. Describe the corresponding motion of the pendulum in each case.
b) Compute the basins for each fixed point. Use a reasonably fine grid of initial conditions, and then integrate from each one until the trajectory has settled down to one of the fixed points. (You'll need to establish a criterion to decide whether convergence has occurred.) Show that the boundary between the basins looks like a fractal.

ANSWERS TO SELECTED EXERCISES

Chapter 2

2.1.1 $\sin x = 0$ at $x^* = n\pi$, for integer n

2.1.3 (a) $\ddot{x} = \frac{d}{dt}(\dot{x}) = \frac{d}{dt}(\sin x) = (\cos x)\dot{x} = \cos x \, \sin x = \frac{1}{2}\sin 2x$

2.2.1 $x^* = 2$, unstable; $x^* = -2$, stable

2.2.10 (a) $\dot{x} = 0$ (b) $\dot{x} = \sin \pi x$ (c) impossible: between any two stable fixed points, there must be an unstable one (assuming that the vector field is smooth). (d) $\dot{x} = 1$

2.2.13 (a) $v = \frac{rm}{k}\left(\frac{e^{rt} - e^{-rt}}{e^{rt} + e^{-rt}}\right)$, where $r = \sqrt{gk/m}$. (b) $\sqrt{mg/k}$ (d) $V_{avg} = 29{,}300/116 \approx 253$ ft/s $= 172$ mph (e) $V \approx 265$ ft/s

2.3.2 $x^* = 0$, unstable; $x^* = k_1 a / k_{-1}$, stable

2.4.5 $x^* = 0$, $f'(x^*) = 0$, half-stable by graphical analysis

2.4.6 $x^* = 1$, $f'(x^*) = 1$, unstable

2.5.1 $(1-c)^{-1}$

2.5.6 (a) Conservation of mass—the volume of water flowing through the hole equals the volume of water lost from the bucket. Equating the time derivatives of these two volumes yields $av(t) = A\dot{h}(t)$. (b) Change in P.E. $= [\Delta m]gh = [\rho A(\Delta h)]gh =$ change in K.E. $= \frac{1}{2}(\Delta m)v^2 = \frac{1}{2}(\rho A \Delta h)v^2$. Hence $v^2 = 2gh$.

2.6.2 On the one hand, $\int_t^{t+T} f(x)\frac{dx}{dt}\,dt = \int_{x(t)}^{x(t+T)} f(x)dx = 0$. The first equality follows from the chain rule, and the second from the assumption that $x(t) = x(t+T)$.

On the other hand, $\int_t^{t+T} f(x)\frac{dx}{dt}\,dt = \int_t^{t+T} \left(\frac{dx}{dt}\right)^2 dt > 0$ by assumption that $T > 0$ and $\frac{dx}{dt}$ does not vanish identically.

2.7.5 $V(x) = \cosh x$; equilibrium $x^* = 0$, stable

2.8.1 The equation is time-independent so the slope is determined solely by x.

2.8.6 (b) From Taylor's series, we find $x + e^{-x} = 1 + \frac{1}{2}x^2 - \frac{1}{6}x^3 + O(x^4)$. Graphical analysis shows that $1 \le \dot{x} = x + e^{-x} \le 1 + \frac{1}{2}x^2$ for all x. Integration then yields $t \le x(t) \le \sqrt{2}\tan(t/\sqrt{2})$. Hence $1 \le x(1) \le \sqrt{2}\tan(1/\sqrt{2}) \approx 1.208$.
(c) A step size of order 10^{-4} is needed, and yields $x_{\text{Euler}}(1) = 1.15361$.
(d) A step size $\Delta t = 1$ gives three-decimal accuracy: $\Delta t = 1 \Rightarrow x_{\text{RK}}(1) = 1.1536059$; $\Delta t = 0.1 \Rightarrow x_{\text{RK}}(1) = 1.1536389$; $\Delta t = 0.01 \Rightarrow x_{\text{RK}}(1) = 1.1536390$.

2.8.7 (a) $x(t_1) = x(t_0 + \Delta t) = x(t_0) + \Delta t\,\dot{x}(t_0) + \frac{1}{2}(\Delta t)^2\,\ddot{x}(t_0) + O(\Delta t)^3 =$ $x_0 + \Delta t\, f(x_0) + \frac{1}{2}(\Delta t)^2 f'(x_0)f(x_0) + O(\Delta t)^3$, where we've made use of $\dot{x} = f(x)$ and $\ddot{x} = f'(x)\dot{x} = f'(x)f(x)$.
(b) $|x(t_1) - x_1| = \frac{1}{2}(\Delta t)^2 f'(x_0)f(x_0) + O(\Delta t)^3$. Hence $C = \frac{1}{2}f'(x_0)f(x_0)$.

Chapter 3

3.1.1 $r_c = \pm 2$

3.2.3 $r_c = 1$

3.2.6 (a) $c = -b$ (c) Choose $b = -a/2R$.

3.3.1 (a) $\dot{n} = \dfrac{Gnp}{f + Gn} - kn$ (c) transcritical

3.4.4 $r_c = -1$, subcritical pitchfork

3.4.11 (b) $x^* = 0$, unstable (c) $r_c = 1$, subcritical pitchfork; infinitely many saddle-node bifurcations occur as r decreases from 1 to 0 (use graphical analysis). (d) $r_c \approx \left[(4n + 1)\frac{\pi}{2}\right]^{-1}$ for integer $n >> 1$.

3.4.15 $r_c = -3/16$

3.5.4 (a) $m\ddot{x} + b\dot{x} + kx(1 - L_0/(h^2 + x^2)^{1/2}) = 0$ (d) $m << b^2/k$

3.5.5 (a) $T_{\text{fast}} = mr/b$

3.5.7 (b) $x = N/K$, $x_0 = N_0/K$, $\tau = rt$

3.6.5 (b) $u = x/a$, $R = L_0/a$, $h = mg\sin\theta/ka$. (c) $R < 1$, unique fixed point; $R > 1$, one, two, or three fixed points, depending on h.

3.7.2 (b) Cusp at $x = \sqrt{3}$

3.7.4 (d) transcritical (e) saddle-node

3.7.5 (b) $r_c = \frac{1}{2}$ (d) saddle-node curves at $r_c = 2x/(1+x^2)^2$, $s_c = x^2(1-x^2)/(1+x^2)^2$

Chapter 4

4.1.1 $a = \text{integer}$. For a well-defined vector field on the circle, we need $\sin(a(\theta+2\pi k)) = \sin(a\theta)$ for *all* integer k. Hence $2\pi k a = 2\pi n$, for some integer n. Thus $ka = \text{integer}$, for all integer k. This is possible only if a is itself an integer.

4.1.3 Unstable fixed points: $\theta^* = 0, \pi$. Stable fixed points: $\theta^* = \pm\pi/2$.

4.2.1 12 seconds

4.2.3 12/11 hours later, i.e., at approximately 1:05 and 27 seconds. This problem can be solved in many ways. One method is based on Example 4.2.1. It takes the minute hand $T_1 = 1$ hr and the hour hand $T_2 = 12$ hrs to complete one revolution around the clockface. Hence the time required for the minute hand to lap the hour hand is $T = (1 - \frac{1}{12})^{-1} = \frac{12}{11}$ hrs.

4.3.2 (a) $d\theta = 2\,du/(1+u^2)$ (d) $T = 2\int_{-\infty}^{\infty} \frac{du}{\omega u^2 - 2au + \omega}$ (e) $x = u - a/\omega$, $r = 1 - a^2/\omega^2$, $T = \frac{2}{\omega}\int_{-\infty}^{\infty} \frac{dx}{r + x^2} = \frac{2\pi}{\omega\sqrt{r}} = \frac{2\pi}{\sqrt{\omega^2 - a^2}}$.

4.3.10 $b = \frac{1}{2n} - 1$, $c = \int_{-\infty}^{\infty} \frac{du}{1+u^{2n}} = \frac{\pi}{n\sin(\pi/2n)}$.

4.4.1 $b^2 >> m^2 g L^3$, approximation valid after an initial transient

4.5.1 (b) $|\omega - \Omega| \le \frac{\pi}{2} A$

4.6.4 (a) $I_b = I_a + I_R$ (c) $V_k = \frac{\hbar}{2e}\dot{\phi}_k$

4.6.5 Let $R_0 = R/N$. Then $\Omega = I_b R_0 / I_c r$, $a = -(R_0 + r)/r$, $\tau = \left[2eI_c r^2/\hbar(R_0 + r)\right]t$.

4.6.6 Kirchhoff's current law gives $\frac{\hbar}{2er}\frac{d\phi_k}{dt} + I_c \sin\phi_k + \frac{dQ}{dt} = I_b$, $k = 1, \ldots, N$, and Kirchhoff's voltage law gives

$$L\frac{d^2Q}{dt^2} + R\frac{dQ}{dt} + \frac{Q}{C} = \frac{\hbar}{2e}\sum_{j=1}^{N}\frac{d\phi_j}{dt}.$$

Chapter 5

5.1.9 (c) $x = y$, stable manifold; $x = -y$, unstable manifold

5.1.10 (d) Liapunov stable (e) asymptotically stable

5.2.1 (a) $\lambda_1 = 2$, $\lambda_2 = 3$, $\mathbf{v}_1 = (1,2)$, $\mathbf{v}_2 = (1,1)$. (b) $\mathbf{x}(t) = c_1\begin{pmatrix}1\\2\end{pmatrix}e^{2t} + c_2\begin{pmatrix}1\\1\end{pmatrix}e^{3t}$. (c) unstable node (d) $x = e^{2t} + 2e^{3t}$, $y = 2e^{2t} + 2e^{3t}$

5.2.2 $\mathbf{x}(t) = C_1 e^t\begin{pmatrix}\cos t\\ \sin t\end{pmatrix} + C_2 e^t\begin{pmatrix}-\sin t\\ \cos t\end{pmatrix}$

5.2.3 stable node

5.2.5 degenerate node

5.2.7 center

5.2.9 non-isolated fixed point

5.3.1 $a > 0, b < 0$: Narcissistic Nerd, Better Latent than Never, Flirting Fink, Likes to Tease but not to Please. $a < 0, b > 0$: Bashful Budder, Lackluster Libido Lover. $a, b < 0$: Hermit, Malevolent Misanthrope (Answers suggested by my students and also by students in Peter Christopher's class at Worcester Polytechnic Institute.)

Chapter 6

6.1.1 saddle point at $(0,0)$

6.1.5 stable spiral at (1,1), saddle point at (0,0), y-axis is invariant.

6.3.3 $(0,0)$, saddle point

6.3.6 $(-1,-1)$, stable node; $(1,1)$, saddle point

6.3.8 (b) unstable

6.3.9 (a) stable node at $(0,0)$, saddle points at $\pm(2,2)$.

6.4.1 Unstable node at $(0,0)$, stable node at $(3,0)$, saddle point at $(0,2)$. Nullclines are parallel diagonal lines. All trajectories end up at $(3,0)$, except those starting on the y-axis.

6.4.2 All trajectories approach $(1,1)$, except those starting on the axes.

6.4.4 (a) Each species grows exponentially in the absence of the other. (b) $x = b_2 N_1 / r_1$, $y = b_1 N_2 / r_1$, $\tau = r_1 t$, $\rho = r_2 / r_1$. (d) saddle point at $(\rho, 1)$. Almost all trajectories approach the axes. Hence one or the other species dies out.

6.5.1 (a) center at $(0,0)$, saddles at $(\pm 1,0)$ (b) $\frac{1}{2}\dot{x}^2 + \frac{1}{2}x^2 - \frac{1}{4}x^4 = C$

6.5.2 (c) $y^2 = x^2 - \frac{2}{3}x^3$

6.5.6 (e) Epidemic occurs if $x_0 > \ell/k$.

6.6.1 Reversible, since equations invariant under $t \to -t$, $y \to -y$.

6.6.10 Yes. The linearization predicts a center and the system is reversible: $t \to -t$, $x \to -x$. A variant of Theorem 6.6.1 shows the system has a nonlinear center.

6.7.2 (e) Small oscillations have angular frequency $(1-\gamma^2)^{1/4}$ for $-1 < \gamma < 1$.

6.8.2 fixed point at $(0,0)$, index $I = 0$.

6.8.7 $(2,0)$ and $(0,0)$, saddles; $(1,3)$, stable spiral; $(-2,0)$, stable node. Coordinate axes are invariant. A closed orbit would have to encircle the node or the spiral. But such a cycle can't encircle the node (cycle would cross the x-axis: forbidden). Similarly, cycle can't encircle the spiral, since spiral is joined to saddle at $(2,0)$ by a branch of saddle's unstable manifold, and cycle can't cross this trajectory.

6.8.9 False. Counterexample: use polar coordinates and consider $\dot{r} = r(r^2-1)(r^2-9)$, $\dot{\theta} = r^2 - 4$. This has all the required properties but there are no fixed points between the cycles $r = 1$ and $r = 3$, since $\dot{r} \neq 0$ in that region.

6.8.11 (c) For $\dot{z} = z^k$, the origin has index k. To see this, let $z = re^{i\theta}$. Then $z^k = r^k e^{ik\theta}$. Hence $\phi = k\theta$ and the result follows. Similarly, the origin has index $-k$ for $\dot{z} = (\bar{z})^k$.

Chapter 7

7.1.8 (b) Period $T = 2\pi$ (c) stable

7.1.9 (b) $R\phi' = \cos\phi - R$, $R' = \sin\phi - k$, where prime denotes differentiation with respect to the central angle θ. (c) The dog asymptotically approaches a circle for which $R = \sqrt{1-k^2} = \sqrt{\frac{3}{4}}$.

7.2.5 (b) Yes, as long as the vector field is smooth everywhere, i.e., there are no singularities.

7.2.9 (c) $V = e^{x^2+y^2}$, equipotentials are circles $x^2 + y^2 = C$.

7.2.10 Any $a, b > 0$ with $a = b$ suffices.

7.2.12 $a = 1$, $m = 2$, $n = 4$

7.3.1 (a) unstable spiral (b) $\dot{r} = r(1 - r^2 - r^2\sin^2 2\theta)$ (c) $r_1 = \frac{1}{\sqrt{2}} \approx .707$

(d) $r_2 = 1$ (e) No fixed points inside the trapping region, so Poincaré–Bendixson implies the existence of limit cycle.

7.3.7 (a) $\dot{r} = ar(1 - r^2 - 2b\cos^2\theta)$, $\dot{\theta} = -1 + ab\sin 2\theta$. (b) There is at least one limit cycle in the annular trapping region $\sqrt{1-2b} \le r \le 1$, by the Poincaré–Bendixson theorem. Period of any such cycle is $T = \oint dt = \oint (\frac{dt}{d\theta})\, d\theta = \int_0^{2\pi} \frac{d\theta}{-1+ab\sin 2\theta} = T(a,b)$.

7.3.9 (a) $r(\theta) = 1 + \mu\left(\frac{2}{5}\cos\theta + \frac{1}{5}\sin\theta\right) + O(\mu^2)$. (b) $r_{\max} = 1 + \frac{\mu}{\sqrt{5}} + O(\mu^2)$, $r_{\min} = 1 - \frac{\mu}{\sqrt{5}} + O(\mu^2)$.

7.4.1 Use Liénard's theorem.

7.5.2 In the Liénard plane, the limit cycle converges to a fixed shape as $\mu \to \infty$; that's not true in the usual phase plane.

7.5.4 (d) $T \approx (2\ln 3)\mu$.

7.5.5 $T \approx 2\left[\sqrt{2} - \ln\left(1+\sqrt{2}\right)\right]\mu$

7.6.7 $r' = \frac{1}{2}r(1 - \frac{1}{8}r^4)$, stable limit cycle at $r = 8^{1/4} = 2^{3/4}$, frequency $\omega = 1 + O(\varepsilon^2)$.

7.6.8 $r' = \frac{1}{2}r(1 - \frac{4}{3\pi}r)$, stable limit cycle at $r = \frac{3}{4}\pi$, $\omega = 1 + O(\varepsilon^2)$

7.6.9 $r' = \frac{1}{16}r^3(6 - r^2)$, stable limit cycle at $r = \sqrt{6}$, $\omega = 1 + O(\varepsilon^2)$

7.6.14 (b) $x(t,\varepsilon) \sim \left(a^{-2} + \frac{3}{4}\varepsilon t\right)^{-1/2}\cos t$

7.6.17 (b) $\gamma_c = \frac{1}{2}$ (c) $k = \frac{1}{4}\sqrt{1 - 4\gamma^2}$ (d) If $\gamma > \frac{1}{2}$, then $\phi' > 0$ for all ϕ, and $r(T)$ is periodic. In fact, $r(\phi) \propto \left(\gamma + \frac{1}{2}\cos 2\phi\right)^{-1}$, so if r is small initially, $r(\phi)$ remains close to 0 for all time.

7.6.19 (d) $x_0 = a\cos\tau$ (f) $x_1 = \frac{1}{32}a^3(\cos 3\tau - \cos\tau)$

7.6.22 $x = a\cos\omega t + \frac{1}{6}\varepsilon a^2(3 - 2\cos\omega t - \cos 2\omega t) + O(\varepsilon^2)$, $\omega = 1 - \frac{5}{12}\varepsilon^2 a^2 + O(\varepsilon^3)$

7.6.24 $\omega = 1 - \frac{3}{8}\varepsilon a^2 - \frac{21}{256}\varepsilon^2 a^4 - \frac{81}{2048}\varepsilon^3 a^6 + O(\varepsilon^4)$

Chapter 8

8.1.3 $\lambda_1 = -|\mu|$, $\lambda_2 = -1$

8.1.6 (b) $\mu_c = 1$; saddle-node bifurcation

8.1.13 (a) One nondimensionalization is $dx/d\tau = x(y-1)$, $dy/d\tau = -xy - ay + b$, where $\tau = kt$, $x = Gn/k$, $y = GN/k$, $a = f/k$, $b = pG/k^2$ (d) Transcritical bifurcation when $a = b$.

8.2.3 subcritical

8.2.5 supercritical

8.2.8 (d) supercritical

8.2.12 (a) $a = \frac{1}{8}$ (b) subcritical

8.3.1 (a) $x^* = 1$, $y^* = b/a$, $\tau = b-(1+a)$, $\Delta = a > 0$. Fixed point is stable if $b < 1+a$, unstable if $b > 1+a$, and linear center if $b = 1+a$. (c) $b_c = 1+a$ (d) $b > b_c$ (e) $T \approx 2\pi/\sqrt{a}$

8.4.3 $\mu \approx 0.066 \pm 0.001$

8.4.4 Cycle created by supercritical Hopf bifurcation at $\mu = 1$, destroyed by homoclinic bifurcation at $\mu = 3.72 \pm 0.01$.

8.4.9 (c) $b_c = \dfrac{32\sqrt{3}}{27}\dfrac{k^3}{F^2}$

8.4.12 $t \sim O(\lambda_u^{-1}\ln(1/\mu))$.

8.6.2 (d) If $|1-\omega| > |2a|$, then $\lim_{\tau\to\infty}\theta_1(\tau)/\theta_2(\tau) = (1+\omega+\omega_\phi)/(1+\omega-\omega_\phi)$, where $\omega_\phi = \left((1-\omega)^2 - 4a^2\right)^{1/2}$. On the other hand, if $|1-\omega| \le |2a|$, phase-locking occurs and $\lim_{\tau\to\infty}\theta_1(\tau)/\theta_2(\tau) = 1$.

8.6.6 (c) Lissajous figures are planar projections of the motion. The motion in the four-dimensional space $(x, \dot{x}, y, \dot{y})$ is projected onto the plane (x, y). The parameter ω is a winding number, since it is a ratio of two frequencies. For rational winding numbers, the trajectories on the torus are knotted. When projected onto the xy plane they appear as closed curves with self-crossings (like a shadow of a knot).

8.6.7 (a) $r_0 = (h^2/mk)^{1/3}$, $\omega_\theta = h/mr_0^2$ (c) $\omega_r/\omega_\theta = \sqrt{3}$, which is irrational. (e) Two masses are connected by a string of fixed length. The first mass plays the role of the particle; it moves on a frictionless, horizontal "air table." It is connected to the second mass by a string that passes through a hole in the center of the table. This second mass hangs below the table, bobbing up and down and supplying the constant force of its weight. This mechanical system obeys the equations given in the text, after some rescaling.

8.7.2 $a < 0$, stable; $a = 0$, neutral; $a > 0$, unstable

8.7.4 $A < 0$

8.7.9 (b) stable (c) $e^{-2\pi}$

Chapter 9

9.1.2 $\frac{d}{dt}(a_n^2 + b_n^2) = 2(a_n \dot{a}_n + b_n \dot{b}_n) = -2K(a_n^2 + b_n^2)$. Thus $(a_n^2 + b_n^2) \propto e^{-2Kt} \to 0$ as $t \to \infty$.

9.1.3 Let $a_1 = \alpha y$, $b_1 = \beta z + q_1/K$, $\omega = \gamma x$, and $t = T\tau$, and solve for the coefficients by matching the Lorenz and waterwheel equations. Find $T = 1/K$, $\gamma = \pm K$. Picking $\gamma = K$ yields $\alpha = Kv/\pi g r$, $\beta = -Kv/\pi g r$. Also $\sigma = v/KI$, Rayleigh $r = \pi g r q_1 / K^2 v$.

9.1.4 (a) degenerate pitchfork (b) Let $\alpha = [b(r-1)]^{-1/2}$. Then $t_{\text{laser}} = (\sigma/\kappa) t_{\text{Lorenz}}$, $E = \alpha x$, $P = \alpha y$, $D = r - z$, $\gamma_1 = \kappa/\sigma$, $\gamma_2 = \kappa b/\sigma$, $\lambda = r - 1$.

9.2.1 (b) If $\sigma < b + 1$, then C^+ and C^- are stable for all $r > 0$. (c) If $r = r_H$, then $\lambda_3 = -(\sigma + b + 1)$.

9.2.2 Pick C so large that $\frac{x^2}{br} + \frac{y^2}{br^2} + \frac{(z-r)^2}{r^2} > 1$ everywhere on the boundary of E.

9.3.8 (a) yes (b) yes

9.4.2 (b) $x^* = \frac{2}{3}$; unstable (c) $x_1 = \frac{2}{5}$, $x_2 = \frac{4}{5}$; 2-cycle is unstable.

9.5.5 (a) $X = \varepsilon x,\ Y = \varepsilon^2 \sigma y,\ Z = \sigma(\varepsilon^2 z - 1),\ \tau = t/\varepsilon$

9.5.6 Transient chaos does not occur if the trajectory starts close enough to C^+ or C^-.

9.6.1 (a) $\dot{V} \le -kV$ for any $k < \min(2, 2b)$. Integration then yields $0 \le V(t) \le V_0 e^{-kt}$.
(b) $\frac{1}{2} e_2{}^2 \le V < V_0 e^{-kt}$, so $e_2(t) < (2V_0)^{1/2} e^{-kt/2}$. Similarly, $e_3(t) \le O(e^{-kt/2})$.
(c) Integration of $\dot{e}_1 = \sigma(e_2 - e_1)$, combined with $e_2(t) \le O(e^{-kt/2})$, implies $e_1(t) \le \max\{O(e^{-\sigma t}),\ O(e^{-kt/2})\}$. So all components of $\mathbf{e}(t)$ decay exponentially fast.

9.6.6 According to Cuomo and Oppenheim (1992, 1993),

$$\dot{u} = \frac{1}{R_5 C_1}\left[\frac{R_4}{R_1} v - \frac{R_3}{R_2 + R_3}\left(1 + \frac{R_4}{R_1}\right) u\right],\ \dot{v} = \frac{1}{R_{15} C_2}\left[\frac{R_{11}}{R_{10} + R_{11}}\left(1 + \frac{R_{12}}{R_8} + \frac{R_{12}}{R_9}\right)\left(1 + \frac{R_7}{R_6}\right) u - \frac{R_{12}}{R_8} v - \frac{R_{12}}{R_9} uw\right],\quad \dot{w} = \frac{1}{R_{20} C_3}\left[\frac{R_{19}}{R_{16}} uv - \frac{R_{18}}{R_{17} + R_{18}}\left(1 + \frac{R_{19}}{R_{16}}\right) w\right].$$

Chapter 10

10.1.1 $x_n \to 1$ as $n \to \infty$, for all $x_0 > 0$

10.1.10 Yes

10.1.13 Differentiation yields $\lambda = f'(x^*) = g(x^*)g''(x^*)/g'(x^*)^2$. Hence $g(x^*) = 0$ implies $\lambda = 0$ (unless $g'(x^*) = 0$ too; this nongeneric case requires separate treatment).

10.3.2 (b) $1+\sqrt{5}$

10.3.7 (d) Any orbit starting at an irrational number x_0 will be aperiodic, since the decimal expansion of an irrational number never repeats.

10.3.12 (a) The maximum of the map occurs at $x = \frac{1}{2}$. A superstable cycle of period 2^n occurs when this point is an element of a 2^n-cycle, or equivalently, a fixed point of $f^{(2^n)}(x,r)$. Hence the desired formula for R_n is $f^{(2^n)}(\frac{1}{2}, R_n) = \frac{1}{2}$.

10.3.13 (a) The curves are $f^k(\frac{1}{2}, r)$ vs. r, for $k = 1,2,\ldots$ Intuitively, points near $x_m = \frac{1}{2}$ get mapped to almost the same value, since the slope equals zero at x_m. So there is a high density of points near the iterates $f^k(\frac{1}{2})$ of the maximum. (b) The corner of the big wedge occurs when $f^3(\frac{1}{2}) = f^4(\frac{1}{2})$, as is clear from the graphs of part (a). Hence $f(u) = u$, where $u = f^3(\frac{1}{2})$. So u must equal the fixed point $1 - \frac{1}{r}$. The solution of $f^3(\frac{1}{2}, r) = 1 - \frac{1}{r}$ can be obtained exactly as $r = \frac{2}{3} + \frac{8}{3}\left(19 + \sqrt{297}\right)^{-1/3} + \frac{2}{3}\left(19 + \sqrt{297}\right)^{1/3} = 3.67857\ldots$

10.4.4 3.8318741...

10.4.7 (b) *RLRR*

10.5.3 The origin is globally stable for $r < 1$, by cobwebbing. There is an interval of marginally stable fixed points for $r = 1$.

10.5.4 The Liapunov exponent is necessarily negative in a periodic window. But since $\lambda = \ln r > 0$ for all $r > 1$, there can be no periodic windows after the onset of chaos.

10.6.1 (b) $r_1 \approx 0.71994$, $r_2 \approx 0.83326$, $r_3 \approx 0.85861$, $r_4 \approx 0.86408$, $r_5 \approx 0.86526$, $r_6 \approx 0.86551$.

10.7.1 (a) $\alpha = -1-\sqrt{3} = -2.732\ldots$, $c_2 = \alpha/2 = -1.366\ldots$ (b) Solve $\alpha = (1 + c_2 + c_4)^{-1}$, $c_2 = 2\alpha^{-1} - \frac{1}{2}\alpha - 2$, $c_4 = 1 + \frac{1}{2}\alpha - \alpha^{-1}$ simultaneously. Relevant root is $\alpha = -2.53403\ldots$, $c_2 = -1.52224\ldots$, $c_4 = 0.12761\ldots$

10.7.8 (e) $b = -1/2$

10.7.9 (b) The steps in the cobweb staircase for g^2 are twice as long, so $\alpha = 2$.

Chapter 11

11.1.3 uncountable

11.1.6 (a) x_0 is rational $\Leftrightarrow$ the corresponding orbit is periodic

11.2.1 $\frac{1}{3}+\frac{2}{9}+\frac{4}{27}+\cdots=\left(\frac{1}{3}\right)\frac{1}{1-\frac{2}{3}}=1$

11.2.4 Measure = 1; uncountable.

11.2.5 (b) Hint: Write $x \in [0,1]$ in binary, i.e., base-2.

11.3.1 (a) $d = \ln 2/\ln 4 = \frac{1}{2}$

11.3.4 $\ln 5/\ln 10$

11.4.1 $\ln 4/\ln 3$

11.4.2 $\ln 8/\ln 3$

11.4.9 $\ln(p^2 - m^2)/\ln p$

Chapter 12

12.1.5 (a) $B(x, y) = (.a_2a_3a_4\ldots, .a_1b_1b_2b_3\ldots)$. To describe the dynamics more transparently, associate the symbol $\ldots b_3b_2b_1.a_1a_2a_3\ldots$ with (x, y) by simply placing x and y back-to-back. Then in this notation, $B(x, y) = \ldots b_3b_2b_1a_1.a_2a_3\ldots$. In other words, B just shifts the binary point one place to the right. (b) In the notation above, ...1010.1010... and ...0101.0101... are the only period-2 points. They correspond to $(\frac{2}{3}, \frac{1}{3})$ and $(\frac{1}{3}, \frac{2}{3})$. (d) Pick $x =$ irrational, $y =$ anything.

12.1.8 (b) $x_n = x_{n+1}\cos\alpha + y_{n+1}\sin\alpha$, $y_n = -x_{n+1}\sin\alpha + y_{n+1}\cos\alpha + (x_{n+1}\cos\alpha + y_{n+1}\sin\alpha)^2$

12.2.4 $x^* = (2a)^{-1}\left[b - 1 \pm \sqrt{(1-b)^2 + 4a}\right]$, $y^* = bx^*$, $a_0 = -\frac{1}{4}(1-b)^2$

12.2.5 $\lambda = -ax^* \pm \sqrt{(ax^*)^2 + b}$

12.2.15 $\det \mathbf{J} = -b$

12.3.3 The Rössler system lacks the symmetry of the Lorenz system.

12.5.1 The basins become thinner as the damping decreases.

REFERENCES

Abraham, R. H., and Shaw, C. D. (1983) *Dynamics: The Geometry of Behavior. Part 2: Chaotic Behavior* (Aerial Press, Santa Cruz, CA).

Abraham, R. H., and Shaw, C. D. (1988) *Dynamics: The Geometry of Behavior. Part 4: Bifurcation Behavior* (Aerial Press, Santa Cruz, CA).

Ahlers, G. (1989) Experiments on bifurcations and one-dimensional patterns in nonlinear systems far from equilibrium. In D. L. Stein, ed. *Lectures in the Sciences of Complexity* (Addison-Wesley, Reading, MA).

Aitta, A., Ahlers, G., and Cannell, D. S. (1985) Tricritical phenomena in rotating Taylor–Couette flow. *Phys. Rev. Lett.* **54**, 673.

Anderson, P. W., and Rowell, J. M. (1963) Probable observation of the Josephson superconducting tunneling effect. *Phys. Rev. Lett.* **10**, 230.

Anderson, R. M. (1991) The Kermack–McKendrick epidemic threshold theorem. *Bull. Math. Biol.* **53**, 3.

Andronov, A. A., Leontovich, E. A., Gordon, I. I., and Maier, A. G. (1973) *Qualitative Theory of Second-Order Dynamic Systems* (Wiley, New York).

Arecchi, F. T., and Lisi, F. (1982) Hopping mechanism generating 1/f noise in nonlinear systems. *Phys. Rev. Lett.* **49**, 94.

Argoul, F., Arneodo, A., Richetti, P., Roux, J. C., and Swinney, H. L. (1987) Chemical chaos: From hints to confirmation. *Acc. Chem. Res.* **20**, 436.

Arnold, V. I. (1978) *Mathematical Methods of Classical Mechanics* (Springer, New York).

Aroesty, J., Lincoln, T., Shapiro, N., and Boccia, G. (1973) Tumor growth and chemotherapy: mathematical methods, computer simulations, and experimental foundations. *Math. Biosci.* **17**, 243.

Arrowsmith, D. K., and Place, C. M. (1990) *An Introduction to Dynamical Systems* (Cambridge University Press, Cambridge, England).

Attenborough, D. (1992) *The Trials of Life.* For synchronous fireflies, see the episode entitled "Talking to Strangers," available on videotape from Ambrose Video Publishing, 1290 Avenue of the Americas, Suite 2245, New York, NY 10104.

Bak, P. (1986) The devil's staircase. *Phys. Today*, Dec. 1986, 38.

Barnsley, M. F. (1988) *Fractals Everywhere* (Academic Press, Orlando, FL).

Belousov, B. P. (1959) Oscillation reaction and its mechanism (in Russian). Sbornik Referatov po Radiacioni Medicine, p. 145. 1958 Meeting.

Bender, C. M., and Orszag, S. A. (1978) *Advanced Mathematical Methods for Scientists and Engineers* (McGraw-Hill, New York).

Benedicks, M., and Carleson, L. (1991) The dynamics of the Hénon map. *Annals of Math.* **133**, 73.

Bergé, P., Pomeau, Y., and Vidal, C. (1984) *Order Within Chaos: Towards a Deterministic Approach to Turbulence* (Wiley, New York).

Borrelli, R. L., and Coleman, C. S. (1987) *Differential Equations: A Modeling Approach* (Prentice-Hall, Englewood Cliffs, NJ).

Briggs, K. (1991) A precise calculation of the Feigenbaum constants. *Mathematics of Computation* **57**, 435.

Buck, J. (1988) Synchronous rhythmic flashing of fireflies. II. *Quart. Rev. Biol.* **63**, 265.

Buck, J., and Buck, E. (1976) Synchronous fireflies. *Sci. Am.* **234**, May, 74.

Campbell, D. (1979) An introduction to nonlinear dynamics. In D. L. Stein, ed. *Lectures in the Sciences of Complexity* (Addison-Wesley, Reading, MA).

Carlson, A. J., Ivy, A. C., Krasno, L. R., and Andrews, A. H. (1942) The physiology of free fall through the air: delayed parachute jumps. *Quart. Bull. Northwestern Univ. Med. School* **16**, 254 (cited in Davis 1962).

Cartwright, M. L. (1952) Van der Pol's equation for relaxation oscillations. *Contributions to Nonlinear Oscillations*, Vol. 2, Princeton, 3.

Cesari, L. (1963) *Asymptotic Behavior and Stability Problems in Ordinary Differential Equations* (Academic, New York).

Chance, B., Pye, E. K., Ghosh, A. K., and Hess, B., eds. (1973) *Biological and Biochemical Oscillators* (Academic Press, New York).

Coddington, E. A., and Levinson, N. (1955) *Theory of Ordinary Differential Equations* (McGraw-Hill, New York).

Coffman, K. G., McCormick, W. D., Simoyi, R. H., and Swinney, H. L. (1987) Universality, multiplicity, and the effect of iron impurities in the Belousov–Zhabotinskii reaction. *J. Chem. Phys.* **86**, 119.

Collet, P., and Eckmann, J.-P. (1980) *Iterated Maps of the Interval as Dynamical Systems* (Birkhauser, Boston).

Cox, A. (1982) Magnetostratigraphic time scale. In W. B. Harland et al., eds. *Geologic Time Scale* (Cambridge University Press, Cambridge, England).

Crutchfield, J. P., Farmer, J. D., Packard, N. H., and Shaw, R. S. (1986) Chaos. *Sci. Am.* **254**, December, 46.

Cuomo, K. M., and Oppenheim, A. V. (1992) Synchronized chaotic circuits and systems for communications. *MIT Research Laboratory of Electronics Technical Report* No. 575.

Cuomo, K. M., and Oppenheim, A. V. (1993) Circuit implementation of synchronized chaos, with applications to communications. *Phys. Rev. Lett.* **71**, 65.

Cuomo, K. M., Oppenheim, A. V., and Strogatz, S. H. (1993) Synchronization of Lorenz-based chaotic circuits, with applications to communications. *IEEE Trans. Circuits and Systems* (in press).

Cvitanovic, P., ed. (1989a) *Universality in Chaos*, 2nd ed. (Adam Hilger, Bristol and New York)

Cvitanovic, P. (1989b) Universality in chaos. In P. Cvitanovic, ed. *Universality in Chaos*, 2nd ed. (Adam Hilger, Bristol and New York).

Davis, H. T. (1962) *Introduction to Nonlinear Differential and Integral Equations* (Dover, New York).

Devaney, R. L. (1989) *An Introduction to Chaotic Dynamical Systems,* 2nd ed. (Addison-Wesley, Redwood City, CA)

Dowell, E. H., and Ilgamova, M. (1988) *Studies in Nonlinear Aeroelasticity* (Springer, New York).

Drazin, P. G. (1992) *Nonlinear Systems* (Cambridge University Press, Cambridge, England).

Drazin, P. G., and Reid, W. H. (1981) *Hydrodynamic Stability* (Cambridge University Press, Cambridge, England).

Dubois, M., and Bergé, P. (1978) Experimental study of the velocity field in Rayleigh–Bénard convection. *J. Fluid Mech.* **85**, 641.

Eckmann, J.-P., and Ruelle, D. (1985) Ergodic theory of chaos and strange attractors. *Rev. Mod. Phys.* **57**, 617.

Edelstein–Keshet, L. (1988) *Mathematical Models in Biology* (Random House, New York).

Epstein, I. R., Kustin, K., De Kepper, P. and Orban, M. (1983) Oscillating chemical reactions. *Sci. Am.* **248**(3), 112.

Ermentrout, G. B. (1991) An adaptive model for synchrony in the firefly *Pteroptyx malaccae*. *J. Math. Biol.* **29**, 571.

Ermentrout, G. B., and Kopell, N. (1990) Oscillator death in systems of coupled neural oscillators. *SIAM J. Appl. Math.* **50,** 125.

Ermentrout, G. B., and Rinzel, J. (1984) Beyond a pacemaker's entrainment limit: phase walk-through. *Am. J. Physiol.* **246**, R102.

Fairén, V., and Velarde, M. G. (1979) Time-periodic oscillations in a model for the respiratory process of a bacterial culture. *J. Math. Biol.* **9**, 147.

Falconer, K. (1990) *Fractal Geometry: Mathematical Foundations and Applications* (Wiley, Chichester, England).

Farmer, J. D. (1985) Sensitive dependence on parameters in nonlinear dynamics. *Phys. Rev. Lett.* **55**, 351.

Feder, J. (1988) *Fractals* (Plenum, New York).

Feigenbaum, M. J. (1978) Quantitative universality for a class of nonlinear transformations. *J. Stat. Phys.* **19**, 25.

Feigenbaum, M. J. (1979) The universal metric properties of nonlinear transformations. *J. Stat. Phys.* **21**, 69.

Feigenbaum, M. J. (1980) Universal behavior in nonlinear systems. *Los Alamos Sci.* **1**, 4.

Feynman, R. P., Leighton, R. B., and Sands, M. (1965) *The Feynman Lectures on Physics* (Addison-Wesley, Reading, MA).

Field, R., and Burger, M., eds. (1985) *Oscillations and Traveling Waves in Chemical Systems* (Wiley, New York).

Firth, W. J. (1986) Instabilities and chaos in lasers and optical resonators. In A. V. Holden, ed. *Chaos* (Princeton University Press, Princeton, NJ).

Fraser, A. M., and Swinney, H. L. (1986) Independent coordinates for strange attractors from mutual information. *Phys. Rev. A* **33**, 1134.

Gaspard, P. (1990) Measurement of the instability rate of a far-from-equilibrium steady state at an infinite period bifurcation. *J. Phys. Chem.* **94**, 1.

Giglio, M., Musazzi, S., and Perini, V. (1981) Transition to chaotic behavior via a reproducible sequence of period-doubling bifurcations. *Phys. Rev. Lett.* **47**, 243.

Glass, L. (1977) Patterns of supernumerary limb regeneration. *Science* **198**, 321.

Glazier, J. A., and Libchaber, A. (1988) Quasiperiodicity and dynamical systems: an experimentalist's view. *IEEE Trans. on Circuits and Systems* **35**, 790.

Gleick, J. (1987) *Chaos: Making a New Science* (Viking, New York).

Goldbeter, A. (1980) Models for oscillations and excitability in biochemical systems. In L. A. Segel, ed., *Mathematical Models in Molecular and Cellular Biology* (Cambridge University Press, Cambridge, England).

Grassberger, P. (1981) On the Hausdorff dimension of fractal attractors. *J. Stat. Phys.* **26**, 173.

Grassberger, P., and Procaccia, I. (1983) Measuring the strangeness of strange attractors. *Physica D* **9**, 189.

Gray, P., and Scott, S. K. (1985) Sustained oscillations and other exotic patterns of behavior in isothermal reactions. *J. Phys. Chem.* **89**, 22.

Grebogi, C., Ott, E., and Yorke, J. A. (1983a) Crises, sudden changes in chaotic attractors and transient chaos. *Physica D* **7**, 181.

Grebogi, C., Ott, E., and Yorke, J. A. (1983b) Fractal basin boundaries, long-lived chaotic transients, and unstable-unstable pair bifurcation. *Phys. Rev. Lett.* **50**, 935.

Grebogi, C., Ott, E., and Yorke, J. A. (1987) Chaos, strange attractors, and fractal basin boundaries in nonlinear dynamics. *Science* **238**, 632.

Griffith, J. S. (1971) *Mathematical Neurobiology* (Academic Press, New York).

Grimshaw, R. (1990) *Nonlinear Ordinary Differential Equations* (Blackwell, Oxford, England).

Guckenheimer, J., and Holmes, P. (1983) *Nonlinear Oscillations, Dynamical Systems, and Bifurcations of Vector Fields* (Springer, New York).

Haken, H. (1983) *Synergetics,* 3rd ed. (Springer, Berlin).

Halsey, T., Jensen, M. H., Kadanoff, L. P., Procaccia, I. and Shraiman, B. I. (1986) Fractal measures and their singularities: the characterization of strange sets. *Phys. Rev. A* **33**, 1141.

Hanson, F. E. (1978) Comparative studies of firefly pacemakers. *Federation Proc.* **37**, 2158.

Hao, Bai-Lin, ed. (1990) *Chaos II* (World Scientific, Singapore).

Hao, Bai-Lin, and Zheng, W.-M. (1989) Symbolic dynamics of unimodal maps revisited. *Int. J. Mod. Phys. B* **3**, 235.

Harrison, R. G., and Biswas, D. J. (1986) Chaos in light. *Nature* **321**, 504.

He, R., and Vaidya, P. G. (1992) Analysis and synthesis of synchronous periodic and chaotic systems. *Phys. Rev. A* **46**, 7387.

Helleman, R. H. G. (1980) Self-generated chaotic behavior in nonlinear mechanics. In E. G. D. Cohen, ed. *Fundamental Problems in Statistical Mechanics* **5**, 165.

Hénon, M. (1969) Numerical study of quadratic area-preserving mappings. *Quart. Appl. Math.* **27**, 291.

Hénon, M. (1976) A two-dimensional mapping with a strange attractor. *Commun. Math. Phys.* **50**, 69.

Hénon, M. (1983) Numerical exploration of Hamiltonian systems. In G. Iooss, R. H. G. Helleman, and R. Stora, eds. *Chaotic Behavior of Deterministic Systems* (North-Holland, Amsterdam).

Hirsch, J. E., Nauenberg, M., and Scalapino, D. J. (1982) Intermittency in the presence of noise: a renormalization group formulation. *Phys. Lett. A* **87**, 391.

Hobson, D. (1993) An efficient method for computing invariant manifolds of planar maps. *J. Comp. Phys.* **104**, 14.

Holmes, P. (1979) A nonlinear oscillator with a strange attractor. *Phil. Trans. Roy. Soc. A* **292**, 419.

Hubbard, J. H., and West, B. H. (1991) *Differential Equations: A Dynamical Systems Approach, Part I* (Springer, New York).

Hubbard, J. H., and West, B. H. (1992) *MacMath: A Dynamical Systems Software Package for the Macintosh* (Springer, New York).

Hurewicz, W. (1958) *Lectures on Ordinary Differential Equations* (MIT Press, Cambridge, MA).

Jackson, E. A. (1990) *Perspectives of Nonlinear Dynamics,* Vols. 1 and 2 (Cambridge University Press, Cambridge, England).

Jensen, R. V. (1987) Classical chaos. *Am. Scientist* **75**, 168.

Jordan, D. W., and Smith, P. (1987) *Nonlinear Ordinary Differential Equations,* 2nd ed. (Oxford University Press, Oxford, England).

Josephson, B. D. (1962) Possible new effects in superconductive tunneling. *Phys. Lett.* **1**, 251.

Josephson, B. D. (1982) Interview. *Omni*, July 1982, p. 87.

Kaplan, D. T., and Glass, L. (1993) Coarse-grained embeddings of time series: random walks, Gaussian random processes, and deterministic chaos. *Physica D* **64**, 431.

Kaplan, J. L., and Yorke, J. A. (1979) Preturbulence: A regime observed in a fluid flow model of Lorenz. *Commun. Math. Phys.* **67**, 93.

Kermack, W. O., and McKendrick, A. G. (1927) Contributions to the mathematical theory of epidemics—I. *Proc. Roy. Soc.* **115A**, 700.

Kocak, H. (1989) *Differential and Difference Equations Through Computer Experiments,* 2nd ed. (Springer, New York).

Kolar, M., and Gumbs, G. (1992) Theory for the experimental observation of chaos in a rotating waterwheel. *Phys. Rev. A* **45**, 626.

Kolata, G. B. (1977) Catastrophe theory: the emperor has no clothes. *Science* **196**, 287.

Krebs, C. J. (1972) *Ecology: The Experimental Analysis of Distribution and Abundance* (Harper and Row, New York).

Lengyel, I., and Epstein, I. R. (1991) Modeling of Turing structures in the chlorite-iodide-malonic acid-starch reaction. *Science* **251**, 650.

Lengyel, I., Rabai, G., and Epstein, I. R. (1990) Experimental and modeling study of oscillations in the chlorine dioxide-iodine-malonic acid reaction. *J. Am. Chem. Soc.* **112**, 9104.

Levi, M., Hoppensteadt, F., and Miranker, W. (1978) Dynamics of the Josephson junction. *Quart. Appl. Math.* **35**, 167.

Lewis, J., Slack, J. M. W., and Wolpert, L. (1977) Thresholds in development. *J. Theor. Biol.* **65**, 579

Libchaber, A., Laroche, C., and Fauve, S. (1982) Period doubling cascade in mercury, a quantitative measurement. *J. Physique Lett.* **43**, L211.

Lichtenberg, A. J., and Lieberman, M. A. (1992) *Regular and Chaotic Dynamics,* 2nd ed. (Springer, New York).

Lighthill, J. (1986) The recently recognized failure of predictability in Newtonian dynamics. *Proc. Roy. Soc. Lond. A* **407**, 35.

Lin, C. C., and Segel, L. (1988) *Mathematics Applied to Deterministic Problems in the Natural Sciences* (SIAM, Philadelphia).

Linsay, P. (1981) Period doubling and chaotic behavior in a driven anharmonic oscillator. *Phys. Rev. Lett.* **47**, 1349.

Lorenz, E. N. (1963) Deterministic nonperiodic flow. *J. Atmos. Sci.* **20**, 130.

Lozi, R. (1978) Un attracteur étrange du type attracteur de Hénon. J. Phys. (Paris) **39** (C5), 9.

Ludwig, D., Jones, D. D., and Holling, C. S. (1978) Qualitative analysis of insect outbreak systems: the spruce budworm and forest. *J. Anim. Ecol.* **47**, 315.

Ludwig, D., Aronson, D. G., and Weinberger, H. F. (1979) Spatial patterning of the spruce budworm. *J. Math. Biol.* **8**, 217.

Ma, S.-K. (1976) *Modern Theory of Critical Phenomena* (Benjamin/Cummings, Reading, MA).

Ma, S.-K. (1985) *Statistical Mechanics* (World Scientific, Singapore).

Malkus, W. V. R. (1972) Non-periodic convection at high and low Prandtl number. *Mémoires Société Royale des Sciences de Liège,* Series 6, Vol. 4, 125.

Mandelbrot, B. B. (1982) *The Fractal Geometry of Nature* (Freeman, San Francisco).

Manneville, P. (1990) *Dissipative Structures and Weak Turbulence* (Academic, Boston).

Marsden, J. E., and McCracken, M. (1976) *The Hopf Bifurcation and Its Applications* (Springer, New York).

May, R. M. (1972) Limit cycles in predator-prey communities. *Science* **177**, 900.

May, R. M. (1976) Simple mathematical models with very complicated dynamics. *Nature* **261**, 459.

May, R. M. (1981) *Theoretical Ecology: Principles and Applications,* 2nd ed. (Blackwell, Oxford, England).

May, R. M., and Anderson, R. M. (1987) Transmission dynamics of HIV infection. *Nature* **326**, 137.

May, R. M., and Oster, G. F. (1980) Period-doubling and the onset of turbulence: an analytic estimate of the Feigenbaum ratio. *Phys. Lett. A* **78**, 1.

McCumber, D. E. (1968) Effect of ac impedance on dc voltage-current characteristics of superconductor weak-link junctions. *J. Appl. Phys.* **39**, 3113.

Metropolis, N., Stein, M. L., and Stein, P. R. (1973) On finite limit sets for transformations on the unit interval. *J. Combin. Theor.* **15**, 25.

Milnor, J. (1985) On the concept of attractor. *Commun. Math. Phys.* **99**, 177.

Milonni, P. W., and Eberly, J. H. (1988) *Lasers* (Wiley, New York).

Minorsky, N. (1962) *Nonlinear Oscillations* (Van Nostrand, Princeton, NJ).

Mirollo, R. E., and Strogatz, S. H. (1990) Synchronization of pulse-coupled biological oscillators. *SIAM J. Appl. Math.* **50**, 1645.

Misiurewicz, M. (1980) Strange attractors for the Lozi mappings. *Ann. N. Y. Acad. Sci.* **357**, 348.

Moon, F. C. (1992) *Chaotic and Fractal Dynamics: An Introduction for Applied Scientists and Engineers* (Wiley, New York).

Moon, F. C., and Holmes, P. J. (1979) A magnetoelastic strange attractor. *J. Sound. Vib.* **65**, 275.

Moon, F. C., and Li, G.-X. (1985) Fractal basin boundaries and homoclinic orbits for periodic motion in a two-well potential. *Phys. Rev. Lett.* **55**, 1439.

Moore-Ede, M. C., Sulzman, F. M., and Fuller, C. A. (1982) *The Clocks That Time Us.* (Harvard University Press, Cambridge, MA)

Munkres, J. R. (1975) *Topology: A First Course* (Prentice-Hall, Englewood Cliffs, NJ).

Murray, J. (1989) *Mathematical Biology* (Springer, New York).

Myrberg, P. J. (1958) Iteration von Quadratwurzeloperationen. *Annals Acad. Sci. Fennicae* A I *Math.* **259**, 1.

Nayfeh, A. (1973) *Perturbation Methods* (Wiley, New York).

Newton, C. M. (1980) Biomathematics in oncology: modelling of cellular systems. *Ann. Rev. Biophys. Bioeng.* **9**, 541.

Odell, G. M. (1980) Qualitative theory of systems of ordinary differential equations, including phase plane analysis and the use of the Hopf bifurcation theorem. Appendix A.3. In L. A. Segel, ed., *Mathematical Models in Molecular and Cellular Biology* (Cambridge University Press, Cambridge, England).

Olsen, L. F., and Degn, H. (1985) Chaos in biological systems. *Quart. Rev. Biophys.* **18**, 165.

Packard, N. H., Crutchfield, J. P., Farmer, J. D., and Shaw, R. S. (1980) Geometry from a time series. *Phys. Rev. Lett.* **45**, 712.

Palmer, R. (1989) Broken ergodicity. In D. L. Stein, ed. *Lectures in the Sciences of Complexity* (Addison-Wesley, Reading, MA).

Pearl, R. (1927) The growth of populations. *Quart. Rev. Biol.* **2**, 532.

Pecora, L. M., and Carroll, T. L. (1990) Synchronization in chaotic systems. *Phys. Rev. Lett.* **64**, 821.

Peitgen, H.-O., and Richter, P. H. (1986) *The Beauty of Fractals* (Springer, New York).

Perko, L. (1991) *Differential Equations and Dynamical Systems* (Springer, New York).

Pianka, E. R. (1981) Competition and niche theory. In R. M. May, ed. *Theoretical Ecology: Principles and Applications* (Blackwell, Oxford, England).

Pielou, E. C. (1969) *An Introduction to Mathematical Ecology* (Wiley-Interscience, New York).

Politi, A., Oppo, G. L., and Badii, R. (1986) Coexistence of conservative and dissipative behavior in reversible dynamical systems. *Phys. Rev. A* **33**, 4055.

Pomeau, Y., and Manneville, P. (1980) Intermittent transition to turbulence in dissipative dynamical systems. *Commun. Math. Phys.* **74**, 189.

Poston, T., and Stewart, I. (1978) *Catastrophe Theory and Its Applications* (Pitman, London).

Press, W. H., Flannery, B. P., Teukolsky, S. A., and Vetterling, W. T. (1986) *Numerical Recipes: The Art of Scientific Computing* (Cambridge University Press, Cambridge, England).

Rikitake, T. (1958) Oscillations of a system of disk dynamos. *Proc. Camb. Phil. Soc.* **54**, 89.

Rinzel, J., and Ermentrout, G.B. (1989) Analysis of neural excitability and oscillations. In C. Koch and I. Segev, eds. *Methods in Neuronal Modeling: From Synapses to Networks* (MIT Press, Cambridge, MA).

Robbins, K. A. (1977) A new approach to subcritical instability and turbulent transitions in a simple dynamo. *Math. Proc. Camb. Phil. Soc.* **82**, 309.

Robbins, K. A. (1979) Periodic solutions and bifurcation structure at high r in the Lorenz system. *SIAM J. Appl. Math.* **36**, 457.

Rössler, O. E. (1976) An equation for continuous chaos. *Phys. Lett.* A **57**, 397.

Roux, J. C., Simoyi, R. H., and Swinney, H. L. (1983) Observation of a strange attractor. *Physica D* **8**, 257.

Ruelle, D., and Takens, F. (1971) On the nature of turbulence. *Commun. Math. Phys.* **20**, 167.

Saha, P., and Strogatz, S. H. (1994) The birth of period three. *Math. Mag.* (in press)

Schmitz, R. A., Graziani, K. R., and Hudson, J. L. (1977) Experimental evidence of chaotic states in the Belousov–Zhabotinskii reaction. *J. Chem. Phys.* **67**, 3040.

Schnackenberg, J. (1979) Simple chemical reaction systems with limit cycle behavior. *J. Theor. Biol.* **81**, 389.

Schroeder, M. (1991) *Fractals, Chaos, Power Laws* (Freeman, New York).

Schuster, H. G. (1989) *Deterministic Chaos*, 2nd ed. (VCH, Weinheim, Germany).

Sel'kov, E. E. (1968) Self-oscillations in glycolysis. A simple kinetic model. *Eur. J. Biochem.* **4**, 79.

Simó, C. (1979) On the Hénon–Pomeau attractor. *J. Stat. Phys.* **21**, 465.

Simoyi, R. H., Wolf, A., and Swinney, H. L. (1982) One-dimensional dynamics in a multicomponent chemical reaction. *Phys. Rev. Lett.* **49**, 245.

Smale, S. (1967) Differentiable dynamical systems. *Bull. Am. Math. Soc.* **73**, 747.

Sparrow, C. (1982) *The Lorenz Equations: Bifurcations, Chaos, and Strange Attractors* (Springer, New York) Appl. Math. Sci. **41**.

Stewart, W. C. (1968) Current-voltage characteristics of Josephson junctions. *Appl. Phys. Lett.* **12**, 277.

Stoker, J. J. (1950) *Nonlinear Vibrations* (Wiley, New York).

Stone, H. A., Nadim, A., and Strogatz, S.H. (1991) Chaotic streamlines inside drops immersed in steady Stokes flows. *J. Fluid Mech.* **232**, 629.

Strogatz, S. H. (1985) Yeast oscillations, Belousov–Zhabotinsky waves, and the non-retraction theorem. *Math. Intelligencer* **7** **(2)**, 9.

Strogatz, S. H. (1986) *The Mathematical Structure of the Human Sleep–Wake Cycle.* Lecture Notes in Biomathematics, Vol. **69**. (Springer, New York).

Strogatz, S. H. (1987) Human sleep and circadian rhythms: a simple model based on two coupled oscillators. *J. Math. Biol.* **25**, 327.

Strogatz, S. H. (1988) Love affairs and differential equations. *Math. Magazine* **61**, 35.

Strogatz, S. H., Marcus, C. M., Westervelt, R. M., and Mirollo, R. E. (1988) Simple model of collective transport with phase slippage. *Phys. Rev. Lett.* **61**, 2380.

Strogatz, S. H., Marcus, C. M., Westervelt, R. M., and Mirollo, R. E. (1989) Collective dynamics of coupled oscillators with random pinning. *Physica D* **36**, 23.

Strogatz, S. H., and Mirollo, R. E. (1993) Splay states in globally coupled Josephson arrays: analytical prediction of Floquet multipliers. *Phys. Rev. E* **47**, 220.

Strogatz, S. H., and Westervelt, R. M. (1989) Predicted power laws for delayed switching of charge-density waves. *Phys. Rev. B* **40**, 10501.

Sullivan, D. B., and Zimmerman, J. E. (1971) Mechanical analogs of time dependent Josephson phenomena. *Am. J. Phys.* **39**, 1504.

Tabor, M. (1989) *Chaos and Integrability in Nonlinear Dynamics: An Introduction* (Wiley-Interscience, New York).

Takens, F. (1981) Detecting strange attractors in turbulence. *Lect. Notes in Math.* **898**, 366.

Testa, J. S., Perez, J., and Jeffries, C. (1982) Evidence for universal chaotic behavior of a driven nonlinear oscillator. *Phys. Rev. Lett.* **48**, 714.

Thompson, J. M. T., and Stewart, H. B. (1986) *Nonlinear Dynamics and Chaos* (Wiley, Chichester, England).

Tsang, K. Y., Mirollo, R. E., Strogatz, S. H., and Wiesenfeld, K. (1991) Dynamics of a globally coupled oscillator array. *Physica D* **48**, 102.

Tyson, J. J. (1985) A quantitative account of oscillations, bistability, and travelling waves in the Belousov–Zhabotinskii reaction. In R. J. Field and M. Burger, eds. *Oscillations and Traveling Waves in Chemical Systems* (Wiley, New York).

Tyson, J. J. (1991) Modeling the cell division cycle: cdc2 and cyclin interactions. *Proc. Natl. Acad. Sci. USA* **88**, 7328.

Van Duzer, T., and Turner, C. W. (1981) *Principles of Superconductive Devices and Circuits* (Elsevier, New York).

Vohra, S., Spano, M., Shlesinger, M., Pecora, L., and Ditto, W. (1992) *Proceedings of the First Experimental Chaos Conference* (World Scientific, Singapore).

Weiss, C. O., and Vilaseca, R. (1991) *Dynamics of Lasers* (VCH, Weinheim, Germany).

Wiggins, S. (1990) *Introduction to Applied Nonlinear Dynamical Systems and Chaos* (Springer, New York).

Winfree, A. T. (1972) Spiral waves of chemical activity. *Science* **175**, 634.

Winfree, A. T. (1974) Rotating chemical reactions. *Sci. Amer.* **230 (6)**, 82.

Winfree, A. T. (1980) *The Geometry of Biological Time* (Springer, New York).

Winfree, A. T. (1984) The prehistory of the Belousov–Zhabotinsky reaction. *J. Chem. Educ.* **61**, 661.

Winfree, A. T. (1987a) *The Timing of Biological Clocks* (Scientific American Library).

Winfree, A. T. (1987b) *When Time Breaks Down* (Princeton University Press, Princeton, NJ).

Winfree, A. T., and Strogatz, S. H. (1984) Organizing centers for three-dimensional chemical waves. *Nature* **311**, 611.

Yeh, W. J., and Kao, Y. H. (1982) Universal scaling and chaotic behavior of a Josephson junction analog. *Phys. Rev. Lett.* **49**, 1888.

Yorke, E. D., and Yorke, J. A. (1979) Metastable chaos: Transition to sustained chaotic behavior in the Lorenz model. *J. Stat. Phys.* **21**, 263.

Zahler, R. S., and Sussman, H. J. (1977) Claims and accomplishments of applied catastrophe theory. *Nature* **269**, 759.

Zaikin, A. N., and Zhabotinsky, A. M. (1970) Concentration wave propagation in two-dimensional liquid-phase self-organizing system. *Nature* **225**, 535.

Zeeman, E. C. (1977) *Catastrophe Theory: Selected Papers 1972–1977* (Addison-Wesley, Reading, MA).

AUTHOR INDEX

SUBJECT INDEX